Nanobiosciences: Current Techniques and Applications

Nanobiosciences: Current Techniques and Applications

Edited by **Oscar Watson**

SYRAWOOD
PUBLISHING HOUSE

New York

Published by Syrawood Publishing House,
750 Third Avenue, 9th Floor,
New York, NY 10017, USA
www.syrawoodpublishinghouse.com

Nanobiosciences: Current Techniques and Applications
Edited by Oscar Watson

International Standard Book Number: 978-1-68286-142-4 (Hardback)

Printed in the United States of America.

Contents

Preface

Nanobioscience is an upcoming field which has a wide range of applications in diverse fields such as in medicine, manufacturing materials, etc. This book outlines the processes and latest applications of nanobiosciences in detail. It aims to shed light on some of the unexplored aspects of nanobiosciences and the recent researches in this field through its extensive content which cover diverse topics like synthesis of nanobiomaterials, scanning probe microscopy, nano-mechanics, nanoelectromechanical systems, etc. Scientists and students actively engaged in this field will find this book full of crucial and innovative concepts.

This book unites the global concepts and researches in an organized manner for a comprehensive understanding of the subject. It is a ripe text for all researchers, students, scientists or anyone else who is interested in acquiring a better knowledge of this dynamic field.

I extend my sincere thanks to the contributors for such eloquent research chapters. Finally, I thank my family for being a source of support and help.

Editor

Chitosan Hydrogel as siRNA vector for prolonged gene silencing

Zhiwei Ma[1†], Chuanxu Yang[2†], Wen Song[3], Qintao Wang[1*], Jørgen Kjems[2*] and Shan Gao[2*]

Abstract

Background: The periodontitis is one of the most prevalent diseases with alveolar resorption in adult people and is the main cause of the tooth loss. To investigate the possibility for protecting the loss of alveolar bone in periodontal diseases, a RNAi-based therapeutic strategy is applied for silencing RANK signaling using thermosensitive chitosan hydrogel as siRNA reservoir and vector.

Results: The thermosensitive chitosan hydrogel was formed from solution (PH = 7.2, at 4°C) at 37°C within 8 minutes. The degradation rates of hydrogel were ~50% and 5% (W remaining/W beginning) in the presence and absence of lysozyme, respectively, over a period of 20 days. The concurrent cumulative *in vitro* release of Cy3-labeled siRNA from the hydrogel was 50% and 17% over 14 days, with or without lysozyme digestion, respectively. High cell viability (>88%) was maintained for cells treated with hydrogel loaded with RANK specific siRNA and RANK knockdown was prolonged for up to 9 days when cells were incubated with siRNA/hydrogel complex. *In vivo* release of siRNA was investigated in a subcutaneous delivery setup in mice. The fluorescent signal from siRNA within hydrogel was remained for up to 14 days compared to less than one day for siRNA alone.

Conclusions: Chitosan hydrogel can potentially serve as a suitable reservoir and vector for local sustained delivery of siRNA in potential therapy.

Background

Most of periodontal diseases are highly destructive and characterized by loss of the periodontal ligament and alveolar bone, eventually causing tooth loss. It is possible to minimalize the disease progression to some extent via routine dental hygiene improvements but no traditional therapy can efficiently recover the reattachment of periodontal ligament tissues and induce the bone regeneration.

The alveolar bone damage is usually caused by abnormal activity of osteoclast. The receptor activator of NF-κB (RANK) and its ligand RANKL are key molecules for differentiation and activation of osteoclasts [1]. When RANKL binds to RANK, it triggers downstream signaling that results in transformation of precursor cells to mature osteoclast. It would therefore be desirable to inhibit the pathway of osteoclast maturation via targeting RANK for prevention of alveolar bone loss.

RNA-interference (RNAi) is a post-transcriptional gene silencing process that is mediated either by synthetic small interfering RNAs (siRNAs) or endogenous microRNAs. RNAi has been established as an efficient tool for investigation of gene function due to its high specificity of gene silencing including genes associated with many diseases [2]. Therapeutic RNAi applications have been well developed over the recent years and significant progress have been made in therapies for cancer [3], neurodegenerative diseases [4], infectious [5] and inflammatory lesions [6-8]. However, the main concern for siRNA therapeutics is the development of a safe delivery system for the siRNA that can protect siRNA from degradation and deliver it into specific cells without side effects. Among non-viral delivery systems, traditional lipid-based transfection system has been developed for decades and widely applied due to its well-controlled pharmacokinetic behavior and high transfection efficiency; however, the drawbacks are cytotoxicity and limited durability, which limit the

* Correspondence: qintaowang@hotmail.com; jk@mb.au.dk; shg@mb.au.dk
†Equal contributors
[1]State Key Laboratory of Military Stomatology, Department of Periodontology and Oral Medicine, The School of Stomatology, Fourth Military Medical University, Xi-an, China
[2]Interdisciplinary Nanoscience Center (iNANO), Department of Molecular Biology and Genetics, Aarhus University, Gustav Wiedsvej 14, 8000 Aarhus C, Denmark
Full list of author information is available at the end of the article

application *in vivo* [9]. Chitosan is a natural cationic polymer that can bind negative charged siRNAs via an electrostatic interaction to form complexes spontaneously in slightly acidic aqueous milieu. It has been comprehensively investigated for drug, plasmid [10] and siRNA delivery [11,12], and considered a suitable transfection vehicle due to its low toxicity, low immunogenicity, low cost, and good biocompatibility [13]. We have also successfully applied chitosan/siRNA derived therapies for arthritis and radiation-induced fibrosis in animal disease models [7,8]. Hydrogel containing siRNA has also been demonstrated effective in cancer therapy [14] and chitosan-based hydrogel has been used as a reservoir for drug loading to enable sustained release of variously therapeutic agents at controlled positions [15,16]. In particular, a chitosan-glycerol phosphate thermosensitive hydrogel, with improved biocompatibility [17,18] and rapid solution-gel state shifting [19], has been employed to deliver ellagic acid [20], collagen composite [21] or ferulic acid [22] for the treatment of various diseases including cancer [23]. In our study the chitosan hydrogel is loaded with RANK siRNA with aim to develop potential RNAi- based therapeutic strategy for periodontitis.

Results
siRNA Validation
Three siRNAs, designed to target mouse RANK, were transfected in murine macrophage cell line, RAW264.7. Two of these siRNAs showed a strong knockdown of RANK mRNA expression (Additional file 1: Figure S1). siRNA2 was selected for further experiments and referred as siR-RANK.

Hydrogel mediated prolonged gene silencing
The bioactivity of hydrogel loaded with siR-RANK was investigated in RAW264.7 cells. The efficiency of RANK gene silencing was evaluated at day 3, 6 and 9 post-transfection. Transfection with siRNA formulated using the commercial reagent TransIT-TKO was included as positive control for transfection efficiency. Compared to the >70% knockdown effect 48 h post transfection by TKO (Additional file 1: Figure S1), the RANK mRNA expression was partly recovered after 72 h (3 days) (Figure 1) (>40%), and completely returned to untreated level 6 and 9 days post transfection (Figure 1). In contrast, cells treated with hydrogel loaded with siR-RANK demonstrated a significant prolonged RANK mRNA knockdown effect gradually increasing from 30%, 50% to 60% at 3 d, 6 d and 9 d post transfection, respectively (Figure 1), with statistical significance of $p < 0.05$ (evaluated by one-way ANOVA).

Cell viability assay
To evaluate the cytotoxicity of the hydrogel alone or hydrogel/siRNA complex, MTT assay was performed for cell treated with hydrogel and chitosan/siRNA hydrogel for 3 days (Figure 2). Compared to the non-treated control, viability of cells incubated with chitosan hydrogel alone or formulated with siRNAs was ~90%, indicating good biocompatibility and relatively low cytotoxicity of the hydrogel with and without siRNA.

siRNA release test *in vitro*
In order to determine the release profile of siRNA from chitosan hydrogel, Cy3-labeled siRNA was formulated with the hydrogel and subsequently incubated at 37°C in PBS. As a reference, a fluorescence intensity-concentration standard curve was determined with multiple dilution of Cy3-labeled siRNA in PBS and PBS alone. Fluorescence intensity was measured for samples collected from various time points and background signal from PBS alone was subtracted. The accumulative release profiles were calculated based on the concentrations obtained (Figure 3). A biphasic release profile was observed: A low initial burst for the first 2–3 days followed by a sustained slow release with a total of 17% released siRNA at day 14. The siRNA was released much faster when the hydrogel was treated with lysozyme, accumulating to a total of 50% release at day 14 (Figure 3).

To investigate whether siRNA is released as free siRNA or siRNA complexed with chitosan, a filter assay was performed. As expected, free siRNA, with a molecular weight of 13.3 kD, passes nearly completely through a membrane with MWCO of 100 kDa (Additional file 2: Figure S2). In contrast, less than 10% siRNA was recovered in the filter assay when formulated with chitosan indicating that the siRNA remains bound positively charged chitosan after release from the hydrogel.

Hydrogel degradation *in vitro*
To understand the degradation behavior of the hydrogel material, FE-SEM imaging was used to visualize the gel structure at different time points after gel-formation. Surface images showed that the non-treated hydrogel formed a complicated sponge-like structure with abundant cross-linking and the average diameter of pores of approximately 50 m (Figure 4A-E). The lysozyme treatment changed the apparent structure dramatically (Figures 4F-J and 5K-O), especially in the presence of 0.5g/L of lysozyme at day 10. Here the frame structure was loosened during the incubation and the cross-linking was destroyed, resulting in enlarged cavities (Figure 4K vs. 4A and F). No significant difference was observed between 0.1g/L and 0.5 g/L after extended incubation (>20 days) (Figure 4G-J and 5I-O). To quantify the hydrogel degradation, the residual chitosan hydrogel was weighed at different time points. The lysozyme clearly enhanced the degradation rate and the mass loss showed a positive correlation with the concentration of

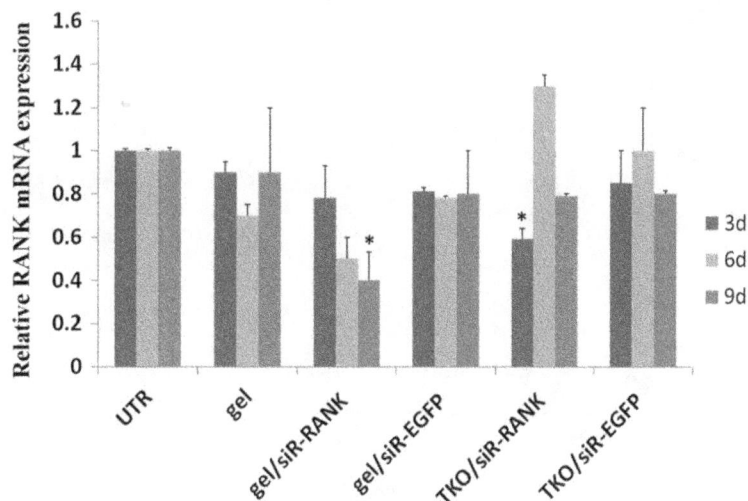

Figure 1 Prolonged gene silencing by siRNA/hydrogel complex. Cells were treated by hydrogel loaded with siR-RANK for 3, 6, and 9 days. Cells transfected via TransIT TKO/siR-RANK was included as positive control, and other controls included untransfected cells (UTR), empty hydrogel (Gel), or hydrogel formulated with siR-EGFP and TKO/siR-EGFP. The knockdown efficiency was evaluated by Real Time RT-PCR. RANK mRNA expression was normalized with β-actin and the relative RANK mRNA expression was calculated with UTR set to 1 and the data was presented as mean ± SD (n = 3). *p < 0.05 compared to untransfected cells.

lysozyme (Figure 5). It is reasonable to speculate that the initial burst of chitosan/siRNA hydrogel in PBS may derive from surface associated siRNA by simple diffusion, while the prolonged release might involve interiorly located siRNA depending on the hydrogel degradation for release.

siRNA release test *in vivo*
In vivo release of siRNA was investigated by optical *in vivo* imaging after subcutaneous injection of Cy5-siRNA loaded hydrogel in mice. Injection of siRNA alone in PBS was included as control. The fluorescent signal from siRNA within hydrogel remained for up to 7 days compared to less than one day for siRNA alone (Figure 6A). A quantitative release profile was provided by measuring the fluorescent signal using Living Image 4.2 Software (Figure 6B). A more sustained siRNA signal was demonstrated for mice injected with siRNA loaded in hydrogel than from mice injected with siRNA alone (~70% *vs* >95% decrease in signal intensity 24 h post-injection, respectively; Figure 6B).

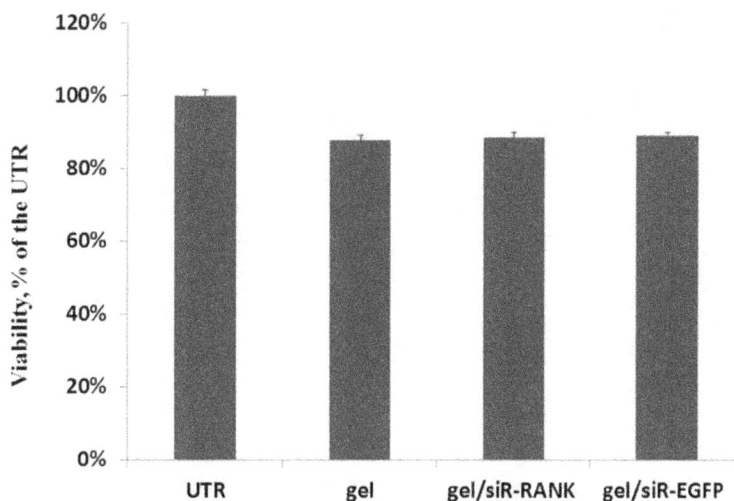

Figure 2 Viability for cells incubated with chitosan hydrogel alone or coupled with siRNA. Cell viability was assessed by MTT assay at 72 h post-transfection. The relative absorbance (at 570 nm) was measured and normalized to the level of non-treated control. Data was presented as mean ± SD (n = 3).

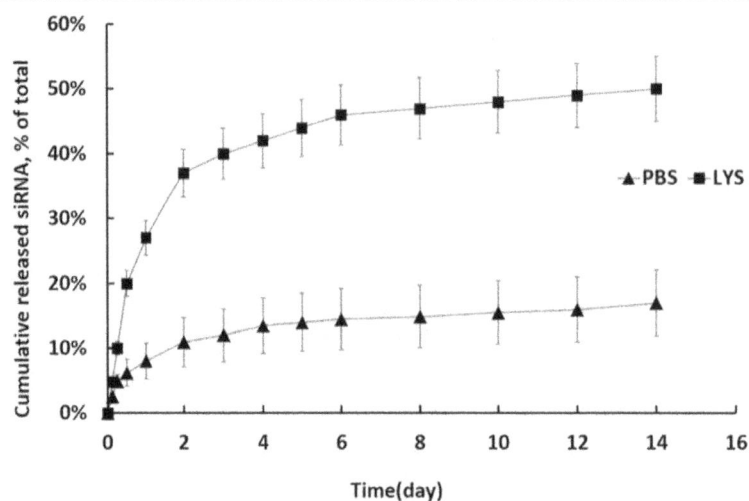

Figure 3 Cumulative release profile of siRNA from chitosan hydrogel/siRNA. The formulated hydrogel was degraded either in PBS alone (marked as PBS) or PBS contained lysozyme (marked as LYS). Fluorescence intensity was measured for samples collected from various time points and background level from blank well was subtracted. The accumulative release profiles were calculated based on concentrations obtained. Data was presented as mean ± SD (n = 3).

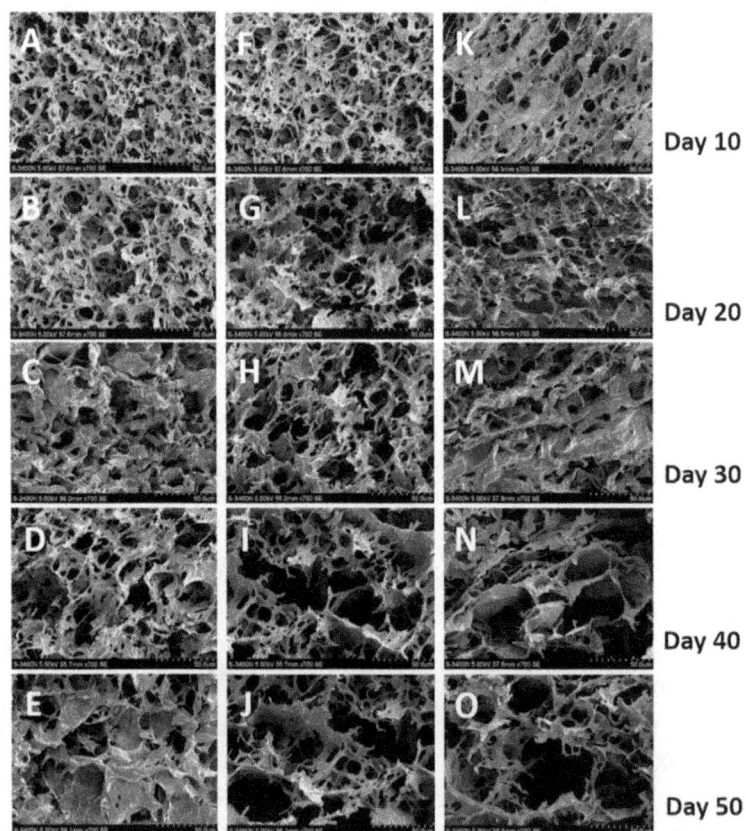

Figure 4 FE-SEM images of hydrogel. Hydrogel incubated in the absence (**A-E**) or presence of, 0.1 g/L and 0.5 g/L lysozyme (**F-J** and **K-O**, respectively; scale bar = 50 μm).

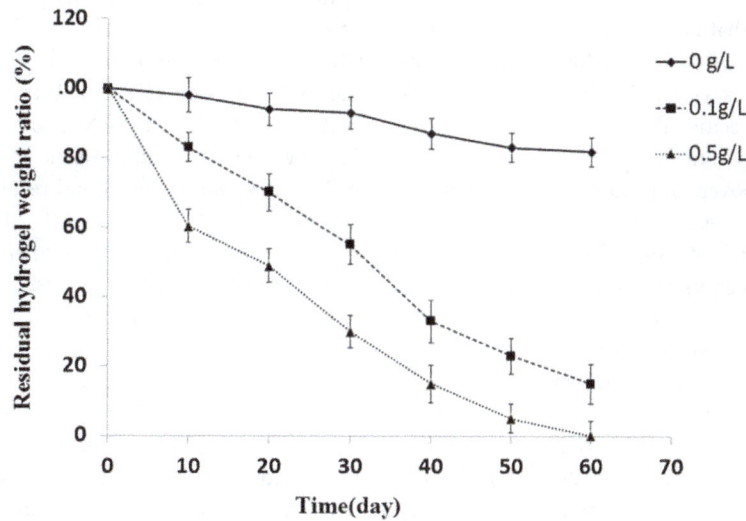

Figure 5 Mass loss of hydrogel treated by lysozyme. Hydrogel was incubated in the absence or presence of lysozyme at various concentrations (0.1 g/L and 0.5 g/L), and the residual weight of dried hydrogel was measured at different intervals. Data were presented as mean ± SD (n = 3).

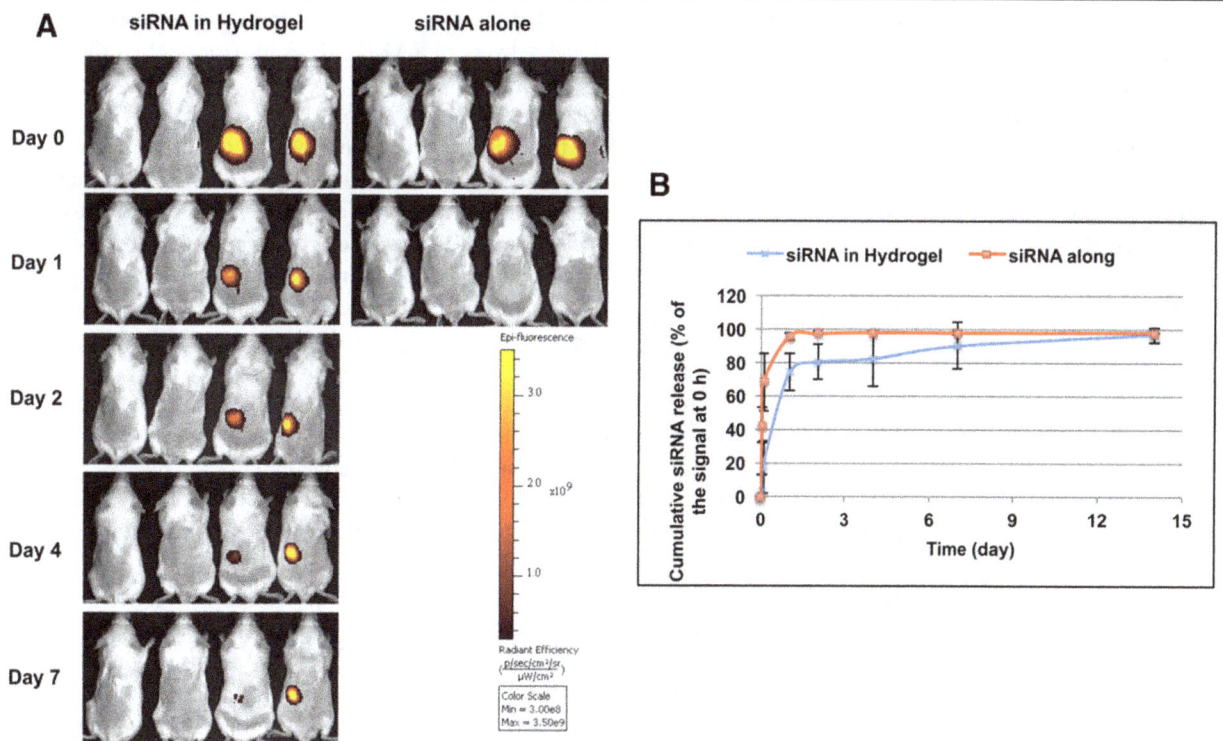

Figure 6 Release profile of Cy5 labeled siRNA complexed within hydrogel in mice. Mice were injected s.c. with chitosan hydrogel/Cy5-siRNA, Cy5-siRNA alone and PBS buffer (n = 4). Fluorescent optical imaging was performed at the indicated time points after injection. **(A)**, Images from 2 mice 0, 1, 2, 4 and 7 days after injection with chitosan hydrogen/Cy5-siRNA and images from 2 mice 0 and 1 day after injection with Cy5-siRNA alone. Two mice injected with buffer (Two mice on the left in each panel) were included as control for both scanning. **(B)**, After quantifying the signal intensity of images, siRNA release was calculated as reduction of fluorescent signal: (RE (0 h) – RE (Designed Time Point)) /RE (0 h) * 100%. Average value from each time point (0, 1, 2 h, and 1, 2, 4, 7, 14 d) was presented (mean ± SD, n = 4).

Discussion

We have demonstrated that siRNA, loaded into chitosan hydrogel, is released in a sustained fashion both *in vitro* and *in vivo*. Using this system, RANK specific siRNA exhibited a prolonged specific silencing effect for up to 9 days *in vitro* without apparent cytotoxicity.

RNAi has recently proven a promising approach for the development of new treatment protocols for human diseases. However, the lack of suitably non-viral systemic delivery vehicles has limited its application in clinic where effective and prolonged silencing effects are needed. Compared to systemic delivery, local delivery systems can provide a relatively high drug concentration in the lesion areas and reduce side effects. For instance, local delivery of siRNA using biodegradable polymer enabled a scaffold to support and repair bone defects [24]. However, the predetermined shape restricts the application for many kinds of lesions encountered in clinic. To address this, injectable hydrogel is widely applied for tissue engineering, also for drug delivery including siRNA delivery [25-27], due to its ability to undergo sol–gel phase transition.

We have successfully applied chitosan/siRNA therapies in several animal disease models with arthritis and radiation-induced fibrosis [7,8]. Among chitosan-based hydrogels, the one cross-linked with β-glycerophosphate has been most extensively employed and already proven effective in several kinds of drug delivery [20-22,25]. In addition, the chitosan-based gels have been used for tissue regeneration in periodontal disease [28]. We have developed an injectable thermosensitive chitosan hydrogel system for siRNA delivery and observed prolonged knockdown efficacy for up to 9 days. The effect is more pronounced than in a previous collagen gel study where substantial knockdown of GFP was observed for only up to 6 days in GFP expressing HEK293 cells [26].

The *in vitro* siRNA release profile suggests that electrostatic interactions between the nucleotides and positively charged polymer slow down the release. In addition, diffusion through the biopolymer pores facilitated by biopolymer degradation may also control the release. This resembles the observation in a previous report for siRNA release from calcium cross-linked alginate with chitosan [26]. In that study siRNA could only be released from the hydrogel when gradually degraded by an enzyme [26]. In another study, gelatin hydrogel, incorporated with cationized gelatin (CG)-siRNA complexes, was enzymatically degraded by the treatment with collagenase to enable the release of siRNA [25]. Similarly, we observed clear degradation in the presence of lysozyme, indicating that this enzyme is able to degrade the chitosan gel *in vitro* [29-31].

Compared to *in vitro*, siRNA was released more quickly *in vivo*, implying the existence of lysozyme or other degradation pathways *in vivo* [32]. The chitosan hydrogel/siRNA system has previously been applied in tumor therapy in mice [27]. They observed labelled siRNA uptake in tumour cells up to 270 µm away from the hydrogel edge 24 h post intra-tumoral injection, however, the half-life of siRNA *in vivo* was not addressed. The hydrogel was administrated twice weekly at a dose of 150 µg/kg body weight and resulted in significant antitumor effect [27]. Similarly, the biodegradable hydrogel, poly-D, L-lactic acid-p-dioxanonepolyethylene glycol block co-polymer (PLA-DX-PEG) has been used as a carrier for Noggin specific siRNA *in vivo*, resulting in efficient silencing of the target mRNA expression and successfully induces the ectopic bone formation [24].

The RANK-RANKL system has been proven to play a vital role in the differentiation and activation of osteoclasts in periodontitis and silencing the RANK expression in osteoclast precursors may provide a new approach to prevent and convert the alveolar absorption in periodontitis. In this study, we demonstrate chitosan hydrogel could prolong gene silencing effect of siRNA up to 9 days *in vitro*. Theoretically, this extended release profile may be provide sufficient siRNA to prevent activation of osteoclasts and maintain the therapeutic effects on periodontitis for an extended period of time. In combination with the excellent biocompatibility we observed, our system may potentially serve as a suitable vector for continuous local delivery of RANK siRNA in periodontitis therapy.

Conclusions

In this study we have developed a chitosan-glycerol phosphate thermosensitive hydrogel enabling to functionally deliver RANK siRNA into cells. The hydrogel exhibited prolonged siRNA release profiles both *in vitro* and *in vivo*, accompanied with sustained gene silencing effect in cell. The next step will be to investigate its efficacy in animal periodontitis model and explore its potential to regenerate periodontal tissues.

Methods

siRNA target RANK screening

Three siRNA duplexes targeting RANK (siRNA1, siRNA2, and siRNA3) were ordered from GenePharma (Shanghai, China). The negative control siRNA (siNC) was provided by Genepharm, and siRNA-EGFP duplex was purchased from Ribotask (Odense, Denmark). All the sequences were listed in Table 1.

Mouse macrophage cell line, RAW264.7 (ATCC® TIB-71™), used as osteoclast precursor, was maintained in RPMI media (Gibco®) supplemented with 10% fetal bovine serum (Sigma), 1% penicillin-streptomycin (Gibco®) at 37°C in 5% CO_2 and 100% humidity. Cells were seeded onto 24-well plate twenty-four hours before transfection. siRNA was transfected using the commercial reagent TransIT-TKO (Mirus Bio Corporation) at final siRNA

Table 1 Sequence of siRNAs

Target	Sense sequence (5'-3')	Antisense sequence (5'-3')
RANK1	GAGCAGAACUGACUCUAUGUU	CAUAGAGUCAGUUCUGCUCUU
RANK2	GCGCAGACUUCACUCCAUAUU	UAUGGAGUGAAGUCUGCGCUU
RANK3	GCGCUGACAGCUAAUUUGUTT	ACAAAUUAGCUGUCAGCGCTT
EGFP	GACGUAAACGGCCACAAGUTC	ACUUGUGGCCGUUUACGUCGC
Negative control	UUCUCCGAACGUGUCACGUTT	ACGUGACACGUUCGGAGAATT

concentration 50 nM. Cells were harvested at 48 hr post-transfection.

mRNA expression level was assessed by Quantitative Real-Time RT-PCR (qRT-PCR) following routine procedures of total RNA purification using TRIzol reagent (Invitrogen, Copenhagen) and reverse transcription using SuperScript® II Reverse Transcriptase (RT) kit (Invitrogen, Copenhagen). The comparative CT (threshold cycle) method described in the manufacturer's protocol (Stratagene, Copenhagen) was used to quantitate the relative RANK mRNA expression level. The β-actin mRNA was amplified as an internal control to normalise the expression level. The primer sequences for RANK gene are, forward: 5'-TAGGACGTCAGGCCAAAGGACAAA -3', reverse: 5'-AGGGCCTACTGCCTAAGTGTGTTT -3', with a product size of 132 bp. Primer sequences for β-actin gene are, forward: 5'-ACACAGTGCTGTCTGGTGGT-3', Reverse: 5'-CTGGAAGGTGGACAGTGAGG-3' with a product size of 172 bp.

Formulation of chitosan hydrogel /siRNA complex

Chitosan (150 kDa, 95% deacetylation) was provided by HEPPE MEDICAL (Germany). Indicated chitosan was dissolved in 1% acetic acid (Sigma) at the concentration of 2% w/v. Sodium glycerophosphate (Sigma) were used as crosslink reagent and added dropwise into chitosan solution while stirring (pH 7.2, 4°C). Pre-calculated siRNA (5 μg) were added into the clear and homogeneous liquid solution (200 μl). The hydrogel spontaneously became solid during the incubation at 37°C for 8 minutes, which was ready for further experiments.

siRNA transfection mediated by hydrogel

The 24-well transwell (Thincert, In Vitro As, Fredensborg, Denmark) loaded with hydrogel was inserted over the cell culture plates where containing pre-seeded cells (RAW264.7, 2×10^4/well), as shown in Figure 7. The transfection was terminated and the RANK gene expression was measured by Quantitative Real Time RT-PCR after transfected for 3, 6, 9 days.

Cytotoxicity of hydrogel/siRNA complex

The cell viability was determined by MTT assay at day 3 of hydrogel mediated transfection as described above (n = 3). MTT reagent (5 mg/ml, Sigma, Copenhagen)

was then added to each well at 1:3 dilution, after removed transwells, rinsed with PBS and exchanged the medium. After incubation of 2 ~ 4 h at 37°C, the reaction was stopped by replacing of DMSO (Sigma, Copenhagen) while the color changes became visible. The relative absorbance (at 570 nm) was measured using a Victor X5 Multilabel Plate reader (PerkinElmer).

siRNA release test in vitro

To assess siRNA release profile, the Cy3-labled siRNA (5 μg) was formulated with chitosan hydrogel as described above. The hydrogel was immersed in 500 μl phosphate buffer (PBS, Gibco®) solution (pH 7.4) with or without lysozyme (0.5 mg/ml, Sigma). At pre-determined intervals, all of the supernatant was drawn and replaced by the fresh buffer accordingly. The fluorescence intensity from the supernatant was measured by fluorescence plate reader (FLUOstar OPTIMA) and PBS solution was used as blank control.

To investigate whether siRNA in release solution as free or chitosan complexed siRNA, Cy3-labled siRNA was encapsulated in chitosan hydrogel (n = 3). After gelation, 500 μl of PBS was added and incubated 2 h at 37°C. The PBS solution was collected and 250 μl the solution was loaded on to a centrifugal device with molecular weight cut off (MWCO) of 100 kDa (Pall Corporation) and centrifuged at 3000 g for 5 min. The Cy3 fluorescent intensity of filtered solution was measured by FLUOstar OPTIMA (BMG Labtechnologies) and normalized to input solution. As a non-particulate control, free Cy3-siRNA was also loaded on a centrifuge device treated similarly.

Hydrogel degradation in vitro

To further understand the mechanism of siRNA release from the hydrogel, the hydrogel degradation was assessed.

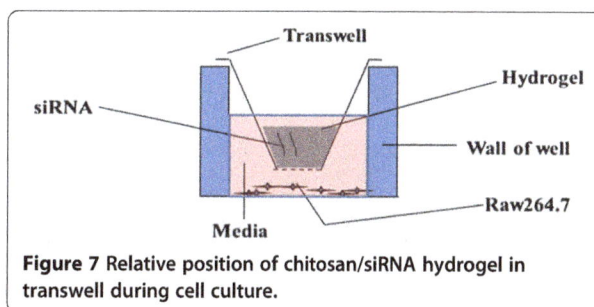

Figure 7 Relative position of chitosan/siRNA hydrogel in transwell during cell culture.

Briefly, 1ml chitosan hydrogel solution was placed into 37°C incubator to form the hydrogel. The hydrogel, treated using three concentration of lysozyme (0, 0.1, 0.5 g/L), was quickly frozen with liquid nitrogen for 5min and totally freeze-dried before precisely weighed by electronic balance. The percentage of mass loss was determined by measuring hydrogel weight at initial time point (Wi) and weight at different time points pre-designed (Wd), and calculated by ((Wi-Wd)/Wi). After gold coating, the freeze-dried sample was also observed with Field Emission Scanning Electron Microscope (FE-SEM).

siRNA release test *in vivo*

To evaluate siRNA release within hydrogel *in vivo*, BALB/c female mice (Taconic Europe, Ll. Skensved, Denmark) were chosen. All procedures of animal work were approved by "The Experimental Animal Inspectorate in Denmark" under The Danish Veterinary and Food Administration, Ministry of Food, Agriculture and Fisheries.

Hair from the back and the abdomen was shaved to avoid autofluorescence during scanning. Cy5-labelled siRNA/chitosan hydrogel formulation in 50 microliter sodium acetate buffer (containing 5 μg siRNA duplex) was subcutaneously injected into the left flank region of mice. The unformulated Cy5-labelled siRNA in PBS with the same volume was administrated as control. The mice were scanned using an IVIS® 200 imaging system (Xenogen, Caliper Life Sciences) before and right after injection followed by 1 hr, 2 hrs, 1 day, 2, 3, 7 and 14 days post injection. The scanning was performed under anesthesia with 2.5% isoflurane. Cy5 excitation (λex = 640 nm) and emission (λem = 700 nm) filters were used. Total emission from inflicted areas (Region of Interest, ROI) of each mouse was quantified using Living Image 4.0 software package. According to the user instruction, the radiant efficiency (RE) of injection sites was measured (photons/sec/cm2/sr)/ (μW/cm2), which is presented radiance/illumination power density [33,34].

Additional files

Additional file 1: Figure S1. RANK gene knockdown efficiency. Three siRNA against murine RANK (siRNA1, siRNA2 and siRNA3) were transfected in triplicate into RAW264.7 cells using TransIT-TKO reagent. Untreated cells (UTR) or cells transfected with siRNA against EGFP (siR-EGFP) and siR-NC were included as controls. Cells were harvested 48 hrs post transfection and RANK mRNA levels were evaluated by quantitative RT-PCR. Data were presented as mean ± SD (n = 3). * siRNA was selected for further experiment.

Additional file 2: Figure S2. Analysis of siRNA released from chitosan hydrogel. The Cy3-labled siRNA was encapsulated in chitosan hydrogel (n = 3) and incubated in PBS for 2h at 37°C, the PBS solution was collected and half of the solution (release solution) was centrifuged through a filter device. The Cy3 fluorescent intensity of filtered solution was measured and normalized to input solution. Unformulated free Cy3-siRNA (free siRNA) was applied as control.

Competing interests
The authors declare that they have no competing interests.

Authors' contributions
ZM designed and carried out all the experiments. CY carried out the hydrogel preparation and joined most of experiments. WS analyzed the data and drafted the manuscript. QW participated in the design of the study. JK analyzed the data and revised the manuscript. SG drafted and revised the manuscript. All authors read and approved the final manuscript.

Acknowledgements
This work was supported by grants from the Nature Science Foundation of China (NSFC 81271137) and the Lundbeck Foundation - Lundbeck Foundation Nanomedicine Center for Individualized Management of Tissue Damage and Regeneration. We thank San Hein for his supervision of chitosan preparation and Menglin Chen for her kindly help in hydrogel release test *in vitro*. We thank Rita Rosendahl and Claus Bus for their excellent technical assistance, as well as Frederik Dagnæs-Hansen for his arrangement of animal works.

Author details
[1]State Key Laboratory of Military Stomatology, Department of Periodontology and Oral Medicine, The School of Stomatology, Fourth Military Medical University, Xi-an, China. [2]Interdisciplinary Nanoscience Center (iNANO), Department of Molecular Biology and Genetics, Aarhus University, Gustav Wiedsvej 14, 8000 Aarhus C, Denmark. [3]State Key Laboratory of Military Stomatology, Department of Prothodontics, The School of Stomatology, Fourth Military Medical University, Xi-an, China.

References
1. Koide M, Kinugawa S, Takahashi N, Udagawa N: Osteoclastic bone resorption induced by innate immune responses. *Periodontol 2000* 2010, 54:235–246.
2. Lopez-Fraga M, Martinez T, Jimenez A: RNA interference technologies and therapeutics: from basic research to products. *BioDrugs* 2009, 23:305–332.
3. Sotillo E, Thomas-Tikhonenko A: Shielding the messenger (RNA): microRNA-based anticancer therapies. *Pharmacol Ther* 2011, 131:18–32.
4. Gonzalez-Alegre P: Therapeutic RNA interference for neurodegenerative diseases: from promise to progress. *Pharmacol Ther* 2007, 114:34–55.
5. Lupberger J, Brino L, Baumert TF: RNAi: a powerful tool to unravel hepatitis C virus-host interactions within the infectious life cycle. *J Hepatol* 2008, 48:523–525.
6. Laroui H, Theiss AL, Yan Y, Dalmasso G, Nguyen HT, Sitaraman SV, Merlin D: Functional TNFalpha gene silencing mediated by polyethyleneimine/TNFalpha siRNA nanocomplexes in inflamed colon. *Biomaterials* 2011, 32:1218–1228.
7. Howard KA, Paludan SR, Behlke MA, Besenbacher F, Deleuran B, Kjems J: Chitosan/siRNA nanoparticle-mediated TNF-alpha knockdown in peritoneal macrophages for anti-inflammatory treatment in a murine arthritis model. *Mol Ther* 2009, 17:162–168.
8. Nawroth I, Alsner J, Behlke MA, Besenbacher F, Overgaard J, Howard KA, Kjems J: Intraperitoneal administration of chitosan/DsiRNA nanoparticles targeting TNFalpha prevents radiation-induced fibrosis. *Radiother Oncol* 2010, 97:143–148.
9. Schroeder A, Levins CG, Cortez C, Langer R, Anderson DG: Lipid-based nanotherapeutics for siRNA delivery. *J Intern Med* 2010, 267:9–21.
10. Duceppe N, Tabrizian M: Advances in using chitosan-based nanoparticles for in vitro and in vivo drug and gene delivery. *Expert Opin Drug Deliv* 2010, 7:1191–1207.
11. Howard KA, Rahbek UL, Liu X, Damgaard CK, Glud SZ, Andersen MO, Hovgaard MB, Schmitz A, Nyengaard JR, Besenbacher F, Kjems J: RNA interference in vitro and in vivo using a novel chitosan/siRNA nanoparticle system. *Mol Ther* 2006, 14:476–484.
12. Holzerny P, Ajdini B, Heusermann W, Bruno K, Schuleit M, Meinel L, Keller M: Biophysical properties of chitosan/siRNA polyplexes: profiling the polymer/siRNA interactions and bioactivity. *J Control Release* 2012, 157:297–304.
13. Baldrick P: The safety of chitosan as a pharmaceutical excipient. *Regul Toxicol Pharmacol* 2010, 56:290–299.

14. Ramakrishnan S: **Hydrogel-siRNA for cancer therapy.** *Cancer Biol Ther* 2011, **11:**849–851.
15. Jo S, Kim S, Cho TH, Shin E, Hwang SJ, Noh I: **Effects of recombinant human bone morphogenic protein-2 and human bone marrow-derived stromal cells on in vivo bone regeneration of chitosan-poly(ethylene oxide) hydrogel.** *J Biomed Mater Res A* 2013, **101:**892–901.
16. Chakavala SR, Patel NG, Pate NV, Thakkar VT, Patel KV, Gandhi TR: **Development and in vivo evaluation of silver sulfadiazine loaded hydrogel consisting polyvinyl alcohol and chitosan for severe burns.** *J Pharm Bioallied Sci* 2012, **4:**S54–S56.
17. Ahmadi R, de Bruijn JD: **Biocompatibility and gelation of chitosan-glycerol phosphate hydrogels.** *J Biomed Mater Res A* 2008, **86:**824–832.
18. Hoemann CD, Chenite A, Sun J, Hurtig M, Serreqi A, Lu Z, Rossomacha E, Buschmann MD: **Cytocompatible gel formation of chitosan-glycerol phosphate solutions supplemented with hydroxyl ethyl cellulose is due to the presence of glyoxal.** *J Biomed Mater Res A* 2007, **83:**521–529.
19. Ruel-Gariepy E, Chenite A, Chaput C, Guirguis S, Leroux J: **Characterization of thermosensitive chitosan gels for the sustained delivery of drugs.** *Int J Pharm* 2000, **203:**89–98.
20. Kim S, Nishimoto SK, Bumgardner JD, Haggard WO, Gaber MW, Yang Y: **A chitosan/beta-glycerophosphate thermo-sensitive gel for the delivery of ellagic acid for the treatment of brain cancer.** *Biomaterials* 2010, **31:**4157–4166.
21. Wang L, Stegemann JP: **Thermogelling chitosan and collagen composite hydrogels initiated with beta-glycerophosphate for bone tissue engineering.** *Biomaterials* 2010, **31:**3976–3985.
22. Cheng YH, Yang SH, Lin FH: **Thermosensitive chitosan-gelatin-glycerol phosphate hydrogel as a controlled release system of ferulic acid for nucleus pulposus regeneration.** *Biomaterials* 2011, **32:**6953–6961.
23. Seth S, Johns R, Templin MV: **Delivery and biodistribution of siRNA for cancer therapy: challenges and future prospects.** *Ther Deliv* 2012, **3:**245–261.
24. Manaka T, Suzuki A, Takayama K, Imai Y, Nakamura H, Takaoka K: **Local delivery of siRNA using a biodegradable polymer application to enhance BMP-induced bone formation.** *Biomaterials* 2011, **32:**9642–9648.
25. Saito T, Tabata Y: **Preparation of gelatin hydrogels incorporating small interfering RNA for the controlled release.** *J Drug Target* 2012, **20:**864–872.
26. Krebs MD, Jeon O, Alsberg E: **Localized and sustained delivery of silencing RNA from macroscopic biopolymer hydrogels.** *J Am Chem Soc* 2009, **131:**9204–9206.
27. Han HD, Mora EM, Roh JW, Nishimura M, Lee SJ, Stone RL, Bar-Eli M, Lopez-Berestein G, Sood AK: **Chitosan hydrogel for localized gene silencing.** *Cancer Biol Ther* 2011, **11:**839–845.
28. Akncbay H, Senel S, Ay ZY: **Application of chitosan gel in the treatment of chronic periodontitis.** *J Biomed Mater Res B Appl Biomater* 2007, **80:**290–296.
29. Lee KY, Ha WS, Park WH: **Blood compatibility and biodegradability of partially N-acylated chitosan derivatives.** *Biomaterials* 1995, **16:**1211–1216.
30. Onishi H, Machida Y: **Biodegradation and distribution of water-soluble chitosan in mice.** *Biomaterials* 1999, **20:**175–182.
31. Tomihata K, Ikada Y: **In vitro and in vivo degradation of films of chitin and its deacetylated derivatives.** *Biomaterials* 1997, **18:**567–575.
32. Lim SM, Song DK, Oh SH, Lee-Yoon DS, Bae EH, Lee JH: **In vitro and in vivo degradation behavior of acetylated chitosan porous beads.** *J Biomater Sci Polym Ed* 2008, **19:**453–466.
33. Iversen F, Yang C, Dagnaes-Hansen F, Schaffert DH, Kjems J, Gao S: **Optimized siRNA-PEG conjugates for extended blood circulation and reduced urine excretion in mice.** *Theranostics* 2013, **3:**201–209.
34. Nawroth I, Alsner J, Deleuran BW, Dagnaes-Hansen F, Yang C, Horsman MR, Overgaard J, Howard KA, Kjems J, Gao S: **Peritoneal macrophages mediated delivery of chitosan/siRNA nanoparticle to the lesion site in a murine radiation-induced fibrosis model.** *Acta Oncol (Stockholm, Sweden)* 2013, **52:**1730–1738.

2

Synthesis and bioactivities of silver nanoparticles capped with 5-Amino-β-resorcylic acid hydrochloride dihydrate

Syeda Sohaila Naz[1*], Muhammad Raza Shah[2†], Nazar Ul Islam[3,4†], Ajmal Khan[2†], Samina Nazir[1†], Sara Qaisar[1†] and Syed Sartaj Alam[5†]

Abstract

Background: Conjugated and drug loaded silver nanoparticles are getting an increased attention for various biomedical applications. Nanoconjugates showed significant enhancement in biological activity in comparison to free drug molecules. In this perspective, we report the synthesis of bioactive silver capped with 5-Amino-β-resorcylic acid hydrochloride dihydrate (AR). The *in vitro* antimicrobial (antibacterial, antifungal), enzyme inhibition (xanthine oxidase, urease, carbonic anhydrase, α-chymotrypsin, cholinesterase) and antioxidant activities of the developed nanostructures was investigated before and after conjugation to silver metal.

Results: The conjugation of AR to silver was confirmed through FTIR, UV–vis and TEM techniques. The amount of AR conjugated with silver was characterized through UV–vis spectroscopy and found to be 9% by weight. The stability of synthesized nanoconjugates against temperature, high salt concentration and pH was found to be good. Nanoconjugates, showed significant synergic enzyme inhibition effect against xanthine and urease enzymes in comparison to standard drugs, pure ligand and silver.

Conclusions: Our synthesized nanoconjugate was found be to efficient selective xanthine and urease inhibitors in comparison to Ag and AR. On a per weight basis, our nanoconjugates required less amount of AR (about 11 times) for inhibition of these enzymes.

Keywords: 5-Amino-β-resorcylic acid hydrochloride dihydrate, Silver nanoparticles, Urease, Xanthine oxidase

Background

In the new research areas of science, the research field of metal nanoparticles is an extensive and the most emerging field. The huge surface area of the metal nanoparticles is responsible for their diverse optical, chemical, magnetic, mechanical, and catalytic properties as compared to large bulk materials [1,2]. Different methods like the chemical, physical, and biological methods are reported for the metal nanoparticles synthesis [3-8]. Silver nanoparticles (AgNPs) have been synthesized using various methods, i.e. polysaccharide method, tollens method, irradiation method, biological, polyoxometalates. Silver is well-known for its antimicrobial activities and

has been utilized for years towards biomedical applications [9].

Current research showed that conjugation of Ag to plant extracts boosted the enzyme inhibition and antimicrobial activities but these reports are not very systematic [10,11]. In this context the aim of our present work is to synthesize AgNPs from a synthetic biocide 5-Amino-β-resorcylic acid hydrochloride dihydrate (AR). The choice of AR towards facile synthesis of AgNPs seems to be original one and may present several advantages in terms of water solubility, possibility to conjugation to Ag metal due to the presence of carboxylic, amino and phenolic hydroxyl groups. To ascertain the potential of the synthesized nanoconjugates for *in vivo* applications, the stability of the suspensions was investigated against several parameters such as pH, temperature and salt concentration. Barron AgNPs (Ag) was prepared by reduction

* Correspondence: syedasohailanaz@yahoo.com
†Equal contributors
[1]Nanoscience and Catalysis Division, National Centre for Physics, Quaid-i-Azam University Campus, Islamabad 44000, Pakistan
Full list of author information is available at the end of the article

of AgNO$_3$ with NaBH$_4$. The *in vitro* antibacterial, antifungal, enzyme inhibition (xanthine oxidase, urease, carbonic anhydrase, α-chymotrypsin, cholinesterase) and antioxidant activities of AgAR nanoconjugates were compared with pure AR, Ag and the commercially available antibiotics, enzyme inhibitors and antioxidants.

Results and discussion

The synthesis of AR (Figure 1) was carried out according to our previously published procedure [12]. When the synthesized AR was added to the aqueous solution AgNO$_3$, we observed a change in color from light brown to dark brown upon slow addition of NaBH$_4$ (Additional file 1: Figure S1). Characterization of AgNPs with UV–vis spectroscopy showed surface plasmon resonance peak at 390 nm and the amount of AR conjugated with the surface of silver was found to be 9% by weight (Figure 2).

FTIR spectra of AR was recorded before and after formation of nanoparticles and reported in Figure 3. The disappearance of the peak at carbonyl region (1639 cm^{-1}) in the spectrum of AR indicated the chelation of carboxylic group with silver. From FTIR characterization, a mechanism has been proposed for the synthesis of AgAR nanoconjugates and reported in Figure 4. This figure showed that NaBH$_4$ has been involved in reduction of AgNO$_3$ while carboxylic group of AR provide stability to AgNPs *via* electrostatic interactions [13]. The formation of silver nanoparticles was finally confirmed from transmission electron micrograph and the mean size of the nanoparticles was found to be 8 nm (Figure 5).

In order to determine the potential of synthesized nanoparticles for *in vivo* applications, it was desired to check its stability against high concentration of NaCl, heat and pH. The synthesized nanoconjugates was found to be basic in nature as its pH was found to be 8.49. The stability of nanoparticles was checked at all pH values ranging from 2–13 (Figure 6) and indicated by observing a change in λ$_{max}$. In comparison to other pH values, as the absorbance of nanoparticles was highest at pH 8–9 therefore, it was established that the stability of the nanoconjugates was good at this pH.

When NaCl was added to the nanoparticles solution, a gradual change in the peak shape is observed; an initial

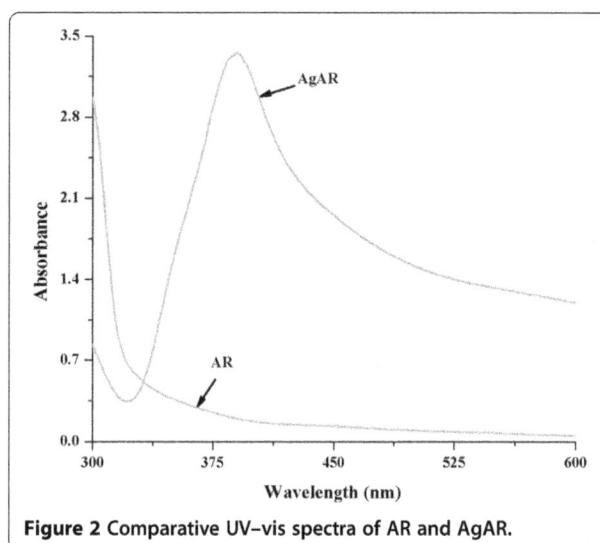

Figure 2 Comparative UV–vis spectra of AR and AgAR.

halide surface layer of unknown structure may form very rapidly (Figure 7). The successive changes in the UV-visible spectra proposed that this layer may have developed into a silver halide layer. For NaCl, the onset concentration for aggregation is considerably lower. This has been discussed in terms of a distinct effect on the nanoparticles surface, in which the surface charge is dropped by nearly a factor of 2. It is not clear that how this is accomplished. One probability is that a chloride layer decreased the number of adsorption sites for the highly charged AR. Instead, the chloride ion may substitute AR entirely but then form AgCl$_2$ rather than AgCl, thereby retaining a negatively charged surface but with a lesser value [14].

Figure 8 showed the absorption spectra of 8 nm AgNPs at 100°C. The result indicated that the temperature effect is negligible, resulting in a very minute reduction in absorbance while a broadening of the plasmon band was not observed.

In the present study, a systematic comparative bioactivity evaluation of synthesized AR, Ag, AgAR and standard drugs was carried out. AgAR nanoconjugates contain 9% of AR as determine from UV–vis study. The purpose of the conjugation of AR to the surface of AgNPs is to determine the change in the activity of AR after attachment to the surface of metal. Selected bioassays

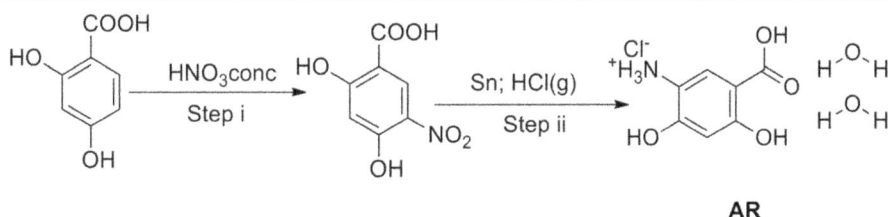

Figure 1 Synthesis of 5-Amino-β-resorcylic acid hydrochloride dihydrate (AR).

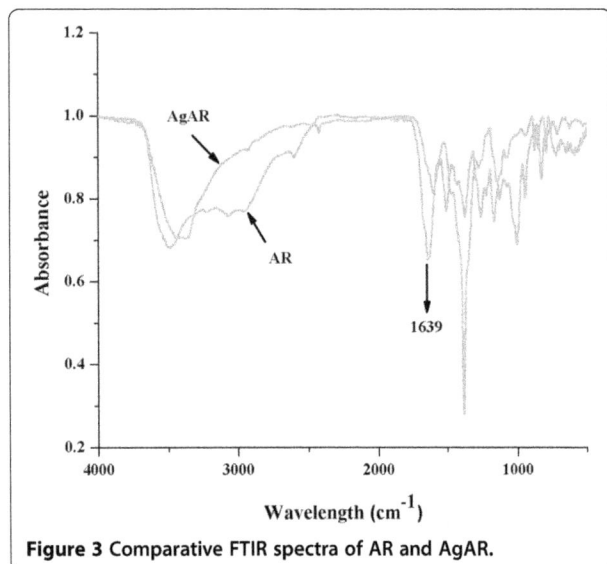

Figure 3 Comparative FTIR spectra of AR and AgAR.

include *in vitro* antimicrobial activities, six enzyme inhibition activities and antioxidant activities.

The antimicrobial studies of Ag, AR and AgAR were determined against three pathogenic microbes namely *Erwinia carotovora* (Figure 9), *Alternaria solani* (Figure 10) *and Fusarium solani* (Figure 11) at three different concentrations (150, 200 and 250 ppm). Silver in the form of bare AgNPs (Ag) showed significant antibacterial activity in comparison to AR and AgAR. However, at the tested concentrations the activity of Ag was lesser than the standard antibiotic streptomycin. Moreover, the antifungal activities of Ag, AR and AgAR were not significant in comparison to the standard fungicide dithane-M45.

The synthesized nanoparticles AgAR, Ag and parent compound AR were then screened for inhibition of xanthine and urease enzymes and the results are reported in Table 1. Ag and AR are found to be inactive against these enzymes. The AgAR has shown significant inhibition against xanthine oxidase and urease enzymes with IC$_{50}$ value of 15.5 $\pm$ 0.9 and 9.2 $\pm$ 0.9 µg/mL,

respectively. As discussed previously our nanoconjugates contains 9% of ligand AR, therefore the conjugation of AR to Ag in the form of nanoconjugates AgAR not only enhanced the enzyme inhibition activity of AR but also reduced the amount of AR by about 11 times. Moreover, the urease inhibition activity of AgAR was found to be more than the standard drug thiourea with IC$_{50}$ values of 9.2 $\pm$ 0.9 and 21.8 $\pm$ 1.2 µg/mL, respectively. Its means the conjugation of AR to Ag had a robust inhibition effect in comparison to pure Ag and AR. In order to evaluate the selectivity of AgAR towards xanthine oxidase and urease enzymes, the activity of AgAR, Ag and AR were also tested against the carbonic anhydrase, α-chymotrypsin, acetylcholinesterase and butylcholinesterase enzymes. Interestingly the AgAR, Ag and AR were found to be inactive against these enzymes, exhibiting their selectivity towards xanthine oxidase and urease enzymes. In case of carbonic anhydrase AR showed inhibitory effect against carbonic anhydrase-II, but this effect was non-significant in comparison to standard drug acetazolamide.

The antioxidant activity of AgAR, Ag and AR was determined using DPPH radical scavenging assay (RSA). Before conjugation, the antioxidant activity of AR was significant with IC$_{50}$ value of 36.55 $\pm$ 2.05 µg/mL but after conjugation the AR lost its antioxidant activity (Table 2). Additionally, the antioxidant effect produced by AR in comparison to AgAR was also greater than all the three standard drugs (ascorbic acid, butylated hydroxyanisole, N-acetyl cystine).

Conclusions

In conclusion, we have developed an efficient strategy for the synthesis of silver nanoparticles capped with AR toward selective inhibition of urease and xanthine enzymes. The urease and xanthine inhibition potency of AR was amplified by conjugation to Ag. On per weight basis, our nanoconjugates AgAR require less amount of AR (about 11 times) for inhibition enzyme of urease and xanthine enzymes. Moreover, our synthesized

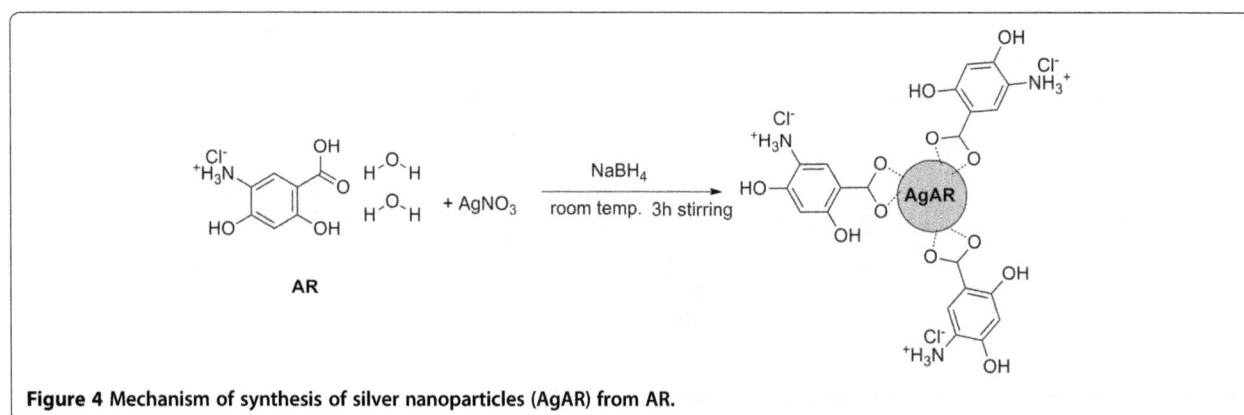

Figure 4 Mechanism of synthesis of silver nanoparticles (AgAR) from AR.

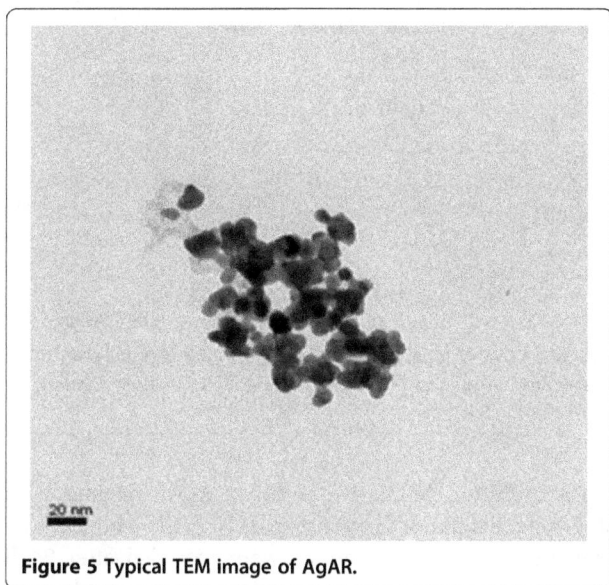

Figure 5 Typical TEM image of AgAR.

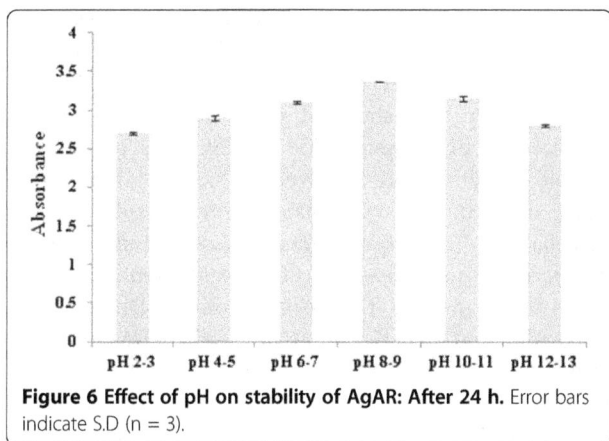

Figure 7 Effect of salt (NaCl) on stability AgAR: After 24 h.

nanoconjugates was found to be stable at all pH vales and was not affected by high concentration of salt solution and heating up to 100°C.

Methods
Materials and instruments
Silver nitrate ($AgNO_3$), sodium borohydride ($NaBH_4$), sodium hydroxide (NaOH), sodium chloride (NaCl), methanol (MeOH), dimethylformamide (DMF) and hydrochloric acid (HCl) were purchased from Merck. Deionized water was used for synthesis of AgNPs. A digital pH meter model 510 (Oakton, Eutech) equipped with a glass working electrode and a reference Ag/AgCl electrode was used. UV–vis spectra were recorded with a Shimadzu UV-240, Hitachi U-3200 spectrometer with a path length of 1 cm. The FTIR spectra were recorded using Shimadzu IR. Microscopic image of synthesized AgNPs was taken with a Zeiss Libra transmission electron microscope (TEM) operated at 120 keV.

Synthesis of ligand and nanoparticles
Synthesis of 5-Amino-β-resorcylic acid hydrochloride dihydrate (AR)
Synthesis of AR illustrated in Figure 1 was carried out by our previously reported method [12]. It was used further as a capping agent for silver nanoparticles.

Synthesis of Silver nanoparticles capped with AR (AgAR)
As AR has low solubility in water therefore a mixture of MeOH, DMF and water (2:3:5) was used as a solvent for its solubility. 1 mM solution of AR was prepared in respective solvent and used for synthesis of AgNPs. 1 mM solution of $AgNO_3$ was prepared in deionized water. Freshly prepared (40 mM in water) solution of $NaBH_4$ was used as a reducing agent. I mL of AR and 10 mL of Ag NO_3 (1 mM) was taken and kept on stirring for

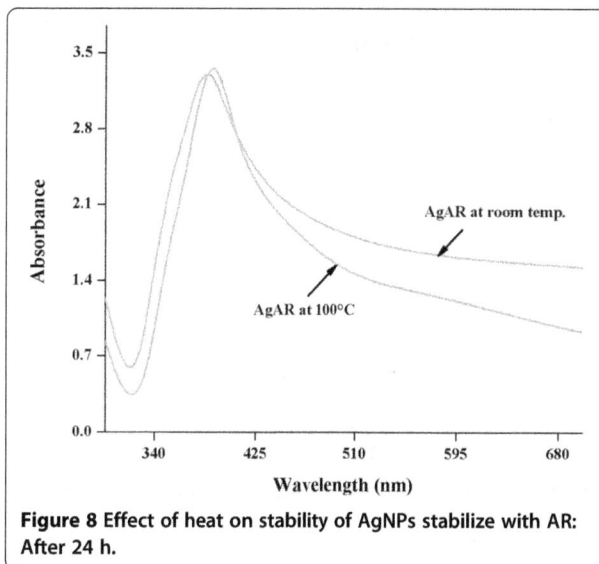

Figure 6 Effect of pH on stability of AgAR: After 24 h. Error bars indicate S.D (n = 3).

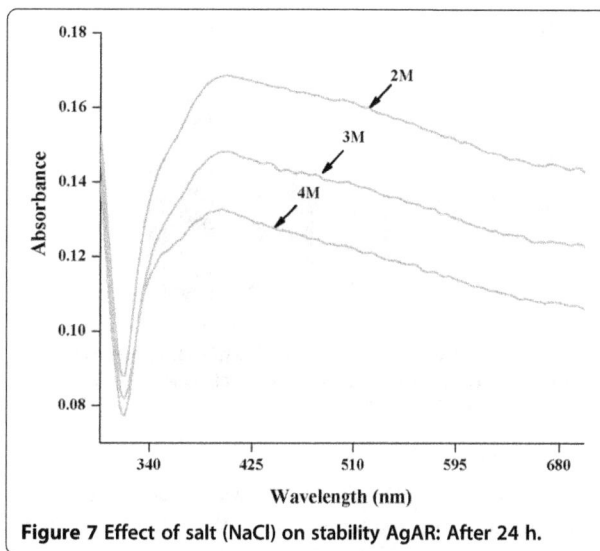

Figure 8 Effect of heat on stability of AgNPs stabilize with AR: After 24 h.

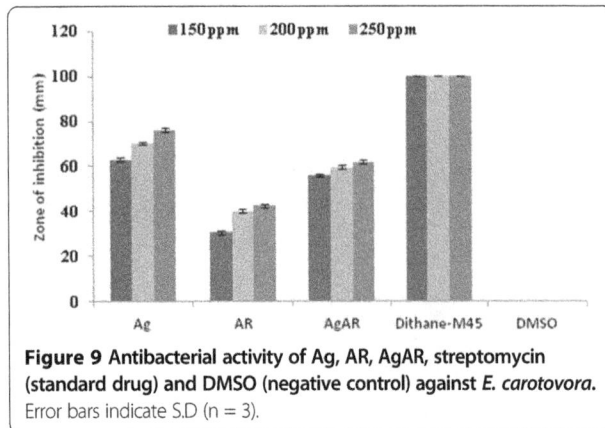

Figure 9 Antibacterial activity of Ag, AR, AgAR, streptomycin (standard drug) and DMSO (negative control) against *E. carotovora*. Error bars indicate S.D (n = 3).

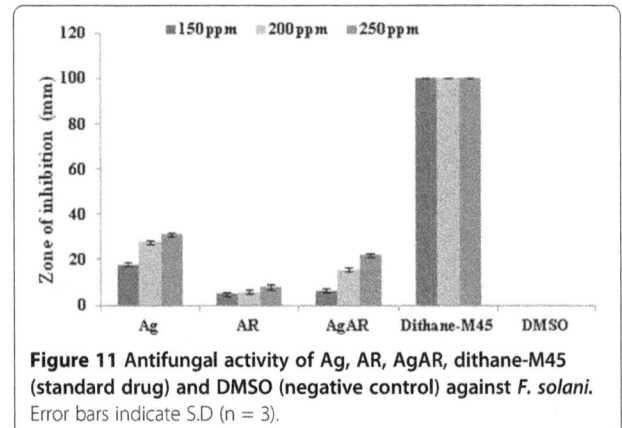

Figure 11 Antifungal activity of Ag, AR, AgAR, dithane-M45 (standard drug) and DMSO (negative control) against *F. solani*. Error bars indicate S.D (n = 3).

30 min. Then 1 mL of $NaBH_4$ was added drop wise. After addition of reducing agent, the color of the reaction mixture generally changed from light brown to dark brown (Additional file 1: Figure S1). After 3 h stirring, the solution was characterized by UV–vis spectroscopy (Figure 1). Similarly different ratios of Ag, AR and $NaBH_4$ were tested (Additional file 2: Table S1). From table S1 the optimum ratio for nanoparticles formation was found to be 10: 1.0: 1.0 (Ag: AR: $NaBH_4$). Solid silver nanoparticles capped with AR (AgAR) were collected by freeze drying for FTIR and TEM characterizations and bioactivity screening.

Bare silver nanoparticles (Ag)

10 mL of Ag NO_3 (1 mM) was taken followed by drop wise addition of $NaBH_4$ and then kept on stirring for 30 min. Solid bare silver nanoparticles (Ag) was collected after freeze drying.

Stability check of silver nanoparticles
pH effect

For pH study, 2 mL of freshly prepared AgAR was taken. Its pH was recorded and found to be 8.49. The pH of AgAR ranging from 10–13 was adjusted by using 1 M

NaOH solution. Similarly, the pH of AgAR ranging from 2–7 was maintained by using 1 N HCl. The UV–vis spectra of resulting solutions were recorded after 24 h (Figure 6).

Salt effect

Effect of high concentration of NaCl (2–4 M) on synthesized AgAR was studied. For this purpose 2 mL of freshly prepared AgAR was taken in a beaker. Then 2 mL of 2 M NaCl solution was added to it. The resulting solution was kept at room temperature for 24 h. Then its UV–vis spectrum was recorded. Similarly, the effect of 3 and 4 M NaCl was monitored by following the same protocol (Figure 7).

Heat effect

Heat effect of synthesized nanoparticles AgAR was studied by taking 10 mL of freshly prepared AgAR in a round bottom flask (25 mL). The solution was heated up to 100°C for 30 min. The nanoparticles was then kept at room temperature for and its UV–vis spectrum was recorded (Figure 8).

Bioactivity screening of silver nanoparticles
Antifungal activity

The antifungal activity was evaluated by the agar-well diffusion method [15]. Dithane-M45 was used as a standard drug. Three different concentration (150, 200 and 250 ppm) of standard drug and tested samples prepared in dimethysulfoxide (DMSO). 10 μL of each sample was uniformly dissolved in PDA (potato dextrose agar) medium on agar plate. Using sterilized borer mycelial plug (5 mm) of *Alternaria solani* and *Fusarium oxysporum* f. sp. *lycopersici* (FOL) were fixed at the center of the agar plate. The positive control (Dithane-M45 fungicide) and negative control (media + DMSO) were also run in parallel for reference. The agar plate cultures were incubated at 25°C for 7 days and growth of the fungal strains were observed on daily basis. After

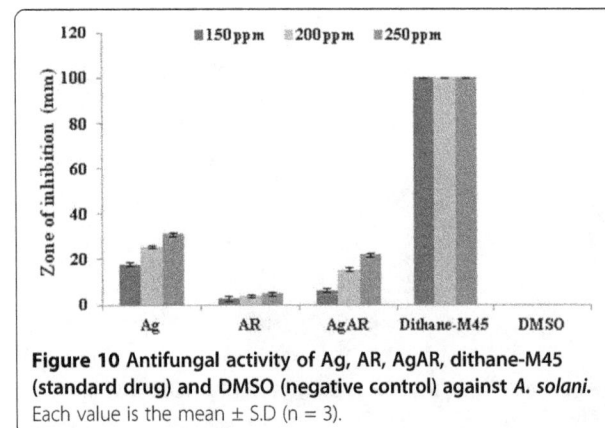

Figure 10 Antifungal activity of Ag, AR, AgAR, dithane-M45 (standard drug) and DMSO (negative control) against *A. solani*. Each value is the mean ± S.D (n = 3).

Table 1 Comparative enzyme inhibition activities of AR, Ag and AgAR

S. No	Enzymes		AR	Ag	AgAR	Standard
1	Xanthine oxidase	% Inhibition	8	9.5	86.8	98.6
		$IC_{50} \pm$ S.E.M* (μg/mL)	°NA	NA	15.5 ± 0.9	2.0 ± 0.1
2	Carbonic anhydrase-II	% Inhibition	71.2	20	0	89.0
		$IC_{50} \pm$ S.E.M (μg/mL)	23.43 ± 1.12	NA	NA	0.21 ± 0.03
3	a-Chymotrypsin	% Inhibition	7	10	0	98.6
		$IC_{50} \pm$ S.E.M (μg/mL)	NA	NA	NA	5.7 ± 0.1
4	Acetylcholi-nesterase	% Inhibition	8	5	0	98.6
		$IC_{50} \pm$ S.E.M (μg/mL)	NA	NA	NA	8.5 ± 0.1
5	Butylcholin-esterase	% Inhibition	12	8	0	98.6
		$IC_{50} \pm$ S.E.M (μg/mL)	NA	NA	NA	8.5 ± 0.1
6	Urease	% Inhibition	5	15	97.0	96.9
		$IC_{50} \pm$ S.E.M(μg/mL	NA	NA	9.2 ± 0.9	21.8 ± 1.2

°NA = Not applicable, *S.E.M = Standard error mean.

incubation, the cultures were compared for reduction in radial colony growth of the fungus in negative, positive and test cultures. Percent growth Inhibition in radial colony growth was determined in triplicate by using the following formula.

$$I = (C - T)/C \times 100$$

I = Percentage of inhibition, C = Diameter of fungal colony in control, T = Diameter of the fungal colony in treatment.

Antibacterial activity

The antibacterial activity of samples was determined according to method described by Yin *et al.* [16]. For this study gram-negative bacteria *Erwinia carotovora* was selected [17]. Streptomycin was used as a positive control. Three different concentrations (150, 200 and 250 ppm) of each sample and standard drug was prepared in DMSO. In sterilized petridishes, 10 mL PDA was poured and bacterial inoculation was carried out by streaking to form different colonies. The disc of 5 mm was kept at the center of agar plate and 10 μL of each sample was

Table 2 DPPH radical scavenging activities of Ag, AR and AgAR

S. No.	Code	$IC_{50} \pm$ S.E.M (μg/mL)	RSA*(%)
1	AR	36.55 ± 2.05	93
2	Ag	NA	10
3	AgAR	NA	5
4	AA'	40.1 ± 1.2	96
5	BHA"	44.6 ± 0.7	95
6	NAC°	105.9 ± 1.1	95

*Radical scavenger activity, 'AA (ascorbic acid), "BHA (butylated hydroxyanisole), °NAC (N-acetyl cystine).

poured on disk. The positive control and negative control (media + DMSO) were also run in parallel for reference. and zone of inhibition was observed after 72 h. Percent growth of inhibition was calculated by following formula and analysis was done in triplicate.

$$IE (\%) = DC\text{–}DS/DC - 5 \times 100$$

DC stands for diameter of control and 5 is the size of disc which was 5 mm.

Xanthine oxidase inhibition assay

The XO inhibitory activity of test compounds was determined by measuring the rate of hydroxylation of the substrate (xanthine) and subsequent formation of uric acid, which is a colorless end product of the reaction and showed absorption at 295 nm. Briefly, the reaction mixture containing 10 μL of 1 mg/L pure compound or 0.2 mg/mL of nanoparticles was dissolved in DMSO, 150 μL of phosphate buffer (0.05 mol/L, pH 7.4), 0.003 units of Xanthine Oxidase dissolved in buffer (20 μL), and 20 μL of 0.1 mmol/L xanthine as substrate for enzyme. After addition of xanthine oxidase, the mixture was incubated for 10 min at room temperature and pre-read in the UV region (λ_{max} 295 nm). The substrate was added to reaction mixture, and continuous reading for 15 min at an interval of 1 min was observed (Spectra MAX-340). The percentage inhibitory activity induced by the samples were determined against a DMSO blank and calculated using the following formula. Inhibition (%) = 100 - [(OD test compound/OD control) × 100]. The IC_{50} of the compounds as calculated using EZ-Fit windows-based software (Perrella Scientific Inc. Amherst, U.S.A.). To compare the inhibitory activities of the compounds, allopurinol was used as standard. The reaction for each compound was performed in triplicate [18].

Urease inhibition assay

Exact 25 μL of enzyme (jack bean urease) solution and 5 μL of test compounds (0.2 mg/mL of nanoparticles) were incubated with 55 μL of buffers containing 100 mM urea for 15 min at 30°C in each well of 96-well plates,. Ammonia production was measured as a urease activity by indophenol method [19]. Final volumes were maintained as 200 μL by adding 45 μL phenol reagent (1% w/v phenol and 0.005% w/v sodium nitroprussside) and 70 μL of alkali reagent (0.5% w/v NaOH and 0.1% active chloride NaOCl) to each well. Using a microplate reader (Molecular Devices, CA, USA), the increase in absorbance was measured at 630 nm after 50 min at pH 6.8. The results (change in absorbance per min) were collected using softMax Pro software (Molecular Devices, CA, USA). Thiourea was used as the standard inhibitor and percentage inhibitions were calculated as follow: $100-(OD_{testwell}/OD_{control}) \times 100$ [20]. The analysis was done in triplicate.

Carbonic anhydrase-II inhibition assay

The experiment was run with the buffer containing HEPES-*Tris* solution at a total concentration of 20 mM and pH (7.2-7.9). For this 140 μL of the HEPES-*Tris* solution was mixed with 20 μL of freshly prepared aqueous solution of purified bovine erythrocyte CA-II (0.1-0.2 mg/2000 μL of demonized water for 96-well), Fluka MP Biomedicals. The test compound was dissolved in 10% DMSO, out of this 20 μL was added in a reaction mixture, followed by the addition of 4-NPA at concentration of 0.8 mM diluted in ethanol [21]. The reaction was initiated by addition of 4-NPA after 15 min incubation of test compound. The compounds were tested in triplicate. In this assay, the reaction was performed by using 96-well plates. To initiate the reaction, the plate was placed in a microplate reader and the amount of reaction product formed was monitored at 1 min interval for 30 min at 400 nm. The reaction temperature was kept between 25–28°C.

α-Chymotrypsin assay

The inhibitory activity of α-chymotrypsin was performed in 50 mM *Tris–HCl* buffer pH 7.6 with 10 mM $CaCl_2$, as mentioned by Cannell *et al.* with the slight modification. The enzyme α-chymotrypsin (12 units/mL prepared in buffer mentioned above) with the 0.5 mM test compound prepared in DMSO, was incubated at 30°C for 25 min. The reaction was initiated by the addition of the chromogenic substrate, *N*-succinyl-L-phenylalanine-*p*-nitroaniline (SP*p*NA; 0.4 mM final concentration prepared in the buffer as above). The change in absorbance by release of *p*-nitroanilide was continuously monitored at 410 nm. The positive control without test compound was replaced by DMSO (final concentration 7%). The

analysis was performed in triplicate. The percentage of inhibition based upon initial velocity and calculated as [22]:

$$\% \text{ Inhibition } = 100 - (OD_{test}/OD_{control}) * 100$$

Cholinesterase Inhibition assays

Acetyl cholinesterase (AChE) and butyl cholinesterase (BChE) inhibiting activities were measured by the spectrophotometric method by using acetylthiocholine iodide and butyrylthiocholine chloride as substrates. The reaction mixture contained 130 μL of (100 mM) sodium phosphate buffer (pH 8.0), 20 μL of DTNB, 10 μL of tested compound solution and 20 μL of AChE or BChE solution, which were mixed and incubated for 15 min at 25°C. The reaction was then initiated by the addition of 20 μL acetylthiocholine or butyrylthiocholine, respectively. The hydrolysis of acetylthiocholine and butyrylthiocholine were monitored at a wavelength of 412 nm (15 min). Absolute ethanol, which becomes 5% in the reaction mixture, was used as solvent for test compounds and the standard inhibitor. All the reactions were performed in triplicate in 96-well microplate using SpectraMax Plus 384 [23,24].

Antioxidant assay

Measurement of superoxide radical scavenging activity was carried out by the modified method used by Ferda [25]. The reaction mixture comprised of 40 mL of 280 mM b-nicotinamide adenine dinucleotide reduced form (NADH), 40 mL of 80 mM nitro blue tetrazolium (NBT), 20 mL phenazine methosulphate (PMS), 10 mL of 0.2 mg/mL sample and 90 mL of 0.1 M phosphate buffer (pH 7.4). Reagents were prepared in buffer solution and the sample in DMSO. The reaction was performed in 96-well microtitre plate at room temperature and absorbance was measured at 560 nm. The formation of superoxide was monitored by measuring the formation of water soluble blue formazan dye. A lower absorbance of reaction mixture indicates a higher scavenging activity of the sample. Percent radical scavenging activity (% RSA) by samples can be determined in comparison with a control.

$$\text{RSA (\%)} = 100 - (OD_{testcompound}/OD_{control}) * 100$$

The reaction mixture containing 5 μL of (0.5 mg/mL) test samples (in DMSO) and 95 mL of DPPH (300 mmol) in ethanol, was taken in a 96-well microtitre plate and incubated in ELISA (multiple reader, Spectra Max – 3400) at 37°C for 30 min. The absorbance was measured at 515 nm. RSA (%) was determined in triplicate by comparison with a DMSO containing controls. Ascorbic acid, butylated hydroxyanisole and N-acetyl cystine were used as the positive controls [26].

Additional files

> **Additional file 1: Figure S1.** Optical recognition of AgNO₃, NaBH₄, AR and AgAR.
>
> **Additional file 2: Table S1.** Optimization of reaction conditions by changing the amount of Ag, AR and NaBH₄.

Abbreviations

AR: 5-Amino-β-resorcylic acid hydrochloride dihydrate; AgAR: Silver nanoparticles capped with AR; AgNPs: Silver nanoparticles; UV-Visible spectra: Ultra violet-visible spectra; FTIR: Fourier transformed infrared; TEM: Transmission electron microscope.

Competing interests

The authors state that they have no competing interests.

Authors' contributions

SSN performed the synthesis of nanoparticles and wrote the manuscript. MRS supervised the whole work. AK and SSA helped in interpretation of biological data. NUI, SN and SQ helped with the analysis. All authors read and approved the final manuscript.

Acknowledgements

We are indebted to the HEJ, Research Institute of Chemistry, ICCBS, University of Karachi and Higher Education Commission of Pakistan for the availability of chemicals and instruments. We thank Dr. Massimo Bertino of Department of Physics, Virginia Commonwealth University for his assistance with acquiring TEM images and Dr. Mohammad Aslam Baig of National Centre of Physics, Quaid-i-Azam University Campus, Islamabad, for his contribution in language correction of the manuscript.

Author details

¹Nanoscience and Catalysis Division, National Centre for Physics, Quaid-i-Azam University Campus, Islamabad 44000, Pakistan. ²H.E.J. Research Institute of Chemistry, International Centre for Chemical and Biological Sciences, University of Karachi, Karachi 75270, Pakistan. ³Institute of Chemical Sciences, University of Peshawar, Peshawar 25120, Pakistan. ⁴Sarhad University of Science and Information Technology, Peshawar 2500, Pakistan. ⁵Department of Plant Pathology, Khyber Pakhtunkhwa Agricultural University Peshawar, Peshawar, Pakistan.

References

1. Corma A, Garcia H: **Supported gold nanoparticles as catalysts for organic reactions.** *Chem Soc Rev* 2008, **37**(9):2096–2126.
2. Sýkora D, Kašička V, Mikšík I, Řezanka P, Záruba K, Matějka P, Král V: **Application of gold nanoparticles in separation sciences.** *J Sep Sci* 2010, **33**(3):372–387.
3. Yu DG: **Formation of colloidal silver nanoparticles stabilized by Na + –poly ([gamma]-glutamic acid)-silver nitrate complex via chemical reduction process.** *Colloids Surf B: Biointerfaces* 2007, **59**(2):171–178.
4. Kéki S, Török J, Deák G, Darkczi L, Zsuga M: **Silver nanoparticles by PAMAM-assisted photochemical reduction of Ag+.** *J Colloid Interface Sci* 2000, **229**(2):550–553.
5. Petit C, Lixon P, Pileni MP: **In situ synthesis of silver nanocluster in AOT reverse micelles.** *J Phys Chem* 1993, **97**(49):12974–12983.
6. Sandmann G, Dietz H, Plieth W: **Preparation of silver nanoparticles on ITO surfaces by a double-pulse method.** *J Electroanal Chem* 2000, **491**(1–2):78–86.
7. Smetana AB, Klabunde KJ, Sorensen CM: **Synthesis of spherical silver nanoparticles by digestive ripening, stabilization with various agents, and their 3-D and 2-D superlattice formation.** *J Colloid Interface Sci* 2005, **284**(2):521–526.
8. Tan Y, Wang Y, Jiang L, Zhu D: **Thiosalicylic acid-functionalized silver nanoparticles synthesized in one-phase system.** *J Colloid Interface Sci* 2002, **249**(2):336–345.
9. Abou El-Nour KMM, Eftaiha A, Al-Warthan A, Ammar RAA: **Synthesis and applications of silver nanoparticles.** *Arab J Chem* 2010, **3**(3):135–140.
10. Vogt RL, Dippold L: **Escherichia coli O157: H7 outbreak associated with consumption of ground beef, June-July 2002.** *Public Health Rep* 2005, **120**(2):174–178.
11. Dar MA, Ingle A, Rai M: **Enhanced antimicrobial activity of silver nanoparticles synthesized by< i> Cryphonectria</i> sp. evaluated singly and in combination with antibiotics.** *Nanomedicine* 2013, **9**(1):105–110.
12. Naz SS, Islam NU, Tahir MN: **5-Carboxy-2,4-dihydroxyanilinium chloride dihydrate.** *Acta Cryst* 2011, **E67**:o299.
13. Sohaila NS, Ul IN, Raza SM, Sartaj AS, Zafar I, Massimo B, Louis F, Afifa A: **Enhanced biocidal activity of Au nanoparticles synthesized in one pot using 2, 4-dihydroxybenzene carbodithioic acid as a reducing and stabilizing agent.** *J Nanobiotechnol* 2013, **11**(1):13.
14. Espinoza MG, Hinks ML, Mendoza AM, Pullman DP, Peterson KI: **Kinetics of Halide-Induced Decomposition and Aggregation of Silver Nanoparticles.** *J Phys Chem B* 2012, **116**:8305–8313.
15. Ahmad B, Azam S, Bashir S, Hussain F, Chaudhary MI: **Insecticidal, brine shrimp cytotoxicity, antifungal and nitric oxide free radical scavenging activities of the aerial parts of Myrsine africana L.** *Afr J Biotechnol* 2011, **10**(8):1448–1453.
16. Tsao S-M, Yin M-C: **In-vitro antimicrobial activity of four diallyl sulphides occurring naturally in garlic and Chinese leek oils.** *J Med Microbiol* 2001, **50**(7):646–649.
17. Toth IK, Bell KS, Holeva MC, Birch PRJ: **Soft rot erwiniae: from genes to genomes.** *Mol Plant Pathol* 2003, **4**(1):17–30.
18. Lee SK, Mbwambo ZH, Chung H, Luyengi L, Gamez EJ, Mehta RG, Kinghorn AD, Pezzuto JM: **Evaluation of the antioxidant potential of natural products.** *Comb Chem High Throughput Screen* 1998, **1**(1):35.
19. Weatherburn M: **Phenol-hypochlorite reaction for determination of ammonia.** *Anal Chem* 1967, **39**(8):971–974.
20. Choudhary MI, Ali M, Wahab A-t, Khan A, Rasheed S, Shyaula SL, Rahman A-u: **New antiglycation and enzyme inhibitors from Parmotrema cooperi.** *Sci China Chem* 2011, **54**(12):1926–1931.
21. Arslan O: **Inhibition of bovine carbonic anhydrase by new sulfonamide compounds.** *Biochem Mosc* 2001, **66**(9):982–983.
22. Cannell RJ, Kellam SJ, Owsianka AM, Walker JM: **Results of a large scale screen for the production of protease inhibitors.** *Planta Med* 1988, **54**(01):10–14.
23. Ellman GL, Courtney KD, Andres V, Featherstone RM: **A new and rapid colorimetric determination of acetylcholinesterase activity.** *Biochem Pharmacol* 1961, **7**(88):95.
24. Gupta A, Thakur SS, Uniyal PL, Gupta R: **A survey of bryophytes for presence of cholinesterase activity.** *Am J Bot* 2001, **88**(12):2133–2135.
25. Pezzuto JM, Kosmeder JW, Park EJ, Lee SK, Cuendet M, Gills J, Bhat K, Grubjesic S, Park H-S, Mata-Greenwood E: **Characterization of natural product as chemopreventive agents.** In *Cancer Chemoprevention.* 2nd edition. Edited by Kelloff GJ, Hawk ET, Sigman CC. Totowa, NJ: Humana Press Inc; 2005:3–37. http://link.springer.com/chapter/10.1007%2F978-1-59259-768-0_1.
26. Thadhani MV, Iqbal CM, Sajjad A, Iman O, Hina S, Veranja K: **Antioxidant activity of some lichen metabolites.** *Nat Prod Res* 2011, **25**:1827–1837.

3

Fluorescent carbon dots as an efficient siRNA nanocarrier for its interference therapy in gastric cancer cells

Qing Wang[1,2†], Chunlei Zhang[2†], Guangxia Shen[2*], Huiyang Liu[2], Hualin Fu[2] and Daxiang Cui[1,2*]

Abstract

Background: Fluorescent carbon dots (Cdots) have attracted increasing attention due to their potential applications in sensing, catalysis, and biomedicine. Currently, intensive research has been concentrated on the synthesis and imaging-guided therapy of these benign photoluminescent materials. Meanwhile, Cdots have been explored as nonviral vector for nucleic acid or drug delivery by chemical modification on purpose.

Results: We have developed a microwave assisted one-step synthesis of Cdots with citric acid as carbon source and tryptophan (Trp) as both nitrogen source and passivation agent. The Cdots with uniform size show superior water solubility, excellent biocompatibility, and high quantum yield. Afterwards, the PEI (polyethylenimine)-adsorbed Cdots nanoparticles (Cdots@PEI) were applied to deliver Survivin siRNA into human gastric cancer cell line MGC-803. The results have confirmed the nanocarrier exhibited excellent biocompatibility and a significant increase in cellular delivery of siRNA, inducing efficient knockdown for Survivin protein to 6.1%. In addition, PEI@Cdots complexes mediated Survivin silencing, the arrested cell cycle progression in G_1 phase as well as cell apoptosis was observed.

Conclusion: The Cdots-based and PEI-adsorbed complexes both as imaging agents and siRNA nanocarriers have been developed for Survivin siRNA delivery. And the results indicate that Cdots-based nanocarriers could be utilized in a broad range of siRNA delivery systems for cancer therapy.

Keywords: Carbon dots, siRNA interference therapy, Gastric cancer, Nanocarriers

Introduction

Over recent decades, great advances have been made in the combination of nanotechnology and medicine, which is paving the way towards the goal of clinic application [1,2]. A large amount of biocompatible fluorescence nanomaterials, such as quantum dots, metal nanoclusters, and fluorescent polymers, have been developed [3-5]. Carbon dots (Cdots) are currently emerging as a class of promising fluorescent probe on account of their low photobleaching, no optical blinking, tunable photoluminescence, versatile surfaces, and excellent biocompatibility [6,7]. Therefore,

fluorescent Cdots possess additional benefits over organic fluorophores and semiconductor quantum dot, which are more or less circumscribed by their photobleaching or intrinsic potential hazards of heavy metal elements (e.g. Cd and Pb) [8]. These excellent properties of Cdots have made bright prospects in the applications of bioimaging, drug delivery, biochemical detection, and sensors [9-11]. Currently, intensive research has been focus on the synthesis of Cdots with high quantum efficiency and the construction of multifunctional systems based on Cdots [10,12,13]. Until now, various precursors including graphite, C_{60}, citric acid, glucose, and silk have been developed for the preparation of Cdots with a wide variety of approaches, processes, and tools [10,12,14-16]. Sun's group has reported a new strategy to prepare core-shell dots based on Cdots doped with inorganic salts with quantum yield around 45% ~ 60%, but the preparation process is quite complicated [17].

---

* Correspondence: gxshen@sjtu.edu.cn; dxcui@sjtu.edu.cn
†Equal contributors
[2]Institute of Nano Biomedicine and Engineering, Key Laboratory for Thin Film and Microfabrication Technology of the Ministry of Education, Department of Instrument Science & Engineering, School of Electronic Information and Electrical Engineering, Shanghai Jiao Tong University, 800 Dongchuan RD, Shanghai 200240, China
[1]School of Life Sciences and Biotechnology, Shanghai Jiao Tong University, Shanghai 200240, China

Recently, our group has developed a green synthetic route for Cdots with high quantum yield around 24.2% using Ribonuclease A (RNase A) as an assisting and passivating reagent *via* microwave assisted one step procedure [18]. Interestingly, the RNase A@Cdots can effectively inhibit the survival rate of cancer cells. But the price of RNase A is not economic enough. Taking into account the mechanism of the photoluminescence (PL) enhancement in RNase A@Cdots, electron-donating effect from neighbor amino acids especially those with benzene rings could play an important role. Therefore, we select tryptophan (Trp), a kind of amino acids with benzene ring while possesses a higher nitrogen content than tyrosine and phenylalanine, for the synthesis of Cdots with lower cost. Besides, we have ever developed a new theranostic platform based on photosensitizer-conjugated Cdots with excellent imaging and tumor homing ability for NIR fluorescence imaging guided photodynamic therapy [19]. However, till now few reports are closely associated with the using of Cdots as gene transfection vector for cancer therapy.

RNA interference (RNAi) has emerged as a valuable research tool to downregulate the expression of specific target proteins in a wide variety of cells [20]. RNAi is a biological process in which RNA molecules inhibit gene expression by causing the destruction of specific mRNA molecules. Viral delivery such as Lentivirus, Adenovirus and Adeno-Associated-Virus, has successfully been used for therapeutic applications, including cancer therapy [21,22]. To deal with concerns over the potential risk of undesired immune and toxic side reactions in virus-mediated nucleic acid delivery systems, non-viral gene delivery systems based on inorganic nanoparticles, cationic liposomes, and cationic polyamidoamine dendrimers have been employed as carriers for gene silencing [23-25]. The nanomaterials-based non-viral gene delivery systems

have benefited from many advantages over viral vectors, as they are simple to prepare, rather stable, easy to modify and relatively safe. Meanwhile, Cdots would be an ideal nanocarrier due to the high biological safety, well-defined structures together with their tunable surface functionalities.

In this work, the Cdots-based and PEI-adsorbed nanocarrier (Cdots@PEI) was developed to delivery siRNA against antiapoptotic protein Survivin in human gastric cancer cell line MGC-803. It is well known that Survivin play an important role in cell division, apoptosis, and checkpoint mechanisms of genomic integrity [26,27]. In addition, Survivin expression is often upregulated in human cancers, as it can be treated as a specific tumor marker with prognostic and therapeutic implications from studies of gastric carcinomas [28]. For the preparation of the Cdots-based nanocarrier, firstly, the Cdots was directly synthesized *via* microwave pyrolysis of citric acid in the presence of tryptophan and purified by gradient centrifugation and dialyzed with MWCO 3500 dialysis bag for 72 h. Secondly, the Cdots@PEI complexes were formed by electrostatic interaction between negatively charged Cdots and positively charged PEI (as depicted in Figure 1). The Cdots@PEI complexes were used to mix with Survivin siRNA (siRNA-Cdots@PEI). Afterwards, we investigated gene transfection efficacy, cellular uptake, and biological effects of the siRNA-Cdots@PEI complexes towards gastric cancer cells MCG-803 by means of reverse transcription polymerase chain reaction (RT-PCR), Western blotting, apoptosis assay and cell cycle analysis. The results showed that siRNA can attach to the surface of the Cdots@PEI complexes and notably enhanced the gene delivery efficiency. The system renders the possibility of Cdots to serve as a universal transmembrane carrier for intracellular gene and drug delivery and imaging applications in cancer gene therapy.

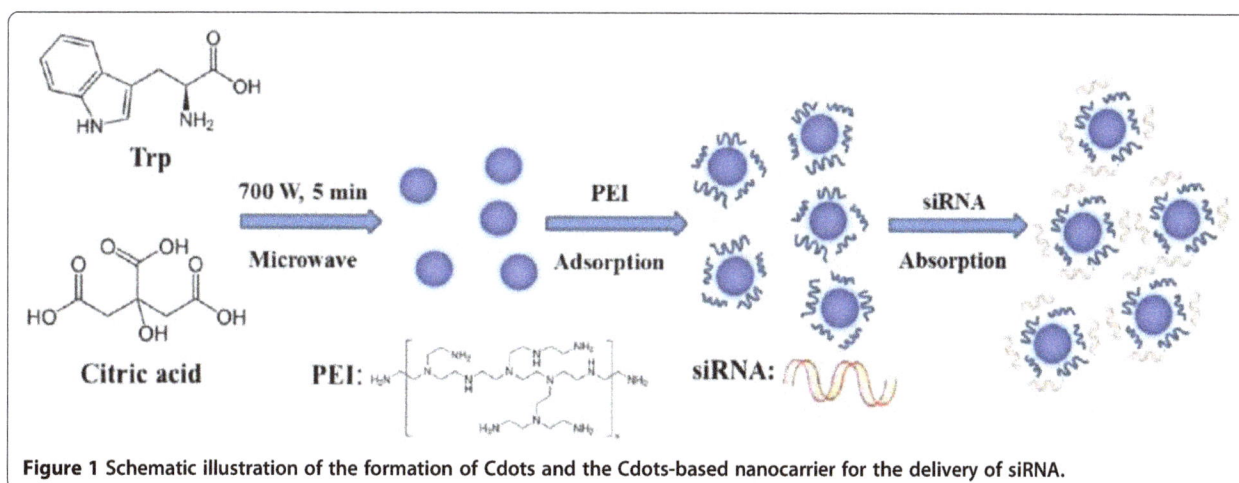

Figure 1 Schematic illustration of the formation of Cdots and the Cdots-based nanocarrier for the delivery of siRNA.

Experimental sections

Materials

Citric acid, L-tryptophan (99%) and polyethylenimine (PEI) with a molecular weight of 1800 Da were purchased from Aladdin Reagent Co. Ltd. (Shanghai, China). Dimethyl sulfoxide (DMSO) was obtained from Sinopharm Chemical Reagent Co., Ltd, China. Hieff™ qPCR SYBR® Green Master Mix and Annexin V-FITC/PI Apoptosis Detection Kit were purchased from Yeasen Corporation (Shanghai, China). 3-[4, 5-dimethylthiazol-2yl]-2,5-diphenyltetrazolium bromide (MTT) was purchased from Invitrogen Corporation (Carlsbad, CA, USA). Random primer and M-MLV reverse transcriptase were purchased from Promega (Madison, WI, USA). RIPA lysis buffer and BCA (bicinchoninic acid) protein assay kit were purchased from Beyotime Biotechnology (Jiangsu, China). Monoclonal rabbit anti-Survivin antibody, polyclonal rabbit anti-β-actin antibody, and horseradish peroxidase (HRP)-conjugated goat anti-rabbit IgG were purchased from Epitomics (Burlingame, CA, USA). Complete protease inhibitor cocktail and BM Chemiluminescence Western Blotting Kit were obtained from Roche (Mannheim, Germany). Human gastric cancer MGC-803 cells and human gastric mucous epithelial GES-1 cells were available in the Cell Bank of Type Culture Collection of Chinese Academy of Sciences. Cell culture products and reagent were purchased from GIBCO. All the above chemicals were used without any further purification. Ultrapure water (Millipore Milli-Q grade) with a resistivity of 18.2 MΩ cm was used in all the preparations.

Synthesis and characterization of Cdots and Cdots@PEI

The luminescent Cdots was synthesized by a one-step green route of microwave assisted pyrolysis method. Briefly, 2 g citrate and 0.01 g L-Trp was dissolved in 30 ml of ultrapure water, and stirring for 1 h to form a homogeneous solution in a 100 ml beaker. Then the beaker containing clear transparent solution was placed at the center of the rotation plate of a domestic microwave oven (700 W) and heated for 3 minutes. When cooled down to room temperature, the Cdots were isolated from the opaque suspension via centrifugation with 10,000 rpm which aims at removing carbon residual. Excessive citric acid and L-Trp were removed via repeated dialysis against deionized water using a low molecular weight cut-off membrane (1000 Da) for 2 days. Finally, different concentration can be got through rotary evaporation, which is used to remove water as it depends. The morphologies of the Cdots were detected using a transmission electron microscope (TEM, 2100F, JEOL, Japan), operating at an accelerated voltage of 200 kV. Photoluminescence (PL) spectra were measured on a Hitachi FL-4600 spectrofluorometer. UV–vis spectra were recorded with a Varian Cary 50 spectrophotometer

(Varian Inc., Palo Alto, CA, USA). X-ray diffraction (XRD) measurement was performed with a D8 Advance (Bruker AXS Corporation, Germany). Fourier transform infrared (FTIR) spectra were conducted on a Nicolet 6700 spectrometer (Thermo Electron Corporation, Madison, WI) using KBr pellets. X-ray photoelectron spectrum (XPS) was acquired with a Kratos Axis Ultra^{DLD} spectrometer (AXIS Ultra, Kratos Analytical Ltd, Japan) using a monochromatic Al Kα source (1486.6 eV). Zeta potential was completed using a NICOMP 380 ZLS Zeta potential/Particle sizer (PSS Nicomp, Santa Barbara, CA, USA) equipped with a He-Ne laser (λ = 633 nm).

Quantum yield measurement

The quantum yield of the Cdots was determined using quinine sulfate in 0.1 M H_2SO_4 (quantum yield: 54%) as the standard sample. According the emission peak area and absorbance of Trp@Cdots and quinine sulfate, the QY of the Trp@Cdots could be calculated from Equation 1 below:

$$\varphi_{sample} = \frac{A_{std}}{A_{sample}} \times \frac{F_{sample}}{F_{std}} \times \frac{n_{sample}^2}{n_{std}^2} \times \varphi_{std}$$

Where Φ_{std} is the known quantum yield of the standard compound, F_{sample} and F_{std} are the integrated areas of fluorescence of the sample and standard in the emission region at 350–600 nm. A_{std} and A_{sample} are the absorbance of the standard and sample at the excitation wavelength (360 nm); n is the refractive index of solvent, for water the refractive index is 1.33, 0.1 M H_2SO_4 is 1.33. All samples were diluted to ensure the optical densities less than 0.10 measured by Varian Cary 50 UV–vis spectrophotometer to minimize re-absorption effects.

Preparation of siRNA-Cdots@PEI complexes and agarose gel electrophoresis

Three couples of siRNA oligonucleotides (noted as Surv-1, Surv-2 and Surv-3, respectively) and a couple of non-silencing-siRNA as negative control (NC) oligonucleotides were chemically synthesized by Shanghai Genechem Co. All siRNAs were annealed with complementary antisense strands with 3′-dTdT overhangs. The siRNA duplexes are as follows:

Surv-1: sense, 5′-CACCGCAUCUCUACAUUCATT (dTdT)-3′; antisense, 5′-UGAAUGUAGAGAUGCG GUGTT (dTdT)-3′.
Surv-2: sense, 5′-GAAGCAGUUUGAAGAAUUATT (dTdT)-3′; antisense, 5′-UAAUUCUUCAAACUGC UUCTT(dTdT)-3′.
Surv-3: sense, 5′-GGUCCCUGGAUUUGCUAAUT (dTdT)-3′; antisense, 5′-AUUAGCAAAUCCAGG GACCTT(dTdT)-3′.

NC: sense, 5′-UUCUCCGAACGUGUCACGUTT (dTdT)-3′; antisense, 5′-ACGUGACACGUUCG GAGAATT(dTdT)-3′.

The Cdots-based complexes were formed by electrostatic interactions between positively charged PEI and the negatively charged Cdots, and then the negatively charged phosphate backbone of siRNA attaching to the Cdots@PEI complexes, resulting the siRNA loaded siRNA-Cdots@PEI complexes. The Cdots@PEI and siRNA-Cdots@PEI complexes were purified by gel filtration over a Sephadex G-50 column equilibrated with 10 mM NaCl to remove unadsorbed PEI or siRNA. The electrophoretic mobility of the siRNA-Cdots@PEI complexes was determined by 1% agarose gel in 1× TBE buffer with a constant voltage of 120 V for 20 min. The siRNA loading amount was determined by measuring the absorption at 260 nm using the relation 1OD duplex = 3.0 nmols after subtracting the absorbance contributed by Cdots@PEI at the same wavelength. In aqueous solution of pH 7.4, the concentration of siRNA was determined to be about 15 nM when the concentration of Cdots@PEI was 100 μg/ml. All data when used for siRNA-Cdots@PEI complexes were expressed as 100 μg/ml Cdots@PEI with about 15 nM siRNA, unless otherwise specified.

Cell culture and MTT assay

Human gastric cancer MGC-803 cells and human gastric epithelial GES-1 cells were available in the Cell Bank of Type Culture Collection of Chinese Academy of Sciences. All the cells were cultured in Dulbecco's modified Eagle's medium (DMEM) plus 10% (vol/vol) fetal bovine serum (Gibco) and penicillin-streptomycin (100 U/ml to 0.1 mg/ml) and incubated in a humidified incubator containing 5% CO_2 at 37°C. MTT assay was carried out to investigate the cytotoxicity of Cdots and Cdots@PEI. MGC-803 and GES-1 cells were first seeded to 96-well plates at a seeding density of 5×10^3 cells per well in 100 μl complete medium, which was incubated at 37°C for 24 h. Then the culture medium in each well was replaced by 100 μl fresh complete medium containing serial concentrations of Cdots and Cdots@PEI. After incubation for 24 h, the medium was replaced with 150 μL fresh medium containing 15 μl MTT (5 mg/ml in PBS) and incubated for another 4 h. Afterwards, the culture medium with MTT was removed and 150 μl/well of DMSO was added, followed by shaking for 10 min at room temperature. The absorbance of each well was measured at 490 nm using a standard micro plate reader (Scientific Multiskan MK3, thermo, USA). The cell viability was calculated according to the equation: cell viability = ($OD_{490\,nm}$ of the experimental group/$OD_{490\,nm}$ of the control group) × 100% and the cell viability of control group was denoted as 100%.

In vitro siRNA transfection and cellular uptake of siRNA-Cdots@PEI complexes

Before transfection, the MGC-803 cells were seeded in 6-well plates and the appropriate transfected number of cells is based on the fact that confluent of cells achieve to 30% to 50% at the time of transfection. Each sample prepared siRNA oligo-Cdots@PEI complexes are as follows: A. siRNA oligo stock solution was diluted to 1 μM before transfection. Then 110 μl 1 μM of siRNA oligo was added to 200 μl serum-free DMEM and mix gently at room temperature for 5 min. B. After incubated for 5 min, 100 μl Cdots@PEI are taken into diluted siRNA oligo (mentioned in a.) with immediate shaking (using a scroll instrument or pipetting more than 10 s). After mild centrifugation, the solution needs to stand still at room temperature for 10 min, to allow the effective formation of siRNA oligo-Cdots@PEI complexes. C. While it is incubated, the medium in the cell culture plates need to be refreshed. Each well was added 1.8 ml of complete medium (containing 10% serum and antibiotics). The siRNA oligo-Cdots@PEI complexes were dropped into each well containing cells and the medium. Gently shake the culture plate after mixing. Complete medium can be changed after 4–6 h transfection. Scrambled siRNA with the transfection reagent of Cdots@PEI was used as the nontargeting control. After incubation with different transfection complexes for 48 h, approximately 2×10^5 MGC-803 cells were collected from each sample and then subjected to quantitative reverse transcription-PCR (qRT-PCR) and Western blot analysis to determine the silencing efficiency against Survivin gene and Survivin protein expression. For evaluation of cellular uptake of siRNA-Cdots@PEI complexes, we tracked the cellular internalization of Cdots-PEI, Cy3-labelled siRNA, or Cy3-siRNA-Cdots@PEI in MGC-803 cells. The cells were plated on 14 mm glass coverslips and allowed to adhere for 24 h. After co-incubation with Cdots-PEI, Cy3-labelled siRNA, or Cy3-siRNA-Cdots@PEI for different times, the cells were washed twice with PBS sufficiently and fixed with 4% paraformaldehyde. Confocal fluorescence images were captured with a TCS SP5 confocal laser scanning microscopy (Leica Microsystems, Mannheim, Germany). Blue and red fluorescence images were acquired using DAPI-specific (excitation, 340–380 nm; emission, 450–490 nm) and Cy3-specific (excitation, 515–560 nm; emission, > 590 nm) sets of filters, respectively.

Survivin expression assay by qRT-PCR analysis

In qRT-PCR experiment, the total RNA was extracted from transfected MGC-803 cells using the TRIzol reagent (Invitrogen, America) according to the manufacturer's instructions. A total of 1 μg of RNA was transcribed into cDNA using random primers and M-MLV reverse transcriptase (Promega). The cDNA templates were stored at −20°C. The qRT-PCR was performed in a final volume of 25 µl containing 12.5 µl of Hieff™ qPCR SYBR® Green Master Mix (Yeasen, Shanghai), and 1 µl of each 10 µM primer, and 1 µl of 1:10-diluted cDNA products. The PCR amplification was carried out in a Bio-Rad iQ5 with one cycle at 95°C for 5 min, followed by 30 cycles at 95°C for 30 sec, at 67°C for 30 sec, and at 72°C for 1 min, and finally at 72°C for 5 min. GAPDH was chosen as the endogenous control in the assay. The following PCR primers were used: GAPDH primers, forward: 5′-CCACCCATGGCAAATTCCATGGCA-3′, reverse: 5′-TCTATCTAGACGGCAGGTCAGGTCCA CC-3′; Survivin primers, forward: 5′-GTGAATTTTT GAAACTGGACAG-3′, reverse: 5′-CCTTTCCTAAGA CATTGCTAA-3′.

Western blot analysis

MGC-803 cells were lysed at 72 h after the transfection using RIPA lysis buffer (20 mM Tris, pH 7.5, 150 mM NaCl, 1% Triton X-100, 2.5 mM sodium pyrophosphate, 1 mM EDTA, 1% Na_3VO_4, 0.5 µg/ml leupeptin, and 1 mM phenylmethanesulfonyl fluoride) in the presence of complete protease inhibitor cocktail (Roche Diagnostics). The homogenate was then subjected to 10,000 rpm centrifugation for 10 min at 4°C. All the above procedures were performed in ice bath. The protein concentration was determined using BCA (bicinchoninic acid) protein assay kit (Beyotime Biotech, Jiangsu, China) and store at −20°C. The cell extracts (20 µg total proteins) were mixed with four times loading buffer (16% glycerol, 20% mercaptoethanol, and 2% SDS, and 0.05% bromophenol blue) (3:1, sample/loading buffer) and boiled for 5–7 min at 100°C. The samples were then subjected to 12% sodium dodecyl sulfate poly-acrylamide gel electrophoresis (SDS-PAGE) at 120 V for 1 h and then transferred onto 0.45 mm polyvinylidene difluoride membrane (PVDF, Immobilon-P 0.45 µm, Millipore, Billerica, MA), using a semi-dry system (Biocraft, Tokyo, Japan) at 300 mA for 150 min. Membranes were blocked with Tris-buffered saline containing 0.1% Tween 20 and 5% dry skim milk powder and then incubated with rabbit anti-human survivin antibody (1:1000, Epitomics) and rabbit anti-human β-actin antibody (1:2500, Epitomics) at 4°C overnight with a gentle shaking. The next day, after four 5 min washes with TBST buffer, the bolts were then incubated with horseradish peroxidase (HRP)-conjugated secondary antibody (goat anti-rabbit IgG, 1:2500, Epitomics) for 1 h at room temperature. Antibody binding was detected by enhanced chemiluminescence (BM Chemiluminescence Western Blotting kit, Roche) and autora-diography (Kodak X-OMAT; Kodak, Rochester, NY).

Apoptosis assay by Annexin V-FITC and propidium iodide (PI) staining

The apoptotic and necrotic cells were analyzed by Annexin V/PI apoptosis detection Kit (Yeasen) according to the manufacturer's protocol. In brief, MGC-803 cells were seed in 6-well plates at 5×10^4 cells/well for 24 h before co-incubated with Surv-1-Cdots@PEI, Surv-2-Cdots@PEI, Surv-3-Cdots@PEI, NC-Cdots@PEI, and Cdots@PEI complexes, respectively. The cells incubated with complete medium only were set as blank control. After 48 h incubation, the cells were harvested, washed with PBS and re-suspended in 200 µL of binding buffer containing 5 µL Annexin V and 10 µL PI. After incubation in dark at room temperature for 15 min, 400 µL of binding buffer was added to each sample, and the cells were immediately analyzed by FACSCalibur (BD Biosciences, Mountain View, CA). The data analysis was performed with Flow Jo 7.6 software. Positioning of quadrants on Annexin-V/PI plots was performed to distinguish living cells (Annexin V^-/PI^-), early apoptotic cells (Annexin V^+/PI^-), late apoptotic/necrotic cells (Annexin V^+/PI^+).

Cell cycle analysis

Cell cycle analysis was performed using Flow cytometry. After the incubation of MGC-803 cells with Surv-1-Cdots@PEI, Surv-2-Cdots@PEI, Surv-3-Cdots@PEI, NC-Cdots@PEI, and Cdots@PEI complexes for 48 h, respectively, cells were harvested, washed with PBS and fixed overnight in 70% ethanol at −20°C. Then the cells were washed with PBS and stained with 50 µg/ml PI and 100 µg/ml RNase A for 30 min in the dark at room temperature. The cell cycle phase distribution was acquired by FACSCalibur and G_1, S, and G_2/M populations were quantified using FlowJo 7.6 software.

Result and discussion

Synthesis and characterization of Cdots

The morphology and structure of the Cdots were characterized by HRTEM and XRD. The HRTEM images indicate that the Cdots have outstanding uniform sizes and spherical shape, with an average diameter of about 2.6 nm (Figure 2a). The XRD pattern measured for Cdots shows a broad peak located at 2θ of around 20°(d = 4.2 Å), which is consistent with the (002) lattice spacing of Graphite, meanwhile, the larger interlayer spacing of 4.2 Å compared to that of bulk graphite which is about 0.33 nm might have resulted from the poor crystallization [7]. As shown in Figure 2c insets, the final product of Cdots is brownish in aqueous solution and emitted intensive blue luminescence under excitation of 365 nm UV light. The UV–vis absorption spectrum of the Cdots shows a clear an absorption feature at about 280 nm, which is ascribed to the function of aromatic rings of Tryptophan. The emission of the Cdots depends on the excitation wavelength.

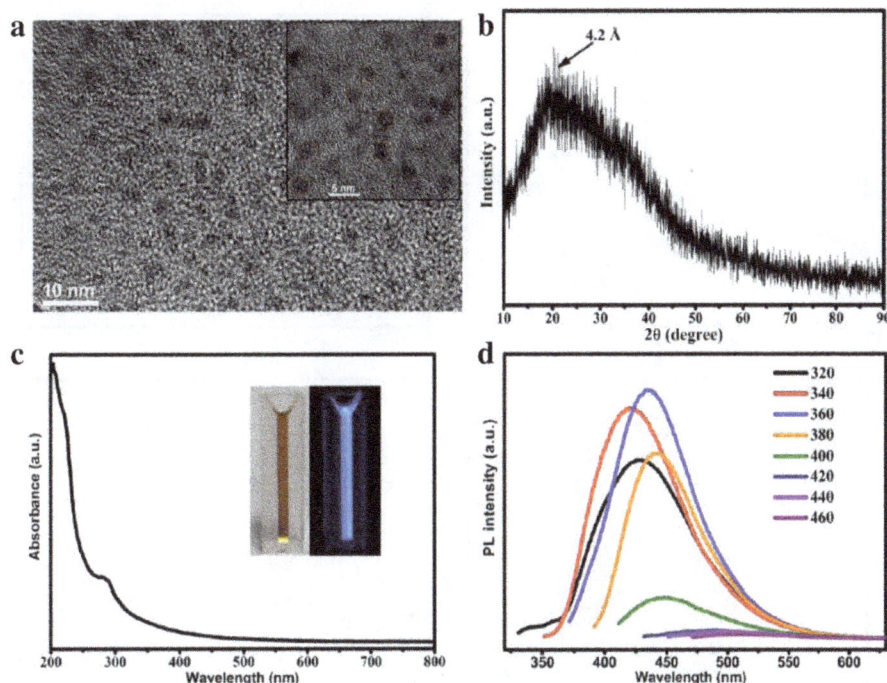

Figure 2 Characterization of the as-synthesized Cdots (a) HRTEM images and inset is the zoom of particles. (b) XRD pattern of the Cdots. The spherical Cdots are marked by circles. **(c)** UV–vis absorption spectrum of the Cdots, insets are digital photos of the Cdots dissolved in water under white-light (left) and UV (365 nm) excitation (right). **(d)** PL spectra of the Cdots when excited at different wavelengths from 320 to 460 nm in a 20 nm increment.

The Cdots show the strongest blue fluorescence under the excitation wavelength of 360 nm, with the highest quantum yield of 20.6%. XPS measurement was performed for the characterization of surface states (Figure 3). Three bands of the XPS survey spectrum at around 531.5, 400.0, and 284.5 eV represent O1s, N1s, and C1s, respectively, which indicating the atomic ratio of O/N/C is 54.6/3.6/41.8 as calculated from the survey spectrum [7,10]. The C1s core level spectrum can be deconvoluted into three contributions at 284.6, 285.2, and 289.0 eV, which are associated with carbon in the states of C-C, C-N, and O-C = O, respectively [7,9]. The O1s peaks at 532.0 and 533.2 eV can be assigned to oxygen in the form of C = O and C-O-C/C-OH, respectively [7,12]. The N1s peaks at 399.5 and 401.2 eV suggest that nitrogen exists mostly in the forms of $(C)_3$-N and N-H, respectively [7,13]. Furthermore, Fourier transform infrared (FTIR) spectral measurement was conducted to determine the surface state of Cdots. As shown in Additional file 1: Figure S1, the as-synthesized Cdots show a main absorption band of O-H/N-H stretching vibration from 3630–2820 cm^{-1} and the existence of carbonyls (C = O) at 1717 cm^{-1}. The bending vibration of C-O/C-N band and stretching peak of the C-O-C bond appear at 1400 cm^{-1} and 1194 cm^{-1}, respectively [9,16,17]. In brief, the as-synthesized Cdots are rich in oxygen and possess mounts of hydroxyl/amine and carboxyl groups, which is useful for further modifications and biological applications.

Formation and characterization of Cdots@PEI and siRNA-Cdots@PEI complexes

The suspension of as-synthesized Cdots was highly stable in aqueous solution, with a zeta potential of −18.9 ± 1.2 mV at pH 7.4, imparting sufficient colloidal stability to the Cdots. Undoubtedly, it's necessary to further exchange exterior charge of the Cdots for the siRNA delivery. So we introduce PEI, which have been reported to be individually considered as nonviral gene carriers with a capability of forming stable complexes by electrostatic interactions with nucleic acids. The zeta potentials of the as-prepared Cdots@PEI and siRNA-Cdots@PEI complexes were 26.6 ± 1.6 and 12.7 ± 0.8 mV, respectively, indicating the successful attachment of siRNA and PEI to the Cdots (Figure 4a). It is noteworthy that even after the load of siRNA; the positively charged siRNA-Cdots@PEI complexes could be conducive to intracellular delivery. The hydrodynamic particle size of siRNA-Cdots@PEI complexes was 4.7 ± 0.8 nm, which was a little larger than that of Cdots (3.9 ± 0.3 nm) (Figure 4b). The results of zeta potentials and hydrodynamic diameters indicated the favorable dispersibility of Cdots@PEI and siRNA-Cdots@PEI complexes in aqueous solution. Furthermore, agarose gel electrophoresis assay was taken to evaluation of capability of the Cdots@PEI complexes. With the attachment of negative charged siRNA, the surface positive charge of siRNA-Cdots@PEI complexes was

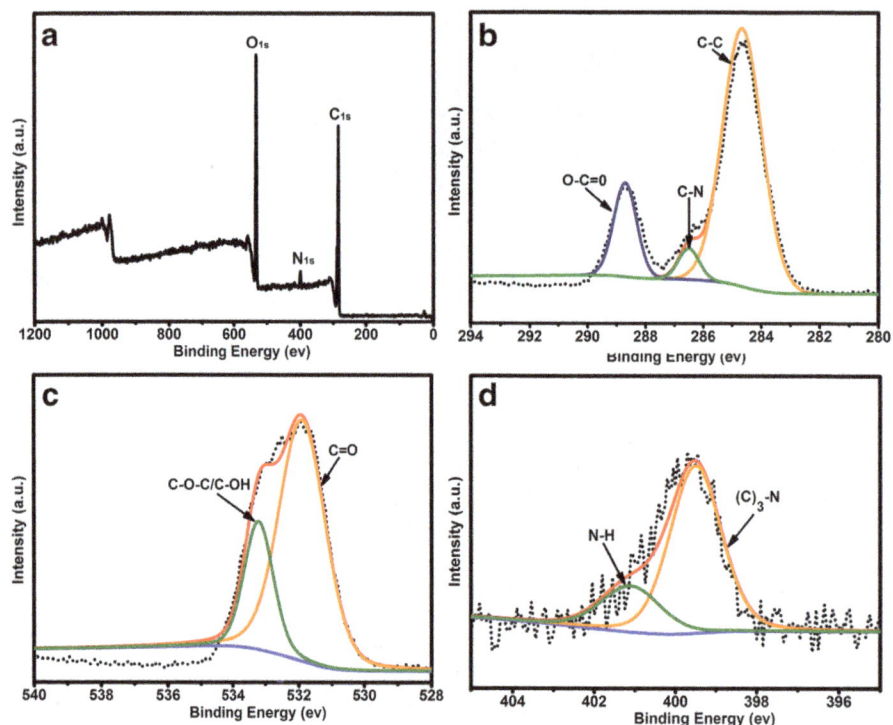

Figure 3 XPS spectra of the Cdots. (a) Survey spectrum of the Cdots with three major peaks of carbon, oxygen, and nitrogen. XPS high resolution survey spectra of **(b)** C1s, **(c)** O1s, and **(d)** N1s region of Cdots.

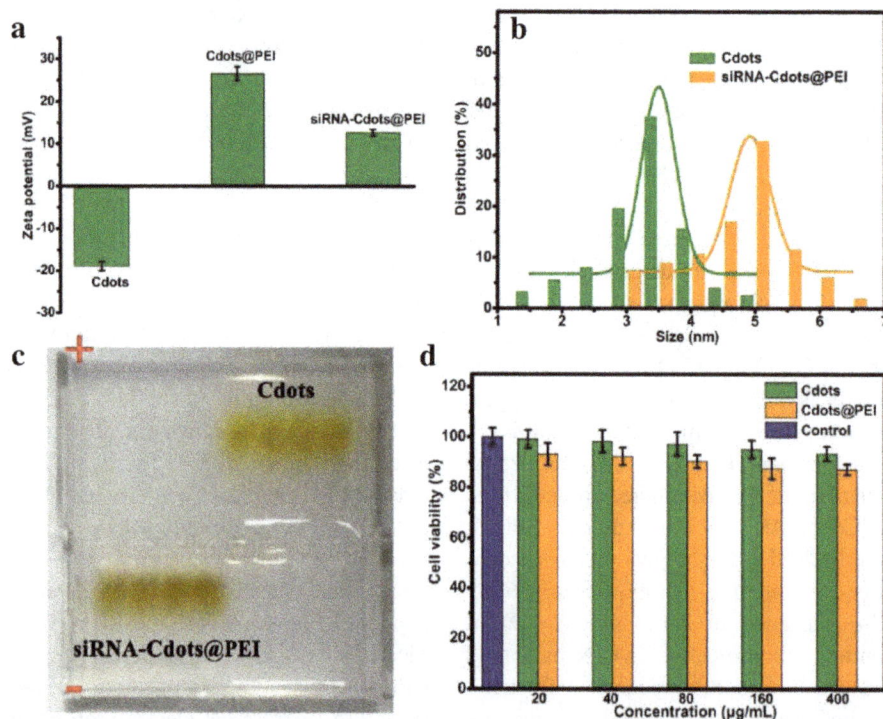

Figure 4 Characterization of the Cdots@PEI and siRNA-Cdots@PEI complexes. (a) Zeta potential of the Cdots, Cdots@PEI and siRNA-Cdots@PEI complexes. **(b)** Hydrodynamic diameters and **(c)** agarose gel electrophoresis analysis of the Cdots and siRNA-Cdots@PEI complexes. There were four paralleled loadings of each example at same time. **(d)** Cell viability of MGC-803 cells after treatment of different concentrations of Cdots and Cdots@PEI complexes.

neutralized and density of the complexes was raised, which caused the opposite direction of migration and the lower electrophoretic mobility when compared with Cdots (Figure 4c). Before the applications of the Cdots-based nanocarrier, the cell toxicity of Cdots and Cdots@PEI complexes was evaluated using MTT assays. Figure 4d showed the comparative viability of the cells incubated with Cdots and Cdots@PEI. It was found that Cdots induced no change in the cell viability when the dose was up to 400 µg/ml, suggesting their excellent biocompatibility. Previous reports have shown that the high content of cationic amine groups in branched polymer PEI mediated the substantial cytotoxicity [29]. The cell viability of MGC-803 cells was not seriously influenced by the addition of Cdots@PEI complexes, as it only slightly decreased to 87.2%. In this regard, the reduced cytotoxicity of Cdots@PEI can be explained by the reduced amounts of primary amine groups exposed at the surface of Cdots@PEI complexes.

Intracellular delivery of siRNA by Cdots@PEI complexes

As known to all, the negatively charged siRNA alone was inhibited from cellular uptake due to its inability to cross cellular membranes [24,25]. Herein, the positively charged siRNA-Cdots@PEI complex was proposed to enhance the cellular uptake and delivery of siRNA intracellularly. The resulting complex was used to transfect MGC-803 cells at pH 7.4 for 2 and 5 h, and the cells treated with

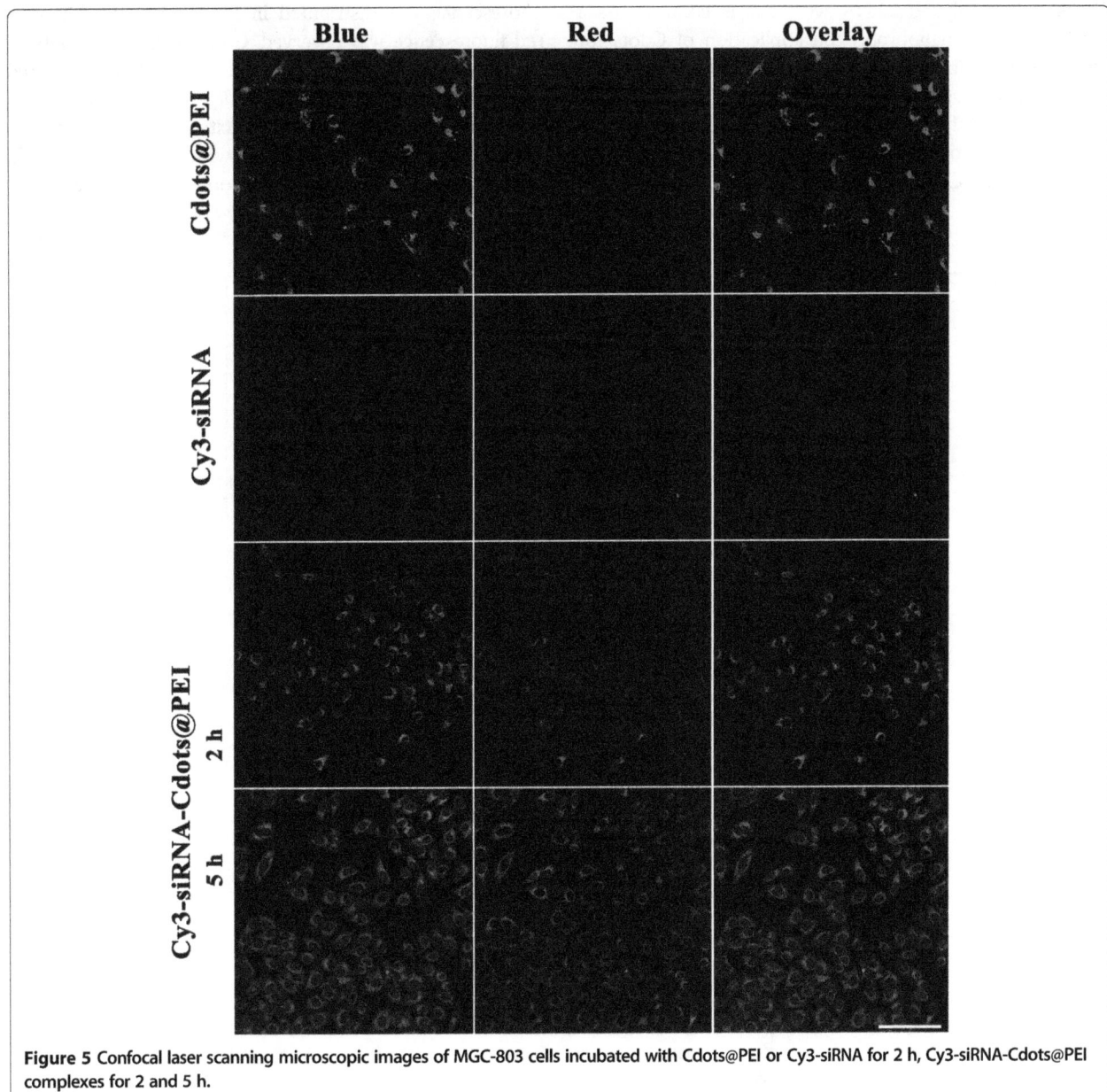

Figure 5 Confocal laser scanning microscopic images of MGC-803 cells incubated with Cdots@PEI or Cy3-siRNA for 2 h, Cy3-siRNA-Cdots@PEI complexes for 2 and 5 h.

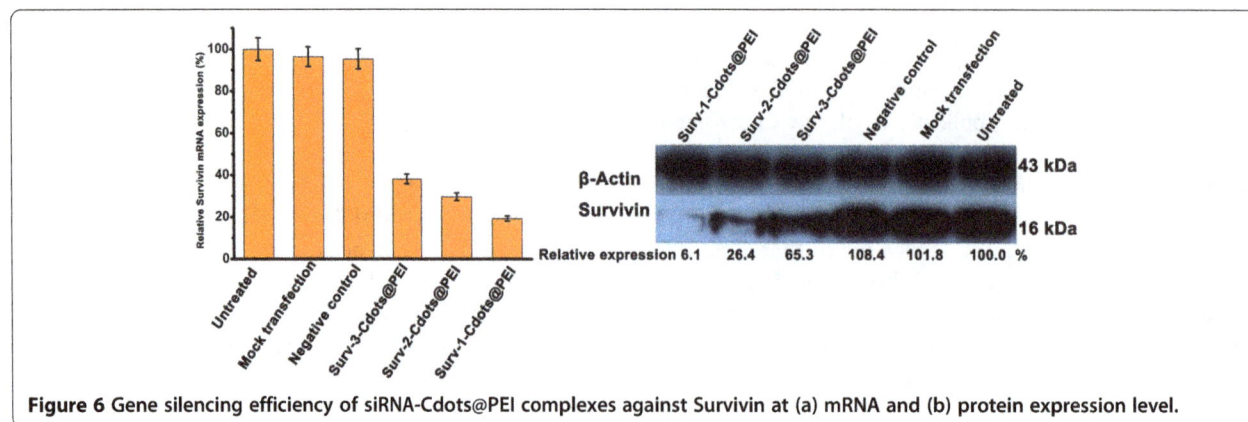

Figure 6 Gene silencing efficiency of siRNA-Cdots@PEI complexes against Survivin at (a) mRNA and (b) protein expression level.

Cdots or Cy3-labeled siRNA were set as controls. As a new fluorescent nanoprobe, the application of Cdots for bioimaging and biolabeling in cancer cells was evaluated. Fluorescence images were acquired using DAPI- and Cy3-specific sets of filters for the blue and red channels of fluorescence, respectively. As shown in Figure 5, after incubation with the Cdots@PEI complexes for 2 h, bright blue

fluorescence was distributed in the cytoplasm while dim red fluorescence was observed. Consistent with the above excitation-dependent fluorescence, the intensity of red fluorescence excited from green light was much weaker than that of blue fluorescence excited from UV light. As anticipated, even after incubation with Cy3-siRNA for 5 h, only a marginal level of red fluorescence was detected

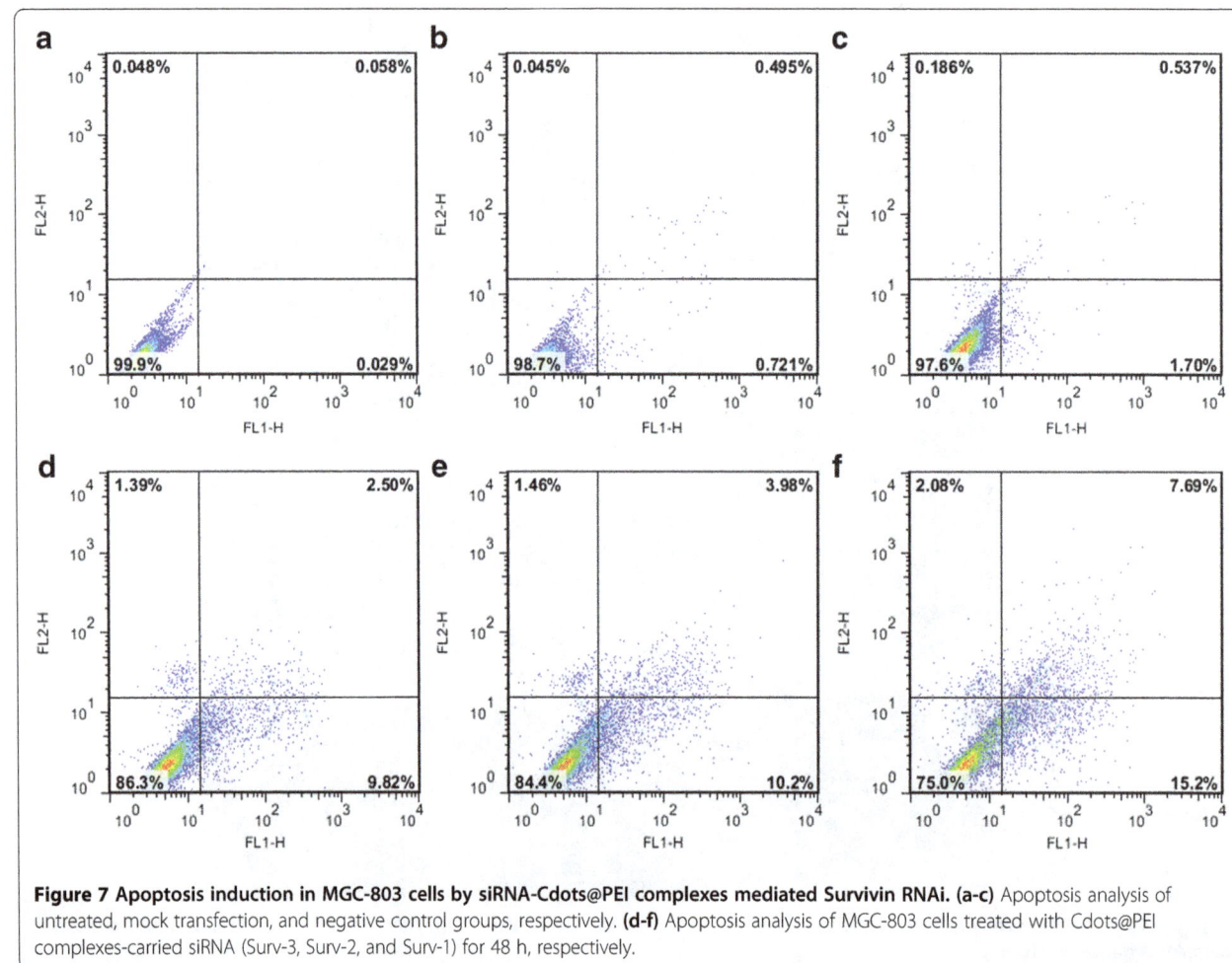

Figure 7 Apoptosis induction in MGC-803 cells by siRNA-Cdots@PEI complexes mediated Survivin RNAi. **(a-c)** Apoptosis analysis of untreated, mock transfection, and negative control groups, respectively. **(d-f)** Apoptosis analysis of MGC-803 cells treated with Cdots@PEI complexes-carried siRNA (Surv-3, Surv-2, and Surv-1) for 48 h, respectively.

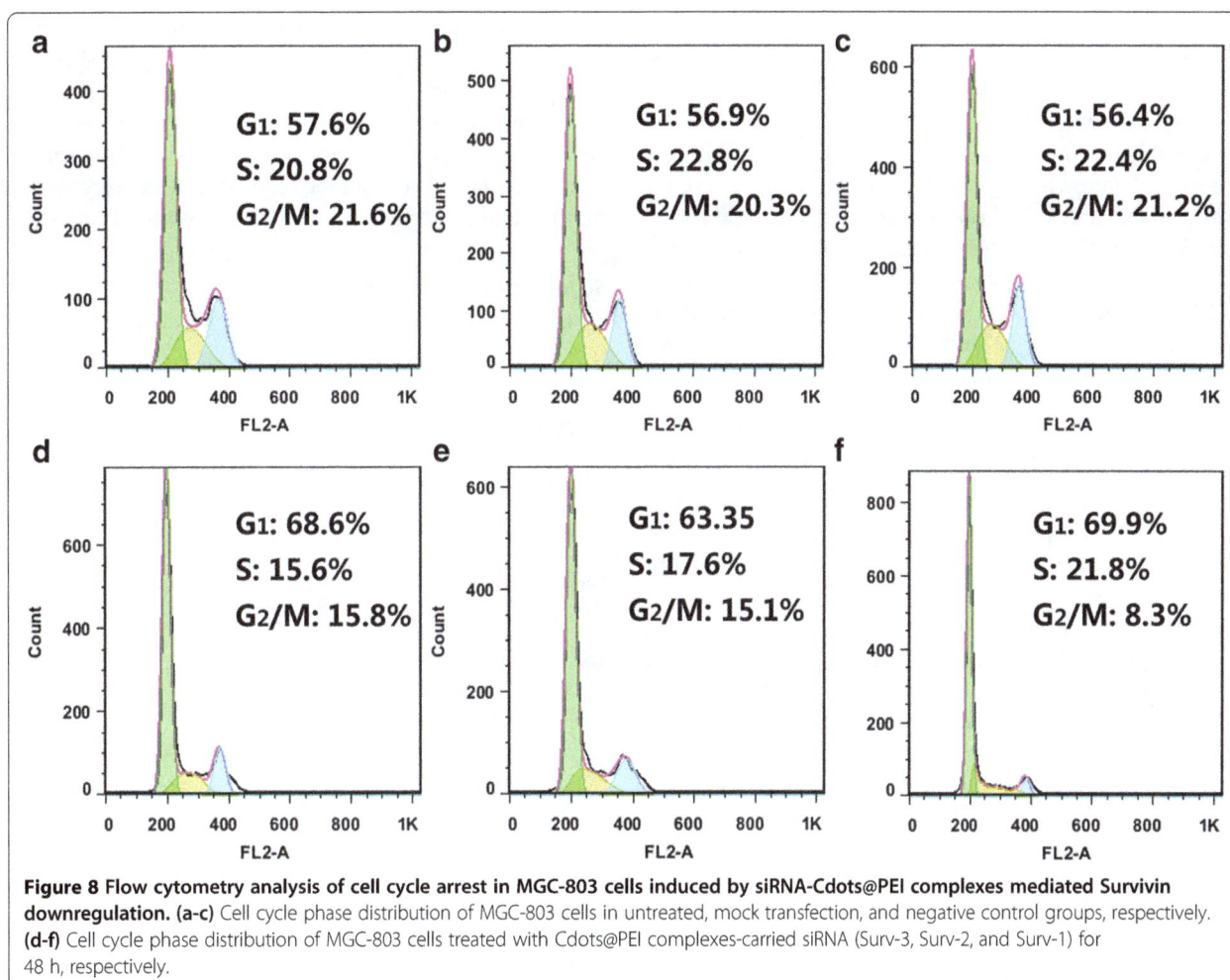

Figure 8 Flow cytometry analysis of cell cycle arrest in MGC-803 cells induced by siRNA-Cdots@PEI complexes mediated Survivin downregulation. **(a-c)** Cell cycle phase distribution of MGC-803 cells in untreated, mock transfection, and negative control groups, respectively. **(d-f)** Cell cycle phase distribution of MGC-803 cells treated with Cdots@PEI complexes-carried siRNA (Surv-3, Surv-2, and Surv-1) for 48 h, respectively.

within the cells, suggesting the low quantity of free siRNA that has entered the cells. In contrast, the siRNA-Cdots@PEI complex was rapidly accumulated into the MGC-803 cells within 2 h. With time growing, blue fluorescence from Cdots and red fluorescence from siRNA increase significantly at a time dependent manner. Hence, it is conceivable that the positively charged Cdots@PEI complexes play an important role in the great enhancement of siRNA internalization. The results indicated that the fluorescent Cdots-based siRNA delivery system not only can be efficient nanocarrier but also intracellular distribution reporter.

Gene silencing efficiency of siRNA-Cdots@PEI complexes

Real-time PCR and Western blot assay was performed to evaluate the gene silencing efficiency of siRNA-Cdots@PEI complexes on the mRNA and protein level, respectively. The GAPDH was selected as a reference gene. The Cdots@PEI complexes functionalized with nontargeting scrambled siRNA served as a negative control. Compared with the blank control group (untreated), the expression level of Survivin mRNA was reduced to $19.3 \pm 1.2\%$, $29.7 \pm 1.8\%$, $38.2 \pm 2.3\%$, $95.3 \pm 6.9\%$, and 96.4 ± 8.7 when the cells were transfected with Surv-1-Cdots@PEI complexes, Surv-2-Cdots@PEI complexes, Surv-3-Cdots@PEI complexes, scrambled siRNA-Cdots@PEI complexes (negative control), and Cdots@PEI complexes (mock transfection), respectively (Figure 6a). The effect of Cdots@PEI complexes-based siRNA delivery on Survivin protein down-regulation was evaluated after 72 h of treatment. As shown in Figure 6b, the expression levels of protein of test group, Survivin in MGC-803 cells exhibited gradually down-regulation compared with the control groups, and the group of Surv-1-Cdots@PEI complexes exhibited the strongest inhibition effect on expression of Survivin protein, which corresponded well with the data given by qRT-PCR.

Apoptosis assay and cell cycle analysis

In order to determine the secondary effects on the context of cell physiology by the downregulation of Survivin protein, the apoptosis levels and cell cycle distributions were evaluated based on flow cytometry analysis. Since Survivin inhibits apoptosis and improves the proliferation of cells, we hypothesized that the performed RNAi by siRNA-Cdots@PEI complexes in MGC-803 cells could induce apoptosis. Using Annexin V-FITC/PI double staining and flow cytometric analyses, rates of apoptosis were quantified as sums of early and late apoptotic cells. As shown in Figure 7a-c, the cells in blank (a), negative control (b), and mock (c) groups showed a large viable cell population with very fewer staining for early apoptotic, late apoptotic and necrosis cells, respectively. However, MGC-803 cells treated with Cdots@PEI complexes-carried siRNA (Surv-3, Surv-2, and Surv-1) for 48 h resulted in a shift from live cells to early apoptotic, late apoptotic, and dead cell populations (Figure 7d-f). Quantitative analysis of the data clearly demonstrated that Surv-3, Surv-2, Surv-1 traetment for 48 h resulted in 9.82%, 10.2%, and 15.2% early apoptotic cells; 2.50%, 3.98%, and 7.69% of late apoptotic cells, respectively. However, the dead cells also increased. It is widely accepted that cell cycle progression can be either negatively or positively regulated by a number of dynamically expressed genes. Previous studies have suggested that Survivin participates in multiple facets of cell division via controlling microtubule stability of the normal mitotic spindle [27,30]. Therefore, secondary effect was further extended to cell cycle analysis by propidium iodide (PI) staining-based flow cytometry cell cycle assay. As shown in Figure 8a-c, there was no significant change in cell cycle distribution in the cells treated with Cdots@PEI complexes (mock group) or scrambled siRNA-Cdots@PEI complexes (negative control group) when compared with the untreated cells (untreated group). However, the cells treated with Cdots@PEI complexes-carried siRNA (Surv-3, Surv-2, and Surv-1) demonstrated an increase in cell population in G_1 phase, indicating that knockdown of Survivin could block the G_1/S transition (Figure 8d-f). Therefore, we hypothesized that the critical role of Survivin in early cell cycle entry would explain cell cycle arrest MGC-803 cells when Survivin was down-regulated.

Conclusion

In summary, the Cdots-based and PEI-adsorbed complexes both as imaging agent and siRNA nanocarrier have been developed for Survivin siRNA delivery. The Cdots were prepared *via* one-step microwave assisted approach with citric acid as carbon source and tryptophan as passivation agent and nitrogen source. The as-synthesized Cdots exhibited excellent water dispersibility, biocompatibility, and high quantum yield. In the elaborately fabricated complexes of siRNA-Cdots@PEI, Cdots acted as a nanocarrier and a fluorescent indicator, while the positively charged PEI acted as the ties attaching negative charged siRNA to Cdots. Furthermore, the confocal fluorescence images indicted the cellular uptake of siRNA-Cdots@PEI complexes, and subsequently qRT-PCR and

Western blot analysis confirmed the successfully entrance of siRNA into MGC-803 cells and superior gene silencing efficiency. Importantly, the siRNA-Cdots@PEI complexes, which target Survivin gene, can induce apoptosis and cell cycle arrest in G_1 phase inhuman gastric cancer cells MGC-803. The resulting Cdots-based delivery system may be used to advance the field of siRNA therapeutics.

Supporting information

Supporting information is available from the XX Online Library or from the author.

Additional file

> **Additional file 1: Figure S1.** FTIR spectra of tryptophan, citric acid, and the Cdots.

Competing interests

The authors declare that they have no competing interests.

Authors' contributions

QW and CZ synthesized and functionalized the Cdots, peformed confocal microscopy imaging, carried out qRT-PCR, western blot, apoptosis assay, cell cycle analysis, and co-drafted the manuscript. HL and HF coordinated the project, characterized the nanocomplexes by TEM, XPS, DLS. GS and DC provided laboratory facilities and revised the manuscript. All authors read and approved the final manuscript.

Acknowledgements

This work was supported by Chinese Key Basic Research Program (973 Project) (No. 2010CB933901), the National Natural Scientific Foundation of China (Grant No. 31170961, 81225010, and81327002), and 863 project of China(no.2012AA022703 and 2014AA020700), Shanghai Science and Technology Fund (No.13NM1401500). Biomedical and Engineering Multidisciplinary Funding of SJTU(No. YG2012MS13), YG2014MS01.

References

1. Peer D, Karp JM, Hong S, FaroKhzad OC, Margalit R, Langer R: **Nanocarriers as an emerging platform for cancer therapy.** *Nat Nanotechnol* 2007, 2:751–760.
2. Wang M, Thanou M: **Targeting nanoparticles to cancer.** *Pharmacol Res* 2010, 62:90–99.
3. Zhang C, Zhou Z, Qian Q, Gao G, Li C, Feng L, Wang Q, Cui D: **Glutathione-capped fluorescent gold nanoclusters for dual-modal fluorescence/X-ray computed tomography imaging.** *J Mater Chem B* 2013, 1:5045–5053.
4. Michalet X, Pinaud FF, Bentolila LA, Tsay JM, Doose S, Li JJ, Sundaresan G, Wu AM, Gambhir SS, Weiss S: **Quantum dots for live cells, in vivo imaging, and diagnostics.** *Science* 2005, 307:538–544.
5. Disney MD, Zheng J, Swager TM, Seeberger PH: **Detection of bacteria with carbohydrate-functionalized fluorescent polymers.** *J Am Chem Soc* 2004, 126:13343–13346.
6. Baker SN, Baker GA: **Luminescent carbon nanodots: emergent Nanolights.** *Angew Chem Int Ed* 2010, 49:6726–6744.
7. Li H, Kang Z, Liu Y, Lee S-T: **Carbon nanodots: synthesis, properties and applications.** *J Mater Chem* 2012, 22:24230–24253.
8. Derfus AM, Chan WCW, Bhatia SN: **Probing the cytotoxicity of semiconductor quantum dots.** *Nano Lett* 2003, 4:11–18.
9. Song Y, Shi W, Chen W, Li X, Ma H: **Fluorescent carbon nanodots conjugated with folic acid for distinguishing folate-receptor-positive cancer cells from normal cells.** *J Mater Chem* 2012, 22:12568–12573.

10. Ruan S, Qian J, Shen S, Zhu J, Jiang X, He Q, Gao H: **A simple one-step method to prepare fluorescent carbon dots and their potential application in non-invasive glioma imaging.** *Nanoscale* 2014, **6:**10040–10047.

11. Liu J-M, Lin L-p, Wang X-X, Lin S-Q, Cai W-L, Zhang L-H, Zheng Z-Y: **Highly selective and sensitive detection of Cu^{2+} with lysine enhancing bovine serum albumin modified-carbon dots fluorescent probe.** *Analyst* 2012, **137:**2637–2642.

12. Wang Q, Liu X, Zhang L, Lv Y: **Microwave-assisted synthesis of carbon nanodots through an eggshell membrane and their fluorescent application.** *Analyst* 2012, **137:**5392–5397.

13. Du F, Zhang M, Li X, Li J, Jiang X, Li Z, Hua Y, Shao G, Jin J, Shao Q, Zhou M, Gong A: **Economical and green synthesis of bagasse-derived fluorescent carbon dots for biomedical applications.** *Nanotechnology* 2014, **25**(31):315702–315712.

14. Peng J, Gao W, Gupta BK, Liu Z, Romero-Aburto R, Ge L, Song L, Alemany L, Zhan X, Gao G, Vithayathil S, Kaipparettu B, Marti A, Hayashi T, Zhu J, Ajayan P: **Graphene quantum dots derived from carbon fibers.** *Nano Lett* 2012, **12**(2):844–849.

15. Lu J, Yeo PSE, Gan CK, Wu P, Loh KP: **Transforming C_{60} molecules into graphene quantum dots.** *Nat Nanotechnol* 2011, **6:**247–252.

16. Wu ZL, Zhang P, Gao MX, Liu CF, Wang W, Leng F, Huang CZ: **One-pot hydrothermal synthesis of highly luminescent nitrogen-doped amphoteric carbon dots for bioimaging from Bombyx mori silk - natural proteins.** *J Mater Chem B* 2013, **1:**2868–2873.

17. Sun Y-P, Wang X, Lu F, Cao L, Meziani MJ, Luo PG, Gu L, Veca L: **Doped carbon nanoparticles as a new platform for highly photoluminescent dots.** *J Phys Chem C* 2008, **112**(47):18295–18298.

18. Liu H, Wang Q, Shen G, Zhang C, Li C, Ji W, Wang C, Cui D: **A multifunctional ribonuclease A-conjugated carbon dot cluster nanosystem for synchronous cancer imaging and therapy.** *Nanoscale Res Lett* 2014, **9:**397.

19. Huang P, Lin J, Wang X, Wang Z, Zhang C, He M, Wang K, Chen F, Li Z, Shen G, Cui D, Chen X: **Light-triggered theranostics based on photosensitizer-conjugated carbon dots for simultaneous enhanced-fluorescence imaging and photodynamic therapy.** *Adv Mater* 2012, **24**(37):5104–5110.

20. Hannon GJ: **RNA interference.** *Nature* 2002, **418:**244–251.

21. Fellmann C, Lowe SW: **Stable RNA interference rules for silencing.** *Nat Cell Biol* 2014, **16:**10–18.

22. Maillard PV, Ciaudo C, Marchais A, Li Y, Jay F, Ding SW, Voinnet O: **Antiviral RNA interference in mammalian cells.** *Science* 2013, **342:**235–238.

23. Wang L, Wang X, Bhirde A, Cao J, Zeng Y, Huang X, Sun Y, Liu G, Chen X: **Carbon-Dot-based Two-photon visible nanocarriers for safe and highly efficient delivery of siRNA and DNA.** *Adv Healthcare Mater* 2014, **3:**1203–1209.

24. Liu C, Zhang P, Zhai X, Tian F, Li W, Yang J, Liu Y, Wang H, Wang W, Liu W: **Nano-carrier for gene delivery and bioimaging based on carbon dots with PEI-passivation enhanced fluorescence.** *Biomaterials* 2012, **33:**3604–3613.

25. Qi L, Shao W, Shi D: **JAM-2 siRNA intracellular delivery and real-time imaging by proton-sponge coated quantum dots.** *J Mater Chem B* 2013, **1:**654–660.

26. Yang D, Welm A, Bishop JM: **Cell division and cell survival in the absence of survivin.** *Proc Natl Acad Sci* 2004, **101:**15100–15105.

27. Suzuki A, Hayashida M, Ito T, Kawano H, Nakano T, Miura M, Akahane K, Shiraki K: **Survivin initiates cell cycle entry by the competitive interaction with Cdk4/p16(INK4a) and Cdk2/Cyclin E complex activation.** *Oncogene* 2000, **19:**3225–3234.

28. Lu CD, Altieri DC, Tanigawa N: **Expression of a novel antiapoptosis gene, survivin, correlated with tumor cell apoptosis and p53 accumulation in gastric carcinomas.** *Cancer Res* 1998, **58:**1808–1812.

29. Parhamifar L, Larsen AK, Hunter AC, Andresen TL, Moghimi SM: **Polycation cytotoxicity: a delicate matter for nucleic acid therapy-focus on polyethylenimine.** *Soft Matter* 2010, **6:**4001–4009.

30. Montazeri Aliabadi H, Landry B, Mahdipoor P, Uludağ H: **Induction of apoptosis by survivin silencing through siRNA delivery in a human breast cancer cell line.** *Mol Pharmaceutics* 2011, **8:**1821–1830.

4

PVM/MA-shelled selol nanocapsules promote cell cycle arrest in A549 lung adenocarcinoma cells

Ludmilla Regina de Souza[1], Luis Alexandre Muehlmann[2], Mayara Simonelly Costa dos Santos[1], Rayane Ganassin[2], Rosana Simón-Vázquez[3], Graziella Anselmo Joanitti[2], Ewa Mosiniewicz-Szablewska[4], Piotr Suchocki[5,6], Paulo César Morais[7,8], África González-Fernández[3], Ricardo Bentes Azevedo[2] and Sônia Nair Báo[2]*

Abstract

Background: Selol is an oily mixture of selenitetriacylglycerides that was obtained as a semi-synthetic compound containing selenite. Selol is effective against cancerous cells and less toxic to normal cells compared with inorganic forms of selenite. However, Selol's hydrophobicity hinders its administration *in vivo*. Therefore, the present study aimed to produce a formulation of Selol nanocapsules (SPN) and to test its effectiveness against pulmonary adenocarcinoma cells (A549).

Results: Nanocapsules were produced through an interfacial nanoprecipitation method. The polymer shell was composed of poly(methyl vinyl ether-co-maleic anhydride) (PVM/MA) copolymer. The obtained nanocapsules were monodisperse and stable. Both free Selol (S) and SPN reduced the viability of A549 cells, whereas S induced a greater reduction in non-tumor cell viability than SPN. The suppressor effect of SPN was primarily associated to the G2/M arrest of the cell cycle, as was corroborated by the down-regulations of the CCNB1 and CDC25C genes. Apoptosis and necrosis were induced by Selol in a discrete percentage of A549 cells. SPN also increased the production of reactive oxygen species, leading to oxidative cellular damage and to the overexpression of the GPX1, CYP1A1, BAX and BCL2 genes.

Conclusions: This study presents a stable formulation of PVM/MA-shelled Selol nanocapsules and provides the first demonstration that Selol promotes G2/M arrest in cancerous cells.

Background

Low therapeutic efficacy and drug resistance are the most common problems related to the currently available chemotherapeutic agents used in tumoral clinical application. In the search for new chemotherapeutic drugs, several selenium (Se) compounds shown anticancer and anticarcinogenic activities [1,2]. In particular, those containing Se at its 4+ oxidation state, namely selenite, present the highest antioxidant and anticancer activities [2]. However, Se(IV)-containing compounds generally present high systemic toxicity, limiting their clinical application. In this context, a selenite-containing compound named Selol, which was first obtained at Warsaw Medical University, Poland [3], has shown antitumor activity and low systemic toxicity [4,5]. Selol is a mixture of different selenitetriacylglycerides

and appears to act primarily through the induction of oxidative stress in cancer cells [5]. Interestingly, Selol was shown to sensitize leukemia cells to the cytotoxicity of vincristine and doxorubicin, insomuch that it was suggested that Selol could be used in combination with other drugs in chemotherapeutic protocols [4].

The potential synergism of Selol with classical anticancer drugs can be exploited to treat tumors, such as non-small cell lung cancer (NSCLC) [4]. NSCLC, whose most frequently observed histological subtype is adenocarcinoma, is particularly aggressive and the leading cause of cancer death worldwide [6,7]. It is estimated that more than 75% of patients with NSCLC present locally advanced or metastatic disease, severely limiting the success of treatments [8,9]. Platinum-based treatment, which is the most recommended first-line therapy, reaches response rates of only 20-40% and mean survivals between 7 and 12 months [8,10,11]. Multidrug protocols and a treatment break with non-platinum-based drugs after a fixed

* Correspondence: snbao@unb.br
[2]Institute of Biological Sciences, University of Brasília, Brasília, DF 70910-900, Brazil
Full list of author information is available at the end of the article

course of initial chemotherapy have been shown to prolong the survival of NSCLC patients [12,13]. Thus, Selol may be a potential candidate for combined anti-NSCLC strategies [14].

Despite the therapeutic potential presented by Selol, its hydrophobicity is a major obstacle to its biological application. For instance, high hydrophobicity often hinders intravenous (iv) administration and may thus confer an undesirable pharmacokinetic profile [15]. This problem can be circumvented through the nanoencapsulation of Selol in an aqueous vehicle to form a nanocapsule-based drug delivery system. The nanoencapsulation of Selol with a polymer presenting highly reactive chemical groups allowing surface modification could bring up new possibilities for delivering this anticancer agent. In this context, the copolymer poly(methyl vinyl ether-co-maleic anhydride) (PVM/MA) has been reported to be a biocompatible and biodegradable material useful for preparing drug delivery systems [16]. Additionally, the copolymer PVM/MA presents a surfactant effect and anhydride groups, which readily react with a series of molecules. On that ground, we report the development and the first *in vitro* efficacy tests of a PVM/MA-shelled Selol nanocapsule formulation intended for the treatment of lung adenocarcinomas.

Results and discussion
Formulation screening
The interfacial precipitation method of preformed polymer through solvent displacement yields nanosized Selol capsules only within a certain range of solute and solvent concentrations. Thus, to identify the best formulation parameters, different concentrations of each component were tested.

First, different Selol-to-PVM/MA ratios were tested (0.1, 0.2, 0.5, 0.7, 1.0, 1.5, 2.0, and 4.0, w:w), and the concentrations of acetone, ethanol and water were fixed to 20, 40 and 40% (v:v), respectively. As shown in Figure 1 (a), the increase in this ratio led to a directly proportional increase in the hydrodynamic diameter (HD) values of the nanocapsules. Using Selol-to-PVM/MA ratios from 0.1 to 1.0, monodisperse nanocapsule populations were obtained, and the polydispersity index (PDI) values remained below 0.1. Formulations prepared with Selol-to-PVM/MA ratios higher than 1.0 showed visible decantation minutes after preparation and were not used for dynamic light scattering analysis. Therefore, this parameter was set to 1.0 for the next steps because this was the highest value that allowed stable nanocapsules to be obtained.

Next, different concentrations of Selol plus PVM/MA were tested. The concentrations of acetone, ethanol and water were set to 20, 40 and 40% (v:v), respectively. As expected, smaller nanocapsules were obtained at the lowest concentrations of Selol plus PVM/MA (Figure 1 (b)). The PDI was not significantly affected by this variable, remaining close to 0.1. The concentration of Selol plus PVM/MA was set to 0.8% (w:v) in further experiments because it provided good colloidal characteristics in addition to a good yield of nanocapsules.

Then, different concentrations of acetone and ethanol were tested in the process of encapsulation with 0.8% Selol plus PVM/MA and a ratio of 1.0 Selol-to-PVM/MA. Different volumes of acetone were used for dissolving a fixed amount of Selol and PVM/MA, and the final volume reached 100% with ethanol:water (1:1, v:v). As shown in Figure 1(c), a major change in the nanocapsule HD was observed with 40% acetone, but the PDI remained below 0.1. When the concentration of acetone was set to 20% and varying volumes of ethanol were added, it was observed that the HD of the nanocapsules decreased with higher concentrations of ethanol (Figure 1 (d)). The highest HD and PDI values were obtained with 20% ethanol.

Given the results described above, the protocol of Selol nanoencapsulation was established as follows: 1) 100 mg of PVM/MA and 100 mg of Selol were dissolved in 5 mL of acetone; 2) 10 mL of ethanol and 10 mL of water were added; and 3) the purification steps were then performed.

The method of nanoprecipitation by solvent displacement yielded monodisperse nanocapsules at almost all of the conditions tested and also allowed modulation of the nanocapsule diameter. Noteworthy, by varying the concentrations of acetone and ethanol, nanocapsules of different HDs were obtained, likely due to differences in solvent diffusion, as previously suggested [17]. Even for a concentration of nanocapsule components (Selol plus PVM/MA) near the upper critical limit of 2%, as noted by Aubry et al. (2009) [18] for this method, stable and monodisperse nanocapsules were obtained. As expected, at higher concentrations of Selol, larger capsules were obtained, which can be attributed to the nucleation-and-growth phenomenon [19,20].

Characterization of Selol nanocapsules
The Selol nanocapsules (SPN) formulation presented a single population of nanocapsules with an HD of 344.4 ± 4.8 nm, a PDI of 0.061 ± 0.005 and a zeta potential (ζ po tential) of −29.3 mV ± 1.5. The transmission electron microscope (TEM) image revealed a population of nanocapsules with an average diameter of 207.9 ± 80.9 nm (Figure 1(e)). These nanometric structures presented a spherical shape and slightly rough surface, as observed with a scanning electron microscope (SEM) (Figure 1(f)). A spherical equilibrium shape is expected with this method due to the three-dimensional primordial droplet nuclei growth conferred by the interfacial tension between the droplets and the dispersant [20,21]. Furthermore, according to the

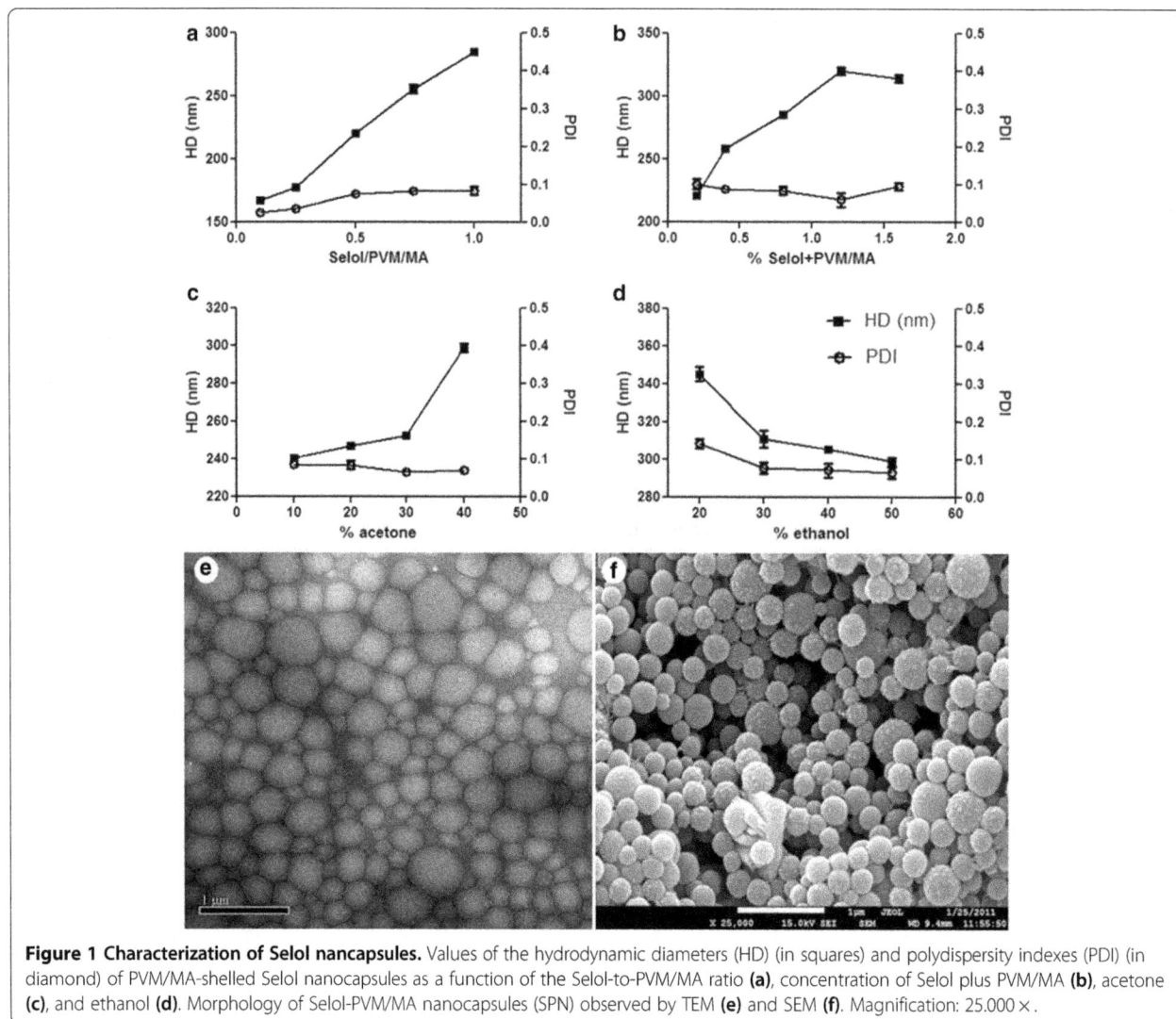

Figure 1 Characterization of Selol nancapsules. Values of the hydrodynamic diameters (HD) (in squares) and polydispersity indexes (PDI) (in diamond) of PVM/MA-shelled Selol nanocapsules as a function of the Selol-to-PVM/MA ratio **(a)**, concentration of Selol plus PVM/MA **(b)**, acetone **(c)**, and ethanol **(d)**. Morphology of Selol-PVM/MA nanocapsules (SPN) observed by TEM **(e)** and SEM **(f)**. Magnification: 25.000 × .

TEM images, the Selol nanocapsules appear to be deformable. Both the deformability and the spherical shape of the nanocapsules are interesting for their administration through parenteral routes [22].

The dosage of the purified Selol nanocapsules showed that the concentration of Selol measured in the nanocapsule was approximately 100% of the initial amount of Selol used in the process. Supported by the low PDI of the Selol composition, this result confirms that Selol was efficiently encapsulated into the PVM/MA shell. However, it was necessary to evaluate the copolymer's capacity for interfacial stabilization after prolonged periods of storage and thermal stress conditions. As observed in Table 1, the SPN was stable at room temperature (RT) for at least 60 days. Under these conditions, there was no statistically significant change in the HD over the evaluated period of time. When stored at 4°C, the SPN average HD increased by approximately 40 nm on the

first day and then remained stable. At both RT and 4°C, the PDI remained well below 0.1, showing that the nanocapsules remained monodisperse. At −20°C, significant variations were observed in both the HD and PDI (>0.3), indicating that this condition is not adequate for storage. After freezing/thawing cycles, no significant changes in both the HD and PDI (p > 0.05) were found up to the eighth cycle. Indeed, PVM/MA presented good capacity for interfacial stabilization in SPN because this system did not present any important changes in its characteristics, neither after 60 days of storage (at RT or 4°C) nor after being subjected to the thermal stress of eight cycles of freezing/thawing. The ζ potential of the nanocapsules was generally between −25 and −30 mV at all of the conditions tested, which is due to the presence of carboxylate groups on their outer surface. A hypothetical scheme of the PVM/MA-shelled Selol nanocapsules is shown in Figure 2. Over the course of encapsulation, the PVM/

Table 1 Thermodynamic stability and dispersibility studies of Selol-PVM/MA nanocapsules

Formulation	Hydrodynamic diameter ± S.D. (nm)	Polydispersity index	Zeta potential (mV)
Room temperature			
Day 0	344.4 ± 4.8[a]	0.061	−29.3
Day 15	349.7 ± 8.2[a]	0.049	−29.7
Day 30	338.9 ± 6.4[a]	0.046	−29.2
Day 45	344.3 ± 2.0[a]	0.034	−26.5
Day 60	336.6 ± 4.6[a]	0.043	−29.4
4°C			
Day 0	344.4 ± 4.8[a]	0.061	−29.3
Day 1	390.2 ± 1.0[b]	0.054	−29.3
Day 15	391.3 ± 7.5[b]	0.044	−25.1
Day 30	373.9 ± 15.0[b]	0.046	−27.8
Day 45	381.6 ± 8.9[b]	0.031	−30.1
Day 60	380.6 ± 7.1[b]	0.039	−27.3
−20°C			
Day 0	344.4 ± 4.8[a]	0.061	−29.3
Day 1	414.2 ± 7.7[c]	0.074	−29.2
Day 15	408.2 ± 10.7[c]	0.113	−27.0
Day 30	471.0 ± 1.3[d]	0.195	−27.4
Day 45	615.2 ± 13.3[e]	0.278	−29.8
Day 60	566.3 ± 14.8[f]	0.196	−26.5
Freezing (−20°C) and thawing cycle			
No cycle	344.4 ± 4.8	0.061	−29.3
I	414.2 ± 7.74	0.074	−29.2
II	414.1 ± 10.0	0.066	−28.9
III	406.7 ± 4.5	0.035	−28.9
IV	410.2 ± 4.7	0.045	−28.1
V	416.5 ± 4.3	0.064	−22.9
VI	405.5 ± 6.0	0.046	−28.5
VII	425.8 ± 0.9	0.079	−28.8
VIII	467.3 ± 9.9	0.143	−27.0
IX	486.5 ± 17.1*	0.106	−28.8
X	495.1 ± 1.0*	0.177	−30.8
XI	509.3 ± 6.6*	0.221	−28.9
XII	566.2 ± 3.5*	0.364	−28.7
XIII	865.1 ± 50.4*	0.581	−27.7
XIV	989.1 ± 26.6*	0.433	−26.9
XV	1027.5 ± 60.7*	0.393	−28.8

All values were expressed as the means ± standard error of the mean. Different letters indicate significant differences between SPN treatment for one day at RT compared with a given time and/or storage condition ($p < 0.05$). *Significantly different compared with the first cycle of freezing (−20°C) and thawing ($p < 0.05$).

MA copolymer, which is poorly soluble in water, becomes amphipathic due to the hydrolysis of some of its anhydride groups exposed to water. The formed polymer shell presents a hydrophilic water-exposed surface, facilitating the stabilization of nanocapsules.

Selol nanocapsules affect cell viability in a concentration- and time-dependent manner

The MTT assay showed that nanoparticles of PVM/MA without Selol (Bl) did not significantly affect the viability of the studied cells (Figure 3). Significant reductions in

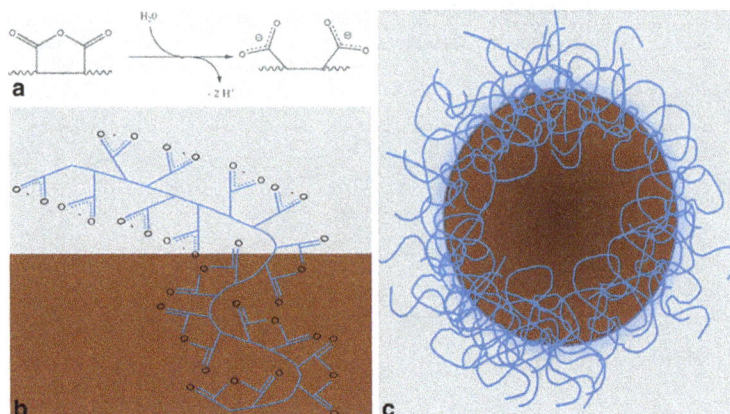

Figure 2 Schematic representation of PVM/MA-shelled Selol nanocapsules. Hydrolysis of an anhydride group yields two carboxylate groups at neutral pH in a PVM/MA strand **(a)**. Partial hydrolysis of the PVM/MA polymer strand exposed to water but not in PVM/MA closer to the oily core **(b)**. The carboxylate-containing parts of PVM/MA are hydrophilic and comprise the nanocapsule shell surface, whereas the anhydride-containing parts are more hydrophobic, closely covering the Selol core **(c)**.

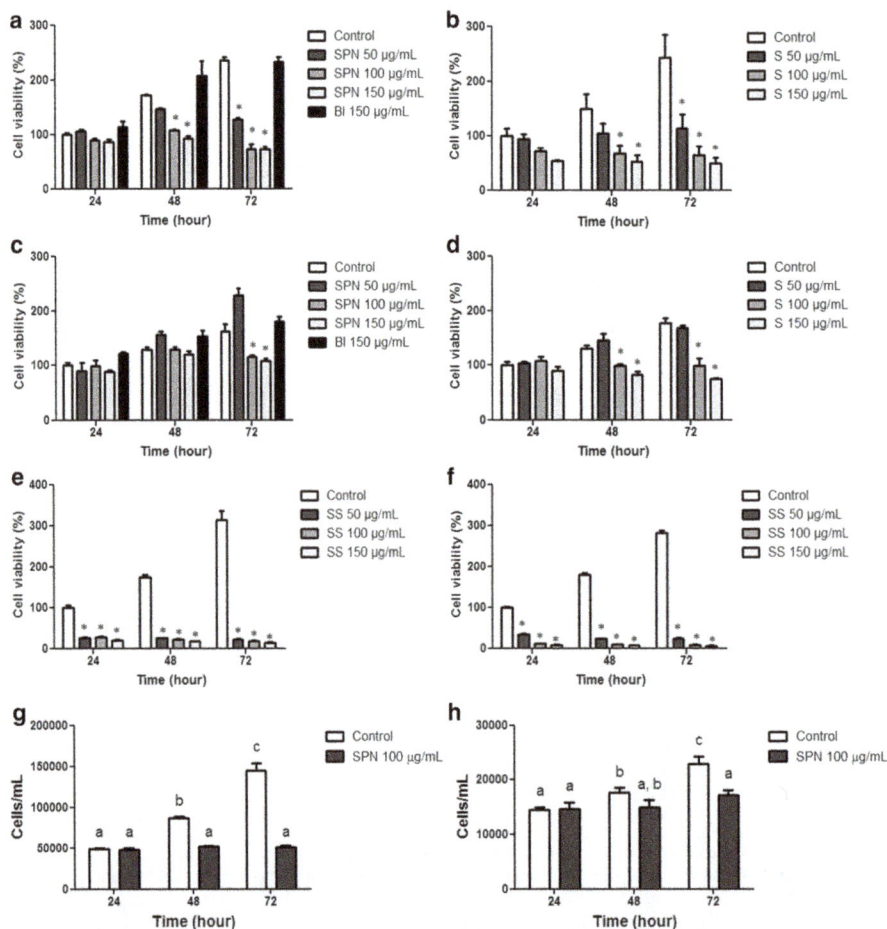

Figure 3 Viability of A549 (a, b, e) and human connective tissue (c, d, f) cells after exposure to Selol-PVM/MA nanocapsules (SPN) (a, c), free Selol (S) (b, d), blank polymeric nanoparticles without Selol (Bl) (a, c) or sodium selenite (SS) (e, f) at 50, 100 and 150 μg/mL Se for 24 h, 48 h and 72 h. All of the values were normalized according to the control group at 24 h (100%). The time-dependent proliferation of control cells was significantly different. *Statistically significant compared with the control group at the corresponding treatment period (p < 0.05). Influence of SPN at a concentration of 100 μg Se/mL on the proliferation of A549 **(g)** and human connective tissue **(h)** cells evaluated after different treatment times. Pairs of means in a same graph identified with different letters are significantly different (p < 0.05).

the viability of human lung adenocarcinoma (A549) cells were observed after exposure to SPN or free Selol (S) at concentrations of 50 µg/mL for 72 h and 100 and 150 µg/mL for 48 and 72 h ($p < 0.05$). In contrast, tissue connective normal cells were shown to be less sensitive to Selol compared with tumor cells. Free Selol treatment reduced the viability of normal cells after 48 h, whereas SPN reduced their viability only after 72 h of incubation. Sodium selenite (SS) was highly toxic for both cell types at all concentrations and times tested. The high toxicity of SS toward normal cells shows that selenite is an effective anticancer agent but is not safe for clinical use in its inorganic form. Therefore, some researchers have aimed to find organic selenium compounds with higher therapeutic indexes [23,24]. Selol, an organic selenite compound, significantly reduced the viability of lung adenocarcinoma A549 cells but was far less toxic on normal cells than sodium selenite. This finding reinforces the previously shown evidence that Se (4+) organified as Selol has a significantly lower potential to exert deleterious effects on non-target tissues than its inorganic form [4,25,15]. Noteworthy, the encapsulation of Selol did not significantly affect its activity against A549 cells and reduced its toxicity toward normal cells. These results encourage further tests of Selol activity in *in vivo* models of pulmonary cancer. Moreover, some characteristics of the investigated nanocapsules point to their potential to act as good drug delivery systems. First, their hydrodynamic diameters can be tuned to values that allow for the passive targeting of tumors via an enhanced permeation and retention effect (EPR) because the tumor microvasculature usually presents pores with diameters of 100 to 780 nm [26]. Second, targeting molecules can be conjugated to the surface of nanocapsules to increase their affinity to cancerous cells. This procedure can be easily performed because PVM/MA has anhydride groups, which can easily react with the hydroxyl or primary amine groups present in most of the available targeting molecules [16].

Subsequent experiments with SS, S and SPN were performed with a concentration of 100 µg/mL Se because this was the lowest concentration in S and SPN treatments that reduced cell viability. Moreover, SPN was always used within 60 days after preparation because its biological activity did not significantly change during storage at 4°C for this period of time ($p > 0.05$) (Figure 4).

SPN reduces cell proliferation and promotes cell cycle arrest

As expected, both untreated normal and A549 cells proliferate in culture, as evidenced by the time-dependent increase in the number of cells ($p < 0.001$ for both 24 vs. 48 h and 48 vs. 72 h for A549; $p < 0.01$ for 24 vs. 48 h, and $p < 0.05$ for 48 vs. 72 h for normal cells) (Figure 3

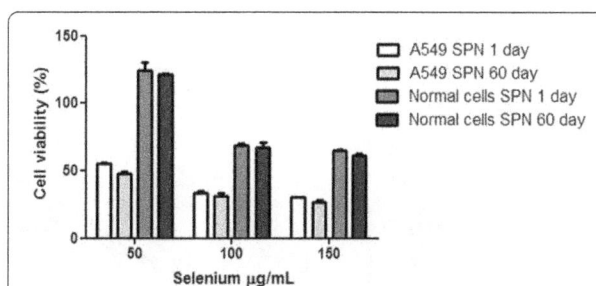

Figure 4 Viability of A549 and human connective tissue cells treated with Selol-PVM/MA nanocapsules (SPN) after 1 and 60 days of storage at 4°C. The cells were treated with 50, 100 and 150 µg/mL Se for 72 h, and the data are expressed as the means ± standard error of the mean of the percentages of viable cells. No statistically significant difference was found between the results for SPN after 1 day and SPN for 60 days ($p > 0.05$).

(g and h)). SPN treatment inhibited the proliferation of both cell types, but the intensity of this effect was cell line-dependent. The number of both normal and tumor cells in the SPN-treated groups did not vary over the treatment time ($p > 0.05$). Reductions of $40.4 \pm 2.7\%$ and $64.7 \pm 1.7\%$ in the number of SPN-treated A549 cells were evidenced at 48 h ($p < 0.001$) and 72 h ($p < 0.001$), respectively, relative to the control A549 cells. However, normal cells were affected to a lesser extent because a significant reduction in their number was observed only at 72 h ($24.8 \pm 6.7\%$, $p < 0.01$) compared with the control normal cells.

The real-time cell index monitoring also showed changes in this parameter after the first hours of SPN treatment on A549 cells, which became more significant for higher concentrations and longer times of incubation (Figure 5). Cell growth was detected up to 20 h, 28 h and 56 h in the presence of SPN at 150 µg/mL, 100 µg/mL and 50 µg/mL concentrations, respectively. Over these times, the cell indexes remained constant and subsequently decreased. Similar to previous results, the cell index in the treatment with 100 µg/mL SPN was almost invariant at 24, 48, and 72 h.

These results were corroborated by light microscopy images, which showed a more intense reduction in the confluence of SPN-treated A549 cells compared with SPN-treated normal cells (Figure 6). A549 cells treated with SPN presented a significantly higher proportion of cells arrested at the G2/M phase compared with that obtained for control cells (Table 2). These results are in good agreement with the decreases in the gene expression levels of CCNB1 (cyclin B1), CDC25C and WEE1 (FC (fold change) > 2.0) and unchanged expression levels of the gene transcripts of CCND1 (cyclin D1) and CCNE1 (cyclin E1) (FC < 2.0) (Figure 7(a)), indicating that Selol acts on the G2/M arrest of the cell cycle and

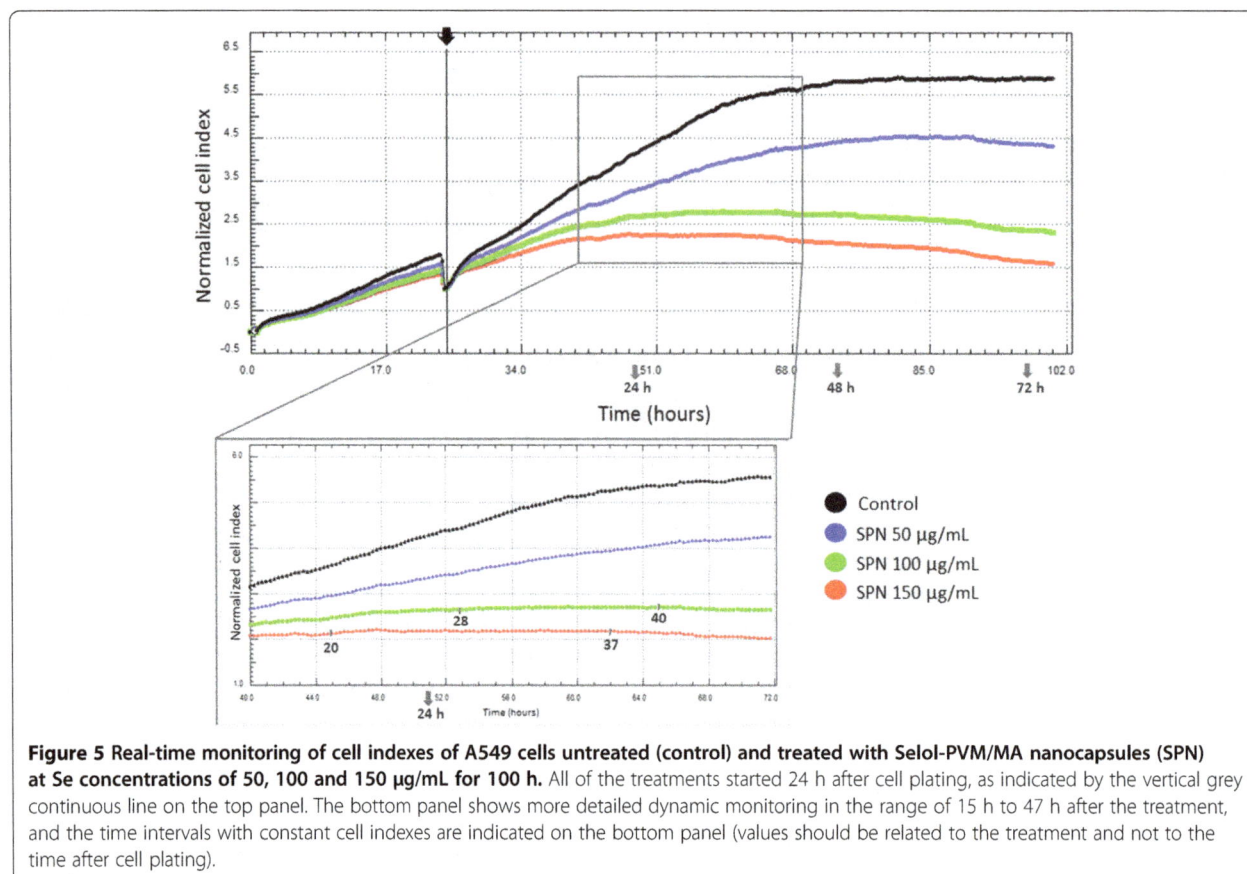

Figure 5 Real-time monitoring of cell indexes of A549 cells untreated (control) and treated with Selol-PVM/MA nanocapsules (SPN) at Se concentrations of 50, 100 and 150 µg/mL for 100 h. All of the treatments started 24 h after cell plating, as indicated by the vertical grey continuous line on the top panel. The bottom panel shows more detailed dynamic monitoring in the range of 15 h to 47 h after the treatment, and the time intervals with constant cell indexes are indicated on the bottom panel (values should be related to the treatment and not to the time after cell plating).

has no activity on the G1 and S checkpoints. In response to SPN treatment, the expression levels of CCNB1 and CDC25C were reduced by half at 24 h and were markedly down-regulated after 48 h (approximately 17-fold for CCNB1 and 21-fold for CDC25C) and 72 h (approximately 25-fold for CCNB1 and 50-fold for CDC25C) compared with the control cells. The expression of WEE1 was reduced only to half at 48 h and 72 h.

In relation to normal cells, A549 cells were more susceptible to Selol, both free and encapsulated, which may be partially due to enhanced endocytic activity [27]. Another possible explanation that is supported by the results of the present study is that tumor cells may be more sensitive to Selol due to their higher proliferation rates because Selol appears to primarily act as an inhibitor of cell proliferation. SPN significantly increased the percentage of A549 cells arrested in the G2/M phase of cell cycle and consequently reduced the number of living cells. In the G2 phase, the wee1 protein inactivates the mitosis promoting factor (cyclin B1/CDK1), and cdc25C is a positive regulator of this complex. As shown in previous studies, G2/M arrest may require activation of wee1 in addition to inactivation of cdc25C [28,29].

Additionally, the decrease in cell proliferation induced by Selol was not associated with a decreased energy metabolism coordinated by the main mitochondrial deacetylase sirtuin 3 [30], as observed by the unchanged transcript levels of the SIRT3 gene (FC < 2.0) (Figure 7(a)). Moreover, subtle changes in the expression of the β-actin gene (ACTB) in SPN-treated cells compared with untreated cells showed that this gene is not suitable as an endogenous control for SPN treatment (Figure 7(a)), likely due to association of the β-actin protein with cytoskeletal components and consequent cell division events [31].

Morphological alterations

Phase contrast microscopy revealed that neither SPN nor free Selol induced morphological changes in A549 cells. SPN treatment did not induce any morphological alterations in normal cells because most of them were shown to be spindle-shaped with cytoplasmic projections, as expected. In contrast, free Selol induced visible cytoplasmic retraction (arrows) in normal cells (Figure 6). Comparatively, pronounced morphological changes were induced by SS in both normal and A549 cells.

Some ultrastructural signs of cell death were observed in A549 cells after treatment with Selol. Mitochondrial

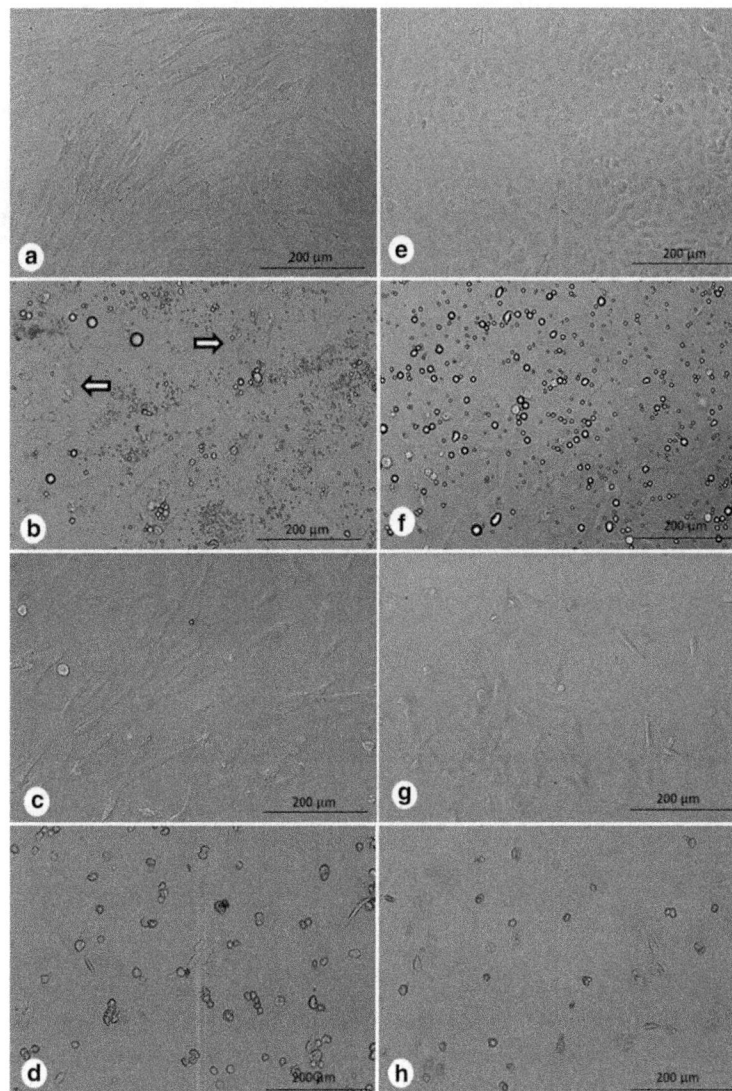

Figure 6 Morphology of human cells from connective tissue (a, b, c, d) and A549 cells (e, f, g, h). The cells were untreated **(a, e)**, exposed to free Selol (S) **(b, f)** or to Selol-PVM/MA nanocapsules (SPN) **(c, g)** for 72 h. Morphologic changes due to sodium selenite (SS) treatment for 24 h **(d, h)**. Decrease in cell confluence by S and SPN treatments. Normal morphology of A549 **(f, g)** and SPN-treated connective tissue cells. Cytoplasmic retraction (arrows) on human cells from connective tissue after treatment with S **(b)**. The bright, black-bordered spherical structures are free Selol microdroplets **(b, f)**. Magnification: 20x.

Table 2 Effect of Selol-PVM/MA nanocapsule (SPN) treatment on the cell cycle distribution of A549 cells

A549	G0/G1 (%)	S (%)	G2/M (%)
Control 48 h	69.6 ± 2.9	18.4 ± 2.3	12.2 ± 1.3
SPN 48 h	56.6 ± 3.8	26.5 ± 0.9	16.9 ± 3.7*
Control 72 h	70.7 ± 3.9	18.8 ± 2.9	10.6 ± 1.7
SPN 72 h	56.2 ± 1.3	25.0 ± 1.9	18.7 ± 2.7*

*$p < 0.05$ compared with the control. The data were obtained from three independent experiments and are presented as the means ± standard error of the mean.

changes and intense vacuolization were apparent (Figure 8). Additionally, cytoplasm swelling suggestive of necrosis was present. These morphological changes suggest that apoptosis and necrosis are induced by treatment with Selol. Most of the normal cells did not present any morphological changes after SPN treatment (Figure 9). However, the cells exposed to free Selol presented large endosomes containing Selol, whereas the SPN-treated cells did not present visible Selol particles. Endocytosis of free Selol droplets in both cell lines caused the compression of adjacent organelles. In addition, an unambiguous identification of Selol nanoparticles was not possible due to similarities with cellular lipids.

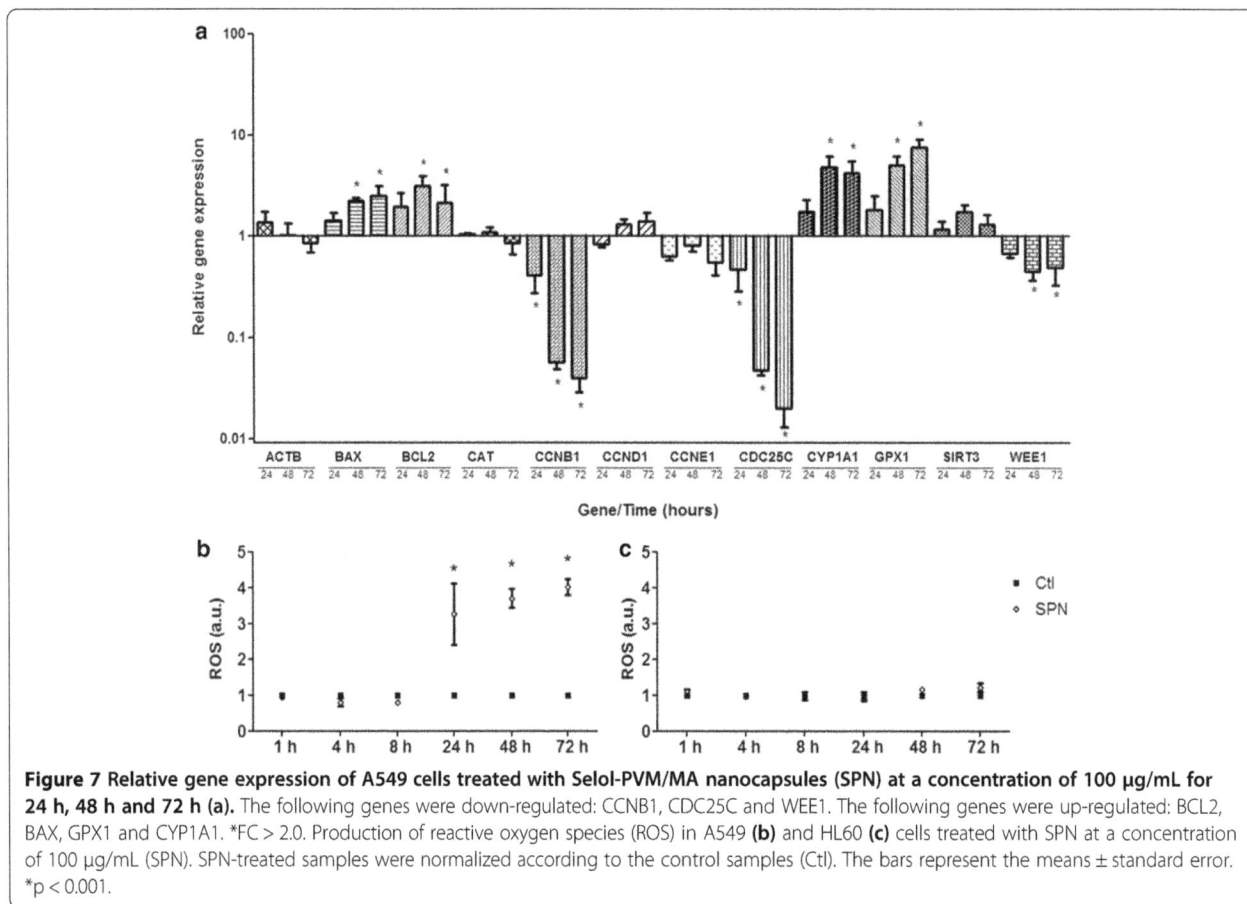

Figure 7 Relative gene expression of A549 cells treated with Selol-PVM/MA nanocapsules (SPN) at a concentration of 100 μg/mL for **24 h, 48 h and 72 h (a).** The following genes were down-regulated: CCNB1, CDC25C and WEE1. The following genes were up-regulated: BCL2, BAX, GPX1 and CYP1A1. *FC > 2.0. Production of reactive oxygen species (ROS) in A549 **(b)** and HL60 **(c)** cells treated with SPN at a concentration of 100 μg/mL (SPN). SPN-treated samples were normalized according to the control samples (Ctl). The bars represent the means ± standard error. *p < 0.001.

Cell death signaling

S and SPN induced the exposure of phosphatidylserine on the outer leaflet of the plasma membrane and impairment of plasma membrane permeability only in a discrete percentage of A549 cells. After 72 h of exposure, only 11.2 ± 0.6% (p < 0.01) and 12.9 ± 2.3% (p < 0.001) of A549 cells were propidium iodide (PI)- and/or annexin V-positives after S and SPN treatments, respectively. In control A549 cells, this percentage was 2.9 ± 0.5% (Figure 10).

SPN treatment caused an increase in the percentage of A549 cells presenting reduced mitochondrial membrane potential ($\Delta\Psi m$) compared with control cells at 48 h (14.5 ± 4.2%, p < 0.05) and 72 h (22.8 ± 5.4%, p < 0.001) (Figure 11(a)). Figure 11(b) also shows the $\Delta\Psi m$ of control and SPN-treated A549 cells at 72 h. These findings are compatible with the mitochondrial damages evidenced by ultrastructure analysis.

DNA fragmentation increased after 48 h (19.8 ± 1.1% vs. 3.6 ± 1.0% for control, p < 0.001) and 72 h (24.0 ± 2.5% vs. 4.6 ± 1.0% for control, p < 0.001) of SPN treatment (Figure 11(c)). As shown in Figure 11c, this treatment resulted in an increase in the sub-G1 cell population, a clear reduction in the percentage of cells in the G1 phase and an increase in the percentage of cells in the G2/M phase (Figure 11(d)).

The levels of transcripts related to apoptosis, namely BAX (BCL2-associated X protein) and BCL-2 (B-cell lymphoma 2), increased after 48 h (2-fold and 3-fold for BAX and BCL-2, respectively) and 72 h (2-fold for both BAX and BCL-2) (Figure 7(a)). These results suggest that Selol triggers damages that are able to activate the apoptosis mechanisms. However, the BAX/BCL2 ratio, which is recognized as an initiator of the caspases activation pathway, was slightly increased only at 72 h. The results presented suggest that the mechanism of action of Selol in A549 cells is not crucially dependent on the direct induction of apoptosis or necrosis. Suchocki et al. (2007) [4] demonstrated mitochondrial changes and DNA fragmentation in a leukemia cell line, and Estevanato et al. (2012) [32] and Wilczynska et al. (2011) [25] observed the translocation of phosphatidylserine in breast and cervix cancer cell lines. In the present study, the exposition of phosphatidylserine, changes in the mitochondrial membrane potential, DNA fragmentation, and/or alterations in the plasma membrane permeability were observed in low percentages of A549 cells after treatment with free or encapsulated Selol. These findings, however, do not exclude the possibility that Selol may induce extensive A549 cell death at the highest concentrations and/or

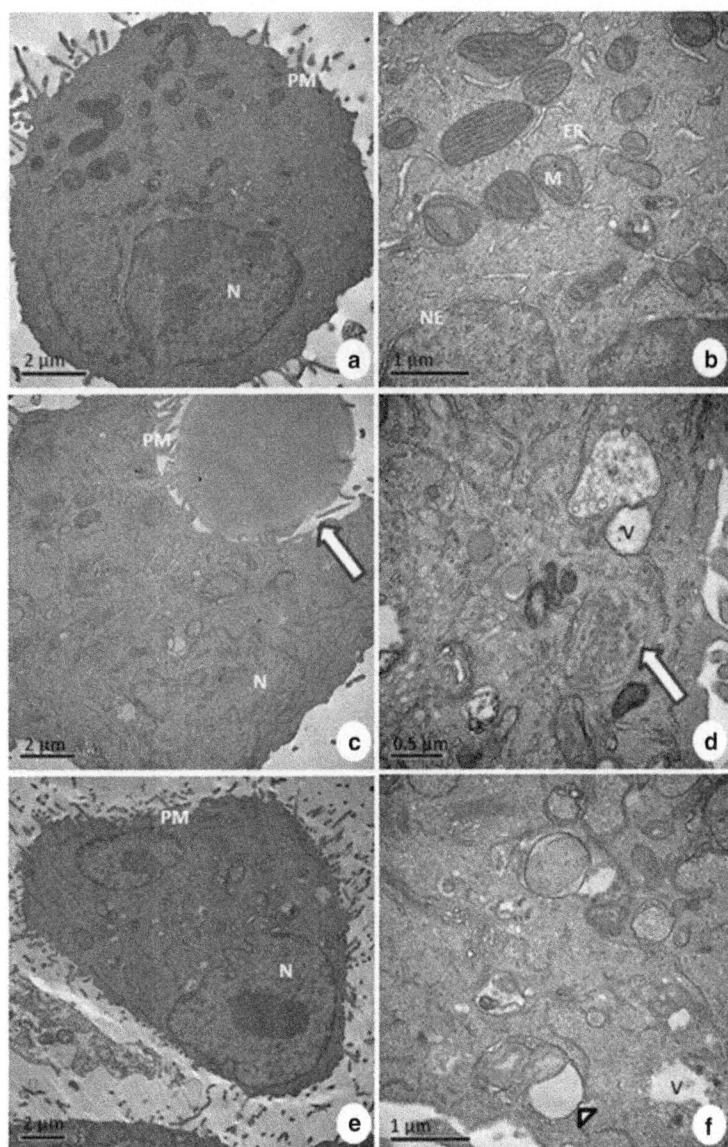

Figure 8 Ultrastructural morphology of A549 cells: control (a, b), treated with free Selol (S) (c, d) and treated with Selol-PVM/MA nanocapsules (SPN) (e, f). Note the integrity of the mitochondria (M), endoplasmic reticulum (ER), plasma membrane (PM), nucleus (N), nuclear envelope (NE), and cytosol of the control cells. An endocytosed Selol droplet is shown with an arrow in **c**. Endosomes, as indicated by an arrow in **d**, were usually identified with the free Selol treatment. Cytoplasm characteristic of necrotic cells can be observed in **e** (arrow). Intense formation of vacuoles (V) containing cellular organelles and changes in the morphology of mitochondria (arrowhead) can be evidenced in **d** and **f**.

after prolonged treatment times. Prolonged cell cycle arrest may elicit apoptosis insomuch that cell death may not only be due to primary damage but also result from accumulation of damages arising from the cell cycle arrest itself [33].

Oxidative stress in A549 cells

The production of reactive oxygen species (ROS) was evaluated in A549 cells compared with HL60 promyelocytic cells. The HL60 cell line is a cell model that is frequently used to access the oxidant or antioxidant potential of different compounds [34,35]. In the current study, Selol surprisingly did not induce an oxidative burst on HL60 cells (Figure 7(b)). Otherwise, ROS accumulation was observed in A549 cells at higher SPN exposure times (24, 48 and 72 h) (Figure 7(b)). These findings suggest that the SPN-induced oxidative burst is variable according to the cell type.

Additionally, increased expression of GPX1 (glutathione peroxidase 1) (5-fold and 7-fold at 48 h and 72 h,

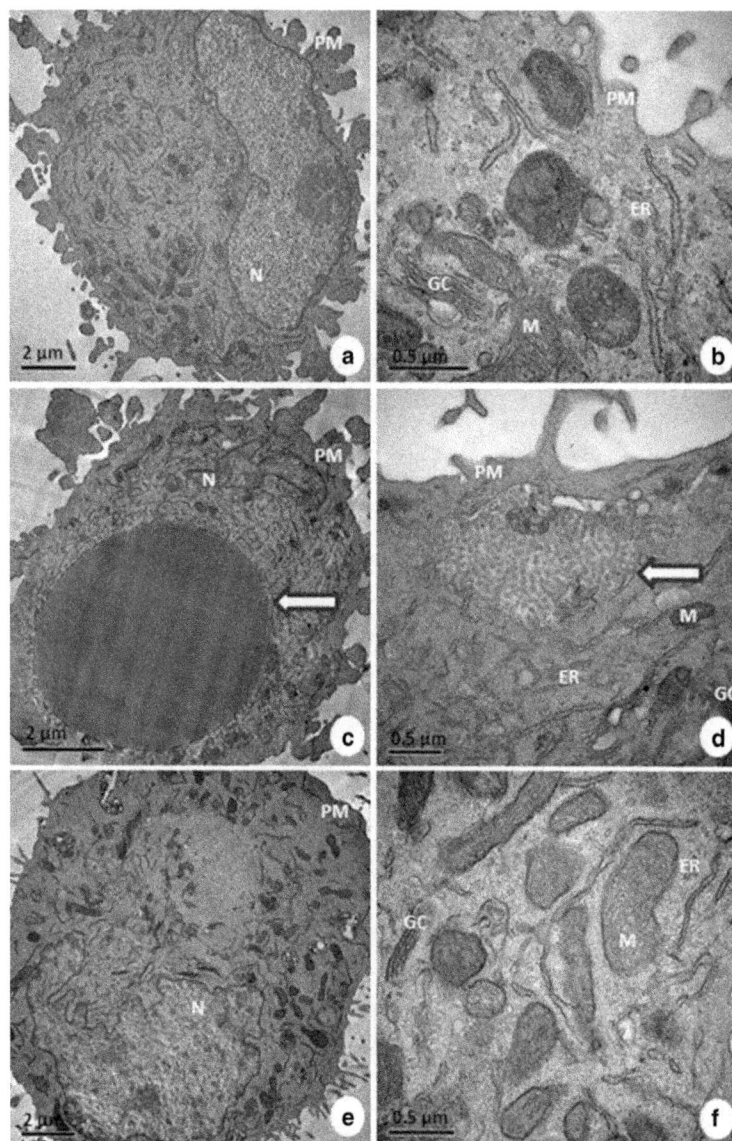

Figure 9 Ultrastructural morphology of human connective tissue cells: control (a, b), treated with free Selol (S) (c, d) or treated with Selol-PVM/MA nanocapsules (SPN) (e, f). Note the integrity of the control cells, as expected. In **c**, a Selol vesicle being taken up by a cell exposed to free Selol is shown by an arrow. In **d**, the arrow points to an endosome. In **e** and **f**, the ultrastructural morphology is unchanged by SPN treatment. N, M, ER, GC, PM, and V: nucleus, mitochondria, endoplasmic reticulum, Golgi complex, plasma membrane and vesicle, respectively.

respectively) was evidenced on A549 cells (Figure 7(a)). Increased levels of ROS coincident with the up-regulation of GPX1 indicate that A549 cells responded in response to exposure to SPN by activation of the antioxidant defense systems. Suchocki et al. (2010) [5] showed the involvement of ROS and the inhibition of CYP1A1 (cytochrome P450) induced by Selol in cervix cancer cell lines. Conversely, in this study, the expression of CYP1A1 (4-fold on both 48 h and 72 h) on A549 cells was enhanced by SPN treatment (Figure 7(a)). The metabolism and detoxification of xenobiotics performed by cytochrome P450 may be associated with SPN metabolism and ROS generation [36]. Therefore, the present study suggests that ROS production on A549 cells may be associated with cell death and inhibition of proliferation, as evidenced by Chen et al. (2013) [37].

Despite the induction of oxidative stress observed on A549 cells, SPN had no effect on the gene expression of catalase at the concentration and times evaluated (Figure 7(b)), corroborating the results of a study on prostate cancer cells treated with Selol for 24 and 48 h [38].

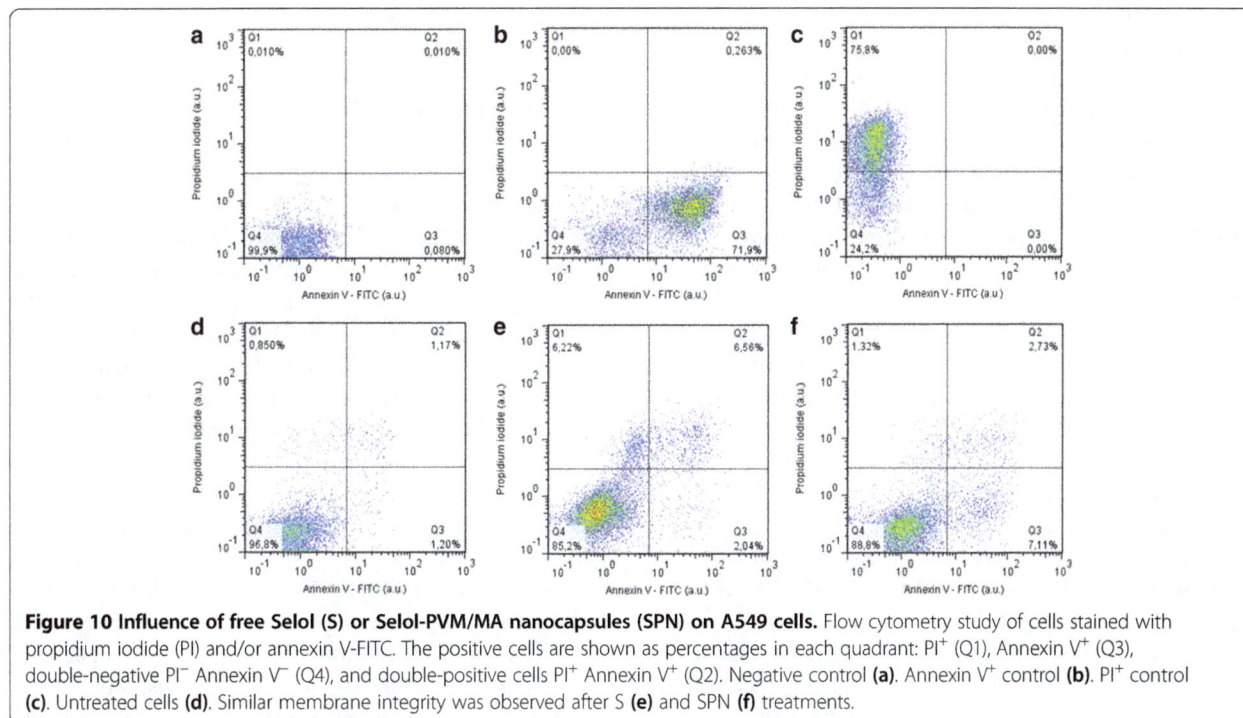

Figure 10 Influence of free Selol (S) or Selol-PVM/MA nanocapsules (SPN) on A549 cells. Flow cytometry study of cells stained with propidium iodide (PI) and/or annexin V-FITC. The positive cells are shown as percentages in each quadrant: PI$^+$ (Q1), Annexin V$^+$ (Q3), double-negative PI$^-$ Annexin V$^-$ (Q4), and double-positive cells PI$^+$ Annexin V$^+$ (Q2). Negative control **(a)**. Annexin V$^+$ control **(b)**. PI$^+$ control **(c)**. Untreated cells **(d)**. Similar membrane integrity was observed after S **(e)** and SPN **(f)** treatments.

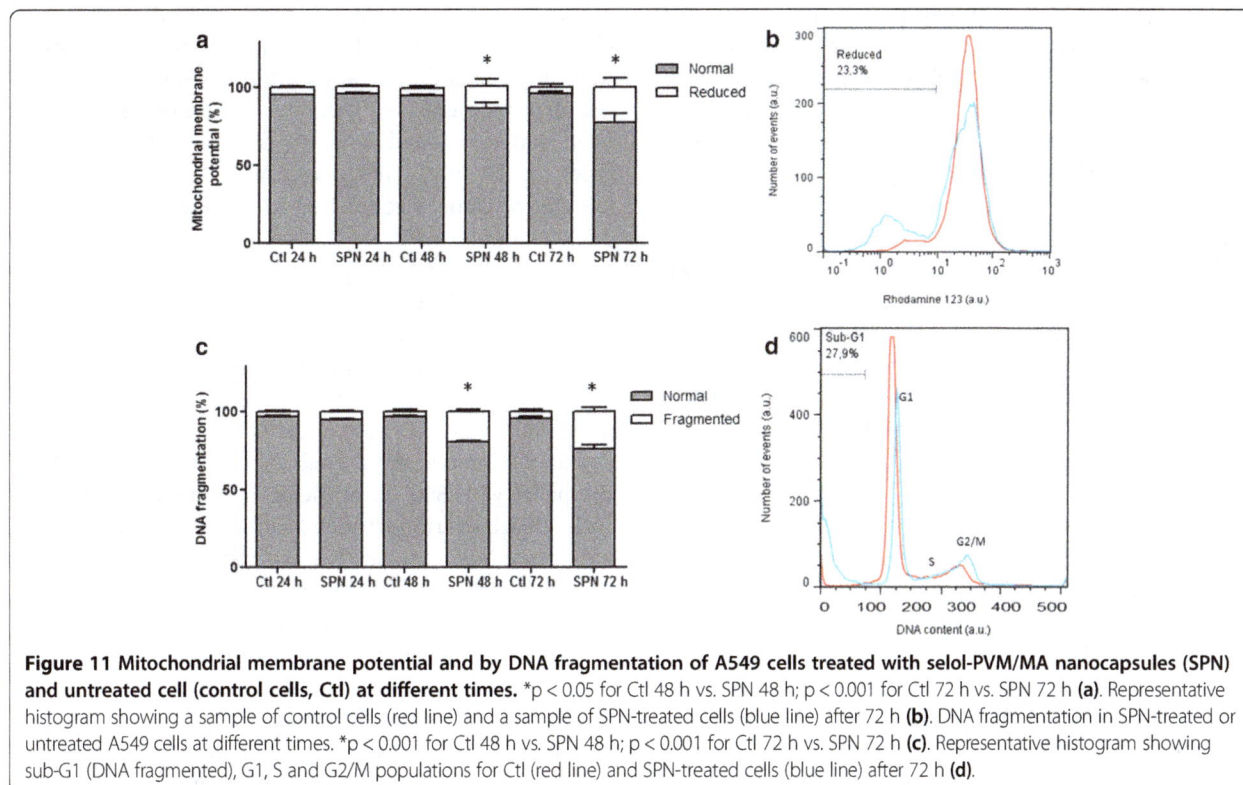

Figure 11 Mitochondrial membrane potential and by DNA fragmentation of A549 cells treated with selol-PVM/MA nanocapsules (SPN) and untreated cell (control cells, Ctl) at different times. *p < 0.05 for Ctl 48 h vs. SPN 48 h; p < 0.001 for Ctl 72 h vs. SPN 72 h **(a)**. Representative histogram showing a sample of control cells (red line) and a sample of SPN-treated cells (blue line) after 72 h **(b)**. DNA fragmentation in SPN-treated or untreated A549 cells at different times. *p < 0.001 for Ctl 48 h vs. SPN 48 h; p < 0.001 for Ctl 72 h vs. SPN 72 h **(c)**. Representative histogram showing sub-G1 (DNA fragmented), G1, S and G2/M populations for Ctl (red line) and SPN-treated cells (blue line) after 72 h **(d)**.

Conclusions

This study demonstrates that Selol can act as a cytostatic agent and corroborates previous reports indicating that it is effective against cancerous cells and safer for clinical applications than sodium selenite. Furthermore, the present study includes the development of a stable and monodisperse aqueous vehicle for Selol delivery, namely PVM/MA-shelled Selol nanocapsules, which are able to maintain the activity of free Selol against pulmonary adenocarcinoma cells, exhibit reduced toxicity to non-tumor cells *in vitro* and are thus potentially suitable for the treatment of some types of lung cancer. Further *in vivo* studies should be performed to evaluate the potential of this formulation for human therapy.

Methods

Materials

Selol composed of 5% selenium (w:w) was provided by Warsaw Medical University (Poland). PVM/MA (Gantrez AN 119) was kindly gifted by ISP Corp. (Brazil). Human lung adenocarcinoma (A549) and promyelocytic leukemia (HL60) cell lines were purchased from American Type Culture Collection (USA). Sodium selenite, trifluoroacetic acid, dimethyl sulfoxide and rhodamine 123 were purchased from Sigma (USA). Dulbecco's modified Eagle's medium, F12 medium, Roswell Park Memorial Institute medium, 3,4,5-dimethylthiazol-2,5 biphenyl tetrazolium bromide (MTT), annexin V-FITC and propidium iodide (PI) were provided by Invitrogen (USA). Fetal bovine serum, penicillin and streptomycin were purchased from Gibco (USA). RNase A was obtained from Promega (USA). Primers and 2',7'-dichlorodihydrofluorescein diacetate (DCFH-DA) were acquired from Life Technology (USA). Phorbol 12-myristate 13-acetate (PMA) was obtained from Abcam (England). GeneJet RNA purification and cDNA Maxima kits were provided by Life Science (USA). DNase I, free-RNase kit was purchased from Thermo Scientific (USA). The TaqMan gene expression assay was obtained from AB Applied Biosystems (USA). Grids and supports of copper and osmium tetroxide were obtained from Electron Microscopy Sciences (USA). Dichloromethane was purchased from Vetec (Brazil). Ethanol and acetone were purchased from J. T. Baker (USA). Phosphate buffer saline was obtained from Laborclin (Brazil).

Preparation of nanocapsules

Nanocapsules were prepared through an interfacial nano-precipitation method. Briefly, Selol (oil phase) and PVM/MA (surfactant) were dissolved in acetone at room temperature (RT). Next, ethanol and distilled water were sequentially added to the acetone solution, under mild stirring, to form a yellowish, opaque suspension. The organic solvents were removed by distillation at 45°C under reduced pressure (80 mbar) in a rotavapor apparatus (Rotavapor

RII®, Buchi Switzerland). Next, the resulting oily nanocapsules were centrifuged at 22,000 × g for 30 min, the transparent aqueous supernatant was removed, and the pellet was resuspended in distilled water. This preparation was immediately characterized and/or stored at 4°C until usage.

Colloidal characterization

The nanocapsules were dispersed in phosphate buffer saline (PBS) (pH 7.4) at a concentration of Selol equivalent to 100 μg/mL. Then, the hydrodynamic diameter (HD) and polydispersity index (PDI) and the zeta potential (ζ potential) were measured at 25°C by dynamic light scattering (DLS) and electrophoretic laser Doppler anemometry (ZetaSizer Nano ZS®, Malvern Instruments), respectively.

Surface morphology and structure

The shape and surface morphology of the capsules were investigated using a field emission scanning electron microscope (SEM) (JEOL JSM 7001-F®, Japan). Before analysis, the composition was diluted with ultrapure water to 5% (v:v), and 20 μL was deposited onto copper supports. Next, the sample was fixed with 1% osmium tetroxide vapor (w:v) for 1 h, left to dry at RT and coated with gold using a Blazers SCD 050® sputter coater (Blazers Union AG, Liechtenstein). The images were digitized using an UltraScan® camera connected to the Digital Micrograph® 3.6.5 computer software (Gatan, USA).

The diameter values of 300 nanocapsules were measured with the Image Pro-Plus® 5.1 software from images captured with a transmission electron microscope (TEM) (JEOL JEM 1011®, Japan). Before this analysis, the sample was diluted with ultrapure water to 3% (v:v) and deposited onto a copper grid. The dried sample was fixed and contrasted with 1% osmium tetroxide vapor (w:v) for 20 min. The images were digitized using an UltraScan® camera connected to the Digital Micrograph 3.6.5® computer software (Gatan, USA).

Efficiency of Selol encapsulation

The Selol concentration in the nanocapsules was estimated according to Suchocki et al. (2003) [39]. Briefly, 300 μL of the nanocapsules dispersion was centrifuged at 22,000 × g and 4°C for 30 min. The pellet was left to dry at RT for two days. Next, Selol was extracted from the pellet with 800 μL of dichloromethane and oxidized with 200 μL of trifluoroacetic acid. The selenium absorbance was measured at a wavelength of 380 nm in a quartz cuvette. The efficiency of Selol encapsulation was calculated considering the ratio of the encapsulated mass to the mass of Selol that was initially used.

Thermodynamic stability studies

Storage at room temperature, 4°C and −20°C. Aliquots of the nanocapsule dispersion were stored at RT, 4°C or −20°C, and their colloidal characteristics (HD, PDI and ζ potential) were evaluated every 15 days.

Freezing and thawing cycles. Fifteen cycles of freezing ($-20°C$) and thawing ($25°C$) were applied to a nanocapsule aliquot. After each cycle, the HD, PDI and ζ potential were evaluated.

Cell culture

The A549 human lung adenocarcinoma cell line was cultured with a 1:1 (v:v) mixture of Dulbecco's modified Eagle's medium (DMEM) and F12 medium supplemented with 10% (v:v) fetal bovine serum (FBS) and 1% (v:v) antibiotic solution (100 units/mL penicillin and 100 mg/mL streptomycin). Human connective tissue cells harvested from the dental pulp of normal teeth were maintained in primary culture and used as non-tumor control cells, namely normal cells [40]. These cells were grown in DMEM supplemented with 10% FBS and 1% antibiotic solution, as described above. Both cells were maintained at $37°C$ in a 5% CO_2 and 80% humidity environment.

Treatment design

The cells were allowed to adhere to culture microplates for 24 h and were then treated as follows: (1) Selol-PVM/MA nanocapsules (SPN), (2) free Selol (S), (3) blank polymeric nanoparticles without Selol (Bl), and (4) sodium selenite (SS). Cells treated with culture medium, culture medium/acetone and culture medium/PBS corresponded to the control groups of SPN, S and SS, respectively. For the free Selol treatments, Selol was dissolved in acetone and then added to the culture medium, as described by Suchocki et al. (2007) [4]. Before the tests, a viability study was performed to ensure that the volumes of acetone required for each Selol concentration would not be cytotoxic themselves. Bl nanoparticles were prepared using the same method with the same concentrations of the components used for SPN but without Selol. Each treatment was performed in triplicate with different Se concentrations (50, 100 and 150 µg/mL) and times of exposure (24, 48 and 72 h). The concentration of PVM/MA in the treatments with Bl was equivalent to that used in the SPN treatment.

To verify whether the biological activity of SPN was maintained during storage at $4°C$, the cell viability was also evaluated on days 1st and 60th after preparation of the SPN.

Cell viability

The cells were seeded and treated as described above. Next, the cells were incubated with 0.5 mg/mL MTT for 2.5 h. The MTT solution was removed, and formazan was extracted from the cells with dimethyl sulfoxide [41]. The absorbance at a wavelength of 595 nm was measured using a spectrophotometer (Spectramax M2, USA) and was used as an index of cell viability. The results were expressed as percentages relative to the control groups after 24 h of treatment.

Cell counting

A549 and normal cells were treated with SPN (100 µg/mL) for 24, 48 and 72 h. Next, the cells were harvested and quantified using a Scepter™ Cell Counter (Millipore, USA).

Morphology and confluence of cells

The morphology and confluence of the cells were analyzed using a phase contrast microscope (Zeiss, Germany) and the AxioVision® software (Zeiss, Germany). For ultrastructural analysis, the cells were fixed, contrasted and dehydrated in agreement with the method described by Carneiro et al. (2011) [42]. Ultrathin sections were observed through TEM, and the images were digitized.

Real-time cell index

Real-time cell analysis was performed with a RTCA instrument (xCelligence, Roche, Switzerland) [43]. Briefly, A549 cells were seeded for 24 h on plates containing microelectronic sensor arrays and later incubated with medium or treated with different concentrations of SPN (50, 100, and 150 µg/mL). The cells were automatically monitored every 15 min for up to 100 h. Two independent experiments were performed.

Reactive oxygen species

The intracellular production of reactive oxygen species (ROS) was measured using 2',7'-dichlorodihydrofluorescein diacetate (DCFH-DA) as an oxidation-sensitive fluorescent probe. The ROS production was evaluated in A549 and HL60 cells [34,35]. The HL60 cells were cultured in RPMI (Roswell Park Memorial Institute medium) supplemented with 10% FBS and 1% antibiotic solution and were maintained at $37°C$, 5% CO_2 and 80% humidity. Then, the cells treated or not treated with SPN (100 µg/mL) for different times (1, 4, 8, 24, 48 and 72 h) were stained with 2.5 µM DCFH-DA for 30 min at $37°C$. HL60 cells treated with 10 mM phorbol 12-myristate 13-acetate (PMA) were used as a positive control. A total of 10,000 events per sample were analyzed using a FC500® cytometer (Beckman Coulter, USA) and the FlowJo 7.6.3 software.

Annexin V-FITC/propidium iodide staining

The externalization of phosphatidylserine and the loss of plasma membrane integrity, which are signs of apoptosis and necrosis, respectively, were assessed with a double-staining kit consisting of FITC-labeled annexin V and propidium iodide (PI). The cells were incubated with 100 µL of binding buffer containing 10 mM HEPES/NaOH (pH 7.4), 140 mM NaCl and 2.5 mM $CaCl_2$. Next, 5 µL of annexin V-FITC and 10 μL of PI (50 μg/mL) were added, and the cells were incubated for 15 min in the dark at RT. The cells were analyzed with a CyFlow® space cytometer (Partec, Germany), and 10,000 events were counted per sample. Cells not incubated with Annexin V-FITC and PI were used as the negative control. Cells incubated with 100 μg/mL SS for 24, 48 and 72 h were used as annexin V staining-positive cells. Cells killed by heating at 60°C for 5 min were used as PI staining-positive cells. All of the cytometry results reported in the CyFlow® space were analyzed using the Windows™ Flow Max® and FlowJo 7.6.3 software programs.

DNA fragmentation and cell cycle analysis

The cell cycle was evaluated by the quantification of total DNA. A549 cells were fixed with 70% ethanol for 2 h at 4°C, rinsed with PBS, incubated with 50 μg/mL RNase A for 30 min at 37°C and stained with 50 μg/mL PI for 30 min at RT [44]. A total of 10,000 events per sample were counted with a CyFlow® cytometer, and the percentage of cells in different phases of the cell cycle was determined. Only those cells presenting DNA content in the range of 2n-4n were considered in the cell cycle analysis. Fragmented DNA was identified in the sub-G1 (DNA content < 2n) population and calculated considering the totality of events.

Mitochondrial membrane potential

The fluorescent cationic substrate rhodamine 123 (Rho123) was used to assess the mitochondrial membrane potential ($\Delta\Psi m$) in A549 cells. The cells were incubated with 5 μg/mL Rho123 for 15 min at RT and washed twice with PBS [45]. A total of 10,000 events were analyzed per sample using a CyFlow® cytometer.

Quantitative RT-PCR (qRT-PCR)

A549 cells were harvested after SPN treatment (100 μg/mL) for 24, 48 and 72 h. The total RNA was extracted using a GeneJet RNA purification kit and was then treated with DNase I, free-RNase kit. cDNA was synthesized from mRNA using cDNA Maxima reverse transcription reagents. qRT-PCR was performed using a TaqMan gene expression assay, and the amplification reactions were performed using a Fast Real-time System 7900HT (Applied BioSystems, USA). The following primers (and their specifications) were used: ACTB (Hs99999903_m1), BAX (Hs00180269_m1), BCL2 (Hs00608023_m1), CAT (Hs00156308_m1), CDC25C (Hs00156411_m1), CCNB1 (Hs01030099_m1), CCND1 (Hs00765553_m1), CCNE1 (Hs01026536_m1), CYP1A1 (Hs00153120_m1), GAPDH (Hs02758991_g1), GPX1 (Hs00829989_gH), SIRT3 (Hs00953477_m1), and WEE1 (Hs00268721_m1). All of the kits were used according to the manufacturer's instructions.

The cDNA dilutions were defined based on threshold cycles (CT) (17 – 20) derived from the amplification of the constitutive gene GAPDH ($R^2 \geq 0.9974$). Each sample was normalized based on the mRNA expression level of GAPDH. The gene expression values were obtained using the $2^{-\Delta\Delta CT}$ equation. A gene was considered to be differentially expressed when the transcript rate (FC, fold change) was at changed by at least twofold compared with the untreated control sample [46].

Statistical analysis

All of the experiments were performed in triplicate and repeated three times. The results are represented as the means ± standard deviation. Significant differences were assessed by one- or two-way analyses of variance followed by Tukey or Bonferroni's post-tests ($\alpha = 0.05$) using the GraphPad Prism 5.0 software.

Competing interests

We confirm that we have given due consideration to the protection of intellectual property associated with this work and that there are no impediments to publication, including the timing of publication, with respect to intellectual property. The authors disclose no potential conflicts of interest.

Authors' contributions

LRS was the principal investigator and took primary responsibility for the paper. LRS, LAM, RBA and SNB participated in the design and coordination of the study. EM and PS conducted the Selol development. LAM, RG and PCM developed and characterized the Selol nanocapsules. LRS, MSCS, RS and GAJ performed the biological assays. AG participated in the coordination of some experiments and helped draft the manuscript. LRS, LAM, RBA and SNB wrote the manuscript and all of the authors helped discuss the results, adding thoughtful insights to the manuscript, and approved the final manuscript.

Acknowledgments

This work was supported by the Conselho Nacional de Desenvolvimento Científico e Tecnológico (CNPq), the Coordenação de Aperfeiçoamento de Pessoal de Nível Superior (CAPES), the Financiadora de Estudos e Projetos (FINEP), the Fundação de Empreendimentos Científicos e Tecnológicos (FINATEC), the Instituto Nacional de Ciência e Tecnologia (INCT), the European Union Seventh Framework Programme [FP7/REGPOT-2012-2013.1] under grant agreement BIOCAPS-316265 and Xunta de Galicia (INBIOMED 2012/273, DXPCTSUG-FEDER), and the National Science Centre in Poland (Grants N N405 360639 and N N202 166440).

Author details

[1]Institute of Biological Sciences, Molecular Biology Programme, University of Brasília, Brasília, DF 70910-900, Brazil. [2]Institute of Biological Sciences, University of Brasília, Brasília, DF 70910-900, Brazil. [3]Biomedical Research Center (CINBIO), Institute of Biomedical Research of Vigo, University of Vigo, Vigo, Pontevedra 36310, Spain. [4]Institute of Physics, Polish Academy of Sciences, Warsaw 02-668, Poland. [5]Department of Bioanalysis and Drugs Analysis, Warsaw Medical University, Warsaw 02-097, Poland. [6]Department of Pharmaceutical Chemistry, National Medicines Institute, Warsaw 00-725, Poland. [7]Institute of Physics, University of Brasília, Brasília, DF 70910-900, Brazil. [8]School of Automation, Huazhong University of Science and Technology, Wuhan, Hubei 430074, China.

References

1.   El-Bayoumy K, Sinha R: Molecular chemoprevention by selenium: a genomic approach. Mutat Res 2005, 591:224–236.

2. Drake E: **Cancer chemoprevention: selenium as a prooxidant, not an antioxidant.** *Med Hypotheses* 2006, 67:318–322.

3. Fitak B, Grabowski M, Suchocki P: **Preparat przeciwnowotworowy i sposób jego wytwarzania.** *Pol Patent* 1999, 1765301994.

4. Suchocki P, Misiewicz I, Skupinska K, Waclawek K, Fijalek Z, Kasprzycka-Guttman T: **The activity of Selol in multidrug-resistant and sensitive human leukemia cells.** *Oncol Rep* 2007, 18:893–900.

5. Suchocki P, Misiewicz-Krzemińska I, Skupińska K, NiedŸwiecka K, Lubelska K, Fijaek Z, Kasprzycka-Guttman T: **Selenitetriglicerydes affect CYP1A1 and QR activity by involvement of reactive oxygen species and Nrf2 transcription factor.** *Pharmacol Rep* 2010, 62:352–361.

6. Bennett VA, Davies EA, Jack RH, Mak V, Moller H: **Histological subtype of lung cancer in relation to socio-economic deprivation in South East England.** *BMC Cancer* 2008, 8:139.

7. Jemal A, Siegel R, Xu J, Ward E: **Cancer statistics, 2010.** *CA-Cancer J Clin* 2010, 60:277–300.

8. Pfister DG, Johnson DH, Azzoli CG, Sause W, Smith TJ, Baker S Jr, Olak J, Stover D, Strawn JR, Turrisi AT, Somerfield MR: **American Society of Clinical Oncology treatment of unresectable non–small-cell lung cancer guideline: Update 2003.** *J Clin Oncol* 2004, 22:330–353.

9. Goldstraw P, Crowley J, Chansky K, Giroux DJ, Groome PA, Rami-Porta R, Postmus PE, Rusch V, Sobin L: **International Association for the Study of Lung Cancer International Staging Committee; Participating Institutions. The IASLC Lung Cancer Staging Project: proposals for the revision of the TNM stage groupings in the forthcoming (seventh) edition of the TNM Classification of malignant tumours.** *J Thorac Oncol* 2007, 2:706–714.

10. Wakelee H, Belani CP: **Optimizing first-line treatment options for patients with advanced NSCLC.** *Oncologist* 2005, 10:1–10.

11. Das A, Bortner J, Desai D, Amin S, El-Bayoumy K: **The selenium analog of the chemopreventive compound S, S'-(1, 4-phenylenebis [1, 2-ethanediyl]) bisisothiourea is a remarkable inducer of apoptosis and inhibitor of cell growth in human non-small cell lung cancer.** *Chem Biol Interact* 2009, 180:158–164.

12. Azzoli CG, Temin S, Aliff T, Baker S Jr, Brahmer J, Johnson DH, Laskin JL, Masters G, Milton D, Nordquist L, Pfister DG, Piantadosi S, Schiller JH, Smith R, Smith TJ, Strawn JR, Trent D, Giaccone G: **2011 focused update of 2009 American Society of Clinical Oncology clinical practice guideline update on chemotherapy for stage IV non–small-cell lung cancer.** *J Clin Oncol* 2009, 29:3825–3831.

13. Paz-Ares L, De Marinis F, Dediu M, Thomas M, Pujol JL, Bidoli P, Molinier O, Sahoo TP, Laack E, Reck M, Corral J, Melemed S, John W, Chouaki N, Zimmermann A, Visseren-Grul C, Gridelli C: **Maintenance therapy with pemetrexed plus best supportive care versus placebo plus best supportive care after induction therapy with pemetrexed plus cisplatin for advanced non-squamous non-small-cell lung cancer (PARAMOUNT): a double-blind, phase 3, randomised controlled trial.** *Lancet Oncol* 2012, 13:247–255.

14. Li Y, Li X, Wong YS, Chen T, Zhang H, Liu C, Zheng W: **The reversal of cisplatin-induced nephrotoxicity by selenium nanoparticles functionalized with 11-mercapto-1-undecanol by inhibition of ROS-mediated apoptosis.** *Biomaterials* 2011, 32:9068–9076.

15. Jastrzębski Z, Czyżewska-Szafran H, Remiszewska M: **Effect of administration route on the dynamics of selol absorption, distribution and elimination.** *Pol J Environ Stud* 1995, 6:64–66.

16. Arbos P, Arangoa M, Campanero M, Irache J: **Quantification of the bioadhesive properties of protein-coated PVM/MA nanoparticles.** *Int J Pharm* 2002, 242:129–136.

17. Fraga M, Bruxel F, Lagranha VL, Teixeira HF, Matte U: **Influence of phospholipid composition on cationic emulsions/DNA complexes: physicochemical properties, cytotoxicity, and transfection on Hep G2 cells.** *Int J Nanomed* 2011, 6:2213–2220.

18. Aubry J, Ganachaud F, Cohen Addad JP, Cabane B: **Nanoprecipitation of polymethylmethacrylate by solvent shifting: 1. Boundaries.** *Langmuir* 2009, 25:1970–1979.

19. Vitale SA, Katz JL: **Liquid droplet dispersions formed by homogeneous liquid-liquid nucleation:"The ouzo effect".** *Langmuir* 2003, 19:4105–4110.

20. Guido S, Villone M: **Measurement of interfacial tension by drop retraction analysis.** *J Colloid Interf Sci* 1999, 209:247–250.

21. Hirsjärvi S, Dufort S, Gravier J, Texier I, Yan Q, Bibette J, Sancey L, Josserand V, Passirani C, Benoit JP, Coll JL: **Influence of size, surface coating and fine chemical composition on the in vitro reactivity and in vivo biodistribution of lipid nanocapsules versus lipid nanoemulsions in cancer models.** *Nanomed-Nanotechnol* 2013, 9:375–387.

22. Morachis JM, Mahmoud EA, Almutairi A: **Physical and chemical strategies for therapeutic delivery by using polymeric nanoparticles.** *Pharmacol Rev* 2012, 64:505–519.

23. Wang L, Yang Z, Fu J, Yin H, Xiong K, Tan Q, Jin H, Li J, Wang T, Tang W, Yin J, Cai G, Liu M, Kehr S, Becker K, Zeng H: **Ethaselen: A potent mammalian thioredoxin reductase 1 inhibitor and novel organoselenium anticancer agent.** *Free Radical Bio Med* 2012, 52:898–908.

24. Suzuki M, Endo M, Shinohara F, Echigo S, Rikiishi H: **Differential apoptotic response of human cancer cells to organoselenium compounds.** *Cancer Chemoth Pharm* 2010, 66:475–484.

25. Wilczynska JD, Ksiazek I, Nowak K, Suchocki P, Flis S, Kiljan M, Anuszewska E: **Study of the effect of Selol and sodium selenite on HeLa cells in vitro.** *CHEMIK* 2011, 65:110–114.

26. Jang SH, Wientjes MG, Lu D, Au JLS: **Drug delivery and transport to solid tumors.** *Pharm Res* 2003, 20:1337–1350.

27. Bareford LM, Swaan PW: **Endocytic mechanisms for targeted drug delivery.** *Adv Drug Deliver Rev* 2007, 59:748–758.

28. Liu X, Wang J, Sun B, Zhang Y, Zhu J, Li C: **Cell growth inhibition, G2M cell cycle arrest, and apoptosis induced by the novel compound Alternol in human gastric carcinoma cell line MGC803.** *Invest New Drug* 2007, 25:505–517.

29. Ghiasi N, Habibagahi M, Rosli R, Ghaderi A, Yusoff K, Hosseini A, Abdullah S, Jaberipour M: **Tumour suppressive effects of WEE1 gene silencing in breast cancer cells.** *Asian Pac J Cancer Prev* 2013, 14:6605–6611.

30. Li H, Feng Z, Wu W, Li J, Zhang J, Xia T: **SIRT3 regulates cell proliferation and apoptosis related to energy metabolism in non-small cell lung cancer cells through deacetylation of NMNAT2.** *Int J Oncol* 2013, 43:1420–1430.

31. Bunnell TM, Burbach BJ, Shimizu Y, Ervasti JM: **β-Actin specifically controls cell growth, migration, and the G-actin pool.** *Mol Biol Cell* 2011, 22:4047–4058.

32. Estevanato LL, Da Silva JR, Falqueiro AM, Mosiniewicz-Szablewska E, Suchocki P, Tedesco AC, Morais PC, Lacava ZGM: **Co-nanoencapsulation of magnetic nanoparticles and selol for breast tumor treatment: in vitro evaluation of cytotoxicity and magnetohyperthermia efficacy.** *Int J Nanomed* 2012, 7:5287.

33. Shinomiya N, Kuno Y, Yamamoto F, Fukasawa M, Okumura A, Uefuji M, Rokutanda M: **Different mechanisms between premitotic apoptosis and postmitotic apoptosis in X-irradiated U937 cells.** *Int J Radiat Oncol* 2000, 47:767–777.

34. Teufelhofer O, Weiss R, Parzefall W, Schulte-Hermann R, Micksche M, Berger W, Elbling L: **Promyelocytic HL60 cells express NADPH oxidase and are excellent targets in a rapid spectrophotometric microplate assay for extracellular superoxide.** *Toxicol Sci* 2003, 76:376–383.

35. Chen H, Zhang B, Yao Y, Chen N, Chen X, Tian H, Wang Z, Zheng Q: **NADPH oxidase-derived reactive oxygen species are involved in the HL-60 cell monocytic differentiation induced by isoliquiritigenin.** *Molecules* 2012, 17:13424–13438.

36. Burczynski ME, Penning TM: **Genotoxic polycyclic aromatic hydrocarbon ortho-quinones generated by Aldo-Keto Reductases Induce CYP 1A1 via Nuclear Translocation of the Aryl Hydrocarbon Receptor.** *Cancer Res* 2000, 60:908–915.

37. Chen T, Chen M, Chen J: **Ionizing radiation potentiates dihydroartemisinin-induced apoptosis of A549 cells via a caspase-8-dependent pathway.** *PLoS One* 2013, 8:e59827.

38. Ksiazek I, Sitarz K, Roslon M, Anuszewska E, Suchocki P, Wilczynska DJ: **The influence of selol on the expression of oxidative stress genes in normal and malignant prostate cells.** *Cancer Genomics Proteomics* 2013, 10:225–232.

39. Suchocki P, Jakoniuk D, Fitak B: **Specific spectrophotometric method with trifluoroacetic acid for the determination of selenium (IV) in selenitetriglycerides.** *J Pharm Biomed* 2003, 32:1029–1036.

40. Pereira LO, Longo JPF, Azevedo RB: **Laser irradiation did not increase the proliferation or the differentiation of stem cells from normal and inflamed dental pulp.** *Arch Oral Biol* 2012, 57:1079–1085.

41. Mosmann T: **Rapid colorimetric assay for cellular growth and survival: application to proliferation and cytotoxicity assays.** *J Immunol Methods* 1983, 65:55–63.

42. Carneiro MLB, Nunes ES, Peixoto RCA, Oliveira RGS, Lourenço LHM, Silva ICR, Simioni AR, Tedesco AC, Souza AR, Lacava ZGM, Báo SN: **Free Rhodium (II)**

citrate and rhodium (II) citrate magnetic carriers as potential strategies for breast cancer therapy. *J Nanobiotechnology* 2011, **9**:11.

43. Xing JZ, Zhu L, Gabos S, Xie L: Microelectronic cell sensor assay for detection of cytotoxicity and prediction of acute toxicity. *Toxicol In Vitro* 2006, **20**:995–1004.

44. Liu L, Ni F, Zhang J, Jiang X, Lu X, Guo Z, Xu R: Silver nanocrystals sensitize magnetic-nanoparticle-mediated thermo-induced killing of cancer cells. *Acta Biochim Biophys Sin (Shanghai)* 2011, **43**:316–323.

45. Joanitti GA, Azevedo RB, Freitas SM: Apoptosis and lysosome membrane permeabilization induction on breast cancer cells by an anticarcinogenic Bowman–Birk protease inhibitor from Vigna unguiculata seeds. *Cancer Lett* 2010, **293**:73–81.

46. Robertson KL, Mostaghim A, Cuomo CA, Soto CM, Lebedev N, Bailey RF, Wang Z: Adaptation of the black yeast Wangiella dermatitidis to ionizing radiation: molecular and cellular mechanisms. *PLoS One* 2012, **7**:e48674.

5

Effects of an 11-nm DMSA-coated iron nanoparticle on the gene expression profile of two human cell lines, THP-1 and HepG2

Ling Zhang[1,2], Xin Wang[1], Jinglu Zou[1], Yingxun Liu[1] and Jinke Wang[1*]

Abstract

Background: Iron nanoparticles (FeNPs) have attracted increasing attention over the past two decades owing to their promising application as biomedical agents. However, to ensure safe application, their potential nanotoxicity should be carefully and thoroughly evaluated. Studies on the effects of FeNPs on cells at the transcriptomic level will be helpful for identifying any potential nanotoxicity of FeNPs and providing valuable mechanistic insights into various FeNPs-induced nanotoxicities.

Results: This study investigated the effects of an 11-nm dimercaptosuccinic acid-coated magnetite nanoparticle on the gene expression profiles of two human cell lines, THP-1 and HepG2. It was found that the expression of hundreds of genes was significantly changed by a 24-h treatment with the nanoparticles at two doses, 50 μg/mL and 100 μg/mL, in the two cell types. By identifying the differentially expressed genes and annotating their functions, this study characterized the general and cell-specific effects of the nanoparticles on two cell types at the gene, biological process and pathway levels. At these doses, the overall effects of the nanoparticle on the THP-1 cells were the induction of various responses and repression of protein translation, but in the HepG2 cells, the main effects were the promotion of cell metabolism, growth and mobility. In combination with a previous study, this study also characterized the common genes, biological processes and pathways affected by the nanoparticle in two human and mouse cell lines and identified *Id3* as a nanotoxicity biomarker of the nanoparticle.

Conclusion: The studied FeNPs exerted significant effects on the gene expression profiles of human cells. These effects were highly dependent on the innate biological functions of cells, i.e., the cell types. However, cells can also show some cell type-independent effects such as repression of *Id3* expression. *Id3* can be used as a nanotoxicity biomarker for iron nanoparticles.

Keywords: Iron nanoparticle, Gene expression profile, THP-1, HepG2

Background

Iron nanoparticles (FeNPs) have attracted increasing attention over the past two decades owing to their promising applications as biomedical agents [1,2]. FeNPs have been the most intensively studied and commercialized nanomaterial in recent years. Despite their generally good biocompatibility relative to other metal nanomaterials [3], their potential nanotoxicity has been recognized [4,5]. For this reason, many studies have investigated the potential nanotoxicity of FeNPs [6]. Some important nanotoxicities of FeNPs were thus discovered, including reduction of cell viability [7,8] and induction of cellular inflammation [9,10], mitochondrial injury [8,11,12], apoptosis [8,13,14], reactive oxygen species (ROS) [8,11,15,16], autophagy [8,11], oxidative stress [14,17,18], cell motility impairment [15], and DNA damage [17,18].

In response to any intracellular and extracellular environmental changes, cells can rapidly change their transcriptomic output, i.e., gene expression profile. In this way, cells adapt to the environmental changes for their survival and function. However, excessive environmental changes can damage the normal physiological activities and biological functions of cells. Therefore, evaluation of gene expression profile changes is helpful in identifying

* Correspondence: wangjinke@seu.edu.cn
[1]State Key Laboratory of Bioelectronics, Southeast University, Nanjing 210096, China
Full list of author information is available at the end of the article

the potential nanotoxicity of nanomaterials [16,19-21]. Identifying all the genes whose expression is affected by a nanomaterial at the cell or tissue levels can provide valuable clues for identifying any potential toxicity and the relevant molecular mechanism [16,19,20,22]. Moreover, current transcriptomic profiling techniques including GeneChip and RNA-seq allow the analysis of global gene expression [20,23,24]. Therefore, increasing numbers of studies have investigated the nanotoxicity of various nanomaterials at the transcriptome level [16,20,24]. Many previously unknown nanotoxicities of nanomaterials were thus uncovered, such as intracellular production of ROS and the resulting cell apoptosis induced by silver, silica and magnetic nanoparticles [19,20,25].

Recent transcriptomic studies have provided valuable mechanistic insights into the various nanotoxicities induced by FeNPs. For example, a transcriptomic analysis found that the transcription of many genes relevant to iron metabolism (*Trf, Tfrc, Lcn2, Hfe*) and osmosis (*Slc5a3, Slc6a12*) was significantly changed by FeNPs in mouse RAW264.7 cells [21], indicating that the iron and osmotic homeostasis of the cells was disturbed by FeNPs. The subsequent measurement of the cellular iron content revealed that the internalized FeNPs were degraded in the acidic environment of the lysosomes and thus released iron ions in the cells, which changed the iron and osmotic homeostasis of the cells. In complementary responses, the cells downregulated the expression of the *Trf, Tfrc*, and *Hfe* genes to prevent the transfer of extracellular Fe^{2+} into the cells, upregulated the expression of the *Lcn2* gene to promote the transfer of intracellular Fe^{2+} out of the cells, and downregulated the expression of the *Slc5a3* gene to inhibit the transfer of extracellular myo-inositol, a very important organic osmolyte, into the cells [21].

Our lab has recently evaluated the effects of a FeNP material deemed to have good biocompatibility, 11-nm magnetite (Fe_3O_4) FeNPs coated with dimercaptosuccinic acid (DMSA) [9], at the transcriptome level. The potential nanotoxicological effects of these FeNPs at doses of 50 and 100 µg/mL on the gene expression profiles of two mouse cell lines (RAW264.7 and Hepa1-6) were examined [10]. This study characterized the general and cell-specific biological processes affected by the FeNPs in these two cell lines by identifying the differentially expressed genes (DEGs) and annotating their functions, providing new insights into the nanotoxicity of the FeNPs. RAW264.7 cells are a blood cell line belonging to monocyte-macrophage system, whereas Hepa1-6 cells are a liver-derived hepatoma cell line. Generally, the former is mainly involved in immune activity, whereas the latter is responsible for detoxification in the living body. The blood and liver cells encounter the greatest exposure to the nanomaterials in vivo due to the use of

intravenous administration and the passive targeting of nanomaterials. Therefore, the two cell lines are suitable for evaluating the nanotoxicity of FeNPs.

The benefit of using mouse cells is that the nanotoxicity observed *in vitro* can be further evaluated *in vivo* by administering the nanomaterials to mice [26]. However, the similar *in vivo* evaluation cannot be performed in humans. Therefore, a feasible strategy is to evaluate the nanotoxicity of a nanomaterial with human cells and their mouse equivalents. If the *in vitro* nanotoxicity of a nanomaterial is similar in cells of two species, its *in vivo* nanotoxicity can be evaluated in the mouse to judge its *in vivo* nanotoxicity in humans. According to this strategy, based on our recent study of the nanotoxicity of a FeNP with two mouse cells [10], this study treated two equivalent human cell lines, human monocytic THP-1 cells and hepatoma HepG2 cells, with the same FeNPs at the same doses (50 and 100 µg/mL) for the same time (24 h), and profiled the global gene expression with genechips. This study thus identified hundreds of DEGs in two cell lines. By comparing the DEGs, their annotated functions and the associated pathways, this study evaluated the general and cell-specific effects the FeNPs on two human cell lines. By comparing these results with the previously characterized effects of the same FeNPs on two mouse cell lines, this study defined the common effects of the FeNPs on human and mouse cells. This study also identified a cell-independent nanotoxicity biomarker for the FeNPs. Together, the results of this study provide new insights into the nanotoxicity of the FeNPs and the underlying molecular mechanisms.

Results and discussion
Characterization of FeNPs and their cellular internalization
The average hydrodynamic size of the FeNPs was 32 nm (Figure 1A). Zeta potential measurements showed that the FeNPs were negatively charged in water (Figure 1B). The average size of the FeNPs measured by TEM was 11 ± 1.24 nm. The FeNPs were monodisperse and of uniform size in water (Figure 1C). Prussian blue staining revealed that the FeNPs were taken up by the cells and more nanoparticles were internalized into the cells at the high dose (Figure 1D). The blue staining of the FeNPs agglomerates was clearer in the HepG2 cells than in the THP-1 cells. The reason for the different appearance is that the former is an adherent cell but the latter is a suspension cell.

Identification of FeNP-responsive genes
The GeneChip analysis identified 287 and 714 genes as DEGs (i.e., FeNP-responsive genes, FeRGs) in the THP-1 cells treated with 50 µg/mL (low dose) and 100 µg/mL (high dose) of FeNPs, respectively. Under the same

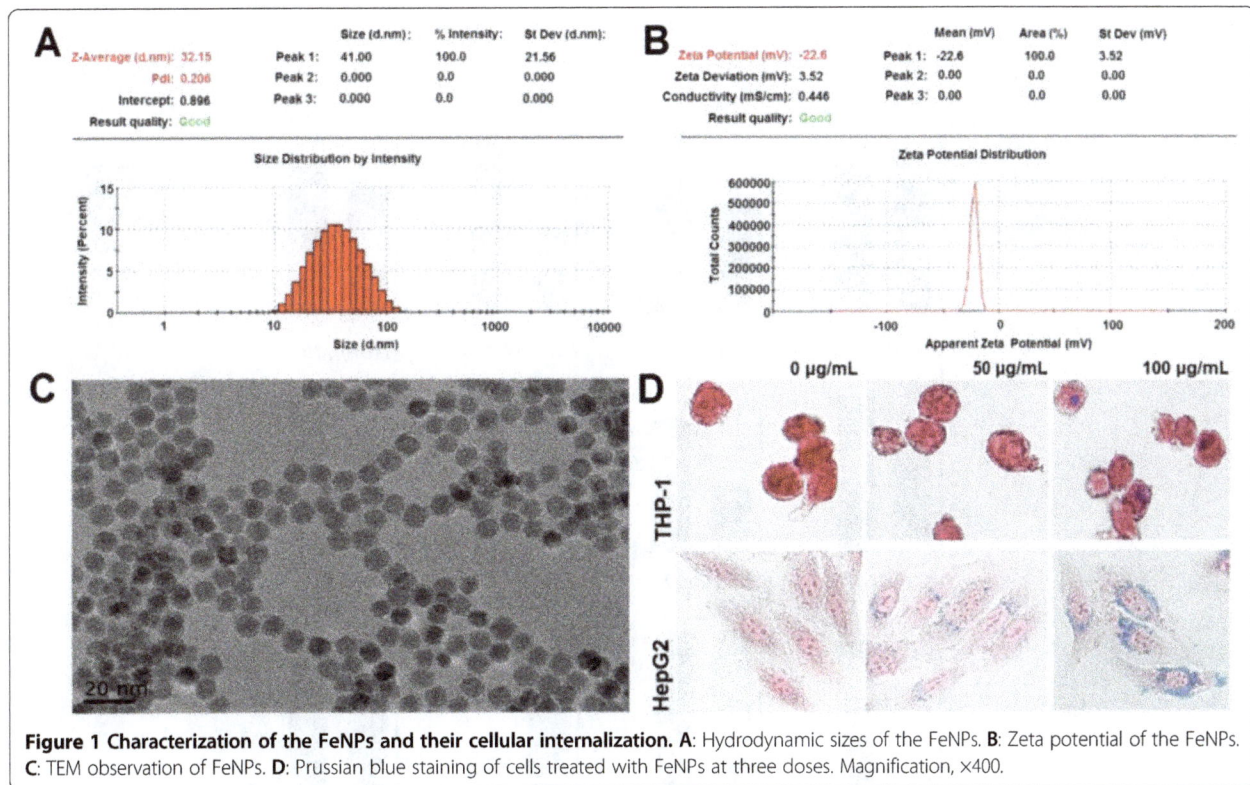

Figure 1 Characterization of the FeNPs and their cellular internalization. A: Hydrodynamic sizes of the FeNPs. **B**: Zeta potential of the FeNPs. **C**: TEM observation of FeNPs. **D**: Prussian blue staining of cells treated with FeNPs at three doses. Magnification, ×400.

conditions, 221 and 265 genes were identified as FeRGs in the HepG2 cells. More genes were regulated by the high-dose FeNPs (hdFeNPs) in the two cell lines, especially in the THP-1 cells. In the THP-1 cells, 229 genes were induced and 58 genes were repressed by the low-dose FeNPs (ldFeNPs), whereas 571 genes were induced and 143 genes were repressed by the hdFeNPs. In the HepG2 cells, 139 genes were induced and 82 genes were repressed by the ldFeNPs, whereas 96 genes were induced and 169 genes were repressed by the hdFeNPs. More genes were induced in the THP-1 cells but repressed in the HepG2 cells. The previous study revealed that more genes were repressed in the RAW264.7 cells but induced in the Hepa1-6 cells [10]. These data indicate that the FeNPs resulted in differential effects on the gene expression patterns of the cells of two species. The expression of some representative genes was confirmed by quantitative PCR (qPCR) analysis (Figure 2).

The genes with highest fold change in transcription revealed the most sensitive responses of the cells to FeNPs at the gene expression level. The top 10 induced and repressed FeRGs in the two cell lines are shown in Figure 3. Clearly, the top FeRGs in two cell lines are completely different. In the THP-1 cells, *Cxcl13* was the gene most significantly induced by both doses of FeNPs. *Cxcl13* is a strong humoral immune response gene in various neuroinflammatory diseases [27]. Other FeRGs including *Adamdec1*, *Ebi3*, *Ifi44l*, *Clec7a* and *Ly96* are also related

to immune responses. *Mmp9* plays a critical role in the positive regulation of the apoptotic process [28]. In the THP-1 cells, most of the most strongly repressed FeRGs encode ribosomal proteins (i.e., *Rps11*, *Rplp2*, *Rpl14*, *Rpl27a*, *Rpl37a* and *Rpl38*). The top FeRGs indicate that the FeNPs resulted in strong activation of defense responses and repression of protein synthesis in the THP-1 cells. Importantly, 7 FeRGs were highly induced, and 7 others were repressed by both ldFeNPs and hdFeNPs in the THP-1 cells (Figure 3), indicating that these effects were stable in the THP-1 cells treated with different doses of FeNPs.

In the HepG2 cells, 3 genes, *Ifi27*, *Ifi6I* and *Tagln*, were highly induced by both the ldFeNPs and hdFeNPs (Figure 3). *Ifi27* was the gene that was most significantly induced by both ldFeNPs and hdFeNPs. Another *Ifi* gene, *Ifi6*, was also significantly induced by both doses of the FeNPs. *Ifi27* and *Ifi6* are associated with immune responses [29,30]. *Ifi27* codes for a mitochondrial protein that contributes to IFN-induced cell death and apoptosis through perturbation of normal mitochondrial function [31]. *Tagln*, which is associated with cell migration, was also induced genes in the HepG2 cells treated with both doses of FeNPs [32]. Seven genes were among the most significantly repressed genes in the HepG2 cells treated with both ldFeNPs and hdFeNPs. *Arnt2* and *Etv5* encode DNA-binding transcription factors (TFs). The TF encoded by *Arnt2* acts as a partner for several sensor

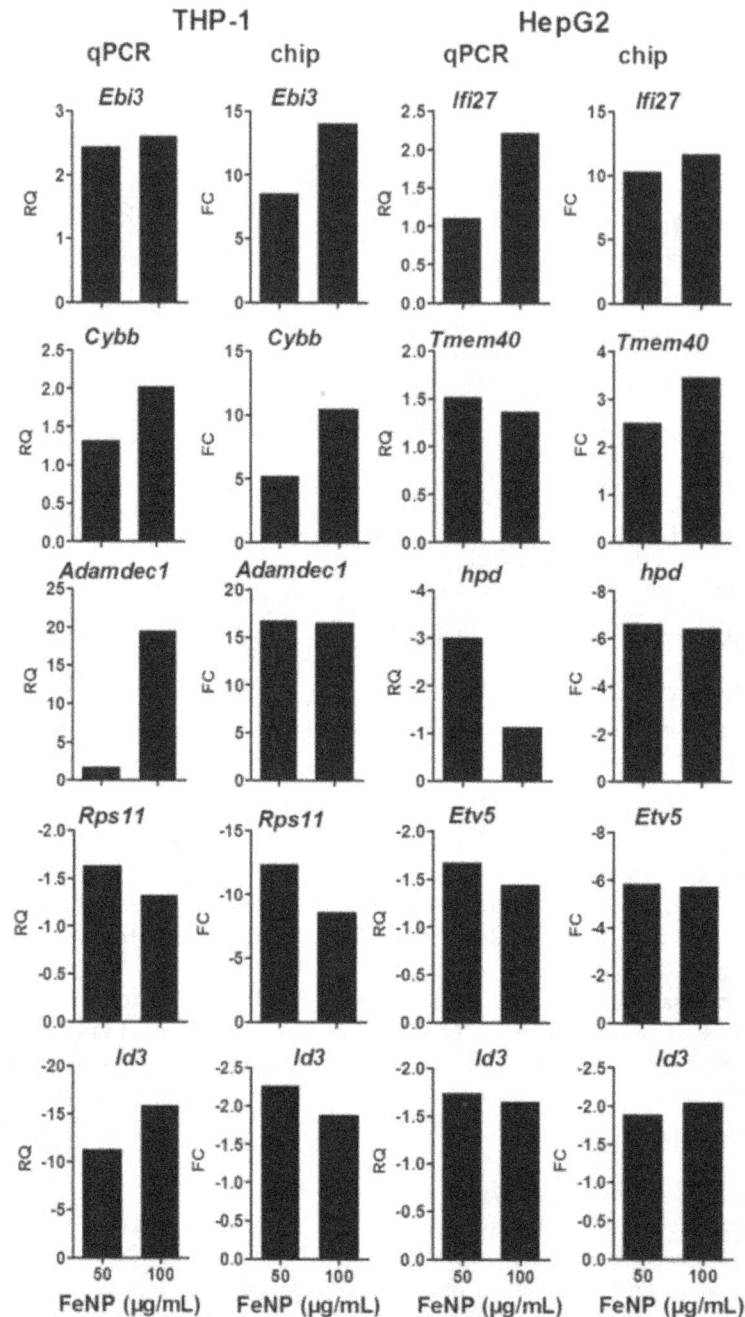

Figure 2 Quantitative PCR (qPCR) detection of the transcription of genes. The transcription of 5 genes in each type of cells was detected with qPCR. The relative quantification (RQ) of the qPCR detection was compared with the fold change (FC) of the GeneChip detection.

proteins that bind the regulatory DNA sequences in genes responsive to developmental and environmental stimuli. The TF coded by *Etv5* is a member of the ETS family. *Hpd* encodes an enzyme in the catabolic pathway of tyrosine; *KIAA1199* (*Cemip*) encodes a cell migration-inducing protein; *Spink6* encodes a serine protease inhibitor selective for kallikreins; and *Frmd3* encodes a putative tumor suppressor protein. *Ccnd1* encodes cyclin D1 of the cyclin family, which functions as regulator of the CDK kinases

CDK4 or CDK6, which are required for the cell cycle G1/S transition. Clearly, most of these top repressed genes are involved in cell growth, proliferation and migration.

The genes commonly regulated in both cell lines revealed the common responses of cells to the FeNPs at the gene expression level. A four-way Venn analysis revealed that 2 genes (*Ifi27* and *Ddx58*) were commonly induced by two doses of FeNPs in these two cell lines (Figure 4A). Eleven genes (*Ifi27*, *Ifi44*, *Ifit3*, *Ddx58*,

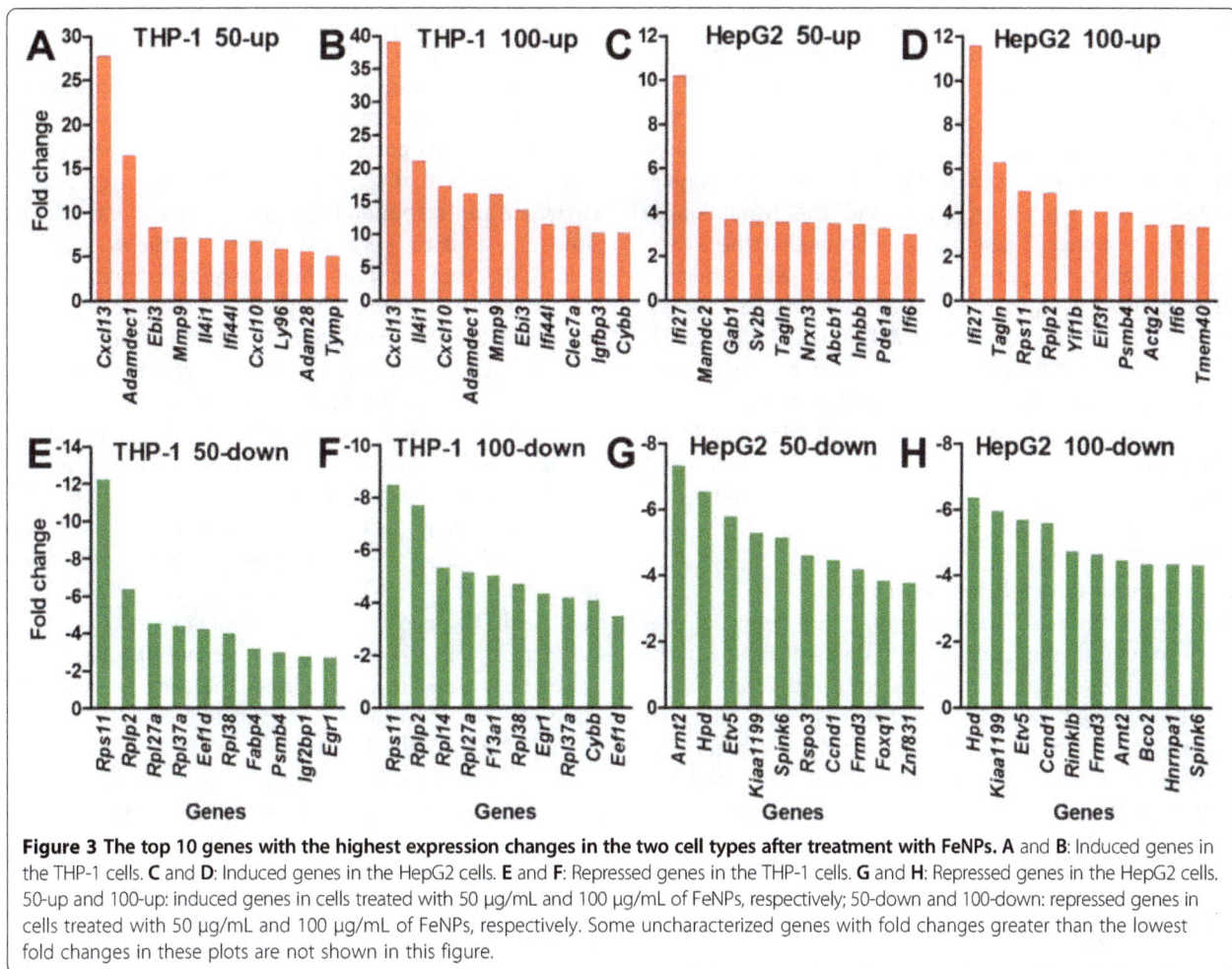

Figure 3 The top 10 genes with the highest expression changes in the two cell types after treatment with FeNPs. A and **B**: Induced genes in the THP-1 cells. **C** and **D**: Induced genes in the HepG2 cells. **E** and **F**: Repressed genes in the THP-1 cells. **G** and **H**: Repressed genes in the HepG2 cells. 50-up and 100-up: induced genes in cells treated with 50 μg/mL and 100 μg/mL of FeNPs, respectively; 50-down and 100-down: repressed genes in cells treated with 50 μg/mL and 100 μg/mL of FeNPs, respectively. Some uncharacterized genes with fold changes greater than the lowest fold changes in these plots are not shown in this figure.

Fbxo16, *Parp9*, *Serpini1*, *Usp25*, *Ccne2*, *Nexn*, and *Rg9mtd2*) were commonly induced by ldFeNPs in both cell lines, and 10 genes (*Ifi27*, *Ifi6*, *Ddx58*, *Akap12*, *Col9a2*, *Nampt*, *Narg1*, *Tmed2*, *Usp16*, and *Zcchc2*) were commonly induced by hdFeNPs in both cell lines

(Figure 4A). However, no genes were commonly repressed by both doses of FeNPs in both cell lines (Figure 4B). Only one gene (*Egr1*) was commonly repressed by the hdFeNPs in both cell lines (Figure 4B). Further identification of genes with fold changes greater

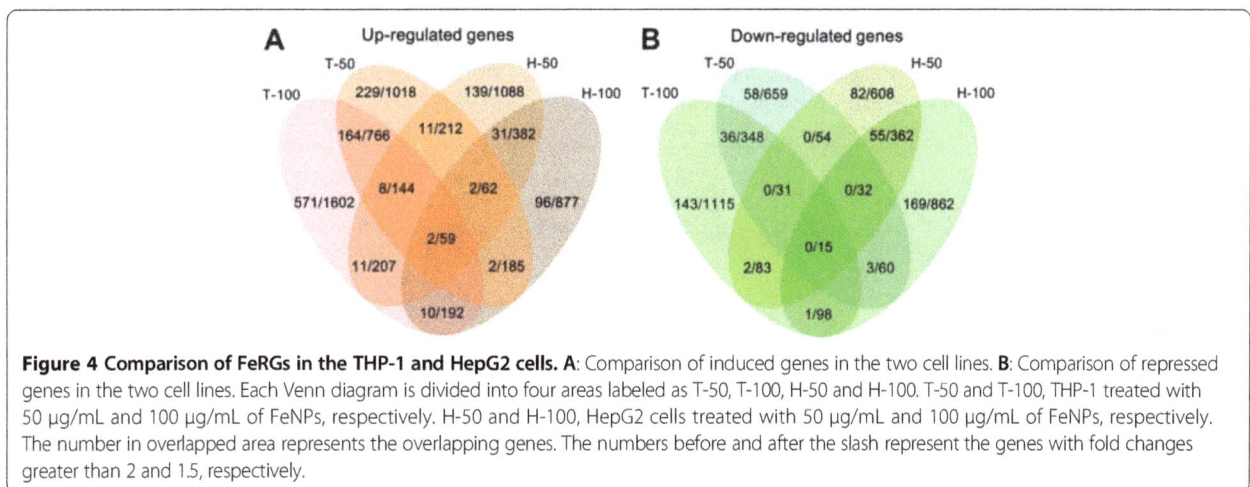

Figure 4 Comparison of FeRGs in the THP-1 and HepG2 cells. A: Comparison of induced genes in the two cell lines. **B**: Comparison of repressed genes in the two cell lines. Each Venn diagram is divided into four areas labeled as T-50, T-100, H-50 and H-100. T-50 and T-100, THP-1 treated with 50 μg/mL and 100 μg/mL of FeNPs, respectively. H-50 and H-100, HepG2 cells treated with 50 μg/mL and 100 μg/mL of FeNPs, respectively. The number in overlapped area represents the overlapping genes. The numbers before and after the slash represent the genes with fold changes greater than 2 and 1.5, respectively.

than 1.5, but at least in one of the cases, greater than 2.0, revealed that 9 and 4 genes were induced and repressed, respectively by both doses of FeNPs in the two cell lines (Figure 5). Among these genes, *Ifi27*, *Ifi44*, *Ifi6* and *Ifit3* express interferon-induced proteins as a defense response to viruses, and *Ddx58* is involved in the viral double-stranded (ds) RNA recognition and the regulation of immune response. These genes indicate that the FeNPs induced cellular responses in the treated cells similar to those induced by viruses [33]. *Parp9* and *Nexn* are associated with cell migration. *Ccne2* belongs to the highly conserved cyclin family and is involved in cell division [34]. *Akap12* encodes a cell proliferation-related protein [35,36]. The repressed genes, *Egr1* and *GLI3*, encode the C_2H_2-type zinc-finger proteins of the EGR family, which play roles in cell proliferation [37,38]. *Id3* is associated with cell growth [39,40].

Cluster analysis of FeNP-responsive genes

The four-way Venn analysis found that only 2 genes were commonly regulated by both doses of FeNPs in the two cell lines (Figure 4). To identify additional commonly regulated genes in these two cell lines, the genes differentially expressed under at least one treatment in each of two cell types were identified. As a result, 55 commonly regulated genes were found (Figure 6). The hierarchical clustering analysis revealed that these genes were classified into four clusters. Clearly, some genes were consistently induced or repressed in both types of cells (Clusters A and D), whereas some genes were inversely regulated in the two cell lines (Clusters B and C). The former reveals the cell-independent effects, whereas the latter reveals the cell-specific effects of the FeNPs on the gene expression in these two cell lines.

Functional annotation of FeNP-responsive genes

The GO analysis revealed that the inflammatory, defense and immune responses, and the responses to stress, wounding, external stimuli, biotic stimuli, viruses, and other organisms were most significantly enriched in the THP-1 cells (Figure 7). The inflammatory and defense responses were the two biological processes most significantly

enriched among the genes induced by both ldFeNPs and hdFeNPs in the THP-1 cells (Figure 7). In combination with the common enrichment of the responses to viruses in the cells treated with both doses of FeNPs and the similar particulate structure of FeNPs and virus, it seems that the THP-1 cells recognized FeNPs as viruses and responded with virus-like cellular effects. Similar responses were also found in the RAW264.7 cells [10], revealing that the virus-like cellular effects are a common cytotoxic response to the FeNPs in the monocyte-macrophage system. In addition, the hdFeNPs activated many more genes of these response-related biological processes in the THP-1 cells (Figure 7), indicating that the hdFeNPs induced more intense virus-like cellular effects. The exacerbated cytotoxicity induced by the hdFeNPs is also indicated by the activation of the biological process of cell death in cells treated with the hdFeNPs (Figure 7), which agrees with the significant apoptosis of the THP-1 cells that resulted from the treatment with 100 µg/mL of the same nanoparticles [7]. In the HepG2 cells, several response-related biological processes were also enriched among the genes induced by ldFeNPs, similar to those observed in the Hepa1-6 cells treated with the hdFeNPs [10].

To evaluate the distribution of FeRGs among various biological processes, the GO terms were classified into five categories at the first level, and the FeRGs belonging to each category were counted. The results demonstrate that the hdFeNPs significantly increased the induced genes in each category in the THP-1 cells but decreased the induced genes in each category in the HepG2 cells (Figure 8). This is completely different from the effects observed in the mouse cell lines. Specifically the hdFeNPs decreased the numbers of the induced genes in these categories in the RAW264.7 cells but increased the numbers of the induced genes in these categories in the Hepa1-6 cells [10]. However, the hdFeNPs increased the numbers of the repressed genes in each category in both two human (Figure 8) and mouse cell lines [10]. These data reveal that two human cell lines responded to the two doses of FeNPs with the different patterns of gene expression. These data also demonstrate that the FeNPs induced

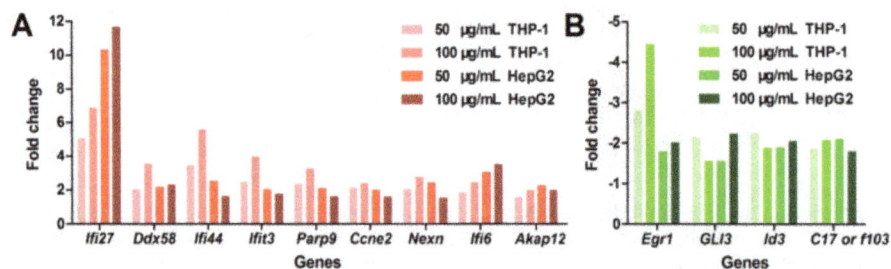

Figure 5 Commonly regulated genes in the THP-1 and HepG2 cells. A: The induced genes and their expression levels in the two types of cells treated with two doses of FeNP. **B**: The repressed genes and their expression levels in the two types of cells treated with two doses of FeNP.

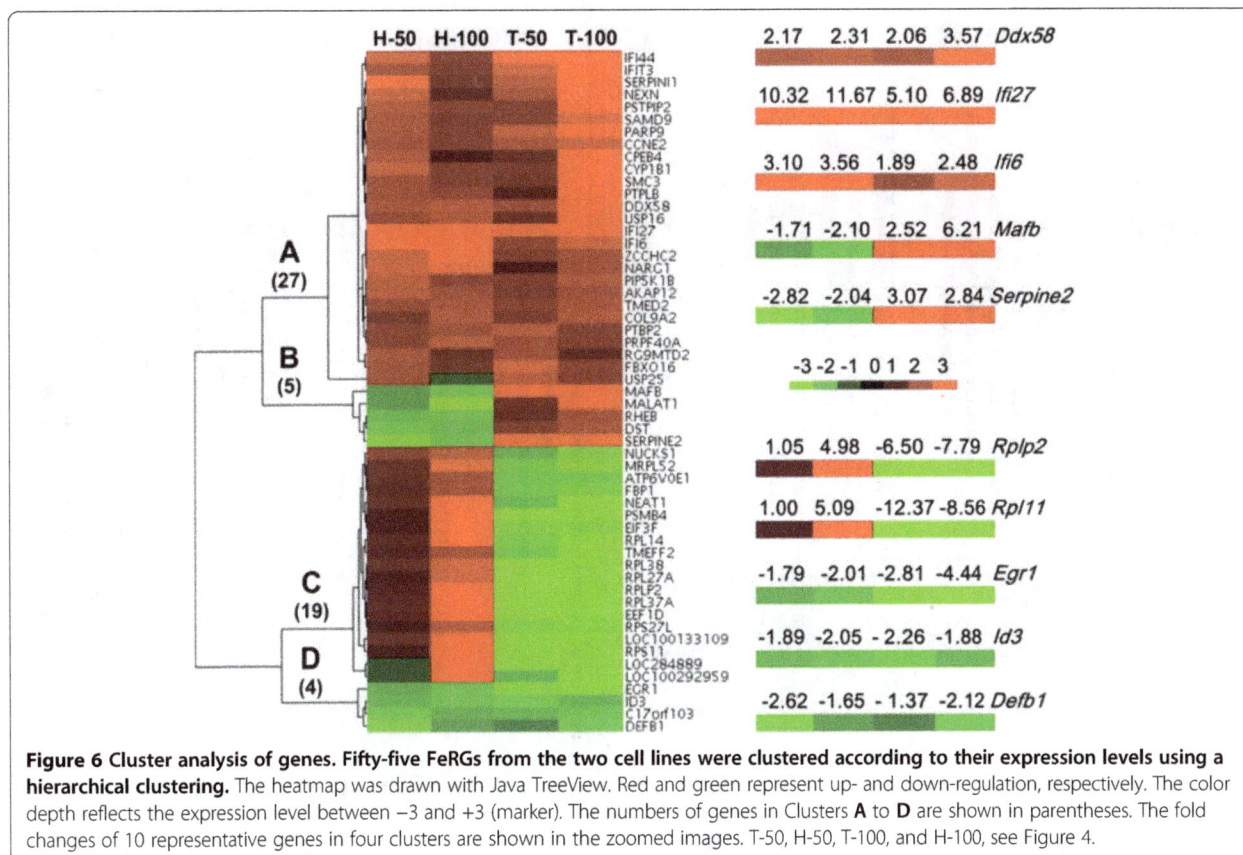

Figure 6 Cluster analysis of genes. Fifty-five FeRGs from the two cell lines were clustered according to their expression levels using a hierarchical clustering. The heatmap was drawn with Java TreeView. Red and green represent up- and down-regulation, respectively. The color depth reflects the expression level between −3 and +3 (marker). The numbers of genes in Clusters **A** to **D** are shown in parentheses. The fold changes of 10 representative genes in four clusters are shown in the zoomed images. T-50, H-50, T-100, and H-100, see Figure 4.

different forms of nanotoxicity in cell types with different biological functions.

To identify the biological processes associated with the genes that were consistently regulated by two doses of FeNPs in two cell lines (clustered in Figure 6), a new GO analysis was performed with these genes. The results revealed that 4 biological processes were enriched mainly by two commonly repressed genes (*Egr1* and *Id3*), whereas 16 biological processes were enriched by 13 genes that were inversely regulated in the two cell lines (Figure 9). The 16 biological processes were all clearly associated with protein translation and were primarily determined by 7 nucleosomal proteins (*Rpl14, Rpl27a, Rplp2, Rpl37a, Rps11, Rpl38,* and *Rps27l*), a translation initiation factor (*Eif3f*) and a translation elongation factor (*Eef1d*). These genes were induced in the HepG2 cells but repressed in the THP-1 cells, indicating that the FeNPs significantly induced protein production in the HepG2 cells but repressed this process in the THP-1 cells. These results are consistent with the results of our recent evaluation of cell viability, which indicated that the ldFeNPs significantly ($p < 0.05$) and hdFeNPs very significantly ($p < 0.01$) decreased the viability of the THP-1 cells, but neither the ldFeNPs nor the hdFeNPs affected the viability of the HepG2 cells (not shown). Interestingly, a gene encoding a mitochondrial ribosomal protein L52 (*Mrpl52*) was also

significantly repressed in the THP-1 cells but induced in the HepG2 cells by the FeNPs (Figure 9), indicating that both cytoplasmic and mitochondrial protein production were significantly affected by the FeNPs. In addition, one biological process (nervous system development) was represented by 3 genes repressed in HepG2 but induced in THP-1 (*Serpine2, Mafb* and *Dst*). Such inverse regulation of a biological process demonstrates the typical cell-specific effects of FeNPs.

Pathway analysis of FeNP-responsive genes

The most significantly ($p < 0.05$) enriched KEGG pathways are shown in Figure 10. In the THP-1 cells, the Toll-like receptor (TLR) signaling pathway is significantly enriched by the FeRGs that were induced by both ldFeNPs and hdFeNPs. TLRs are membrane-bound receptors expressed on innate immune cells such as macrophages and dendritic cells, which generate innate immune responses [41,42]. The TLR signaling pathway was reported to be activated by the FeNPs in RAW264.7 cells [9], and by ceramic (TiO_2, SiO_2 and ZrO_2) and metallic (cobalt) nanoparticles in a human myelomonocytic cell line (U-937) [41]. In addition, the RIG-I-like receptor and chemokine signaling pathways were significantly enriched by the FeRGs induced by both ldFeNPs and hdFeNPs. The cytosolic DNA-sensing pathway and the

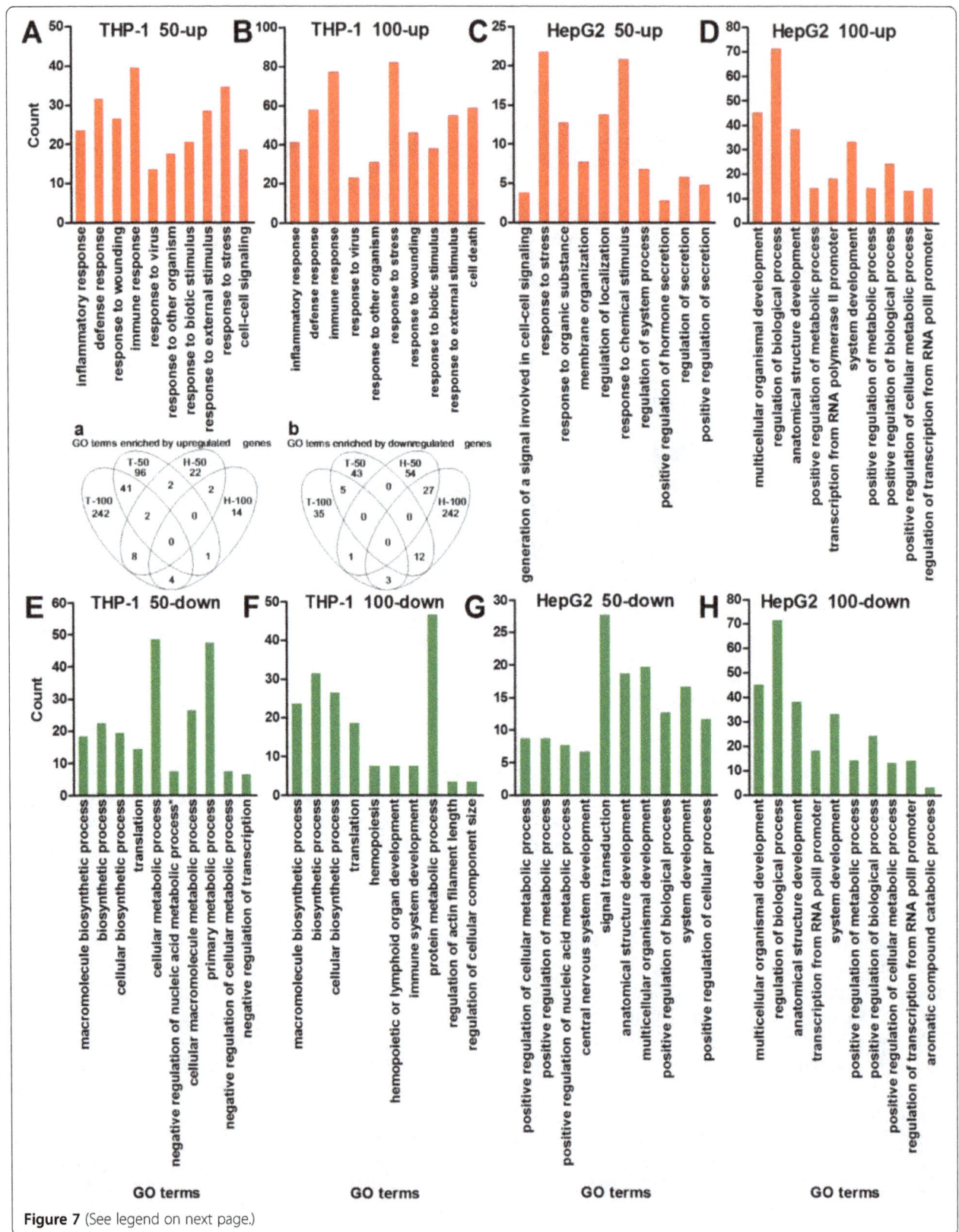

Figure 7 (See legend on next page.)

Figure 7 Top 10 GO terms enriched by induced and repressed genes in the two cell lines. A and **B**: GO terms enriched by the induced genes in the THP-1 cells. **C** and **D**: GO terms enriched by the induced genes in the HepG2 cells. **E** and **F**: GO terms enriched by the repressed genes in the THP-1 cells. **G** and **H**: GO terms enriched by the repressed genes in the HepG2 cells. *Negative regulation of nucleobase, nucleoside, nucleotide and nucleic acid metabolic process. PolII, polymerase II. The p values for all GO terms are less than 0.05. In each plot, the GO terms were aligned from left to right according to their p values from low to high. 50-up, 100-up, 50-down, and 100-down, see Figure 3. Inset: Venn analysis of all GO terms enriched by the induced genes (a) and repressed genes (b) in the THP-1 and HepG2 cell lines. T-50, H-50, T-100, and H-100, see Figure 4. The numbers in overlapped areas represent the overlapping GO terms.

NOD-like receptor signaling pathway were significantly enriched by the FeRGs induced by hdFeNPs. All these pathways are known to play critical roles in immunological responses [43,44]. Additionally, more genes in these pathways were induced by hdFeNPs. These data indicate that the hdFeNPs induced more intense immunological responses than the ldFeNPs in the THP-1 cells. This is consistent with the results of the GO analysis that revealed that the biological process of immune response was consistently highly enriched by the FeRGs induced by both ldFeNPs and hdFeNPs; however, the ldFeNPs induced 40 immune response-related FeRGs, but the hdFeNPs induced 78 immune response-related FeRGs (Figure 7). The more intense reactions of the cells

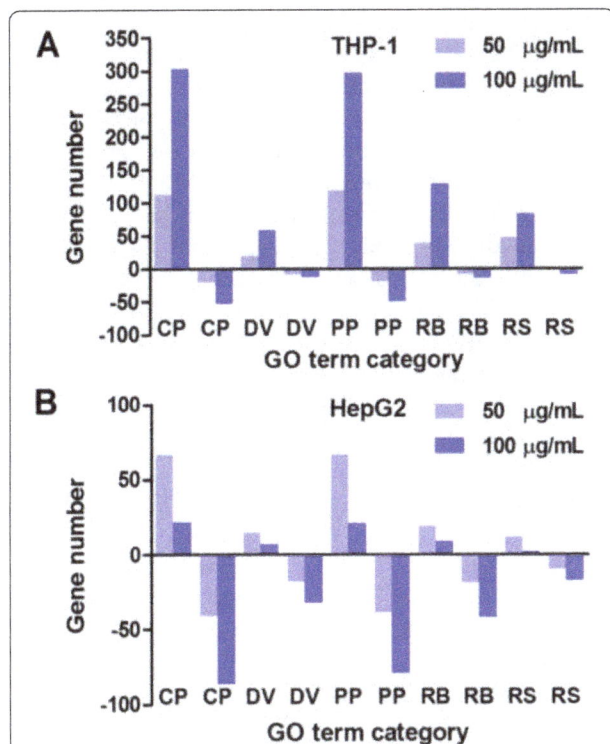

Figure 8 Distribution of FeRGs in the first-level GO category. A: FeRGs in the THP-1 cells. **B**: FeRGs in the HepG2 cells. GO terms (p < 0.01) were classified into five groups at the first level of Biological Processes. CP, cellular process; DV, development; PP, physiological process; RB, regulation of biological process; RS, response to stimulus. The bars above and below the abscissa represent the induced and repressed genes, respectively.

to the hdFeNPs were also demonstrated by the significant activation of five human disease pathways by hdFeNPs, including the Leishmaniasis, Rheumatoid arthritis, Staphylococcus aureus infection, Toxoplasmosis, and Malaria pathways (Figure 10).

It is interesting that the hepatitis C pathway is significantly enriched by the FeRGs induced by both the ldFeNPs and hdFeNPs in the THP-1 cells (Figure 10). Moreover, more genes in this pathway were induced by the hdFeNPs (Figures 10 and 11). In this pathway, the gene Oas1, Oas2 and Oas3 were induced by both doses of FeNPs (Figure 11). These genes encode the 2′,5′-oligoadenylate (2-5A) synthetases, enzymes that play essential roles in the innate immune response to viral infection [45,46]. In this pathway, the gene Ddx58 (coding RIG-I) was also induced by both ldFeNPs and hdFeNPs (Figure 11). RIG-I functions as a pattern recognition receptor that is a sensor for viruses such as hepatitis C virus. The activation of RIG-I-like receptor signaling pathway can induce the production of interferon [47], which is supported by the significant overexpression of many interferon-related genes, including Ifi27, Ifi6, Ifi16, Ifi35, Ifi44, Ifi44l, Ifit1, Ifit2, Ifit3, Ifit5, Ifitm1, Ifih1, Isg20, and Irf7. The GO analysis also revealed that the biological process of response to viruses was significantly enriched by the FeRGs induced by the two doses of FeNPs in the THP-1 cells (Figure 7). These data indicate that the FeNPs act on the THP-1 cells similarly to the responses to viruses including the hepatitis C virus. This is also in consistent with the fact the hydrodynamic size of the FeNPs (41 nm) (Figure 1) used in this study is similar to the size of the hepatitis B virus particles (42 nm).

In the THP-1 cells, only the ribosomal pathway was significantly enriched by the FeRGs repressed by both ldFeNPs and hdFeNPs (Figure 10E and F). In this pathway, 10 genes coding for ribosome proteins, including Rps27l, Rplp2, Rps11, Rps19, Rpl14, Rpl27, Rps10, Rpl38, Rpl37a, and Rpl27a, were commonly repressed by both doses of FeNPs. The extensive repression of these genes suggests that the FeNPs significantly inhibited protein production in this cell. The similar repression of ribosomal pathway was also found in the zebra fish embryos exposed to silver nanoparticles [22]. However, the ribosomal pathway was also the most significantly enriched

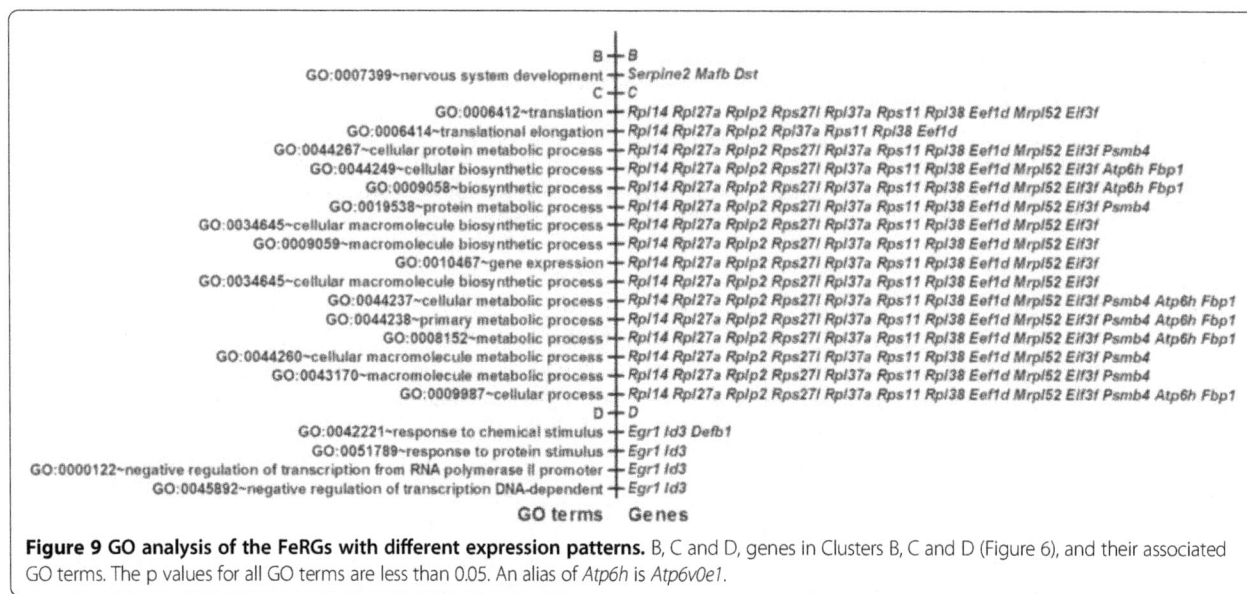

Figure 9 GO analysis of the FeRGs with different expression patterns. B, C and D, genes in Clusters B, C and D (Figure 6), and their associated GO terms. The p values for all GO terms are less than 0.05. An alias of *Atp6h* is *Atp6v0e1*.

by the FeRGs induced by the hdFeNPs in the HepG2 cells (Figure 10D). In this cell type, 11 genes coding ribosome proteins, including *Rps27l*, *Rplp2*, *Rps11*, *Rps19*, *Rpl14*, *Rpl27*, *Rps10*, *Rpl38*, *Rpl37a*, *Rpl27a*, and *Rps6*, were significantly induced by hdFeNPs. Except for *Rps6*, these genes are identical to those altered in the THP-1 cells. These data indicate that the ribosomal pathway was inversely regulated by the FeNPs in two cell lines due to the opposite regulation of a same set of genes coding ribosome proteins. Such differential regulation of a set of genes and their involved pathway demonstrate a typical form of cell-specific toxicology of a nanomaterial.

In the HepG2 cells, no pathway was significantly enriched by the genes induced by the ldFeNPs (Figure 10C). However, 8 pathways were significantly enriched by the genes induced by the hdFeNPs (Figure 10D). These pathways are mainly associated with cellular metabolism. The ribosome pathway was highly enriched by the FeRGs induced by the hdFeNPs. Proteasomes are responsible for protein metabolism (proteolysis). Three pathways are responsible for the metabolism of other nitrogen-containing materials, including the nitrogen, nicotinate and nicotinamide, and alanine, aspartate and glutamate pathways. The insulin signaling pathway is responsible for glucose metabolism. In addition to these metabolism-related pathways, the ECM-receptor interaction and calcium signaling pathways were also significantly enriched by the genes induced by hdFeNPs in HepG2. Two extracellular matrix (ECM) macromolecules, vitronectin (encoded by *Vtn*) and collagen (encoded by *Col6a1*), which benefit cell proliferation and migration, were significantly induced by the hdFeNPs. Four genes (*P2rx4*, *Tnnc1*, *Pde1a*, and *Erbb4*) in the calcium signaling

pathway, which is closely associated with cell adhesion, were significantly induced by hdFeNPs. Therefore, the pathways enriched by the induced genes in the HepG2 cells were mainly associated with cellular metabolism, proliferation, migration and adhesion.

In the HepG2 cells, two pathways, TGF-beta signaling and focal adhesion, were significantly enriched by the FeRGs repressed by both the ldFeNPs and hdFeNPs (Figure 10G and H). Furthermore, more genes in these two pathways are repressed by the hdFeNPs, indicating that the hdFeNPs exert greater effect on the HepG2 cells than the ldFeNPs. This dose-dependent effect can be observed in the THP-1 cells (Figure 10A and B). Importantly, the TGF-beta signaling pathway was most significantly enriched by the genes repressed by both doses of the FeNPs in the HepG2 cells. This pathway is the most important pathway responsible for cell viability. Furthermore, a wide spectrum of cellular functions such as proliferation, apoptosis, differentiation and migration are regulated by TGF-β family members [48]. In the HepG2 cells, 5 (including *Id1*, *Id2*, *Bmp6*, *Smad9*, and *Tgfb1*) and 7 (including *Id1*, *Id2*, *Id3*, *Bmp6*, *Smad6*, *Smad7*, and *Smad9*) genes in this pathway were repressed by the ldFeNPs and hdFeNPs, respectively. *Smad6*, *Smad7* and *Smad9* encode proteins of the SMAD family that act as signal transducers and transcriptional modulators [49]. *Id1*, *Id2* and *Id3* encode helix-loop-helix (HLH) proteins that function as dominant-negative regulators of basic HLH (bHLH) transcription factors by forming inactive heterodimers with intact bHLH. ID proteins play important roles in control of cell growth, differentiation and tumorigenesis [50,51]. The down-regulation of these genes can promote cell growth and increase the risk of tumorigenesis [52]. These data suggest that the FeNPs

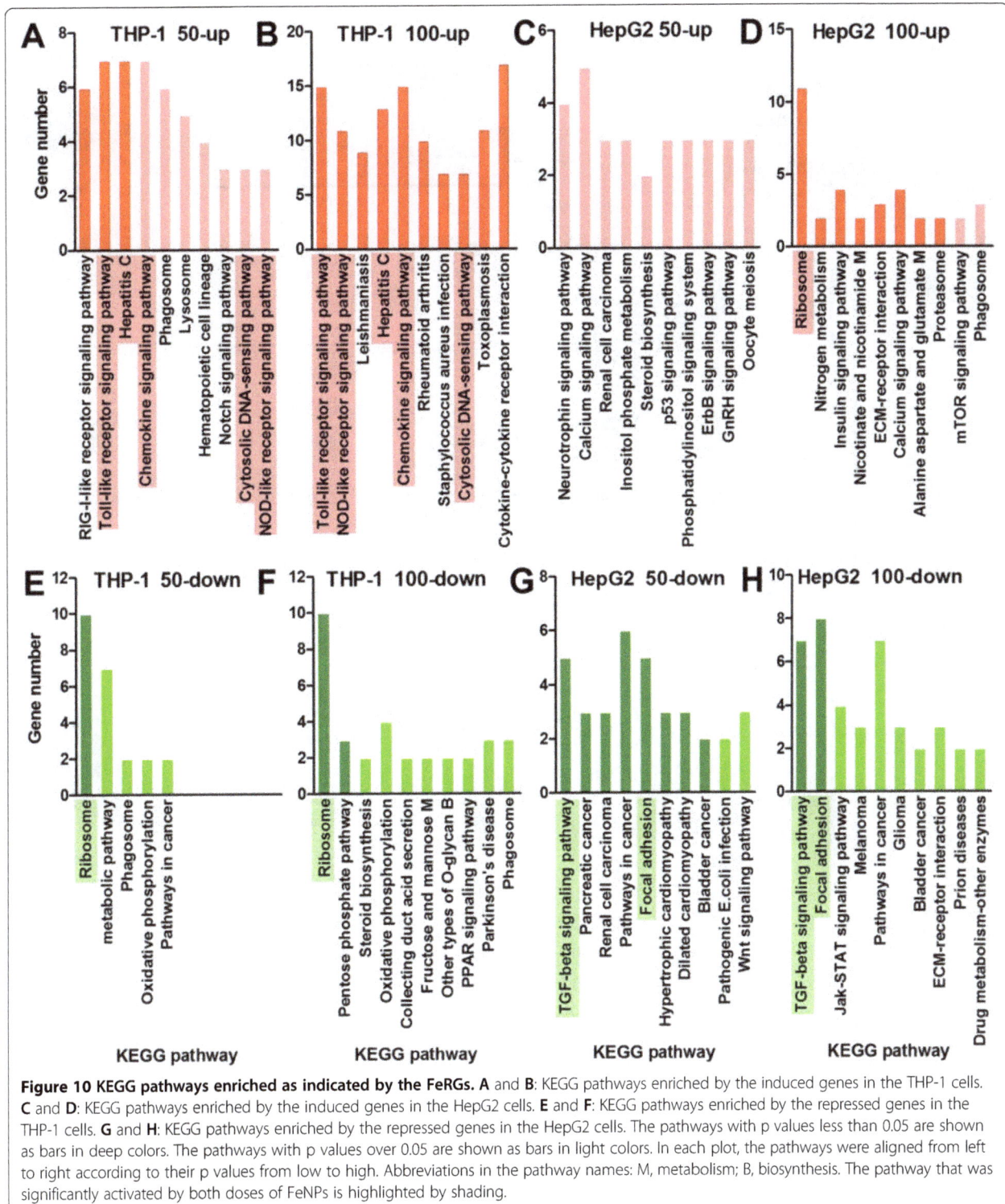

Figure 10 KEGG pathways enriched as indicated by the FeRGs. A and **B**: KEGG pathways enriched by the induced genes in the THP-1 cells. **C** and **D**: KEGG pathways enriched by the induced genes in the HepG2 cells. **E** and **F**: KEGG pathways enriched by the repressed genes in the THP-1 cells. **G** and **H**: KEGG pathways enriched by the repressed genes in the HepG2 cells. The pathways with p values less than 0.05 are shown as bars in deep colors. The pathways with p values over 0.05 are shown as bars in light colors. In each plot, the pathways were aligned from left to right according to their p values from low to high. Abbreviations in the pathway names: M, metabolism; B, biosynthesis. The pathway that was significantly activated by both doses of FeNPs is highlighted by shading.

may stimulate the growth of the HepG2 cells. This is supported by the results of cell viability assay, which revealed that neither the ldFeNPs nor the hdFeNPs affected the viability of the HepG2 cells, and the hdFeNPs even enhanced cell viability (106.4% of the untreated control cells) (not shown). This is also in agreement with

the activation of the ribosomal pathway and several metabolism-related pathways by the hdFeNPs in this cell.

The focal adhesion pathway was also significantly enriched by the FeRGs repressed by both the ldFeNPs and hdFeNPs in the HepG2 cells. The pathway plays an

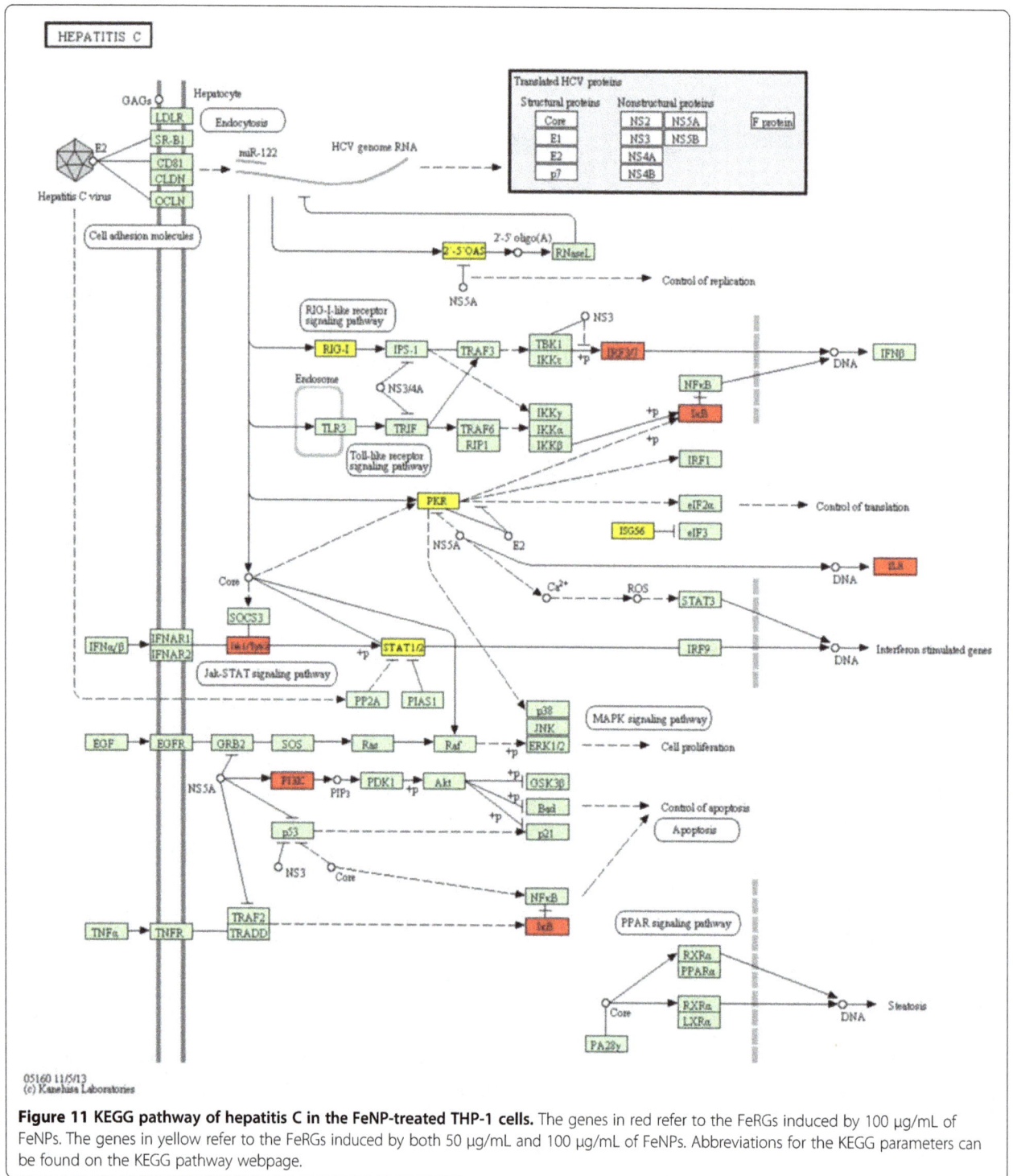

Figure 11 KEGG pathway of hepatitis C in the FeNP-treated THP-1 cells. The genes in red refer to the FeRGs induced by 100 µg/mL of FeNPs. The genes in yellow refer to the FeRGs induced by both 50 µg/mL and 100 µg/mL of FeNPs. Abbreviations for the KEGG parameters can be found on the KEGG pathway webpage.

essential role in important biological processes including cell motility, cell proliferation, cell differentiation, regulation of gene expression, and cell survival [53]. In the HepG2 cells, 4 (*Ccnd1*, *Fyn*, *Fn1*, *Itgb5*, and *Vegfc*) and 8 (*Ccnd1*, *Fyn*, *Fn1*, *Itgb5*, *Col3a1*, *Vav3*, *Igf1r*, and *Egfr*) genes of this pathway were significantly repressed by the ldFeNPs and hdFeNPs, respectively. The down-regulation

of these important cell adhesion-related genes suggests that the growth and mobility of the cells were enhanced by the FeNPs, which is consistent with the slight increase of cell viability and activation of ribosomal pathway. The enhancement of cell growth and mobility may also increase the risk of tumorigenesis. This is supported by the enrichment of the TGF-beta signaling pathway and several cancer-related pathways including those associated with pancreatic cancer, renal cell carcinoma, and bladder cancer in the HepG2 cells (Figure 10G). In the HepG2 cells, these cancer-related pathways are mainly enriched by the genes *Ccnd1*, *Vegfc*, *Tgfb1*, *Arnt2*, *Fn1*, and *Fzd10*.

In the pathway analysis, it was found that the phagosome pathway was also enriched by several genes in both the THP-1 and HepG2 cells. In the THP-1 cells, this pathway was enriched by 6 genes induced by ldFeNPs (*Atp6v1h*, *Hla-Dra*, *Clec7a*, *Cd14*, *Cd36*, and *Cybb*), 2 genes repressed by ldFeNPs (*Calr* and *Atp6v0e1*), and 3 genes repressed by hdFeNPs (*Calr*, *Atp6v0e1*, *Atp6v1c2*) (Figure 9D and E). In the HepG2 cells, this pathway was enriched by 3 genes induced by hdFeNPs (*Calr*, *Atp6v0e1*, and *Ctss*) (Figure 10F). Clearly, the genes *Calr* and *Atp6v0e1* were commonly regulated by the FeNPs in both cell lines; however, the expression of these genes was inversely regulated in the two cell lines. This indicates once again the inverse toxicity of the FeNPs in cells of different functions. However, these genes demonstrated the endocytotic activity of two cells to the FeNPs, which agrees with the previous reports that the FeNPs were internalized into cells by endocytosis [54-57]. In addition to the phagosome pathway, the lysosome and notch signaling pathways were also enriched by the genes induced by ldFeNPs in the THP-1 cells (Figure 10A). In this cell line, the lysosomal pathway was enriched by 5 genes (*Atp6v1h*, *Laptm4b*, *Ap1s3*, *Ctsh*, and *Lamp3*), and the notch signaling pathway was enriched by 3 genes (*Maml2*, *Lfng*, and *Dtx4*) (Figure 10A). These results agree with many previous reports that FeNPs are accumulated into the lysosomes [13,42].

Comparison of human cells with mouse cells

The effects of the same FeNPs at the same two doses on the gene expression profiles of two mouse cell lines (RAW264.7 and Hepa1-6) were recently investigated by our lab [10]. To identify the effects of the FeNPs on the cells of different species, this study used the human equivalents of mouse cells. THP-1 and RAW264.7 are cell lines of monocyte-macrophage system, whereas HepG2 and Hepa1-6 are hepatoma cell lines. The genes with changes ≥ 1.5-fold, the GO terms, and the KEGG pathways for the four cell types were systematically compared. The genes with one or two changes ≥ 2.0-fold among the compared cells were identified as the common genes. The common genes, GO terms, and KEGG pathways identified in the two monocyte-macrophage cell lines, the two hepatoma cell lines, and all four cell lines were shown in Figure 12.

Although some pathways were significantly ($p < 0.05$) enriched in a manner common to both human and mouse cells, the FeRGs involved in these pathways were quite different. For example, in 3 commonly enriched pathways of the two liver cells (Figure 12B), only the focal adhesion pathway shared a gene (*Col3a1*) between the human and mouse cells. In the 8 pathways enriched in common to the two types of blood cells (Figure 12A), only two genes (*Ccl4* and *Ccl5*) were shared by the human and mouse cell lines. The *Ccl4* and *Ccl5* genes shared by two cell lines are involved in the TLR signaling pathway, chemokine signaling pathway, and cytosolic DNA-sensing pathway. The gene *Ccl4*, shared by two cells, is in the cytokine-cytokine receptor interaction pathway. The gene *Ccl5* shared by two cells is in the rheumatoid arthritis pathway. Additionally, it was found that the most common pathways were enriched by the induced or repressed genes in the cells of both species. For example, the TLR signaling pathway is enriched by the induced genes in both human and mouse cells, but the focal adhesion is enriched by the repressed genes in both human and mouse cells (Figure 12A and B). However, two other pathways were inversely regulated in cells of two species. Specifically, the alanine, aspartate and glutamate metabolism and Leishmaniasis pathways were enriched by repressed genes in mouse cells but the induced genes in human cells (Figure 12A and B).

The GO analyses of the common genes revealed that four biological processes were significantly ($p < 0.05$) enriched in the two blood cell types. These biological processes are mainly related to the responses of the cells to stimuli and stress (Figure 12D). The GO analyses of the common genes revealed that no biological processes were significantly ($p < 0.05$) enriched in the two liver cells. However, the biological processes of cell adhesion and cell-matrix adhesion were enriched by three genes, *Ctgf*, *Tgfbi* and *Vtn*, in the two liver cells. It seems that the THP-1 cells are more sensitive to the FeNPs than the RAW264.7 cells because in the 8 common pathways (Figure 12A), many more genes were induced by the FeNPs in the THP-1 cells (Figure 12E). For instance, in the TLR signaling pathway, only two genes (*Ccl4* and *Ccl5*) were significantly induced by the FeNPs in the RAW264.7 cells, however, as many as 15 genes (*Ccl4*, *Ccl5*, *Cxcl10*, *Cxcl11*, *Cd14*, *Cd40*, *Cd86*, *Il8*, *Il1b*, *Nfkbia*, *Stat1*, *Irf7*, *Ly96*, *Tlr8*, and *Pik3r1*) were significantly induced by the FeNPs in the THP-1 cells (Figure 12E).

Identification of the common genes revealed that only the *Id3* gene was commonly regulated by the FeNP in all four cell lines (Figure 12F). *Id3* (inhibitor of DNA binding 3) was the second most significantly repressed gene in the RAW264.7 cells [10]. Previous studies have demonstrated that ID3 is a redox-sensitive signaling molecule [39,40,58]. *Id3* and *Gklf* were identified as two differentially regulated redox-sensitive genes in vascular smooth muscle cells (VSMC) [40]. *Id3* was induced by xanthine/xanthine oxidase (X/XO) but repressed by Fe^{3+}NTA (H-Fe). Conversely, *Gklf* was repressed by X/XO but

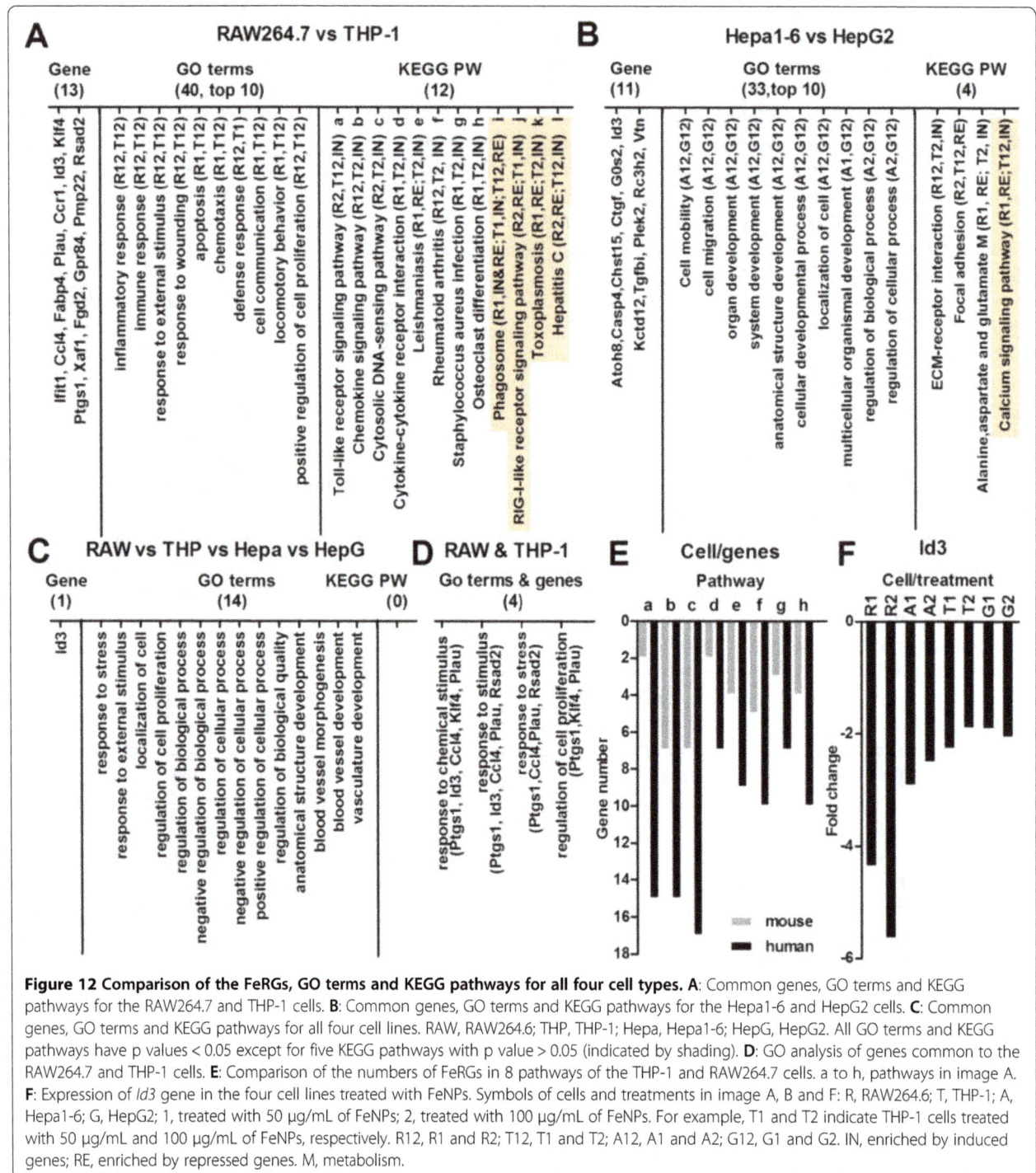

Figure 12 Comparison of the FeRGs, GO terms and KEGG pathways for all four cell types. A: Common genes, GO terms and KEGG pathways for the RAW264.7 and THP-1 cells. **B**: Common genes, GO terms and KEGG pathways for the Hepa1-6 and HepG2 cells. **C**: Common genes, GO terms and KEGG pathways for all four cell lines. RAW, RAW264.6; THP, THP-1; Hepa, Hepa1-6; HepG, HepG2. All GO terms and KEGG pathways have p values < 0.05 except for five KEGG pathways with p value > 0.05 (indicated by shading). **D**: GO analysis of genes common to the RAW264.7 and THP-1 cells. **E**: Comparison of the numbers of FeRGs in 8 pathways of the THP-1 and RAW264.7 cells. a to h, pathways in image A. **F**: Expression of *Id3* gene in the four cell lines treated with FeNPs. Symbols of cells and treatments in image A, B and F: R, RAW264.6; T, THP-1; A, Hepa1-6; G, HepG2; 1, treated with 50 µg/mL of FeNPs; 2, treated with 100 µg/mL of FeNPs. For example, T1 and T2 indicate THP-1 cells treated with 50 µg/mL and 100 µg/mL of FeNPs, respectively. R12, R1 and R2; T12, T1 and T2; A12, A1 and A2; G12, G1 and G2. IN, enriched by induced genes; RE, enriched by repressed genes. M, metabolism.

induced by H-Fe [40]. It was found that *Gklf* could reduce the *Id3* expression by binding to its promoter [40]. It is well known that FeNPs have redox activity and could result in oxidative stress in cells [59-63]. Therefore, the redox-sensitive down-regulation of *Id3* in VSMC is perfectly in agreement with the significant down-regulation in common among the four cell lines treated with two doses of FeNPs. These data suggest that the FeNPs used in this study induced the common down-regulation of *Id3* gene in the four cell lines by its redox activity in cells. These data also show that the down-regulation of the *Id3* gene is a general and sensitive biomarker of the nanotoxicity of the FeNPs, which demonstrates that the redox activity of FeNPs and the resulted oxidative stress are the most significant and prevalent form of the cellular toxicity of the FeNPs. In addition, because *Id3* codes for the ID3 protein of the HLH transcription factor family, which inhibits transcription by forming nonfunctional dimers with other bHLH transcription factors, its significant down-regulation by FeNP suggests that FeNP may disturb the normal biological processes, including cell growth [39], cell differentiation [51], cell apoptosis [64,65], and tumorigenesis [66-68] that are controlled by this transcription factor.

Conclusion

This study investigated the effects of an 11-nm DMSA-coated magnetite FeNP on the gene expression profiles of two human cell lines, THP-1 and HepG2. It was found that the expression of hundreds of genes was significantly changed by a 24-h treatment of the FeNPs at two doses, 50 μg/mL and 100 μg/mL in the two types of cells. By identifying the FeRGs and annotating their functions, this study characterized the general and cell-specific effects of the FeNPs on two cells at the gene, biological process and KEGG pathway levels. At the doses used, the overall effects of the FeNPs in the THP-1 cells was the induction of various responses and repression of protein translation, but in the HepG2 cells these particles promoted cell metabolism, growth and mobility. This study also characterized the common genes, biological processes and pathways effected by the FeNPs in two human and two mouse cell lines, and identified *Id3* as a nanotoxicity biomarker of FeNPs.

Methods
Reagents and cells
The DMSA-coated magnetite FeNPs were supplied by the Gu's lab of Southeast University (Nanjing, China) [69]. HEPES and glutaraldehyde were purchased from Sigma Aldrich (St. Louis, MO, USA). The Trizol reagent and the DMEM cell culture medium were purchased from Invitrogen Gibco (Carlsbad, CA, USA). The Human Genome U133 Plus 2.0 GeneChips® microarrays were purchased from Affymetrix (Santa Clara, CA, USA). The Reverse Transcriptase Kit was purchased from TaKaRa (Dalian, China). The Fast SYBR Master Mix was purchased from Applied Biosystem (Grand Island, NY, USA). The THP-1 and HepG2 cells were purchased from the China Center for Type Culture Collection (Shanghai, China).

Characterization of nanoparticles
The DMSA-coated magnetite FeNPs were synthesized by thermal decomposition [69]. The size and dispersibility of the FeNPs were evaluated using a transmission electron microscope (JEM-2100). The hydrodynamic size distribution of the FeNPs was analyzed with a submicron particle analyzer (Beckman Coulter N4 Plus). The surface charge of the FeNPs was measured with a zeta potential analyzer (Beckman Coulter Delsa 440SX).

Exposure of cells to nanoparticles
The suspension of the FeNPs in water was sterilized by filtration through a 0.22-μm membrane. The human cells were cultured in the DMEM cell culture medium supplemented with 10% fetal calf serum, 100 units/mL penicillin, 100 μg/mL streptomycin and 10 mM HEPES in a humidified 5% CO_2 atmosphere at 37°C. To treat the cells with FeNPs, the cells were seeded in plates or flasks and cultivated overnight, and the culture medium was then replaced with fresh medium containing one of two concentrations (50 and 100 μg/mL) of FeNPs. The cells were cultivated for 24 h longer. To detect the cellular internalization of FeNPs, the cells were stained with Prussian blue as previously described [7].

Detection of gene expression with GeneChip microarray
Total RNA was extracted from the FeNP-treated and untreated control cells with Trizol reagent. The total RNA was quantified using a UV spectrophotometer (NanoDrop 1000) and stored at –80°C for later use. To profile the global gene expression, the RNA samples were analyzed using Affymetrix Human Genome U133 Plus 2.0 GeneChips® microarrays according to standard Affymetrix protocols. Briefly, biotin-labeled cDNA was generated from the RNA using an in vitro transcription reaction in the presence of biotin-labeled ribonucleotides. The labeled cDNA was fragmented and hybridized with the Affymetrix Human Genome U133 Plus 2.0 GeneChips® microarrays at 45°C for 16 h in an Affymetrix hybridization oven. After hybridization, the arrays were washed, stained with streptavidin-phycoerythrin, and scanned with the Affymetrix GeneChip Scanner3000 7G. Image quantitation was performed using GeneChip® Operating Software.

Data analysis of the GeneChip microarray
Following normalization and background filtration, the signal intensity data of the treated cells were compared with those of the control cells. The transcripts with intensity ratios ≥ 2 or ≤ –2 were identified as DEGs [70]. The functional annotation analysis was performed by uploading on the DEGs to DAVID (v6.7 [71]). The enriched GO functions including the biological processes, cellular components and molecular function with p values less than 0.05 were considered significant. The enriched KEGG pathways were determined with DAVID.

Detection of gene expression with qPCR
One μg of total RNA was reverse transcribed into cDNA by using the Reverse Transcriptase Kit. The cDNA was used as template to quantify the transcription of genes with quantitative PCR on a StepOne Plus instrument (Applied Biosystem) using Fast SYBR Master Mix. The primers were as follows (5′ to 3′): *Adamdec1*: AGA

CTG TGA TTG TGG CTC TCC T, TTG TCC TGG CAA GGT AGC ATC T; *Ebi3*: GCT CAG GAC CTC ACA GAC TAC G, GCA GCA GCA AAG CAA GGA CTC; *Cybb*: AGG GTC AAG AAC AGG CTA AGG A, AGC AGG ACT AGA TGA GCC AGA G; *Rps11*: ACA TTC AGA CTG AGC GTG CC, GGA GCT TCT CCT TGC CAG TT; *Id3*: CAC CTT CCC ATC CAG ACA GCC, GCT TCC GGC AGG AGA GGT TC; *Etv5*: CCT GAG AGA CTG GAA GGC AAA, TCA TAG TTC ATG GCT GGC CG; *Tmem40*: ACC GTA TCC ACA GCG TCC TC, ATT GGT CTG GCT TGG TCT CCT; *Hpd*: GCC TCT AGT CCC AGT AGG AG, TCG CCC ATC TCT TTG TTC CA; *Ifi27*: TCC TTC TTT GGG TCT GGC TGA A, CAT GGG CAC AGC CAC AAC TC; *Gapdh*: ATT TGG TCG TAT TGG GCG, CTC GCT CCT GGA AGA TGG. A melting curve analysis was performed at the end of the PCR programs to confirm the specificity of PCR amplification. The relative quantification (RQ) was calculated using the $2^{-\Delta\Delta Ct}$ method.

Abbreviations

FeNP: Iron nanoparticle; DEG: Differentially expressed gene; TEM: Transmission electron microscopy; DMEM: Dulbecco's modified Eagle medium; DMSA: m-2,3-dimercaptosuccinic acid; qPCR: Quantitative PCR; ROS: Reactive oxygen species; CNS: Central nervous system; ldFeNPs: Low-dose FeNPs; hdFeNPs: High-dose FeNPs; MMP: Matrix metalloproteinase; FeRGs: FeNP-responsive genes; DAVID: Database for Annotation, Visualization and Integrated Discovery; GO: Gene ontology; KEGG: Kyoto Encyclopedia of Genes and Genomes; HEPES: 4-(2-Hydroxyethyl)-1-piperazineethanesulfonic acid.

Competing interests

The authors declare that they have no competing interests.

Authors' contributions

LZ, XW, JZ and YL performed the experiments; LZ analyzed data; LZ and JW wrote the manuscript. All authors read and approved the final manuscript.

Acknowledgements

This work was partially supported by the grants from the National Important Science Research Program of China (2011CB933503, 2006CB933205), the National Natural Science Foundation of China (61171030), and the Technology Support Program of Jiangsu (BE2012741).

Author details

[1]State Key Laboratory of Bioelectronics, Southeast University, Nanjing 210096, China. [2]School of Biomedical Engineering, Hubei University of Science and Technology, Xianning 437000, China.

References

1. Sun C, Lee JS, Zhang M. Magnetic nanoparticles in MR imaging and drug delivery. Adv Drug Deliv Rev. 2008;60:1252–65.
2. Salata O. Applications of nanoparticles in biology and medicine. J Nanobiotechnol. 2004;2:3.
3. Karlsson HL, Cronholm P, Gustafsson J, Moller L. Copper oxide nanoparticles are highly toxic: a comparison between metal oxide nanoparticles and carbon nanotubes. Chem Res Toxicol. 2008;21:1726–32.
4. Soenen SJ, De Cuyper M. Assessing iron oxide nanoparticle toxicity in vitro: current status and future prospects. Nanomedicine (Lond). 2010;5:1261–75.
5. Singh N, Jenkins GJ, Asadi R, Doak SH. Potential toxicity of superparamagnetic iron oxide nanoparticles (SPION). Nano Rev. 2010;1:5358.
6. Brayner R. The toxicological impact of nanoparticles. Nano Today. 2008;3:48–55.
7. Liu Y, Chen Z, Wang J. Systematic evaluation of biocompatibility of magnetic Fe$_3$O$_4$ nanoparticles with six different mammalian cell lines. J Nanopart Res. 2010;13:199–212.
8. Yang FY, Yu MX, Zhou Q, Chen WL, Gao P, Huang Z. Effects of iron oxide nanoparticle labeling on human endothelial cells. Cell Transplant. 2012;21:1805–20.
9. Liu Y, Chen Z, Gu N, Wang J. Effects of DMSA-coated Fe$_3$O$_4$ magnetic nanoparticles on global gene expression of mouse macrophage RAW264.7 cells. Toxicol Lett. 2011;205:130–9.
10. Liu Y, Zou J, Wang X, Wang T, Wang J. Effects of 2,3-Dimercaptosuccinic Acid-Coated Fe$_3$O$_4$ Nanoparticles on Genes in Two Mouse Cell Lines. J Biomed Nanotechnol. 2014;10:1574–87.
11. Khan MI, Mohammad A, Patil G, Naqvi SA, Chauhan LK, Ahmad I. Induction of ROS, mitochondrial damage and autophagy in lung epithelial cancer cells by iron oxide nanoparticles. Biomaterials. 2012;33:1477–88.
12. Baratli Y, Charles AL, Wolff V, Ben Tahar L, Smiri L, Bouitbir J, et al. Age modulates Fe$_3$O$_4$ nanoparticles liver toxicity: dose-dependent decrease in mitochondrial respiratory chain complexes activities and coupling in middle-aged as compared to young rats. Biomed Res Int. 2014;2014:474081.
13. Liu Y, Huang J, Wang D, Wang J. Identification of DMSA-coated Fe$_3$O$_4$ nanoparticles induced-apoptosis response genes in human monocytes by cDNA microarrays. Adv Mater Res. 2013;749:377–83.
14. Zhu MT, Wang Y, Feng WY, Wang B, Wang M, Ouyang H, et al. Oxidative stress and apoptosis induced by iron oxide nanoparticles in cultured human umbilical endothelial cells. J Nanosci Nanotechnol. 2010;10:8584–90.
15. Diana V, Bossolasco P, Moscatelli D, Silani V, Cova L. Dose dependent side effect of superparamagnetic iron oxide nanoparticle labeling on cell motility in two fetal stem cell populations. PLoS One. 2013;8:e78435.
16. Kedziorek DA, Muja N, Walczak P, Ruiz-Cabello J, Gilad AA, Jie CC, et al. Gene expression profiling reveals early cellular responses to intracellular magnetic labeling with superparamagnetic iron oxide nanoparticles. Magn Reson Med. 2010;63:1031–43.
17. Alarifi S, Ali D, Alkahtani S, Alhader MS. Iron oxide nanoparticles induce oxidative stress, DNA damage, and caspase activation in the human breast cancer cell line. Biol Trace Elem Res. 2014;159:416–24.
18. Ahamed M, Alhadlaq HA, Alam J, Khan MA, Ali D, Alarafi S. Iron oxide nanoparticle-induced oxidative stress and genotoxicity in human skin epithelial and lung epithelial cell lines. Curr Pharm Des. 2013;19:6681–90.
19. Li X, He Q, Shi J. Global gene expression analysis of cellular death mechanisms induced by mesoporous silica nanoparticle-based drug delivery system. ACS Nano. 2014;8:1309–20.
20. Foldbjerg R, Irving ES, Hayashi Y, Sutherland DS, Thorsen K, Autrup H, et al. Global gene expression profiling of human lung epithelial cells after exposure to nanosilver. Toxicol Sci. 2012;130:145–57.
21. Liu Y, Wang J. Effects of DMSA-coated Fe$_3$O$_4$ nanoparticles on the transcription of genes related to iron and osmosis homeostasis. Toxicol Sci. 2013;131:521–36.
22. van Aerle R, Lange A, Moorhouse A, Paszkiewicz K, Ball K, Johnston BD, et al. Molecular mechanisms of toxicity of silver nanoparticles in zebrafish embryos. Environ Sci Technol. 2013;47:8005–14.
23. Waters KM, Masiello LM, Zangar RC, Tarasevich BJ, Karin NJ, Quesenberry RD, et al. Macrophage responses to silica nanoparticles are highly conserved across particle sizes. Toxicol Sci. 2009;107:553–69.
24. Simon DF, Domingos RF, Hauser C, Hutchins CM, Zerges W, Wilkinson KJ. Transcriptome sequencing (RNA-seq) analysis of the effects of metal nanoparticle exposure on the transcriptome of Chlamydomonas reinhardtii. Appl Environ Microbiol. 2013;79:4774–85.
25. Shim W, Paik MJ, Nguyen DT, Lee JK, Lee Y, Kim JH, et al. Analysis of changes in gene expression and metabolic profiles induced by silica-coated magnetic nanoparticles. ACS Nano. 2012;6:7665–80.
26. Tiwari DK, Jin T, Behari J. Dose-dependent in-vivo toxicity assessment of silver nanoparticle in Wistar rats. Toxicol Mech Methods. 2011;21:13–24.
27. Kowarik MC, Cepok S, Sellner J, Grummel V, Weber MS, Korn T, et al. CXCL13 is the major determinant for B cell recruitment to the CSF during neuroinflammation. J Neuroinflammation. 2012;9:93.
28. Yamamori H, Hashimoto R, Ishima T, Kishi F, Yasuda Y, Ohi K, et al. Plasma levels of mature brain-derived neurotrophic factor (BDNF) and matrix metalloproteinase-9 (MMP-9) in treatment-resistant schizophrenia treated with clozapine. Neurosci Lett. 2013;556:37–41.

29. Hsiao CP, Araneta M, Wang XM, Saligan LN. The association of IFI27 expression and fatigue intensification during localized radiation therapy: implication of a para-inflammatory bystander response. Int J Mol Sci. 2013;14:16943–57.

30. Koldehoff M, Cierna B, Steckel NK, Beelen DW, Elmaagacli AH. Maternal molecular features and gene profiling of monocytes during first trimester pregnancy. J Reprod Immunol. 2013;99:62–8.

31. Rosebeck S, Leaman DW. Mitochondrial localization and pro-apoptotic effects of the interferon-inducible protein ISG12a. Apoptosis. 2008;13:562–72.

32. Dos Santos Hidalgo G, Meola J, Rosa ESJC, Paro de Paz CC, Ferriani RA. TAGLN expression is deregulated in endometriosis and may be involved in cell invasion, migration, and differentiation. Fertil Steril. 2011;96:700–3.

33. Liu H. The immunotherapeutic effect of Fe_3O_4 nanoparticles as adjuvants on mice He22 live cancer. J Nanosci Nanotechnol. 2010;10:514–9.

34. Lauper N, Beck AR, Cariou S, Richman L, Hofmann K, Reith W, et al. Cyclin E2: a novel CDK2 partner in the late G1 and S phases of the mammalian cell cycle. Oncogene. 1998;17:2637–43.

35. Gelman IH. Emerging roles for SSeCKS/Gravin/AKAP12 in the control of cell proliferation, cancer malignancy, and barriergenesis. Genes Cancer. 2010;1:1147–56.

36. Gelman IH. Suppression of tumor and metastasis progression through the scaffolding functions of SSeCKS/Gravin/AKAP12. Cancer Metastasis Rev. 2012;31:493–500.

37. Fukuda N. Cigarette smoking induces vascular proliferative disease through the activation of Egr-1. Cardiovasc Res. 2010;88:207–8.

38. Mayer SI, Rossler OG, Endo T, Charnay P, Thiel G. Epidermal-growth-factor-induced proliferation of astrocytes requires Egr transcription factors. J Cell Sci. 2009;122:3340–50.

39. Felty Q, Porther N. Estrogen-induced redox sensitive Id3 signaling controls the growth of vascular cells. Atherosclerosis. 2008;198:12–21.

40. Nickenig G, Baudler S, Muller C, Werner C, Werner N, Welzel H, et al. Redox-sensitive vascular smooth muscle cell proliferation is mediated by GKLF and Id3 in vitro and in vivo. FASEB J. 2002;16:1077–86.

41. Takeda K, Akira S. Toll-like receptors in innate immunity. Internat Immunol. 2005;17:1–14.

42. Liu YX, Chen ZP, Wang JK. Internalization of DMSA-coated Fe_3O_4 magnetic nanoparticles into mouse macrophage cells. Adv Mater Res. 2012;455–456:1221–7.

43. Wong MM, Fish EN. Chemokines: attractive mediators of the immune response. Semin Immunol. 2003;15:5–14.

44. Nibbs RJB, Graham GJ. Immune regulation by atypical chemokine receptors. Nat Rev Immunol. 2013;13:815–29.

45. Lee YS, Dutta A. MicroRNAs in cancer. Annu Rev Pathol. 2009;4:199–227.

46. Babashah S, Soleimani M. The oncogenic and tumour suppressive roles of microRNAs in cancer and apoptosis. Eur J Cancer. 2011;47:1127–37.

47. Chung RT, Gale Jr M, Polyak SJ, Lemon SM, Liang TJ, Hoofnagle JH. Mechanisms of action of interferon and ribavirin in chronic hepatitis C: summary of a workshop. Hepatology. 2008;47:306–20.

48. Mourskaia AA, Northey JJ, Siegel PM. Targeting aberrant TGF-beta signaling in pre-clinical models of cancer. Anticancer Agents Med Chem. 2007;7:504–14.

49. Kang YJ, Shin JW, Yoon JH, Oh IH, Lee SP, Kim SY, et al. Inhibition of erythropoiesis by Smad6 in human cord blood hematopoietic stem cells. Biochem Biophys Res Commun. 2012;423:750–6.

50. Guo Q, Guo P, Mao Q, Lan J, Lin Y, Jiang J, et al. ID1 affects the efficacy of radiotherapy in glioblastoma through inhibition of DNA repair pathways. Med Oncol. 2013;30:325.

51. Norton JD. ID helix-loop-helix proteins in cell growth, differentiation and tumorigenesis. J Cell Sci. 2000;113(Pt 22):3897–905.

52. Sikder HA, Devlin MK, Dunlap S, Ryu B, Alani RM. Id proteins in cell growth and tumorigenesis. Cancer Cell. 2003;3:525–30.

53. Petit V, Thiery JP. Focal adhesions: structure and dynamics. Biol Cell. 2000;92:477–94.

54. Raynal I, Prigent P, Peyramaure S, Najid A, Rebuzzi C, Corot C. Macrophage endocytosis of superparamagnetic iron oxide nanoparticles: mechanisms and comparison of ferumoxides and ferumoxtran-10. Invest Radiol. 2004;39:56–63.

55. Lunov O, Zablotskii V, Syrovets T, Rocker C, Tron K, Nienhaus GU, et al. Modeling receptor-mediated endocytosis of polymer-functionalized iron oxide nanoparticles by human macrophages. Biomaterials. 2011;32:547–55.

56. Canete M, Soriano J, Villanueva A, Roca AG, Veintemillas S, Serna CJ, et al. The endocytic penetration mechanism of iron oxide magnetic nanoparticles with positively charged cover: a morphological approach. Int J Mol Med. 2010;26:533–9.

57. Osman O, Zanini LF, Frenea-Robin M, Dumas-Bouchiat F, Dempsey NM, Reyne G, et al. Monitoring the endocytosis of magnetic nanoparticles by cells using permanent micro-flux sources. Biomed Microdevices. 2012;14:947–54.

58. Mueller C, Baudler S, Welzel H, Bohm M, Nickenig G. Identification of a novel redox-sensitive gene, Id3, which mediates angiotensin II-induced cell growth. Circulation. 2002;105:2423–8.

59. Singh N, Jenkins GJ, Nelson BC, Marquis BJ, Maffeis TG, Brown AP, et al. The role of iron redox state in the genotoxicity of ultrafine superparamagnetic iron oxide nanoparticles. Biomaterials. 2012;33:163–70.

60. Auffan M, Achouak W, Rose J, Roncato MA, Chaneac C, Waite DT, et al. Relation between the redox state of iron-based nanoparticles and their cytotoxicity toward Escherichia coli. Environ Sci Technol. 2008;42:6730–5.

61. Murray AR, Kisin E, Inman A, Young SH, Muhammed M, Burks T, et al. Oxidative stress and dermal toxicity of iron oxide nanoparticles in vitro. Cell Biochem Biophys. 2013;67:461–76.

62. Keenan CR, Goth-Goldstein R, Lucas D, Sedlak DL. Oxidative stress induced by zero-valent iron nanoparticles and Fe(II) in human bronchial epithelial cells. Environ Sci Technol. 2009;43:4555–60.

63. Sevcu A, El-Temsah YS, Joner EJ, Cernik M. Oxidative stress induced in microorganisms by zero-valent iron nanoparticles. Microbes Environ. 2011;26:271–81.

64. Simbulan-Rosenthal CM, Daher A, Trabosh V, Chen WC, Gerstel D, Soeda E, et al. Id3 induces a caspase-3- and –9-dependent apoptosis and mediates UVB sensitization of HPV16 E6/7 immortalized human keratinocytes. Oncogene. 2006;25:3649–60.

65. Kee BL. Id3 induces growth arrest and caspase-2-dependent apoptosis in B lymphocyte progenitors. J Immunol. 2005;175:4518–27.

66. Wilson JW, Deed RW, Inoue T, Balzi M, Becciolini A, Faraoni P, et al. Expression of Id helix-loop-helix proteins in colorectal adenocarcinoma correlates with p53 expression and mitotic index. Cancer Res. 2001;61:8803–10.

67. Yang HY, Liu HL, Ke J, Wu H, Zhu H, Liu JR, et al. Expression and prognostic value of Id protein family in human breast carcinoma. Oncol Rep. 2010;23:321–8.

68. Castanon E, Bosch-Barrera J, Lopez I, Collado V, Moreno M, Lopez-Picazo JM, et al. Id1 and Id3 co-expression correlates with clinical outcome in stage III-N2 non-small cell lung cancer patients treated with definitive chemoradiotherapy. J Transl Med. 2013;11:13.

69. Chen ZP, Zhang Y, Zhang S, Xia JG, Liu JW, Xu K, et al. Preparation and characterization of water-soluble monodisperse magnetic iron oxide nanoparticles via surface double-exchange with DMSA. Colloid Surf A Physicochem Eng Asp. 2008;316:210–6.

70. Vasil'eva LL, Kochan VA, Obelevskaya KM. Extending the field of application of IPS-020 and IPS-101 power supplies. Meas Tech. 1972;15:1719.

71. Dennis Jr G, Sherman BT, Hosack DA, Yang J, Gao W, Lane HC, et al. DAVID: Database for Annotation, Visualization, and Integrated Discovery. Genome Biol. 2003;4:P3.

6

The application of scanning near field optical imaging to the study of human sperm morphology

Laura Andolfi[1], Elisa Trevisan[2], Barbara Troian[3], Stefano Prato[3], Rita Boscolo[4], Elena Giolo[4], Stefania Luppi[4], Monica Martinelli[4], Giuseppe Ricci[4,5†] and Marina Zweyer[5*†]

Abstract

Background: The morphology of spermatozoa is a fundamental aspect to consider in fertilization, sperm pathology, assisted reproduction and contraception. Head, neck, midpiece, principal and terminal part of flagellum are the main sperm components to investigate for identifying morphological features and related anomalies. Recently, scanning near-field optical microscopy (SNOM), which belongs to the wide family of nanoscopic techniques, has opened up new routes for the investigation of biological systems. SNOM is the only technique able to provide simultaneously highly resolved topography and optical images with a resolution beyond the diffraction limit, typical of conventional optical microscopy. This offers the advantage to obtain complementary information about cell surface and cytoplasmatic structures.

Results: In this work human spermatozoa both healthy and with morphological anomalies are analyzed by SNOM, to demonstrate the potentiality of such approach in the visualization of sperm morphological details. The combination of SNOM topography with optical (reflection and transmission) images enables to examine typical topographic features of spermatozoa together with underlying cytoplasmic structures. Indeed the head shape and inner components as acrosome and nucleus, and the organization of mitochondria in the midpiece region are observed. Analogously for principal tract of the tail, the ridges and the columns are detected in the SNOM topography, while their internal arrangement can be observed in the corresponding SNOM optical transmission images, without requiring specific staining procedures or invasive protocols.

Conclusions: Such findings demonstrate that SNOM represents a versatile and powerful tool to describe topographical and inner structural details of spermatozoa simultaneously. This analysis could be helpful for better characterizing several morphological anomalies, often related to sperm infertility, which cannot be examined by conventional techniques all together.

Keywords: Spermatozoa, Scanning near-field optical microscopy, Morphology

Background

Conventional semen parameters such as sperm concentration, motility and morphology are generally used to assess male fertility [1-3]. However, a significant percentage of males with normal semen parameters, according to WHO guidelines [4], fails to conceive [5]. In these cases, the presence of ultrastructural defects could be hypothesized [6-11]. Several morphological studies have been carried out to define the normal form of sperm (see reference [9] for review) and the last WHO semen manual provides objective criteria to assess the sperm morphology [4]. However, there are many factors that may influence the results of the morphology assessment, including the technician's concept of the definition of normality [12,13] and the staining procedures [14,15]. The most common technique to assess sperm morphology is conventional optical microscopy (OM), usually performed on fixed and stained specimens. In this case,

* Correspondence: zweyer@units.it
†Equal contributors
5Department of Medicine, Surgery and Health Sciences, University of Trieste, Italy
Full list of author information is available at the end of the article

sample preparation is rather easy but the resolution is limited to micron resolution (0.2 μm: Abbe diffraction limit). Moreover, it has been shown that different staining techniques may cause not uniform changes in sperm head dimensions [15]. To reduce the subjectivity of sperm morphology assessment, computer-assisted sperm morphometry analysis (CASMA or ASMA) have been developed in the '90s [16,17]. Subsequently, this analysis has been improved by using special objectives for relief contrast observation [18]. A further progress has been achieved by combining fluorescence microscopy and image analysis with open-access software [19,20]. This method allows automatic determination of sperm morphometry in a reduced time [19,20]. To avoid artefacts induced by fixation and staining, a new optical technique (Trumorph system) has been recently described [21]. This technique allows the analysis of living unmodified spermatozoa that are immobilized in narrow chambers and examined by negative contrast phase microscopy [21]. In the last decades, other fundamental studies have been carried out by transmission and scanning electron microscopies (TEM and SEM), which have enabled to detect ultrastructural details and have provided a huge amount of information concerning morphological organization of spermatozoa [22-25]. However, SEM imaging usually needs harsh sample preparation as drying at critical point, metal coating and vacuum environment (during coating procedure and measurements) that may modify some structural features of cells. Analogously, complex preparation of specimens (as resin embedding and sectioning by ultramicrotomy) are required for performing TEM imaging, which provides two-dimensional view of cell internal arrangement. Recently scanning probe microscopy (SPM) techniques have been exploited to improve the knowledge of spermatozoa morphology. Particularly, atomic force microscopy (AFM) has allowed investigating the topography of spermatozoa with nanometric resolution and without needing invasive sample preparation [26-29]. Moreover, either fixed or living cells can be imaged by this technique. Among scanning probe microscopies (SPM), scanning near-field optical microscope (SNOM) has been demonstrated to be a powerful and versatile tool for the observation of biological samples [30,31]. As in AFM, a sharp probe scans the sample surface to generate a topographic image with similar resolution and sensitivity [32]. The SNOM probe consists of a tapered optical fibre that scans the surface and acts as a point light source with a subwavelength-aperture at the terminal part [32,33], which creates a near-field light that decays exponentially with the distance of 100 nm and carries highly localized information [34]. As result, subdiffraction resolution optical images (reflection, transmission, back-reflection and fluorescence) and high-resolution (nanometer level) topographic image are simultaneously generated. Such

technique has been successfully applied for the investigation of cell membranes at molecular level [35-38], subcellular structures [39,40], thin-sections [41] and chromosomes [42]. In the present work, we exploit the advantage of SNOM technique to describe simultaneously different superficial and internal morphological details of human spermatozoa from normozoospermia and teratozoospermia, without needing invasive or expensive sample preparation.

Results and discussion

Morphological analysis of the sperm structures provides relevant information, necessary part to clinical data to evaluate the ability to fertilization. Numerous cellular structures characterizing the spermatozoa have been elucidated thanks to SEM and TEM studies at ultrastructural level, and more recently by AFM at nanometer scale. In this study, the potentialities of SNOM have been exploited to investigate the morphology of fixed human spermatozoa deriving from samples of patients with normozoospermia and teratozoospermia.

SNOM microscopy offers many distinct advantages over other approaches: it is non-destructive and the sample preparation is simple and inexpensive. SNOM technique exploits the properties of the evanescent field that exists only near the surface, between the probe (optical fibre) and the sample (Figure 1). The probe consists of a single mode tapered optical fibre metal coated with an aperture having a diameter of 50–100 nm at the end of the fibre (Figure 1A). This small aperture, brought at nanometer distance (about 10 nm) from the sample, creates an evanescent component of the incident light whose intensity decays exponentially and allows obtaining optical images with a resolution that overcome the diffraction limit ($\lambda/2$) (Figure 1B). The fibre, positioned within few nanometers from the sample surface, is scanned parallel to the surface and the x,y,z data at each point are acquired. The fibre is attached to a tuning fork that oscillates at its resonance frequency; such amplitude can change upon interaction of the probe with the sample during scanning. Hence, a feedback system regulates the probe-sample distance to maintain constant the oscillation amplitude and the probe-sample distance. As a result, an accurate three-dimensional map of the specimen down to the nanometer level can be obtained. This real topographical map of the sample surface, as for AFM, allows measuring the height profile (i.e. vertical size) of the structures investigated [30]. Simultaneously, the light delivered by the fibre that hints the sample locally is scattered (reflected) or adsorbed according to the different optical properties of the structures encountered. The light reflected from sample surface is collected by an upper detector, while that transmitted one (i.e. the light that pass through and not adsorbed by the sample)

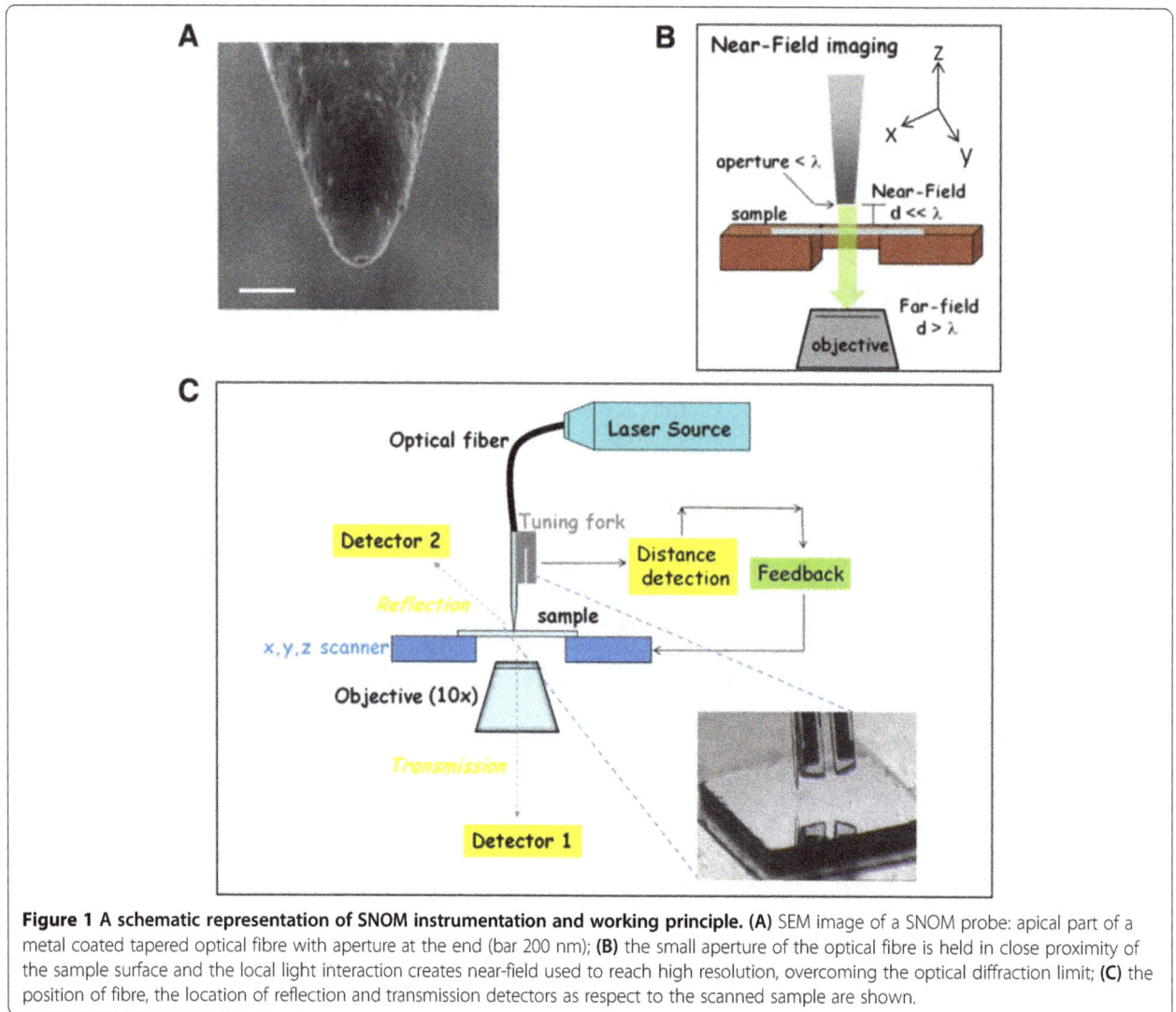

Figure 1 A schematic representation of SNOM instrumentation and working principle. (A) SEM image of a SNOM probe: apical part of a metal coated tapered optical fibre with aperture at the end (bar 200 nm); **(B)** the small aperture of the optical fibre is held in close proximity of the sample surface and the local light interaction creates near-field used to reach high resolution, overcoming the optical diffraction limit; **(C)** the position of fibre, the location of reflection and transmission detectors as respect to the scanned sample are shown.

is collected by a second detector placed below the sample (Figure 1C). These detectors generate SNOM reflection and transmission images simultaneously with high optical resolution in xy axes (<100 nm) higher than conventional optical microscopies (i.e. confocal, phase contrast, differential interference contrast) [39,40]. As result, SNOM allows us to detect simultaneously different signals:

1) Topography: it is an accurate real three-dimensional image of the sample surface. The acquisition of signals in x,y,z axes allows measuring surface details at nanometric level. Differently from traditional microscopies (2D methods), SNOM produces a real 3D topography image, since z dimensions are directly measured point by point on the vertical axis, analogously to AFM images.

2) Optical reflection: the SNOM reflection image derives from local interactions between near field

light and the surface of the specimen, in particular is due to the light scattered from a thin layer within 30–100 nanometers below the sample surface (depending on tip aperture). Hence this image provides mainly morphological information regarding structures immediately below the sample surface.

3) Optical transmission: the light transmitted through the whole thickness of the specimen and not adsorbed by the sample produces the SNOM transmission image. In conventional optical microscopy the entire sample is illuminated. On the contrary, in near-field the light interacts locally, producing signals point by point to reconstruct a whole image of the scanned area. Hence, SNOM imaging is quite similar to traditional optical transmission imaging, but its lateral resolution is more than 10 times higher.

SNOM optical reflection and transmission images can put into evidence the organization of cellular structures positioned at cell surface (i.e. membrane and structures immediately below) or internal ones (i.e. cytoskeleton, nucleus, mitochondria). Therefore together with the topography they can provide a complete picture of the cell morphology.

Before SNOM imaging the semen was purified from other cells and agglomerated proteins to reduce the amount of impurities that can deposit on glass cover slip together with sperm cells. This is an important aspect to get high quality SNOM images and avoid that tip gets dirty during scanning the surface. Moreover, the sequence of purification, fixation and drying procedures, used for SNOM imaging, did not cause remarkable cell damaging that could lead to the formation of artefacts.

As a matter of fact, sperm were firstly fixed, then partially dehydrated with alcohol and dried just before imaging. This procedure should minimize the morphological artefacts, since they mainly appear when cells are dried before fixation [43,44].

To have a clear and immediate picture of the peculiar structures of the spermatozoon, we report in Figure 2 a sketch where the position of such structures is illustrated: head, acrosome, neck, midpiece, anulus, mitochondria that elicoidally surround the axoneme, principal and terminal part of the tail with ultrastructural details (dense fibres and axoneme).

In Figure 3 a reconstruction of a single sperm imaged by SNOM is shown. Figure 3A displays the SNOM topography of the main regions of a sperm cell head, neck, midpiece, anulus, principal and terminal part of flagellum, while SNOM optical reflection and transmission are shown in Figure 3B and C, respectively. The optical images point into evidence that each single region of the sperm has different optical contrast along the cell, likely due to different optical density of the cell structures. SNOM optical reflection images are shown only for the head and principal part, since the rest of the flagellum did not highlight any optical reflection signal.

The analysis of the topographical image is shown in Figure 3D and E. Figure 3D shows a longitudinal profile of the midpiece, between the head and anulus along the red arrow direction of Figure 3A. The profile points into evidence four bumps (indicated by the black arrows), which likely are due to the presence of mitochondria helicoidally surrounding axonema (Figure 2), as also observed in previous AFM work [26]. The total length of this region is found to be 4.5 μm in accordance with the range of values obtained in literature [26]. In Figure 3E horizontal cross section profiles taken on the different part of the sperm are shown (white arrows in Figure 3A). These data display the difference in height of the tail: profile 1, made in the initial part of the midpiece provides height of 450 nm; profile 2, in midpiece region height of

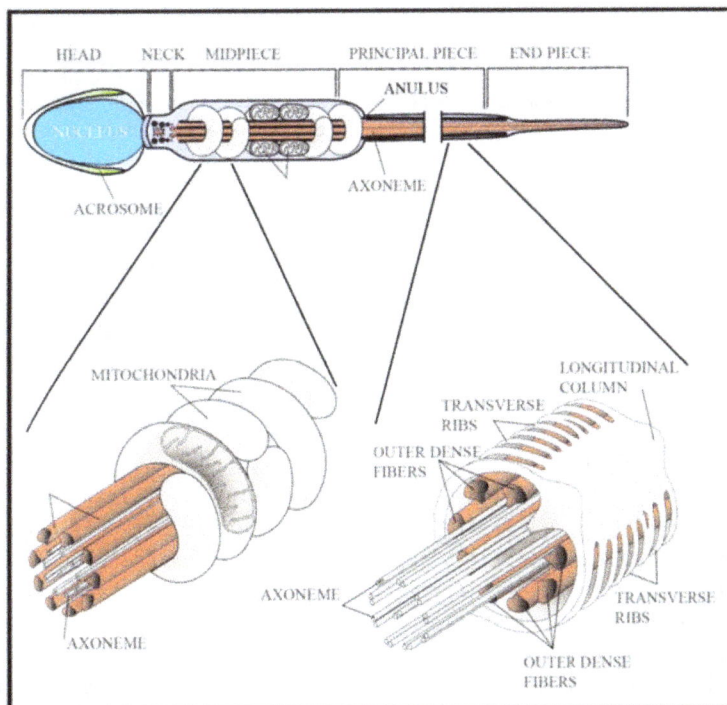

Figure 2 Spermatozoa morphology. Sketch illustrating the sperm cellular structures as modified from text book "Embriologia Umana" Morfogenesi - Processi Molecolari - Aspetti Clinici di M. De Felici et al., Piccin. Gentle permission.

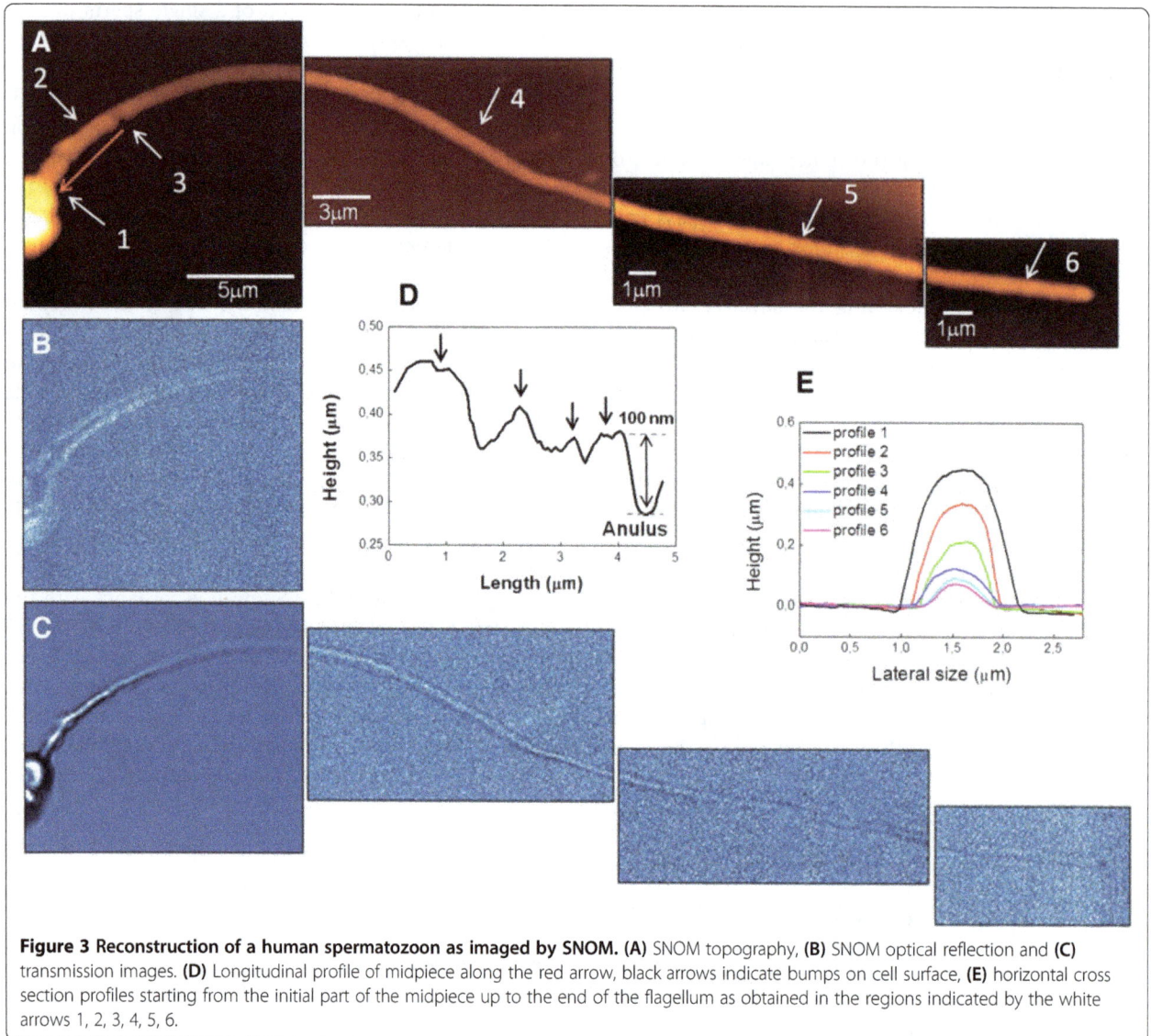

Figure 3 Reconstruction of a human spermatozoon as imaged by SNOM. (A) SNOM topography, **(B)** SNOM optical reflection and **(C)** transmission images. **(D)** Longitudinal profile of midpiece along the red arrow, black arrows indicate bumps on cell surface, **(E)** horizontal cross section profiles starting from the initial part of the midpiece up to the end of the flagellum as obtained in the regions indicated by the white arrows 1, 2, 3, 4, 5, 6.

340 nm; profile 3, in the anulus region height of 220 nm. The rest of the flagellum in three points is measured as shown in Figure 3A: profile 4, height of 120 nm; profile 5, height of 90 nm; profile 6, height of 70 nm. These results, in agreement with literature reported values, point out that the scanning of the spermatozoa by SNOM probe does not introduce artifacts and does not damage cell integrity [26,29].

Afterward, we closely analyze all parts of the spermatozoon. A representative SNOM image acquired on the midpiece region is shown in Figure 4. In this case the topography (Figure 4A and B) shows structural superficial features at higher resolution respect those previously described. The SNOM optical reflection image in Figure 4C shows differences in optical contrast between the external region of the cell (bright area in the reflection image) and the inner region of the cell (dark in the reflection image).

The external bright area mainly involves cell membrane and the cytoplasm region placed immediately beneath. Instead the SNOM optical transmission image (Figure 4D) shows that along the longitudinal axis the midpiece region appears brighter than the rest of the principal part of the spermatozoon. This variation in optical contrast is indicated by the white arrow. In this area some bright ovoid elements can be distinguished (see red arrows) having a length of 0.3-0.5 nm as evaluated by longitudinal profile (Figure 4E), which likely correspond to the mitochondria helicoidally distributed around the axoneme. Moreover, additional discontinuities can be observed as indicated by the arrowheads: the first is in the middle of the midpiece where mitochondria could be not tightly packed, while the second point, close the neck, can identify the different inner organization between midpiece and neck. These data suggest that SNOM optical imaging can allow

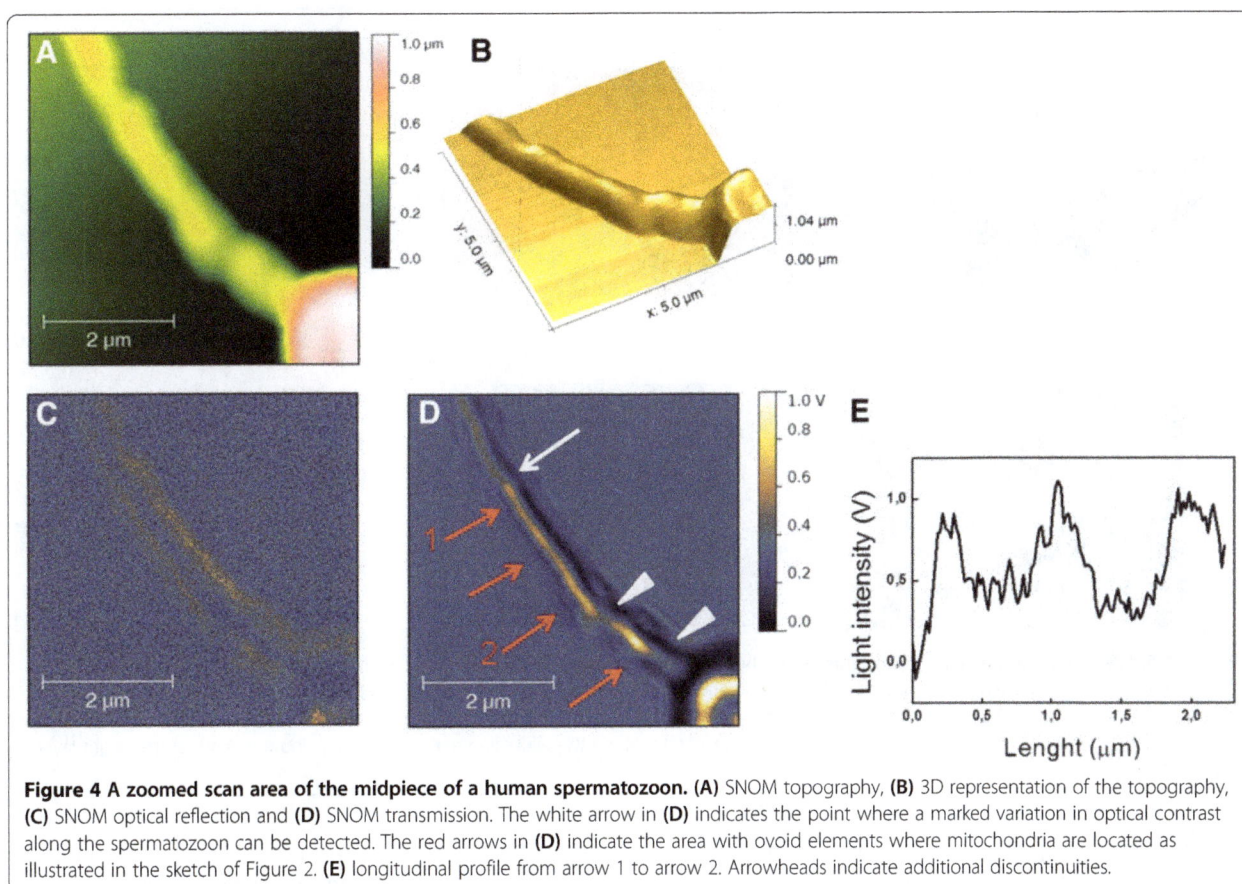

Figure 4 A zoomed scan area of the midpiece of a human spermatozoon. (A) SNOM topography, **(B)** 3D representation of the topography, **(C)** SNOM optical reflection and **(D)** SNOM transmission. The white arrow in **(D)** indicates the point where a marked variation in optical contrast along the spermatozoon can be detected. The red arrows in **(D)** indicate the area with ovoid elements where mitochondria are located as illustrated in the sketch of Figure 2. **(E)** longitudinal profile from arrow 1 to arrow 2. Arrowheads indicate additional discontinuities.

distinguishing different cytoplasmic structures and components thanks to their different optical density.

In Figure 5, a closer SNOM imaging of the sperm head is shown. Figure 5A and B display the SNOM 3D representation (A) and topography (B) of the head. In Figure 5C the head longitudinal profile provides a length of 5.2 µm, in agreement with the range of values obtained in literature with the size of integral sperm head [15,26,45-47]. Head appears flattened on the frontal part (250 nm on average in thickness) respect to the back (950 nm). The analysis of head shape has been taken into account in previous AFM characterization, performed in physiological condition, where the head shape of healthy sperm is compared to that observed in globozoospermia disease [29], which is a pathology consisting in round acrosomeless head leading to infertility [48,49]. In this case the differences of sperm head profiles in the presence and in the absence of acrosoma is investigated. In our case the front head depression can be ascribed to the partial dehydration process following the fixation procedure. This feature, also observed in other AFM investigation performed in air on fixed spermatozoa, does not cause structural variations of the head [29]. In the SNOM reflection image (Figure 5D) a bright layer surrounding the head can be observed. This bright layer, that appears more evident in the SNOM

transmission image (Figure 5E), could be ascribed to region between the plasma membrane and the external nuclear envelope membrane including the acrosome vesicle, which is a cap-like structure partially surrounding the nucleus as indicated in the sketch of Figure 2. In Figure 5F, the two arrows indicate that this bright external layer cover only 2/3 of the head, while the rest is a very dark external part (see also Figure in Additional file 1). The thickness (i.e. width) of this bright layer, evaluated by horizontal cross section profile along the white line, is found to be 230 nm (as evaluated at half width of the maximum height) (see also Figure in Additional file 1). Moreover in these images, immediately below the acrosomal vescicle, the nucleus can be also recognized.

A high magnification imaging of the principal piece is shown in Figure 6. In Figure 6A red arrows indicate a series of lateral thin bumps that likely can be ascribed to the transverse ribs (sketch of Figure 2). Moreover, in this topography image a longitudinal ridge can be observed (white arrow), which can be attributed to the longitudinal column illustrated in the sketch of Figure 2. This feature can be recognized also in the SNOM transmission image of Figure 6C, where the difference in optical contrast allows following the column position within the tail. In this case the SNOM reflection image (Figure 6B)

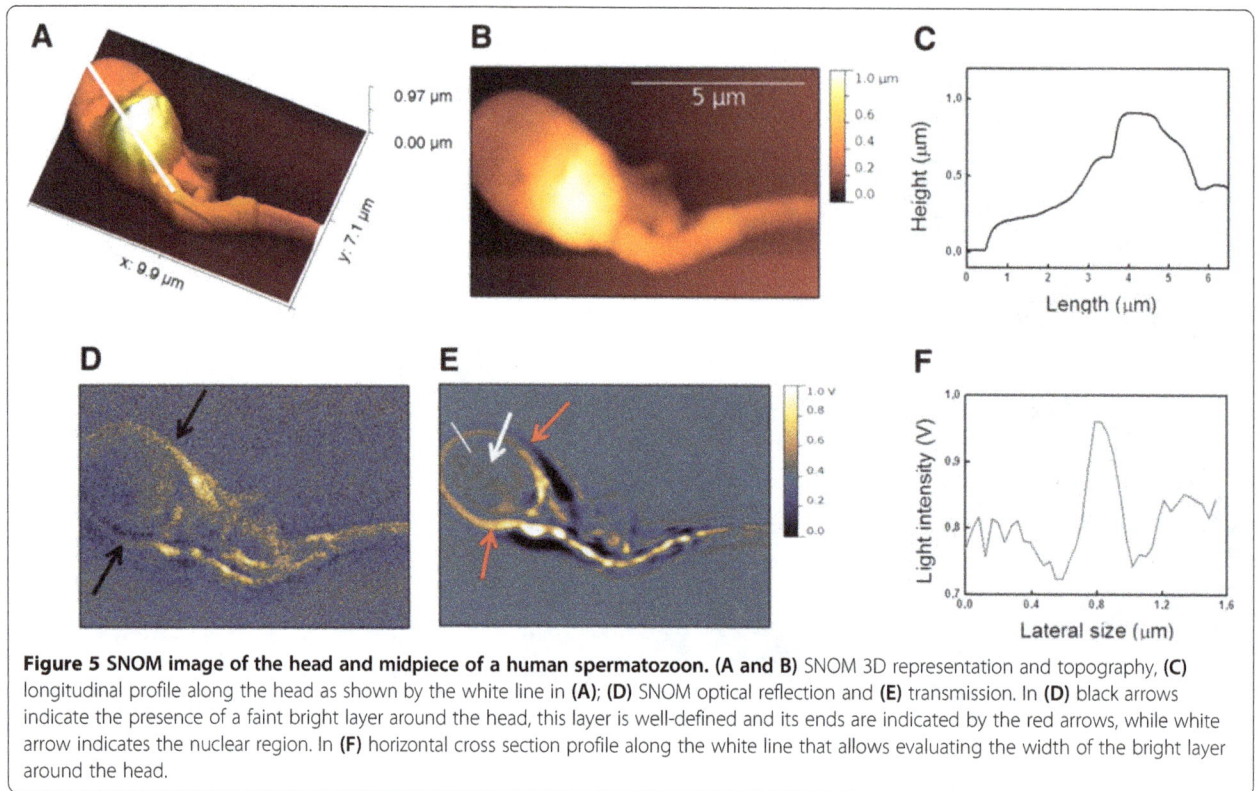

Figure 5 SNOM image of the head and midpiece of a human spermatozoon. (A and B) SNOM 3D representation and topography, **(C)** longitudinal profile along the head as shown by the white line in **(A)**; **(D)** SNOM optical reflection and **(E)** transmission. In **(D)** black arrows indicate the presence of a faint bright layer around the head, this layer is well-defined and its ends are indicated by the red arrows, while white arrow indicates the nuclear region. In **(F)** horizontal cross section profile along the white line that allows evaluating the width of the bright layer around the head.

does not highlight particular differences in optical contrast. These structures may be positioned in the tail likely more deeply than 50–100 nm below the cell membrane, and consequently undetectable by reflection, as already described for SNOM reflection imaging [39].

To understand the potentiality of SNOM technique, imaging is also performed on spermatozoa having morphological anomalies. Figure 7 displays SNOM images acquired on anomalous sperm cells. In this case we take into consideration two sperm anomalies: double tail and coiled tail. In double tail sperm the SNOM topography image shows a head with a reduced acrosome area and a very large post-acrosome area (Figure 7A red arrow). This aspect seems to be also detectable in SNOM reflection

image where a limit between a vescicle-like-structure of the head rostral part and the remnant predominantly component can be observed (red arrow of Figure 7B). In the SNOM topography the midpiece is also well distinguishable (Figure 7A and D) and the blue arrowhead highlights the transition to the principal part of the two emerging flagella (Figure 7A). In the corresponding SNOM optical reflection and transmission images the ovoid elements distinguishable in the case of healthy sperm are poorly delineated and the internal part appears to have an irregular organization as compared to healthy one (white arrows Figure 7B, C, E and F). These images suggest that a defective mitochondria organization could be present.

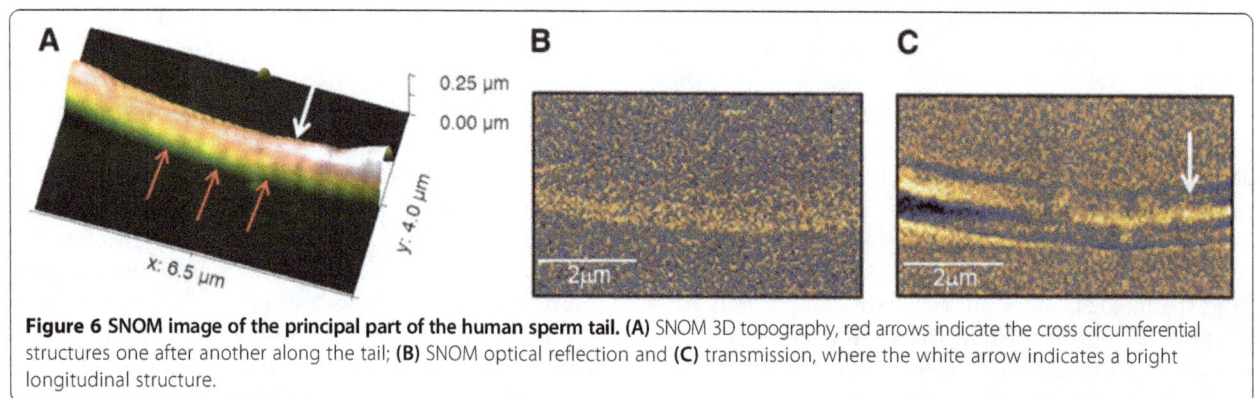

Figure 6 SNOM image of the principal part of the human sperm tail. (A) SNOM 3D topography, red arrows indicate the cross circumferential structures one after another along the tail; **(B)** SNOM optical reflection and **(C)** transmission, where the white arrow indicates a bright longitudinal structure.

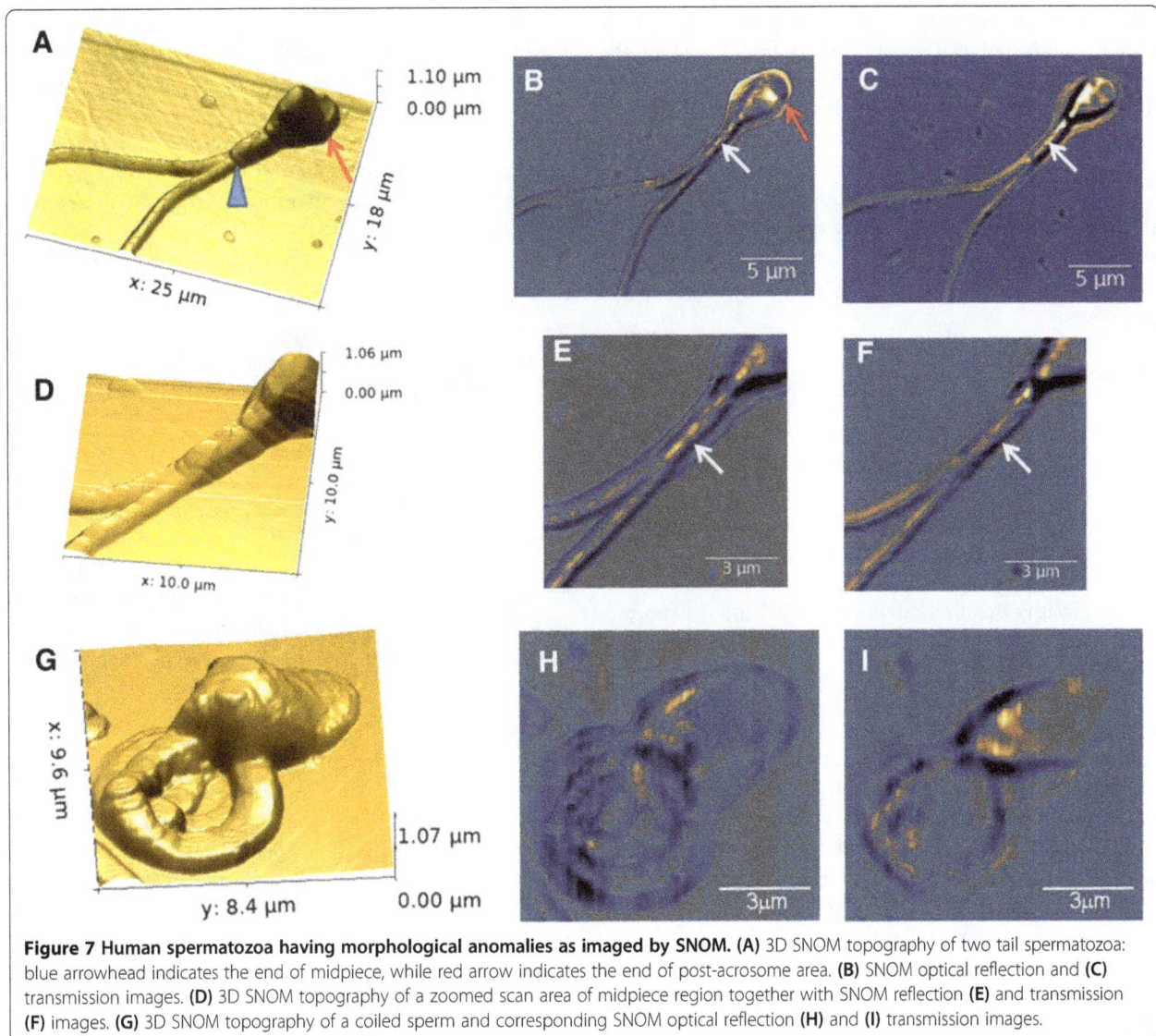

Figure 7 Human spermatozoa having morphological anomalies as imaged by SNOM. (A) 3D SNOM topography of two tail spermatozoa: blue arrowhead indicates the end of midpiece, while red arrow indicates the end of post-acrosome area. **(B)** SNOM optical reflection and **(C)** transmission images. **(D)** 3D SNOM topography of a zoomed scan area of midpiece region together with SNOM reflection **(E)** and transmission **(F)** images. **(G)** 3D SNOM topography of a coiled sperm and corresponding SNOM optical reflection **(H)** and **(I)** transmission images.

In the case of coiled tail sperm, the SNOM topography (Figure 7G) shows that the midpiece region consists of repetitive structures. Such arrangement can be also detected in the SNOM optical reflection image likely related to an arrangement of the membrane and the layer immediately below (Figure 7H). Whereas in the SNOM optical transmission image, due to the overlapping of the coiled tail, it is difficult to detect and describe single components (Figure 7I).

Conclusions

In summary, we demonstrate that SNOM technique enables to describe the cell surface and the arrangement of the spermatozoa internal structures both in healthy and anomalous sperm by acquiring topographic and optical (transmission and reflection) images simultaneously. These structural details of spermatozoa can be detected by using rapid, inexpensive and non-harsh procedures for sample preparation. Specifically, the acrosome region located around the head is revealed and evaluated by SNOM optical transmission images. Analogously, the surface of midpiece region can be observed together with the organization of the mitochondria below the plasma membrane, also in comparison with cell presenting anomalies as the spermatozoon with two tails. Indeed in such case SNOM optical images suggest different organization of mitochondria. Finally the organization and arrangement of circumferential ribs and longitudinal column could be followed below the cell surface thanks to SNOM optical transmission images. This capability can be considered very interesting and useful also for the study of morphological and structural characteristics related to other anomalies as for instance: round acrosomeless sperm head that lead to infertility [48,49]; pathologies that involve the mitochondria (number and assembling), which usually lead to reduced sperm motility [50], flagellum abnormalities. [51-54]. Moreover, SNOM may be very helpful also in male contraception investigations, when sperm damages are intentionally provoked [55-57].

In conclusion SNOM can represent a very promising tool to analyze at high magnification (down to nanometer scale) the morphology of cell membrane and underlying structures both in normozoospermia and teratozoospermia without specific staining or harsh sample preparation procedures.

Methods
Semen analysis
Ten semen samples from men undergoing routine infertility investigations were obtained. Semen samples were collected by masturbation into sterile containers after 3–6 days of sexual abstinence. After complete liquefaction, routine morphological semen analysis was performed using an optical microscope according to World Health Organization guidelines [4] and the morphology was evaluated by precoloured slides Testsimplets® Waldeck-Muenster/Germany. Five samples showed normozoospermia and five samples displayed teratozoospermia, according to WHO 2010 criteria [4]. In order to test semen quality a leukocyte count was carried out by using standard peroxidase test, as described in the WHO laboratory manual. Leukocytospermia was defined as the presence of $>1 \times 10^6$ leukocytes per milliliter of semen. For the detection of antisperm antibodies the reagent kit SpermMar® IgG Test from FertiPro N.V. (Sint-Martens-Latem, Belgium) was used.

A very clean sample is fundamental in all kinds of microscopies to obtain high quality images. Particularly in SPM techniques, it is important to avoid that debris attach to the tip during scanning. This could lead to the probe damaging and/or the formation of topographical artifacts. In the case of spermatozoa, the raw semen was processed and purified (washing steps) to get a clean sperm solution free from agglomerated proteins and other cell populations.

Semen preparation
Semen preparation was performed by Swim-up technique. Such methodology allows obtaining both a spermatozoa selection and a specimen purification. One or more aliquots of 0.5 ml of semen was washed with 1 ml of medium (Quinn's Advantage Medium w/HEPES, SAGE BioPharma™, Bedminster, NJ, USA, supplemented with 0.5% human serum albumin, SAGE Assisted Reproduction Products™, CooperSurgical, Trumbull, CT, USA) into a 5 ml Falcon conical tube (Becton Dickinson Labware, Meylan, France) and then centrifuged at 300×g for 10 min. The excess of supernatant was discarded and the pellet

was resuspended in 0.5 ml of medium. Then, 0.5 ml of medium was gently layered on sperm suspension and the tube was inclinated at an angle of 45° and incubated at 37°C for at least 45 min. The tube was gently set upright and the upper interface was then gently aspirated with a Pasteur pipette. A small aliquot was examined for sperm concentration and sperm motility, the remaining part used for the experiments.

Sample preparation for SNOM imaging
For further sample cleaning, sperm was washed two times with Dulbecco's modified phosphate buffered saline (PBS) (Sigma-Aldrich, USA) by centrifugation at 500×g for 5 min and resuspended in PBS (3×10^6 cell/ml). One aliquot of the cell suspension was deposited on poly-L-lysine coated coverslips (18 mm diameter) (Menzler-Gläser) and incubated at 37°C, 5% CO_2 atmosphere for 1 h to improve the spermatozoa adherence to the substrate. Cells were fixed with 4% paraformaldehyde (PFA) for 15 min at room temperature, washed three times in PBS, two times in water, partially dehydrated (up to 70% ethanol) and allowed to dry before SNOM imaging.

SNOM set-up
Near-field measurements were performed by using a TriA-SNOM microscope (A.P.E. Research, Trieste, Italy), equipped with a flexure scanning stage with a maximum xyz scan range of 100 μm × 100 μm ×10 μm, with a strain gauge sensors to obtain an absolute positioning of the probe. The TriA-SNOM setup was provided with interchangeable laser sources, coupled with a single mode optical fibre. In this work a laser wavelength of 532 nm was used. This laser source power is 20 mW, but after coupling into the fiber, the radiation power emitted by the aperture of SNOM tip is reduced to 2 μW- 20nW on the sample depending on the aperture diameter and shape of SNOM tip [58]. Beside topography images, SNOM optical reflection and optical transmission signals were simultaneously detected with two photomultipliers (R74000, Hamamatsu Photonics Italia S.r.l., Milano, Italy). For these measurements we used a narrow band-pass interference filter at 532 nm wavelength (full width-half maximum = 10 ± 2 nm). SNOM probe was an aluminum-coated tapered pulled optical fiber with a nominal tip aperture of 50 nm (Lovalite, Troyes, France). Two optical view systems were integrated within the SNOM head to control the probe position and select the scan area. An upper optical vision system was used to monitor probe approach to the sample, and one transmission camera (with interchangeable achromatic objectives) was utilized for a bottom view of the sample. The SNOM images are acquired with: 256×256 pixel and 0.1-0.4 msec acquisition time for each point.

SNOM images were processed by A.P.E. Research SPM control software (A.P.E. Research, Trieste, Italy), and analysed by Gwyddion (free software).

Additional file

Additional file 1: A scan area of head and midpiece region of human sperm cell. (A) 3D SNOM topography, (B) SNOM reflection and (C) transmission. (D) longitudinal profile along the head. Black arrows in (B) indicate the ends of post-acrosomial part, while red arrows in (C) indicate the ends of the bright layer partially surrounding the head and that includes the acrosome. In (E) cross section profile along the white line in (C) allows evaluating the width of this layer (266 nm).

Competing interests

The authors declare that they have no competing interests.

Authors' contributions

LA performed SNOM imaging, image analysis and wrote the paper. BT and SP participated in image analysis and data interpretation. ET and RB prepared the samples. EG, SL, MM technically supported ET and RB in the earlier stage of preparation. ET, RB, EG, SL, MM supplied information for writing the final manuscript. MZ analyzed and interpreted the images. MZ and GR contributed in work conception, design and paper writing. All authors read and approved the final manuscript.

Acknowledgements

This work was supported by a grant from the Institute for Maternal and Child Health IRCCS Burlo Garofolo, Trieste, Italy (RC 35/08). We are very grateful to Giovanna Baldini and Roberta Bortul for technical support.

Permission

We acknowledge Piccin publisher for gentle permission to use and modify picture shown in Figure 2A of this paper. The picture has been taken out from textbook "Embriologia Umana" Morfogenesi - Processi molecolari - Aspetti clinici di M. De Felici et al., ed. Piccin.

Author details

[1]IOM-CNR, Area Science Park, Basovizza, Trieste, Italy. [2]Department of Life Sciences University of Trieste, Trieste, Italy. [3]A.P.E. Research Srl, AREA Science Park, Basovizza, Trieste, Italy. [4]Institute for Maternal and Child Health, IRCCS Burlo Garofolo, Trieste, Italy. [5]Department of Medicine, Surgery and Health Sciences, University of Trieste, Italy.

References

1. Redmon JB, Thomas W, Ma W, Drobnis EZ, Sparks A, Wang C, et al. Semen parameters in fertile US men: the study for future families. Study for future families research group. Andrology. 2013;1:806–14.
2. Cooper TG, Noonan E, von Eckardstein S, Auger J, Baker HW, Behre HM, et al. World Health Organization reference values for human semen characteristics. Hum Reprod Update. 2010;16:231–45.
3. Mortimer D, Menkveld R. Sperm morphology assessment—historical perspectives and current opinions. J Androl. 2001;22:192–205.
4. World Health Organization. WHO Laboratory Manual for the Examination and Processing of Human Semen. 5th edn. Geneva; 2010.
5. Hamadaa A, Estevesb SC, Agarwala A. Unexplained male infertility: potential causes and management. Human Andrology. 2011;1:2–16.
6. Rawe VY, Terada Y, Nakamura S, Chillik CF, Olmedo SB, Chemes HE. A pathology of the sperm centriole responsible for defective sperm aster formation, syngamy and cleavage. Hum Reprod. 2002;17:2344–9.
7. Chemes HE, Rawe VY. Sperm pathology: a step beyond descriptive morphology. Origin, characterization and fertility potential of abnormal sperm phenotypes in infertile men. Hum Reprod Update. 2003;9:405–28.
8. Yu JJ, Xu YM. Ultrastructural defects of acrosome in infertile men. Arch Androl. 2004;50:405–9.
9. Menkveld R, Holleboom CA, Rhemrev JP. Measurement and significance of sperm morphology. Asian J Androl. 2011;13:59–68.
10. Chemes E, Sedo CA. Tales of the tail and sperm head aches changing concepts on the prognostic significance of sperm pathologies affecting the head, neck and tail. Asian J Androl. 2012;14:14–23.
11. Escalier D, Tourè A. Morphological defects of sperm flagellum implicated in human male infertility. Sciences. 2012;28:503–11.
12. Cooper TG, Atkinson AD, Nieschlag E: Experience with external quality control in spermatology. 1999, 14:765–769.
13. Bellastella G, Cooper TG, Battaglia M, Ströse A, Torres I, Hellenkemper B, et al. Dimensions of human ejaculated spermatozoa in Papanicolaou-stained seminal and swim-up smears obtained from the Integrated semen Analysis System (ISAS). Asian J Androl. 2010;12:871–9.
14. Henkel R, Schreiber G, Sturmhoefel A, Hipler UC, Zermann DH, Menkveld R. Comparison of three staining methods for the morphological evaluation of human spermatozoa. Fertil Steril. 2008;89:449–55.
15. Maree L, du Plessis SS, Menkveld R, van der Horst G. Morphometric dimensions of the human sperm head depend on the staining method used. Human Reprod. 2010;25:1369–82.
16. Davis RO, Bain DE, Siemers RJ, Thal DM, Andrew JB, Gravance CG. Accuracy and precision of the Cell Form-Human Automated sperm morphometry instrument. Fertil Steril. 1992;58:763–9.
17. Kruger TF, DuToit TC, Franken DR, Acosta AA, Oehninger SC, Menkveld R, et al. A new computerized method of reading sperm morphology (strict criteria) is as efficient as technician reading. Fertil Steril. 1993;59:202–9.
18. Yániz JL, Capistrós S, Vicente-Fiel S, Soler C, Núñez de Murga M, Santolaria P. Use of Relief Contrast(®) objective to improve sperm morphometric analysis by Isas(®) casa system in the ram. Reprod Domest Anim. 2013;48:1019–24.
19. Yániz JL, Vicente-Fiel S, Capistrós S, Palacín I, Santolaria P. Automatic evaluation of ram sperm morphometry. Theriogenology. 2012;77:1343–50.
20. Yániz JL, Capistrós S, Vicente-Fiel S, Soler C, Nuñez de Murga J, Santolaria P. Study of nuclear and acrosomal sperm morphometry in ram using a computer-assisted sperm morphometry analysis fluorescence (CASMA-F) method. Theriogenology. 2014;82:921–4.
21. Soler C, Garcia-Molina A, Sancho M, Contell J, Nuñez M, Cooper TG: A new technique for analysis of humen sperm morphology in unstained cells from raw semen. Reprod Fertil Dev 2014.
22. Zamboni L. The ultrastructural pathology of the spermatozoon as a cause of infertility: the role of electron microscopy in the evaluation of semen quality. Fertil Steril. 1987;48:711–34.
23. Dadoune JP. Ultrastructural abnormalities of human spermatozoa. Hum Reprod. 1988;3:311–8.
24. Courtade M, Lagorce C, Bujan L, Caratero C, Mieusset R. Clinical characteristics and light and transmission electron microscopic sperm defects of infertile men with persistent unexplained asthenozoospermia. Fertil Steril. 1998;70:297–304.
25. Mobberley MA. Electron microscopy in the investigation of asthenozoospermia. Br J Biomed Sci. 2010;67:92–100.
26. Joshi NV, Medina H, Colasante C, Osuna A. Ultrastructural investigation of human sperm using atomic force microscopy. Arch Androl. 2000;44:51–7.
27. Ierardi V, Niccolini A, Alderighi M, Gazzano A, Martelli F, Solaro R. AFM characterization of rabbit spermatozoa. Microsc Res Tech. 2008;71:529–35.
28. Kumar S, Chaudhury K, Sen P, Guha SK. Atomic force microscopy: a powerful tool for high-resolution imaging of spermatozoa. J Nanobiotechnol. 2005;3:1–6.
29. Joshi N, Medina H, Crúz I, Osuna J. Determination of the ultrastructural pathology of human sperm by atomic force microscopy. Fertil Steril. 2001;75:961–5.
30. De Lange F, Cambi A, Huijbens R, de Bakker B, Rensen W, Garcia-Parajo M, et al. Cell biology beyond the diffraction limit: near-field scanning optical microscopy. J Cell Sci. 2001;114:4153–60.
31. Rasmussen A, Deckert V. New dimension in nano-imaging: breaking through the diffraction limit with scanning near-field optical microscopy. Anal Bioanal Chem. 2005;381:165–72.
32. Nakamura H, Sawada K, Kambe H, Saiki T, Sato T. Spatial resolution of near-field scanning optical DOI:10.1143/PTPS.138.173#_blank microscopy with sub-wavelength aperture. Prog Theor Phys Suppl. 2000;138:173–4.
33. Betzig E, Isaacson M, Lewis A. Collection mode near-field scanning optical Microscopy. Appl Phys Lett. 1987;51:2088.
34. Pohl DW, Fischer UC, Durig U. Scanning near-field optical microscopy (SNOM). J Microsc. 1988;152:853–61.

35. Nagy P, Jenei A, Kirsch AK, Szöllosi J, Damjanovich S, Jovin TM. Activation-dependent clustering of the erbB2 receptor tyrosine kinase detected by scanning near-field optical microscopy. J Cell Sci. 1999;112:1733–41.

36. Lapin ZJ, Höppener C, Gelbard HA, Novotny L. Near-field quantification of complement receptor 1 (CR1/CD35) protein clustering in human erythrocytes. J Neuroimmune Pharmacol. 2012;7:539–43.

37. Ianoul A, Street M, Grant D, Pezacki J, Taylor RS, Johnston LJ. Near-field scanning fluorescence microscopy study of ion channel clusters in cardiac myocyte membranes. Biophys J. 2004;87:3525–35.

38. Qiao W, Shang G, Lei FH, Trussardi-Regnier A, Angiboust JF, Millot JM, et al. Imaging of P-glycoprotein of H69/VP small-cell lung cancer lines by scanning near-field optical microscopy and confocal laser microspectrofluorometer. Ultramicroscopy. 2005;105:330–5.

39. Trevisan E, Fabbretti E, Medic N, Troian B, Prato S, Vita F, et al. Novel approaches for scanning near-field optical microscopy imaging of oligodendrocytes in culture. Neuroimage. 2010;49:517–24.

40. Andolfi L, Trevisan E, Zweyer M, Prato S, Troian B, Vita F, et al. The crocidolite fibres interaction with human mesothelial cells as investigated by combining electron microscopy, atomic force and scanning near-field optical microscopy. J Microsc. 2013;249:173–83.

41. Zweyer M, Troian B, Spreafico V, Prato S. SNOM on cell thin sections: observation of Jurkat and MDAMB453 cells. J Microsc. 2008;229:440–6.

42. Hausmann M, Liebe B, Perner B, Jerratsch M, Greulich K, Scherthan H. Imaging of human meiotic chromosomes by scanning near-field optical microscopy (SNOM). DOI:10.1016/S0968-4328(03)00021-0#_blank. Micron. 2003;34:441–7.

43. Yeung CH, Pérez-Sánchez F, Soler C, Poser D, Kliesch S, Cooper TG. Maturation of human spermatozoa (from selected epididymides of prostatic carcinoma patients) with respect to their morphology and ability to undergo the acrosome reaction. Hum Reprod Update. 1997;3:205–13.

44. Cooper TG. Comment of the morphology of spermatozoa in air-dried seminal smears. Int J Andr. 2012;35:105–6.

45. Gergely A, Kovanci E, Senturk L, Cosmi E, Vigue L, Huszar G. Morphometric assessment of mature and diminished-maturity human spermatozoa: sperm regions that reflect differences in maturity. Human Reprod. 1999;14:2007–14.

46. Mossman JA, Pearson JT, Moore HD, Pacey AA. Variation in mean human sperm length is linked with semen characteristics. Human Reprod. 2013;28:22–32.

47. Iglesias I, Vargas-Martin F. Quantitative phase microscopy of transparent samples using a liquid crystal display. J Biomed Opt. 2013;18:26015.

48. Dirican EK, Isik A, Vicdan K, Sozen E, Suludere Z. Clinical pregnancies and livebirths achieved by intracytoplasmic injection of round headedacrosomeless spermatozoa with and without oocyte activation in familial globozoospermia: case report. Asian J Androl. 2008;10:332–4.

49. Harbuz R, Zouari R, Pierre V, Ben Khelifa M, Kharouf M, Coutton C, et al. A recurrent deletion of DPY19L2 causes infertility in man by blocking sperm head elongation and acrosome formation. Am J Hum Genet. 2011;88:351–61.

50. Folgerø T, Bertheussen K, Lindal S, Torbergsen T. Mitochondrial disease and reduced sperm motility. Hum Reprod. 1993;8:1863–8.

51. Chemes HE, Olmedo SB, Carrere C, Oses R, Carizza C, Leisner M, et al. Ultrastructural pathology of the sperm flagellum: association between flagellar pathology and fertility prognosis in severely asthenozoospermic men. Hum Reprod. 1998;13:2521–6.

52. Chemes HE. Phenotypes of sperm pathology: genetic and acquired forms in infertile men. J Androl. 2000;21:799–808.

53. Collodel G, Federico MG, Pascarelli NA, Geminiani M, Renieri T, Moretti E. A case of severe asthenozoospermia: a novel sperm tail defect of possible genetic origin identified by electron microscopy and immunocytochemistry. Fertil Steril. 2011;95:289.e11–6.

54. Moretti E, Geminiani M, Terzuoli G, Renieri T, Pascarelli N, Collodel G. Two cases of sperm immotility: a mosaic of flagellar alterations related to dysplasia of the fibrous sheath and abnormalities of head-neck attachment. Fertil Steril. 2011;95:1787.e19–23.

55. Kumar S, Chaudhury K, Sen P, Guha SK. Topological alterations in human spermatozoa associated with the polyelectrolytic effect of RISUG. Micron. 2006;37:526–32.

56. Kumar S, Chaundhury K, Sen P, Guha SK. Quantitative analysis of surface micro-roughness alterations in human spermatozoa using atomic force microscopy. J Microsc. 2007;227:118–23.

57. Kumar S, Roy S, Chaudhury K, Sen P, Guha SK. Study of the micro-structural properties of RISUG—a newly developed male contraceptive. J Biomed Mater Res B Appl Biomater. 2008;86:154–61.

58. Saiki T, Narita Y. Recent Advances in Near-field Scanning Optical Microscopy. JSAP Int. 2002;5:22–9.

7

Bioimprinted polymer platforms for cell culture using soft lithography

Lynn M Murray[1], Volker Nock[1], John J Evans[2] and Maan M Alkaisi[1*]

Abstract

Background: It is becoming recognised that traditional methods of culture *in vitro* on flat substrates do not replicate physiological conditions well, and a number of studies have indicated that the physical environment is crucial to the directed functioning of cells *in vivo*. In this paper we report the development of a platform with cell-like features that is suitable for *in vitro* investigation of cell activity. Biological cells were imprinted in hard methacrylate copolymer using soft lithography. The cell structures were replicated at high nanometre scale resolution, as confirmed by atomic force microscopy. Optimisation of the methacrylate-based co-polymer mixture for transparency and biocompatibility was performed, and cytotoxicity and chemical stability of the cured polymer in cell culture conditions were evaluated. Cells of an endometrial adenocarcinoma cell line (Ishikawa) were cultured on bioimprinted substrates.

Results: The cells exhibited differential attachment on the bioimprint substrate surface compared to those on areas of flat surface and preferentially followed the pattern of the original cell footprint.

Conclusions: The results revealed for the first time that the cancer cells distinguished between behavioural cues from surfaces that had features reminiscent of themselves and that of flat areas. Therefore the imprinted platform will lend itself to detailed studies of relevant physical substrate environments on cell behaviour. The material is not degraded and its permanency allows reuse of the same substrate in multiple experimental runs. It is simple and does not require expensive or specialised equipment. In this work cancer cells were studied, and the growth behaviour of the tumour-derived cells was modified by alterations of the cells' physical environment. Implications are also clear for studies in other crucial areas of health, such as wound healing and artificial tissues.

Keywords: Bioimprint, Cell culture platform, Cancer cell, Soft lithography, Surface topography, Cell microenvironment

Background

Understanding the control of cell growth and proliferation are central to many health issues, including treatment of cancer [1], implantation of artificial tissues [2], and wound repair [3]. The role of the microenvironment is now well-recognised. In this regard a number of studies have investigated interaction of cells with substrates *in vitro*. Substrate modification has included the plating of small molecules or macromolecules, sometimes applied in patterns. For example molecularly imprinted polymer studies have been undertaken with proteins [4] and an independent role for topography has been suggested [5]. Advances in nanotechnology, such as nanoimprint lithography [6,7], produce topographical surface features down to the nanometre scale and allow for investigation of biomaterial interfaces without chemical variation. Topographically-modified substrates, with wide ranging pattern magnitudes and geometries, have been shown to affect the growth characteristics of cultured cells [8,9]. This hypothesis has led to a number of investigations involving manufacturing physical patterns on substrates in the form of pits, pillars or gratings. These structures are often of smaller dimensions than those of the cells that would constitute a physiological neighbourhood and the relevance of these structures to *in vivo* conditions is uncertain. While these geometric patterns have provided substantial pointers to the importance of the physical environment, they do not contain features that would be recognised by a cell *in vivo*. In this study we report development of a method that

* Correspondence: maan.alkaisi@canterbury.ac.nz
[1]The MacDiarmid Institute for Advanced Materials and Nanotechnology, Department of Electrical and Computer Engineering, University of Canterbury, Christchurch 8140, New Zealand
Full list of author information is available at the end of the article

replicates cell shapes in a polymer and thus contains features of similar size and shape to that of a cell's microenvironment.

We employed Bioimprint methodology [10] in this study. This technique is inspired by nanoimprint lithography and was initially developed in our studies to circumvent deficiencies in high-resolution live cell imaging. Atomic force microscopy (AFM) imaging of live cell cultures was difficult due to the elasticity of the cell membrane and electron microscopy techniques require sacrificial cell samples. A replication protocol was developed to mould the cell surface features into a more rigid and tear-resistant material. The resulting methacrylate co-polymer imprint contained high resolution cell-like features, accurate to 5–20 nm [11-14]. Although other groups have investigated the use of polymeric imprints of cells to obtain information on cell morphology [15,16] this study extends the methodology to enable investigation of cell function.

In this study the biocompatibility of the polymer is confirmed, and we have adapted the imprinted polymer for use as a cell culture platform. We demonstrated a preferential adherence of the cells for the imprinted regions compared to flat areas. These biocompatible bioimprinted templates will provide a platform with potential for investigating localised variation and specific cell adhesion.

Results and discussion
Bioimprint substrates
Bioimprint is a technology we developed for replicating biological cells at high resolution in hard polymer for the purpose of imaging or formation of cell culture platforms. To produce an imprinted substrate it was necessary that the substrate for this initial culture could be separated from the cured polymethacrylate in which the initial culture was moulded. Hence glass was chosen for the initial substrate. Glass provided good cell adhesion and growth environment with minimal adhesive interaction to cured methacrylate co-polymer. Polystyrene, a common surface for cell culture, was not suitable substrate for the initial culture because it formed an inseparable adhesive bond with the cured polymethacrylate.

PDMS-defined borders on culture wells were found to be ideal for confining cultures because of their inexpensive and fast fabrication, adaptability to different size requirements, the reversible but stable conformal seal of PDMS to glass, and fabricated assemblies could be autoclaved to maintain sterile conditions in culture. Circular chamber structures (Figure 1), as opposed to rectangular chambers, were found to minimise the stress induced on the polymethacrylate during UV curing. Due to the short, high intensity UV exposure, chamber designs containing corner regions showed increased mechanical stress in those regions and induced a concavity across the substrate.

The optimal ratio for the liquid methacrylate co-polymer mixture was determined to be 600 µL EGDMA: 300 µL MAA: 100 µL IRGAcure 2022 because of the balance required between the optical and stress properties. More equal ratios of the monomer groups produced a cloudy to opaque white polymer depending on the monomer concentration. Larger ratios (as similar as 600:200:100) caused fatal cracking during the curing phase due to the increased relative quantity of EGDMA cross-linker.

Using the optimised ratio, bioimprint substrates consistently cured into rigid, transparent substrates which were easily separated from the underlying glass microscope slide used for initial cell culture. For assurance of complete

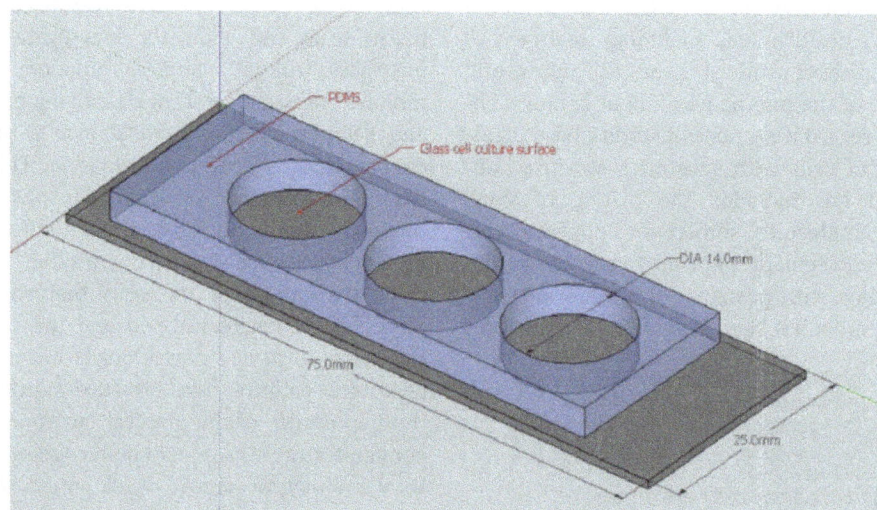

Figure 1 Polydimethylsiloxane [PDMS] with 14 mm circular cut-outs conformally sealed to a glass microscope slide for use as a cell culture substrate for the bioimprinting protocol.

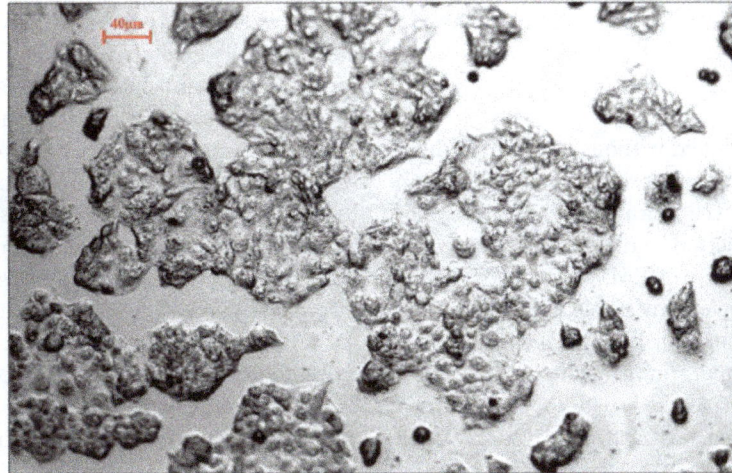

Figure 2 Differential interference micrograph showing a bioimprinted substrate surface containing replica features of Ishikawa cells in polymethacrylate.

curing, chambers containing liquid pre-polymer were exposed to UV for 240 seconds but most bulk curing was complete after as little as 30 seconds.

The bioimprinting protocol successfully produced high resolution replicas of Ishikawa cell features into permanent polymer substrates. Using a pipetting application method instead of spin-coating, which consequently allowed for the removal of triglyme as a thickening agent, did not appear to affect the replication resolution of the methacrylate. Differential interference contrast (DIC) (Figure 2) and AFM (Figure 3)

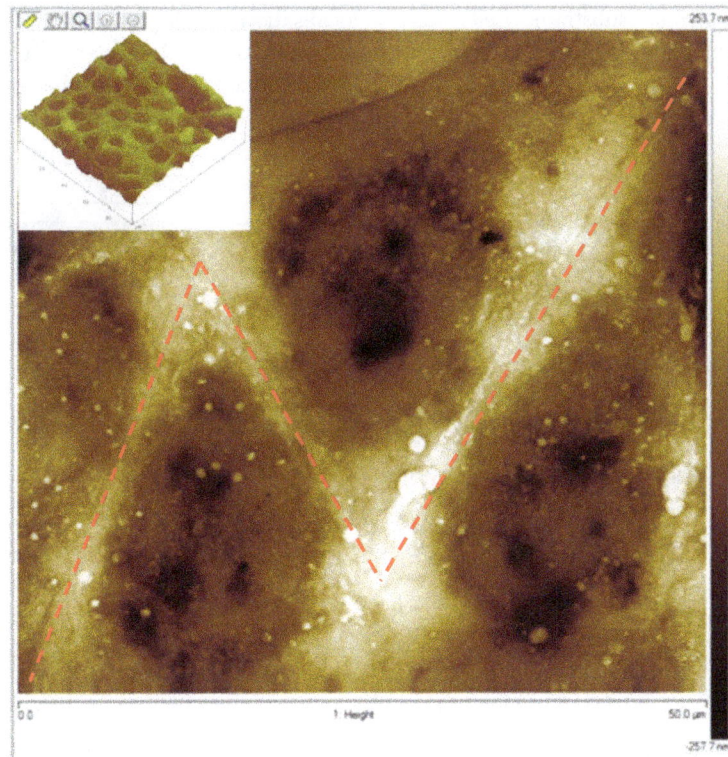

Figure 3 Atomic force microscopy image of multiple cells showing the replication fidelity of bioimprinted Ishikawa cell features in polymethacrylate. Because it is a negative mould indentations and pores on the cell surface appear as protrusions on the AFM. Similarly, the nuclear envelope appears as an impression into the polymer surface. Insert: Low magnification AFM image of Bioimprint in polymethacrylate of a culture of Ishikawa cells. Red dotted lines identify cell borders, dark areas are the replicated nuclei.

showed there was high fidelity feature replication where micron and nanometre scale details are evident.

Biocompatibility

Bioimprint substrates, with and without triglyme, diffused acidic solutes beyond the limiting capacity of the sodium bicarbonate buffer of the cell culture medium. The substrates were therefore subjected to different washing treatments prior to immersion in α-MEM. The phenol red pH indicator included in the medium revealed residual acid in wells in which the substrate had been washed only once with α-MEM prior to immersion. Medium of bioimprint substrates that had been washed with both deionised water and α-MEM leached less acid. Substrates washed with 0.1 M NaOH in addition to deionised water and media washes had no effect on the phenol red pH indicators and thus were satisfactory as culture substrates. Further, methacrylate substrates including triglyme affected pH only slightly less than substrates cured from mixtures not containing triglyme. An extended water wash (>24 hrs) was added to the protocol to provide a tolerance step. This was followed by an additional wash in fresh, sterile medium prior to the application of cells in medium to the substrate. Excluding triglyme from the polymerisation mixture improved optical translucency and the elasticity of the cured polymer substrate. Cells were successfully grown on the prepared platforms as shown in Figure 4.

Secondary cell culture

When cells were incubated on imprinted surfaces the cultured cells exhibited different attachment and growth on the Bioimprint patterned substrate surface compared to those cells on areas of flat surface as shown in Figure 4. Thereby it was revealed that cells distinguish surfaces that had features reminiscent of themselves. The observation therefore indicated that the physical nature of the substrate influenced the cells' behaviour.

It was observed that in cultures that had longer period of proliferation (48 hrs) some of the periphery of the cell culture had lifted during staining. We were able, in these cases, to note that growth of the cells followed the pattern of the imprinted areas (Figure 5). Coomassie Brilliant Blue staining clearly shows cells adhering and spreading across the bioimprinted surface and adhering preferentially to the imprinted surface. Therefore the polymethacrylate substrate will lend itself to detailed studies of behavioural cues generated by relevant physical environments. It is remarkable to observe for the first time how cells reacted to patterns that resemble themselves by following the footprint of the bioimprinted features.

The differential attachment was confirmed by manufacturing imprinted patterns in defined areas using stencils. By this means whether the cells were on flat or imprinted areas could be readily determined by their localisation within the chamber. The results (Figure 6) indicated that cells preferentially adhered and grew on imprinted areas and grew closer to each other.

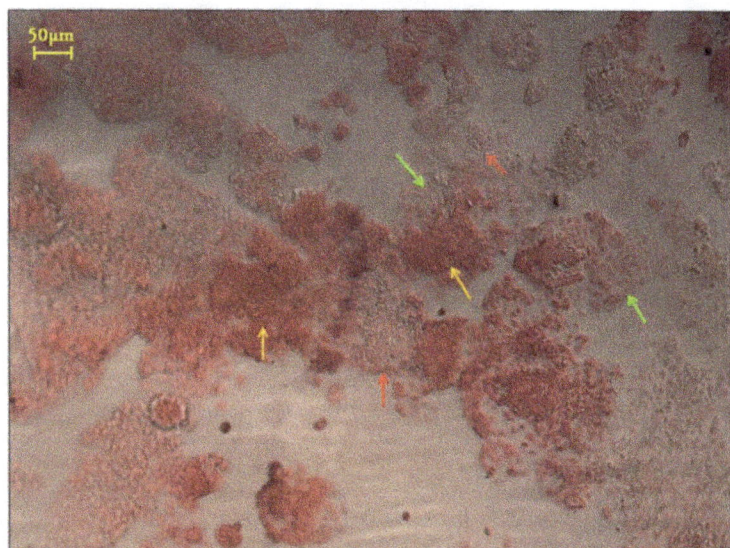

Figure 4 Cell cultured on bioimprinted platform with cell like features. Eosin-stained Ishikawa cells grown for 24 hrs on bioimprinted substrates taken at 24 hrs of initial culture (shown at 50x magnification). Arrows indicate areas of high density cell growth (yellow), cell growth away from bioimprinted regions on flat areas of substrate (green), and identifiable bioimprint regions which are not coverd by cells of the secondry culture (red).

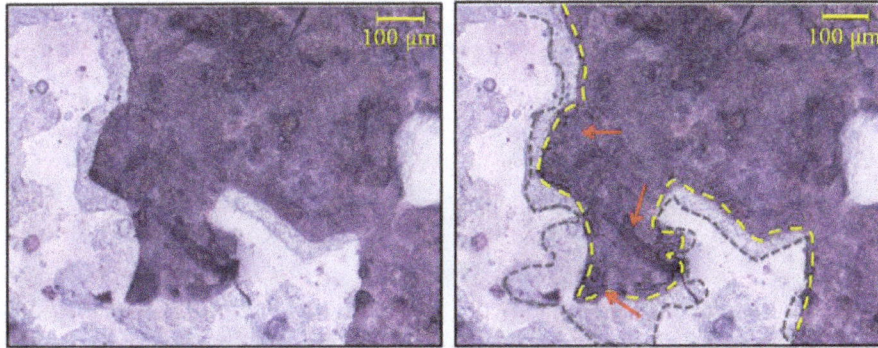

Figure 5 Ishikawa cells grown on bioimprinted substrates for 48 hours, cells are stained with Commassie blue. Left image, unannotated. Right image, the bioimprint features are outlined in grey, the secondary cell culture (dark blue) is outlined in yellow. Arrows (red) note regions of the cell monolayer that peeled off and folded back. This figure illustrates that the growth of the secondary cell culture has been guided by the footprint of the replicated cell patterns.

Lithographically defined micro patterns of pillars and holes were prepared on platforms and cells were cultured on these substrates for comparison with the bioimprint. The dimensions of the lithographically defined patterns were chosen relative to typical Ishikawa cell size which is between 10–50 microns.

Low magnification bright field imaging showed groups of cells clustered across the lithography manufactured substrate, but there was no evidence for differential preference between patterned and unpatterned surfaces (Figure 7). This is in contrast to the observation using Bioimprints. Figure 7 illustrates an array of 5 μm diamond shaped pillars prepared on polymethacrylate substrate; Ishikawa cells were cultured and stained with Coomassie blue. The lithographic patterns without cell-like features had no effect on cell spread between patterned and flat regions.

Interest in topography as an influence on cell behaviour and thence potentially on a range of aspects of health and treatments, independent of biochemical factors, has recently increased. Thus it is important to obtain information on the contribution of cell environment to pathologies such as cancer, as studied in this project, and vascular disease and to interventions that include implants and wound repair.

To improve the effectiveness of inserted medical devices which is expected to be used for monitoring, detections and diagnostics of our health status the interface between the body and foreign materials must be examined and characterised. However we noted the absence of models for *in vitro* investigations that incorporated physical environments similar to those experienced *in vivo*. Here we have, for the first time, developed solid, robust

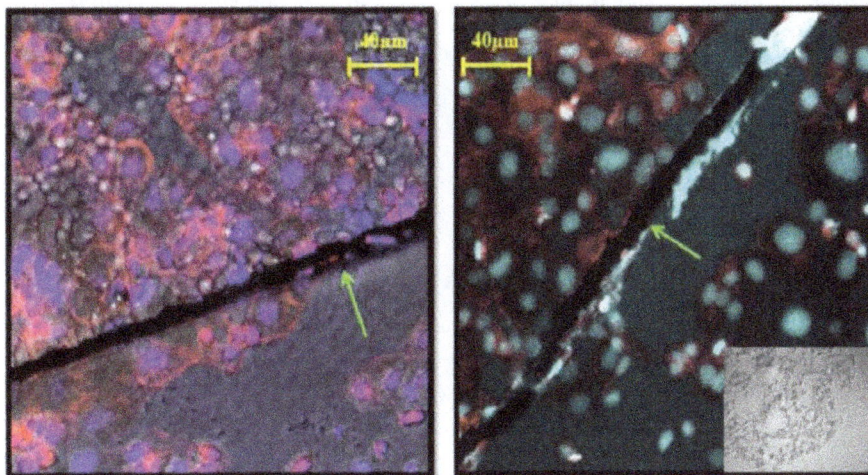

Figure 6 Stencilled Ishikawa Bioimprints showing designated regions of bioimprint features. Left: Confocal fluorescence image of counterstained Ishikawa cells growing on stencilled substrates. Blue, nuclei; Red, cytoplasm. Right Confocal imaging using bright field background channel (which allowed for a single focal plane). Green arrows indicate the border between the flat and bioimprinted (upper) areas and flat (lower) areas. Insert showing a wide view of the border between the flat and imprinted areas of a stencil.

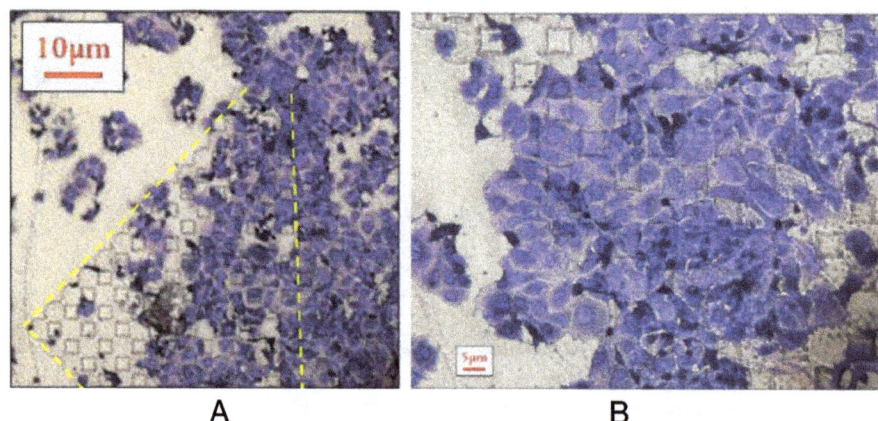

Figure 7 Ishikawa cells grown on substrates with lithographically-manufactured patterns of similar overall size, cells are stained with Commassie blue. **(A)** - illustrates an array of 5 μm diamond shaped pillars prepared on polymethacrylate substrate, Ishikawa cell were cultured and stained with Coomassie blue. The cells show no evidence of preferential spread between patterned and flat regions; **(B)** is a high magnification image.

substrates that have features similar to the cells being studied.

Mixed methacrylate co-polymer was able to reliably replicate high resolution Ishikawa cell surface features to less than 50 nm through fast, high intensity UV exposure methodology [17]. We have used Bioimprints directly for cell imaging studies previously [10,18]. In this study, modifications to the Bioimprint pre-polymer composition and protocol (i.e. removal of triglyme and improving its biocompatability) allow bioimprinted samples to be used as cell culture platforms.

The high light intensity required for the fast curing produced two notable effects on the curing polymer: heat generation and induced stress. Heat generation did not alter the bioimprint quality unless the refrigerated templates of fixed cells were not allowed to acclimate to room temperature prior to exposure and thence introduce defects, possibly resulting from bubbles at the bioimprint interface. The problem of induced crosslinking stress was minimised by using circular sample geometry instead of rectangular geometry. In the chosen geometry the radial distribution of stress induced by high speed (30 seconds), high intensity (100w) UV source curing minimised the curving of the polymer and improved planarization of the bioimprinted surfaces.

The Bioimprint provides a simple and readily-adaptable platform to investigate cell behaviour by methods commonly used for traditional *in vitro* cell culture. The method produces a substrate with nanometre resolution of cell surface features that has attributes that are not provided by the soft surfaces of other cell imprinted templates [19,20]. Importantly the topography-related structures, obtained by the overlay imprinting of the method described here, are more comprehensive than, for example, those obtained from tissue sections [21]. We observed adherence and growth patterns of the cancer

cells on imprinted areas that were distinct from behaviour on flat polymethacrylate surface. Growth is recognised as occurring in areas on the culture platform where cell viability and attachment is high. These results indicate that the cells identified differences in physical topography (flat compared to imprinted) since the substrates, being on the same culture chamber, had been treated identically. We suggest that investigations of guiding cell growth in areas that are currently receiving extensive attention, such as stem cells development and tissue engineering, will also benefit from the method. Other advantages are the ability to store cell details in a hard polymer and prepare cell culture platforms for controlled cell behaviour.

The Bioimprint methodology provides a means of studying cell behaviour in a physical environment that has features of the order of those found *in vivo* and provides a three-dimensional component to the cells' environment. This development is a step increase in biomimicry over that provided by geometrically manufactured substrates. It will be possible, when technical issues are optimised, to extend the concept to imprints in other formats such as cells replicated with structures convex to the base, flexible substrates and in a variety of polymers. It is likely to become possible to manufacture a series of identical substrates from a master mould so that pharmacological treatments of cells on the same imprinted structures can be undertaken. Additionally bioimprinted surfaces may be modified using techniques already developed such as with protein [4] or DNA [22], or adapted to be employed with particulate entities such as viruses [23] to further increase their functionality.

The method produces relevant topography in relation to a cell's micro and nanoenvironment *in vivo*. The resolution of the chosen polymethacrylate polymer is very high (nanometre) and the role of these features that are replicated at this level remain to be defined. The polymer with imprint is permanent and so can be potentially reused within an experiment, incorporated into a later study, or shared with other research laboratories. The process is easy, requires only simple equipment, is inexpensive and the substrate does not require molecular modification. Thus the method provides unique platforms on which the effects of the physical shapes and topography can be investigated. The role of mechanotransduction, the effects on cell behaviour of altered morphology, the cues by which the physical environment either induces tumorigenesis or maintains homeostasis in cells, can all be subjects of study using this method. Importantly this study reported observations on cancer cells of morphological alteration and differential adherence characteristics induced by cues provided by culturing cells on flat and on bioimprinted cell-like patterned platforms.

Determining the effects of the micro-scale patterns allowed us to separate observations of cell growth on flat, micro-patterned and bioimprinted substrates including the nano-scale topographical features. When cells were cultured on the lithographically defined substrate, the pattern showed no effect on the overall culture organization and growth. Cell clusters were visible across the diameter of the substrate irrespective of whether there are patterns or not.

Conclusions

We report development of a unique technique for printing a biological cell in hard polymer that provides high resolution replication and offers a cell culture environment with cell-like features. This enabled us to observe for the first time how cells develop growth characteristics in response to an environment patterned with features that resemble themselves. This methodology has high potential for applications in tissue engineering, medical implants and in studying the influence of physical environment on cell behaviour.

Methods

Cell culture protocol

Ishikawa endometrial cancer cells were cultured in circular chambers of polydimethylsiloxane (PDMS) on glass substrates. To fabricate the PDMS wells, liquid PDMS (Dow Corning, Midland, MI) was mixed at 10:1 elastomer to cross linker ratio, stirred thoroughly, and deaerated before curing. PDMS was poured into polystyrene dishes, which were levelled on a hot plate for curing at 80°C for 2 hrs. Circular chambers were punched into cured PDMS sheets using a 14 mm cork borer. PDMS sheets were then cut to fit a microscope slide and conformally sealed to the slide. PDMS/glass slide constructs were sterilised before use as cell culture substrates.

Ishikawa cells were seeded into the PDMS-bordered wells at 5.0×10^4 cells/cm^2 in α-minimum essential medium (α-MEM) supplemented with 2.2 g/L sodium bicarbonate, 10% fetal bovine serum, 1% GlutaMAX, and 1% penicillin/streptomycin. Ishikawa cells were incubated at 37°C and 5% CO$_2$ for 24 hrs before medium was aspirated and replaced with 4% paraformaldehyde in PBS for at least 30 minutes for cell fixation prior to bioimprinting (all purchased from Life Technologies Co., Carlsbad, CA). Fixative was removed and cultures were rinsed thoroughly in separate PBS and water washes before being placed in 4°C storage for at least 2 hrs to encourage drying of excess water before bioimprinting. Fixed Ishikawa cell cultures were removed from refrigerated storage prior to polymer mixing to bring the samples to room temperature before UV exposure to minimise condensation and bubble artefacts at the cured bioimprint-cell interface.

Bioimprint substrates

The liquid methacrylate co-polymer used for bioimprint substrate fabrication was adapted from previous work [18]. Ethylene glycol dimethacrylate (EGDMA) and methacrylic acid (MAA) (both purchased from Sigma Aldrich, St. Louis, MO) were mixed at the optimised ratio of 600 μL to 300 μL with ~100 μL IRGAcure 2022 (CIBA Specialty Chemicals Basel, Switzerland) added as a photoinitiator. Triglyme was added to the mixture as a thickening agent. The liquid methacrylate solution was mixed for at least 30 seconds with a vortex mixer before being pipetted into the PDMS-defined cell culture wells. The liquid polymer solution was allowed to settle for 10–20 seconds before UV exposure to ensure maximum resolution of small-scale cell features. Slides were placed 15 cm directly beneath a UV light (Omni Cure series 1000 UV, 100w Hg arc lamp, 250-450 nm filter, EXFO Photonic Solutions Inc. Singapore) guide and exposed to UV at 40% aperture opening for 240 seconds. Cured imprints were removed from the PDMS/glass assembly to a water bath and manually agitated to remove larger cell debris. The cured bioimprint was then transferred to an ultrasonic sodium dodecyl sulphate bath (10% w/v in .01 M hydrochloric acid solution) and a 30 minute trypsin soak (0.05% trypsin in PBS) in order to minimise cell material remaining on the bioimprinted polymer surface.

Patterned substrate fabrication

To directly compare the effects of geometrically patterned lithography with those of the bioimprint, different patterned substrates were fabricated. Patterned substrates consisted of regular geometric arrays of pillar or hole patterns of 5–15 μm comparable to the size of the cells under study. The patterns were initially fabricated in SU-8 photoresist (MicroChem SU-8 2100) on silicon wafers using photolithography processes and inverse PDMS moulds were made using soft lithography. The PDMS patterned platforms were replicated in polymethacrylate substrates for cell culture experiments. Ishikawa endometrial cancer cells were cultured in the same conditions as the bioimptinted platforms.

Biocompatibility

To neutralise leaching of methacrylic acid the quenching effect of different washing techniques on the polymethacrylate (EGDMA) substrates prior to use in cell culture was investigated. Bioimprint samples were placed in 12 wells of a 24-well polystyrene tissue culture plate. Bioimprints were washed with (i) deionised water followed by α-MEM medium, (ii) only α-MEM medium, (iii) 0.1 M NaOH followed by deionised water and α-MEM medium, or (iv) left untreated. Washes were pipetted into each well, agitated for approximately 30 seconds, and aspirated. After removal of wash conditions, each well was filled with fresh α-MEM (without cells present) containing phenol red pH indicator.

Cytotoxicity of bioimprinted polymethacrylate samples was investigated by placing a cured bioimprint substrate at the bottom of 3 wells on a polystyrene 6-well plate; the 3 wells without bioimprint samples were maintained as control cultures. Ishikawa cells were seeded at 5.0×10^4 cells/well in all 6 wells and incubated in accordance with the previously outlined protocol for 24 hours, at which point the substrates were removed for imaging.

Secondary cell culture

Ishikawa cells were grown on bioimprinted substrates to verify the biocompatibility of the substrate and determine the topographical influence of bioimprinted features on cell attachment and growth. Ishikawa cells were seeded and cultured on bioimprinted polymethacrylate substrates placed on the bottom of 24-well polystyrene plates. These cells were referred to as secondary cell cultures in order to distinguish them from the initial cell cultures required for bioimprint substrate fabrication. Bioimprints were placed template-side-up and Ishikawa cells were seeded at 5.0×10^4 cells/well and maintained in supplemented α-MEM at 37°C and 5% CO_2 for 24 hrs. At 24 hrs medium was aspirated and cells were fixed with 4% paraformaldehyde in PBS for at least 30 minutes and then washed with PBS several times to remove trace fixative and salts. Cells were stained with Coomassie brilliant blue (Life Technologies Co., Carlsbad, CA) for 5 minutes and washed at least twice with PBS until wash solutions did not contain leached stain.

Competing interest

The authors declare that they have no competing interests.

Authors' contributions

LMM did the experimental work as part of her PhD project. VN was a post doctorate fellow and contributed to the fabrication processes. MMA and JJE were the supervisors of the project who provided the funding, facilities and guidance. All authors helped in the writing of the manuscript, discussions, and interpretation of data and approved the final version of the manuscript.

Acknowledgments

The authors wish to acknowledge the Marsden Fund for funding the project and supporting the post doctorate position and the MacDiarmid Institute for Advanced Materials and Nanotechnology for providing a Ph.D scholarship.

Author details

[1]The MacDiarmid Institute for Advanced Materials and Nanotechnology, Department of Electrical and Computer Engineering, University of Canterbury, Christchurch 8140, New Zealand. [2]The MacDiarmid Institute for Advanced Materials and Nanotechnology, and Centre for Neuroendocrinology, Department of Obstetrics and Gynaecology, University of Otago, Christchurch 8011, New Zealand.

References

1. Mills AM, Longacre TA: **Endometrial hyperplasia.** *Semin Diagn Pathol* 2010, **27**:199–214.
2. Bhattacharya N: **Fetal cell/tissue therapy in adult disease: a new horizon in regenerative medicine.** *Clin Exp Obstet Gynecol* 2004, **31**:167–173.
3. Longmate WM, Dipersio CM: **Integrin regulation of epidermal functions in wounds.** *Adv Wound Care* 2014, **3**:229–246.
4. Lindstrom S, Andersson-Svahn H: **Overview of single-cell analyses: microdevices and applications.** *Lab Chip* 2010, **10**:3363–3372.
5. Kraning-Rush CM, Reinhart-King CA: **Controlling matrix stiffness and topography for the study of tumor cell migration.** *Cell Adh Migr* 2012, **6**:274–279.
6. Blattler T, Huwiler C, Ochsner M, Stadler B, Solak H, Voros J, Grandin HM: **Nanopatterns with biological functions.** *J Nanosci Nanotechnol* 2006, **6**:2237–2264.
7. Lan H, Liu H: **UV-nanoimprint lithography: structure, materials and fabrication of flexible molds.** *J Nanosci Nanotechnol* 2013, **13**:3145–3172.
8. Biggs MJ, Richards RG, Dalby MJ: **Nanotopographical modification: a regulator of cellular function through focal adhesions.** *Nanomed* 2010, **6**:619–633.
9. Nikkhah M, Edalat F, Manoucheri S, Khademhosseini A: **Engineering microscale topographies to control the cell-substrate interface.** *Biomaterials* 2012, **33**:5230–5246.
10. Muys JJ, Alkaisi MM, Evans JJ: **Cancer imaging by atomic force microscopy using a bioimprint cellular transfer technique.** *J Nanobiotechnol* 2005, **4**(1):10.
11. Muys JJ, Alkaisi MM, Evans JJ: **Bioimprint: nanoscale analysis by replication of cellular topography using soft lithography.** *J Biomed Nanotechnol* 2006, **2**:1–5.
12. Murray LM, Nock V, Alkaisi MM, Lee JJM, Woodfield TBF: **Fabrication of polymeric substrates with micro- and nanoscale topography bioimprinted at progressive cell morphologies.** *J Va Sci & Technol B: Microelectron and Nanometer Struct* 2012, **30**:06F902–906F902–906.
13. Nock V, Murray L, Samsuri F, Alkaisi MM, Evans JJ: **Microfluidics-assisted photo nanoimprint lithography for the formation of cellular bioimprints.** *J Vac Sci & Technol B* 2010, **28**:C6K17–C16K22.
14. Nock V, Murray L, Samsuri F, Alkaisi MM, Evans JJ: **Microfluidic arrays for bioimprint of cancer cells.** *Microelectron Eng* 2011, **88**:1828–1831.
15. Wang X, Zhang Y, Du K, Fang X: **Atomic force microscope observation on ultrastructures in plant cells.** *J Nanosci Nanotechnol* 2010, **10**:6624–6628.
16. Li JJ, Zhou XT, Shi J, Zhang F, Li X, Jiang LM, Chen Y: **Upside and downside views of adherent cells on patterned substrates: three-dimensional image reconstruction.** *Microelectron Eng* 2013, **110**:365–368.
17. Samsuri F, Mitchell JS, Alkaisi MM, Evans JJ: **Formation of Nanoscale Bioimprints of Muscle Cells Using UV-Cured Spin-Coated Polymers.** *J Nanotech* 2009, **2009**:6.

18. Samsuri F, Alkaisi MM, Evans JJ, Chitcholtan K, Mitchell JS: **Detection of changes in cell membrane structures using the Bioimprint technique.** *Microelectron Eng* 2011, **88**:1871–1874.

19. DePorter SM, Luib I, McNaughton BR: **Programmed cell adhesion and growth on cell-imprinted polyacrylamide hydrogels.** *Soft Matter* 2012, **8**:10403–10408.

20. Jeon H, Kim G: **Effects of a cell-imprinted poly(dimethylsiloxane) surface on the cellular activities of MG63 osteoblast-like cells: preparation of a patterned surface, surface characterization, and bone mineralization.** *Langmuir* 2012, **28**:13423–13430.

21. Tong WY, Shen W, Yeung CW, Zhao Y, Cheng SH, Chu PK, Chan D, Chan GC, Cheung KM, Yeung KW, Lam YW: **Functional replication of the tendon tissue microenvironment by a bioimprinted substrate and the support of tenocytic differentiation of mesenchymal stem cells.** *Biomaterials* 2012, **33**:7686–7698.

22. Spivaka DA, Shea KJ: **Investigation into the scope and limitations of molecular imprinting with DNA molecules.** *Anal Chim Acta* 2001, **435**:65–74.

23. Bolisay LD, Culver JN, Kofinas P: **Molecularly imprinted polymers for tobacco mosaic virus recognition.** *Biomaterials* 2006, **27**:4165–4168.

8

Carbohydrate functionalization of silver nanoparticles modulates cytotoxicity and cellular uptake

David C Kennedy[1,5], Guillermo Orts-Gil[1*], Chian-Hui Lai[1], Larissa Mller [2], Andrea Haase[3], Andreas Luch[3] and Peter H Seeberger[1,4]

Abstract

Background: Increasing use of silver nanoparticles (Ag-NPs) in various products is resulting in a greater likelihood of human exposure to these materials. Nevertheless, little is still known about the influence of carbohydrates on the toxicity and cellular uptake of nanoparticles.

Methods: Ag-NPs functionalized with three different monosaccharides and ethylene glycol were synthesized and characterised. Oxidative stress and toxicity was evaluated by protein carbonylation and MTT assay, respectively. Cellular uptake was evaluated by confocal microscopy and ICP-MS.

Results: Ag-NPs coated with galactose and mannose were considerably less toxic to neuronal-like cells and hepatocytes compared to particles functionalized by glucose, ethylene glycol or citrate. Toxicity correlated to oxidative stress but not to cellular uptake.

Conclusions: Carbohydrate coating on silver nanoparticles modulates both oxidative stress and cellular uptake, but mainly the first has an impact on toxicity. These findings provide new perspectives on modulating the bioactivity of Ag-NPs by using carbohydrates.

Keywords: Silver, Nanoparticles, Carbohydrates, Nanotoxicology, Bio-interfaces

Introduction

Nanoparticles are playing an increasing role in the development of novel diagnosis methods and in the advanced design of drug delivery systems [1,2]. Silver nanoparticles (Ag-NPs) in particular, show excellent anti-microbial properties and therefore are rapidly being incorporated into a wide array of consumer products such as textiles, cosmetics or packaging materials, increasing the likelihood of human and environmental exposure [3,4]. Moreover, due to their optical properties Ag-NPs are attracting more attention in the fields of biological and chemical sensors [5]. However, Ag-NPs exist in variety of different sizes and shapes but also, very important, with different coatings. Recently, among surface coatings there is an increasing interest in using carbohydrates as biomimetic functional

molecules on the surface of nanoparticles [2,6-8] for the diagnosis and treatment, for instance, of brain diseases like glioma and stroke [9,10]. Glycan functionalised NPs offer several advantages: (i) their synthesis can be performed under biomimetic conditions resulting later on in nanoparticles without traces of chemicals responsible for adverse cellular responses. (ii) the carbohydrates on the surface can serve as targeting molecules and trigger cellular uptake via specific receptors or mediate specific cellular responses [10]. Concurrently, the importance of carbohydrates in cellular signalling and in the regulation of cellular processes continues to emerge [11]. The inherently weak interactions between carbohydrates and proteins or other biomolecules makes these interactions difficult to study. However, because these interactions tend to be multivalent in nature, the use of nanoparticles to mimic the multivalent presentation of carbohydrates found on biomolecular surfaces make carbohydrate-functionalized nanoparticles important systems to study [12].

* Correspondence: guillermo.orts-gil@mpikg.mpg.de
[1]Department of Biomolecular Systems, Max Planck Institute of Colloids and Interfaces (MPIKG), 14476 Potsdam, Germany
Full list of author information is available at the end of the article

Several factors like surface charge and particle size can contribute to the selective binding and uptake of nanomaterials [13,14]. In addition to labelling with a targeting molecule, nanoparticles can induce multivalent effects by clustering antigens on the surface of the particle. Thereby, the binding of relatively weak targeting agents can be enhanced.

Nevertheless, despite the importance of carbohydrates in biology and the vast array of literature on functionalized nanomaterials, little is known about the effects of carbohydrates on the uptake and toxicity of nanoparticles by different type of cells. Although it has been reported that polysaccharides can reduce the toxicity of silver nanoparticles [15] less is known about the influence of monosaccaharides [16] thus, the different results are difficult to rationalise.

Moreover, as pointed out by Johnston et al. [17], the increasing importance of Ag-NPs in the development of novel consumer materials intended for human exposure requires more in depth studies on toxicity mechanisms, as well as, on how silver particles interact with biological molecules and how different surface modifications can be used to reduce or eliminate possible toxic effects.

Here, we discuss the toxicity and the cellular uptake of different silver nanoparticles functionalized with citrate, three different monosaccharides as well as ethylene glycol on two different cell lines. It was found that toxicity correlates with oxidative stress rather than with cellular uptake.

Experimental

Materials
Silver nitrate, sodium citrate, D-glucose, D-mannose, D-galactose and ethylenglycol with MW = 200 (EG-3) with purity >99% were purchased from Sigma-Aldrich.

Synthesis of silver nanoparticles
Citrate-capped silver nanoparticles were synthesized using a standard method [18]: A solution of $AgNO_3$ (10^{-3} M) in deionized water was heated until boiling point. Then, sodium citrate solution in water was added dropwise. The color of the solution slowly turned into gray-yellow after few minutes, indicating the formation of nanoparticles. Heating was continued for an additional 10 min. The solution was then cooled to room temperature and stored in dark before functionalization.

Synthesis of carbohydrate ligands
Thiol-functionalized carbohydrates were synthesized using the following process: glucose, galactose and mannose were each reacted with thiopropionic acid to form the corresponding glycoside in approximately 50% yield. The products were isolated as DMAP-H+ salts. Solutions of these thiolate salts were then directly added to solutions of silver nanoparticles to prepare the carbohydrate functionalized particles.

Functionalization of nanoparticles
To a glass vial charged with a stir bar was added 1 mL of citrate capper silver nanoparticles (1 nM) and 60 µL of a 2 mM stock solution of the corresponding ligand. The solutions were stirred for 3 h at which time each solution was transferred to a 1.5 mL Eppendorf tube and centrifuged for 10 min at 8000 rpm. The supernatant was discarded and the pellet was resuspended in 1 mL of H_2O. These nanoparticle solutions were then used for subsequent reactions and analyses. Scheme of synthesized nanoparticles is shown in Figure 1.

Nanoparticles characterization
Dynamic light scattering (DLS)
Most of DLS measurements were carried out at 37C (to simulate physiological temperature) by use of a Malvern Zeta Nanosizer in water and in cell culture medium (DMEM + 10% FCS). This instrument operates at 4-mW He-Ne laser power, scattering angle of 173 and a wavelength of 633 nm. The intensity correlation functions were fitted by the method of cumulants and by using the Non-Negative Least Squares algorithms (NNLS) included in the Zeta Nanosizer software. The zeta potentials of the samples were obtained from laser Doppler electrophoresis, converting electrophoretic mobilities to zeta potentials. Each sample was prepared in triplicate and measured six times. Experiments consisted of 60 runs per measurement.

Transmission electron microscopy (TEM) and energy dispersive X-ray spectroscopy (EDS)
TEM investigations were performed on a Jeol JEM 2200-FS operating at 200 kV. At high magnification, the in-column Ω-filter was used to improve the contrast. Samples were prepared by immersion of grids of S-160-3 type (Cu coated with carbon film, Plano GmbH) in a small volume (0.5 mL) of solutions followed by solvent evaporation in a dust-protected atmosphere. Particle size distributions were obtained by analysing at least 200 NPs from TEM images using ImageJ software [19]. Energy dispersive X-Ray analysis (EDX) was performed in STEM mode using a spot size between 0.5 and 1.5 nm.

Sugar quantification
Sugar densities were evaluated for Glu, Gal and Mann by a previously reported method [2,7]. Briefly, Glu-Ag-NPs, Gal-Ag-NPs and Mann-Ag-NPs were dispersed in deionized water (0.5 mL) in an ice bath. A freshly prepared 0.5% (w/w) solution of anthrone in sulfuric acid (1 mL) was added slowly to this solution. The resulting solution was gently mixed and heated to 80C for 10 min.

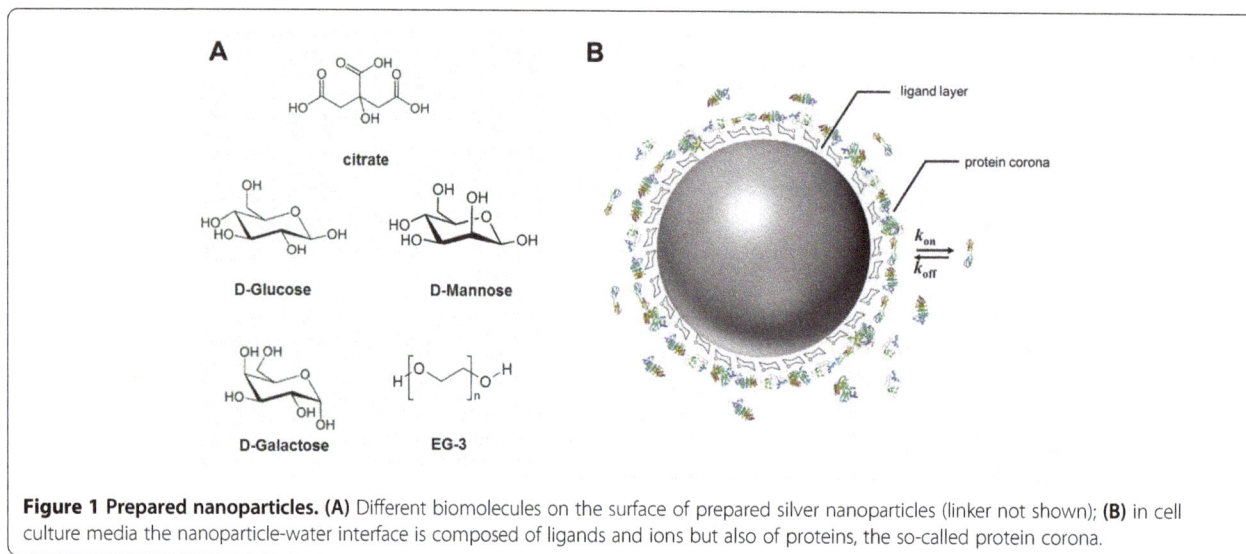

Figure 1 Prepared nanoparticles. (A) Different biomolecules on the surface of prepared silver nanoparticles (linker not shown); **(B)** in cell culture media the nanoparticle-water interface is composed of ligands and ions but also of proteins, the so-called protein corona.

The absorption of the solution was measured at 620 nm and compared with those that were obtained from a standard curve to determine the amount of sugars on the Ag-NP surface.

Cell culture

Neuro-2A and HepG2 cells (American Tissue Culture Center) were each grown in Dulbeccos modified Eagles medium supplemented with 10% fetal bovine serum (PAN-Biotech), 1% penicillin-streptomycin (50 μg/mL, PAN-Biotech), and l-glutamine (2 mM, PAN-Biotech), under standard culture conditions (37C, 5% CO2).

MTT assay

Cells were seeded into wells in a 96-well plate (1 105 cells/mL, 100 μL per well) to cover a 9 6 grid, filling 54 wells. Remaining wells were filled with 200 μL of PBS. After 24 hours, 100 μL volumes of dilutions of nanoparticles in water spanning from 1 nM to 0.01 nM were added to the seeded wells (final concentrations spanning 5 pM to 500 pM). For each functionalized particle, eight dilutions were prepared and for each dilution six replicates were performed. In the remaining 6 wells, 100 μL of PBS was added as a control. Cells were then incubated with complexes for 72 h. After 72 h, 50 μL of a PBS solution of MTT (2.5 mg/mL) was added to each well and then incubated for 3 h. After 3 h, media was aspirated from all wells, leaving purple formazan crystals in those wells with viable cells. To each well, 150 μL of DMSO was added. Plates were then agitated for 10 s and analyzed using a plate reader (NanoQuant Infinite M200 instrument by Tecan Group Ltd.) to determine the absorbance of each well at 570 nm. This reading divided by the average from the reading of the six control wells

was plotted to determine the IC50 value of each complex for each cell line.

Analysis of protein carbonylation as a read-out for intracellular oxidative stress development

HepG2 were seeded in 6-well plates and treated with nanoparticles (final concentrations 2.5 pM and 5pM) for 6 h (induction of protein carbonylation). Cells were washed with PBS three times and lysed by adding a modified RIPA buffer (50 mM Tris/HCl pH 7.4; 150 mM NaCl, 1 mM EDTA, 1% Igepal, 0.25% Na-deoxycholate). Protein concentrations were determined via Bradford assay according to manufacturer instructions (BioRad, Mnchen, Germany). For detection of protein carbonyls OxyBlot kit (Millipore, Schwalbach, Germany) was used according to manufacturer instructions. Briefly, protein carbonyls are labeled by adding 2,4-dinitrophenyl (DNP) hydrazine, which becomes covalently attached as DNP hydrazone and can be detected with the respective DNP antibody. SDS-PAGE was performed according to standard protocols. Gels were transferred onto nitrocellulose membranes with a semidry blotting system. Tubulin antibody was obtained from Abcam (Abcam, Cambridge, UK) and used as a loading control. Images were obtained with GelDoc system (BioRad, Mnchen, Germany) and quantified with ImageLab (BioRad, Mnchen, Germany). The assay was repeated in three independent experiments and results were statistically evaluated.

Confocal microscopy

For imaging, cells were grown on cover slips seeded in 6-well plates. Sterilized cover slips were placed in each well followed by addition of cell suspensions (1 10 5 cells/mL, 2 mL per well). After 24 h, cells were treated with either CuSO4 or CuHis to a final concentration of

25 μM. Cells were incubated with copper complexes for 1, 3, 24 or 72 h at which times the media was removed and cells were fixed by adding 1 mL of fixing solution (3.7% formaldehyde, 4% glucose in PBS). The fixing solution was then removed and 1 mL of PBS was added to cells in each well. To each sample, 4 μL of anti-PRP antibody (Abcam EP1802Y rabbit monoclonal antibody against prion protein) were then added to the PBS and the cells were placed in the fridge to be treated for 12 h at 4C. After 12 h, the PBS was removed, and cells were rinsed 3 times with 1 mL PBS. Cells were then treated with 1 mL PBS and 4 μL of a secondary Goat anti-Rabbit IgG FITC (Invitrogen) and covered with foil. Cells were left at room temperature for 3 h and then treated with DAPI (3 μL Invitrogen). DAPI was added to each well and the plates were then covered with foil again and left at room temperature for 20 min. Finally, the PBS was aspirated, and cells were rinsed 3 times with 1 mL PBS. Cover slips were then removed from the wells. To prepare slides, PBS (20 μL) was added to the surface of each microscope slide and then the removed coverslips were inverted and placed on the PBS. Coverslips were then sealed using nail polish and dried in the dark for 10 min. Slides were imaged using a confocal laser scanning microscope (LSM 700, Zeiss). Z-stack plots (1 micron thick layers) were taken for 6 unique cell clusters from each sample. Stacks were compressed into two-dimensional images using ImageJ software to create a single image showing the entire cell surface. This image was then analyzed using voxel analysis to determine the number of fluorescently labeled pixels, and thus, the level of prion protein at the cell surface. Changes in surface expression and localization were noted and reported.

ICP-MS

Cell samples were digested in 100 μL nitric acid and stored at –20C until analysis. A 50 μL volume of the samples was diluted 1:10000 with 3.5% nitric acid for analysing the cellular uptake of NPs. Lanthanum (10 ppb) was added as internal standard. An external calibration series from 0.5 ppb to 50 ppb was prepared using a silver standard solution. A sample volume of 3 mL was needed for analysis. For this purpose an ICAP-Q (Thermo Fisher Scientific GmbH, Dreieich, Germany) was connected to a concentric nebulizer with a cyclone spray chamber. The working parameters are in Additional file 1.

Results

All nanoparticles were characterized by TEM, DLS, ZP and EDX (Figure 2). Particle sizes computed from statistical analysis of TEM images were around 54 nm (Figure 2-A and Additional File 1). EDX confirms the absence of impurities (Figure 2-B). DLS of particles in water after 24 h show some degree of agglomeration while particles in cell culture media were more stable probably due to the formation of a protein corona (Figure 1-D and Additional File 1). This in good agreement with findings

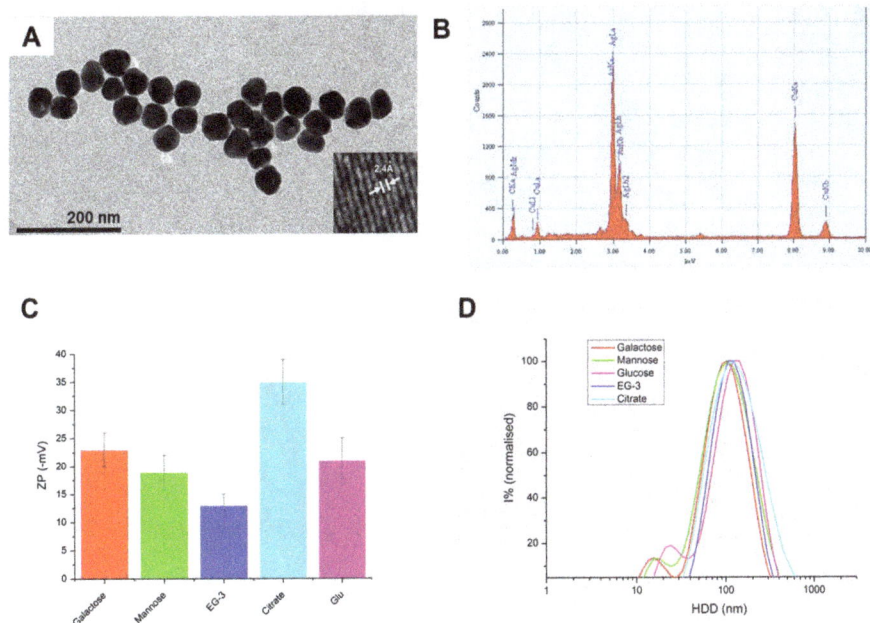

Figure 2 Physico-chemical characterization of nanoparticles. (A) TEM. Inset corresponds to high resolution image showing d lattices; **(B)** Energy dispersive X-ray analysis showing Ag but no impurities; **(C)** zeta potential of different prepared samples; **(D)** DLS of samples in cell culture medium showing well-dispersed nanoparticles.

from Kittler et al. [20] and excludes, in this case, agglomeration as a factor affecting toxicity in cell culture medium [21,22]. ZP shows a change in surface charge for functionalized nanoparticles compared with citrate silver in good agreement with expected values for glycosylated nanoparticles [10] (Figure 2-C). Amount of sugar on nanoparticles was determined using the anthrone/H_2SO_4 method in a similar way as in our previous contribution [23] (see Additional file 1). Values found were between 3.2 and 3.9 molecules sugar/nm^2.

The toxicity of the functionalized silver nanoparticles was tested against two cell lines, a neuronal-like cell line (Neuro-2A) and a hepatocyte cell line (HepG2) by using an MTT assay (Figure 3-A and B). Here, a clear influence of the coating on the toxicity of the particles was observed. While particles functionalized with EG-3, glucose and citrate coated nanoparticles show a similar toxicity, galactose and mannose functionalized nanoparticles were significant less toxic towards both cell lines.

In order to elucidate the mechanism leading to observed coating-dependent toxicities we analyzed the formation of protein carbonyls as an indirect read-out for the oxidative stress inducing activity of nanoparticles. Proteins can become carbonylated either as a direct or an indirect consequence of reactive oxygen species

(ROS) formation. Experiments were performed at particles concentrations 2.5 pM and 5 pM (see Additional file 1). At both concentrations a strong correlation between carbonyls formation and EC_{50} values can be observed (Figure 3-C), suggesting that the toxicity is mainly caused by oxidative stress related to ROS formation.

This may be related to ion release as has already been shown that silver ions can trigger oxidative stress. Many authors have argued that in fact the toxicity of nanosilver is only caused by the ionic form [24]. Therefore, in our case this could mean that either the different types of nanosilver are related to different release rates of ionic silver from the various different coated NP. Dissolution of silver nanoparticles can vary from less than 10% to up to nearly 100%, depending on the coating [25]. Since production of protein carbonyls rather simulates intracellular oxidative stress, the release of silver ions in cell culture media was also measured by ICP-MS in order to also evaluate potential extracellular oxidative stress (see Additional file 1). Nevertheless, no free silver ions were detected in the supernatant of cell culture media, probably due to precipitation of ionic silver in form of AgCl and protein complexes. Therefore, under the studied conditions, intracellular release of silver ions may be the only responsible for cellular damage.

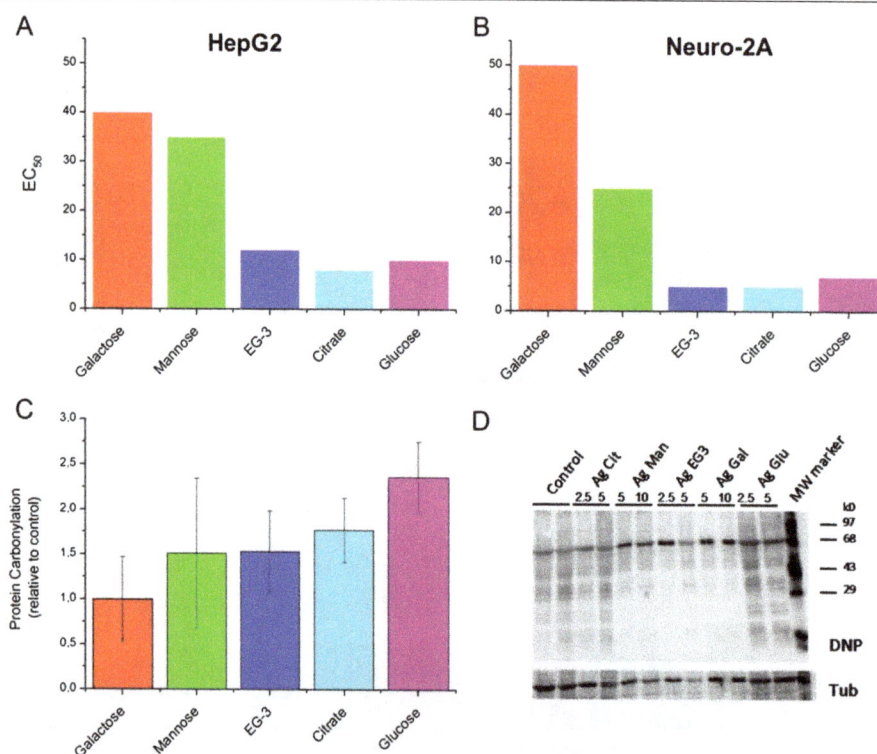

Figure 3 Toxicity results in vitro. (A) EC50 values from MTT assay using silver nanoparticles with different coatings and HepG2 cells; **(B)** Analogous with Neuro-2 cells; **(C)** Detection of oxidative stress from Ag-NPs (concentration 5 pM) via formation of protein carbonyls incubated with HepG2 cells. **(D)** Protein carbonyls were detected at different concentrations (2.5, 5, 10 pM) as (DNP) hydrazone adducts via immunoblots with a DNP antibody.

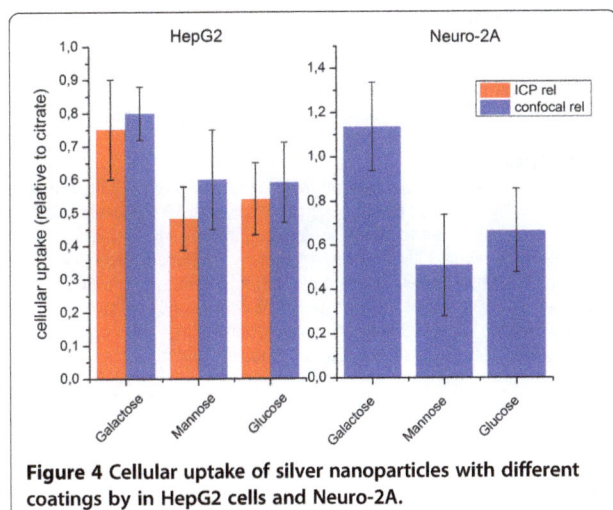

Figure 4 Cellular uptake of silver nanoparticles with different coatings by in HepG2 cells and Neuro-2A.

On the other hand, toxicity may also potentially be influenced by different cellular uptake rates. Here, a Trojan-horse mechanism has been often discussed in literature as a responsible for toxicity of silver nanoparticles. According to this, nanoparticles represent carrier vehicles which penetrate into cells, and then release toxic silver ions by dissolution [26]. To get further insights on main factors leading to toxicity, we analyzed the cellular uptake of the different functionalized silver NPs by ICP-MS and by confocal microscopy. ICP-MS and confocal microscopy showed for both cell lines that the less toxic galactose-functionalized nanoparticles are taken up even more efficiently compared to mannose- or glucose- functionalized particles (Figure 4). Moreover, although mannose and glucose-functionalized nanoparticles present similar cellular uptakes, observed toxicities were considerably different. Thus, particles which are largely internalized into cells do

not necessarily present the highest toxicity. Actually, in this study, glucose-capped nanoparticles present the highest toxicity as well as protein carbonylation, despite their moderate cellular uptake, compared with other nanoparticles. Interestingly, Vaseem et al. showed that glucose reduces the toxicity of nickel nanoparticles towards A549 cells [27]. Thus, intracellular oxidative stress depending on particles coating was the deciding factor leading to toxicity.

Confocal microscopy images show that cellular localization in Neuro-2A cells for the galactose-coated particles are mainly clustered inside the cytoplasm. Therefore, most likely they are contained inside vesicular structures, such as endosomes or lysosomes. Nevertheless, they apparently do not enter the nucleus (Figure 5). Higher density of particles clusters were observed on one side of each cell. Interestingly, for mannose- and glucose-functionalized particles, clusters seem to be spread more evenly through the cell and intracellular clusters tend to be smaller than particles with other functionalities.

Uptake of nanoparticles depending on surface charge has been discussed by other authors. For instance, Badawy et al. showed that negatively charged silver nanoparticles did not overcome electrostatic repulsion barrier towards similar charged bacillus species [28]. As a result, highly negatively charged citrate silver nanoparticles induced less toxicity than H_2-Ag nanoparticles. In our case, we also observe a similar correlation between uptake and surface charge for mannose, glucose or EG3 coated Ag-NP, which are more negatively charged and taken up less efficiently. This is consistent with the fact that cells lines used here are also negatively charged due to various carbohydrate moieties. However, in our case, different uptake rates

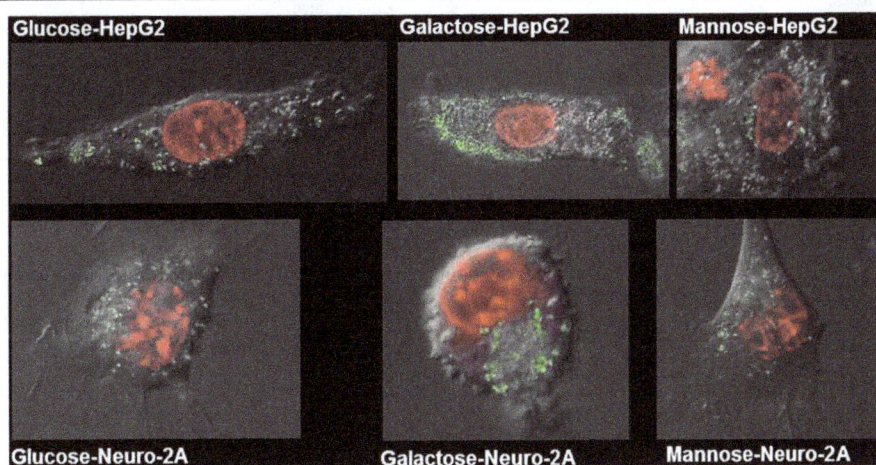

Figure 5 Confocal microscopy images of cells incubated with prepared nanoparticles. Cell nuclei are stained in red. Green dots represent fluorescently labelled nanoparticles.

are not related to different toxicities as actually galact-ose coated NP, which are taken up most efficiently show least toxicity. Eventually this highly increase uptake for galactose coated NP is due to a specific galactose receptor on the surface of cells [23,29].

In fact, internalization of the prepared nanoparticles may depend on various factors, one being surface charge of nanoparticles, another one being the presence of specific receptors on cell surface and finally, it will also depend on the composition of the protein corona [30-32]. We measured the zeta potential of nanoparti-cles after incubation, showing similar overall negative charges for all particles, which confirmed the forma-tion of the protein corona. Although in the last years, more efforts were invested in order to elucidate the detailed composition of the protein corona, this still remains a challenging question which needs exhaustive analysis and techniques. Nevertheless, based on previ-ous studies, the composition of the protein corona on the surface of functionalized particles presented here is expected to be different depending on the particle coating [33-35].

Conclusions

Functionalisation of silver nanoparticles with mono-saccharides modulates their cellular uptake and tox-icity. Galactose and mannose-coated nanoparticles were considerably less toxic to both neuronal-like cells Neuro-2A and hepatocytes, compared to parti-cles functionalized with glucose, ethylene glycol or cit-rate. Observed toxicity was strongly correlated with intracellular oxidative stress, measured as protein car-bonylation, but not to cellular uptake. Summarising, a clear correlation between particle coating, oxidative stress and toxicity has been shown. These results open new perspectives to modulate the bioactivity of Ag-NPs by using carbohydrates.

Additional file

Additional file 1: DLS results/TEM histogram/Calibration curve for sugar quantification/ICP-MS conditions/Carbonyl formation at different concentrations/Extracellular quantification of silver ions.

Competing interests
The authors declare that they have no competing interests.

Authors contributions
DK synthesized and functionalized the nanoparticles, performed confocal microscopy studies and co-wrote the manuscript. GOG coordinated the project, characterized the nanoparticles by TEM, DLS, ZP and EDX and wrote the manuscript. CHL made sugar quantification. LM measured ICP-MS in cells. AH and AL carried out protein carbonylation experiments, organize extracellular ICP-MS experiments and co-wrote the manuscript. PHS provided laboratory facilities at MPI and revised the manuscript. All authors read and approved the final manuscript.

Acknowledgments
Authors thank the Max Planck Society for generous financial support. GOG thanks the Federal Institute for Materials Research and Testing for TEM images of silver nanoparticles. In addition, the authors thank Julian Tharmann and Philipp Reichardt for excellent technical assistance.

Author details
[1]Department of Biomolecular Systems, Max Planck Institute of Colloids and Interfaces (MPIKG), 14476 Potsdam, Germany. [2]Division 1.1 Inorganic Trace Analysis, Federal Institute for Materials Research and Testing (BAM), Richard-Willsttter-Strae 11, 12489 Berlin, Germany.   [3]Departments Chemical and Product Safety, German Federal Institute for Risk Assessment (BfR), 10589 Berlin, Germany. [4]Institute for Chemistry and Biochemistry, Free University Berlin, Arnimallee 22, 14195 Berlin, Germany. [5]National Research Council Canada (CNRC), 100 Sussex Drive, Ottawa, Ontario, Canada.

References
1. Gao J, Gu H, Xu B: **Multifunctional magnetic nanoparticles: design, synthesis, and biomedical applications.** *Acc Chem Res* 2009, **42**(8):1097 1107.
2. Lai C-H, Chang T-C, Chuang Y-J, Tzou D-L, Lin C-C: **Stepwise orthogonal click chemistry toward fabrication of paclitaxel/galactose functionalized fluorescent nanoparticles for HepG2 cell targeting and delivery.** *Bioconjug Chem* 2013, **24**(10):1698 1709.
3. Reidy B, Haase A, Luch A, Dawson K, Lynch I: **Mechanisms of silver nanoparticle release, transformation and toxicity: a critical review of current knowledge and recommendations for future studies and applications.** *Materials* 2013, **6**(6):2295 2350.
4. Sun TY, Gottschalk F, Hungerbhler K, Nowack B: **Comprehensive probabilistic modelling of environmental emissions of engineered nanomaterials.** *Environ Pollut* 2014, **185**:69 76.
5. Pastoriza-Santos I, Liz-Marzan LM: **Colloidal silver nanoplates: state of the art and future challenges.** *J Mater Chem* 2008, **18**(15):1724 1737.
6. Marradi M, Chiodo F, Garcia I, Penades S: **Glyconanoparticles as multifunctional and multimodal carbohydrate systems.** *Chem Soc Rev* 2013, **42**(11):4728 4745.
7. Lai C-H, Lai N-C, Chuang Y-J, Chou F-I, Yang C-M, Lin C-C: **Trivalent galactosyl-functionalized mesoporous silica nanoparticles as a target-specific delivery system for boron neutron capture therapy.** *Nanoscale* 2013, **5**(19):9412 9418.
8. Chiodo F, Marradi M, Calvo J, Yuste E, Penads S: **Glycosystems in nanotechnology: gold glyconanoparticles as carrier for anti-HIV prodrugs.** *Beilstein J Org Chem* 2014, **10**:1339 1346.
9. Zhang C, Wan X, Zheng X, Shao X, Liu Q, Zhang Q, Qian Y: **Dual-functional nanoparticles targeting amyloid plaques in the brains of Alzheimer's disease mice.** *Biomaterials* 2014, **35**(1):456 465.
10. Farr TD, Lai C-H, Grnstein D, Orts-Gil G, Wang C-C, Boehm-Sturm P, Seeberger PH, Harms C: **Imaging early endothelial inflammation following stroke by core shell silica superparamagnetic glyconanoparticles that target selectin.** *Nano Lett* 2014, **14**(4):2130 2134.
11. Seeberger PH, Werz DB: **Synthesis and medical applications of oligosaccharides.** *Nature* 2007, **446**(7139):1046 1051.
12. Mammen M, Choi S-K, Whitesides GM: **Polyvalent interactions in biological systems: implications for design and use of multivalent ligands and inhibitors.** *Angew Chem Int Ed* 1998, **37**(20):2754 2794.
13. Yin Win K, Feng S-S: **Effects of particle size and surface coating on cellular uptake of polymeric nanoparticles for oral delivery of anticancer drugs.** *Biomaterials* 2005, **26**(15):2713 2722.
14. Chithrani BD, Chan WCW: **Elucidating the mechanism of cellular uptake and removal of protein-coated gold nanoparticles of different sizes and shapes.** *Nano Lett* 2007, **7**(6):1542 1550.
15. Miao A-J, Schwehr KA, Xu C, Zhang S-J, Luo Z, Quigg A, Santschi PH: **The algal toxicity of silver engineered nanoparticles and detoxification by exopolymeric substances.** *Environ Pollut* 2009, **157**(11):3034 3041.
16. Sur I, Cam D, Kahraman M, Baysal A, Culha M: **Interaction of multi-functional silver nanoparticles with living cells.** *Nanotechnology* 2010, **21**(17):175104.
17. Johnston HJ, Hutchison G, Christensen FM, Peters S, Hankin S, Stone V: **A review of the in vivo and in vitro toxicity of silver and gold**

particulates: particle attributes and biological mechanisms responsible for the observed toxicity. *Crit Rev Toxicol* 2010, **40**(4):328 346.

18. Pillai ZS, Kamat PV: What factors control the size and shape of silver nanoparticles in the citrate ion reduction method? *J Phys Chem B* 2003, **108**(3):945 951.

19. Woehrle GH, Hutchison JE, zkr S, Finke RG: Analysis of nanoparticle transmission electron microscopy data using a public- domain image-processing program, image. *Turk J Chem* 2006, **30**:1 13.

20. Kittler S, Greulich C, Gebauer JS, Diendorf J, Treuel L, Ruiz L, Gonzalez-Calbet JM, Vallet-Regi M, Zellner R, Kller M, Epple, M: The influence of proteins on the dispersability and cell-biological activity of silver nanoparticles. *J Mater Chem* 2010, **20**:512 518.

21. Drescher D, Orts-Gil G, Laube G, Natte K, Veh RW, Oesterle W, Kneipp J: Toxicity of amorphous silica nanoparticles on eukaryotic cell model is determined by particle agglomeration and serum protein adsorption effects. *Anal Bioanal Chem* 2011, **400**(5):1593 1604.

22. Orts-Gil G, Natte K, Drescher D, Bresch H, Mantion A, Kneipp J, sterle W: Characterisation of silica nanoparticles prior to in vitro studies: from primary particles to agglomerates. *J Nanoparticle Res* 2011, **13**(4):1593 1604.

23. Lai C-H, Lin C-Y, Wu H-T, Chan H-S, Chuang Y-J, Chen C-T, Lin C-C: Galactose encapsulated multifunctional nanoparticle for HepG2 cell internalization. *Adv Funct Mater* 2010, **20**(22):3948 3958.

24. Kittler S, Greulich C, Diendorf J, Kller M, Epple M: Toxicity of silver nanoparticles increases during storage because of slow dissolution under release of silver ions. *Chem Mater* 2010, **22**(16):4548 4554.

25. Loza K, Diendorf J, Sengstock C, Ruiz-Gonzalez L, Gonzalez-Calbet JM, Vallet-Regi M, Koller M, Epple M: The dissolution and biological effects of silver nanoparticles in biological media. *J Mater Chem B* 2014, **2**(12):1634 1643.

26. Park E-J, Yi J, Kim Y, Choi K, Park K: Silver nanoparticles induce cytotoxicity by a Trojan-horse type mechanism. *Toxicol in Vitro* 2010, **24**(3):872 878.

27. Vaseem M, Tripathy N, Khang G, Hahn Y-B: Green chemistry of glucose-capped ferromagnetic hcp-nickel nanoparticles and their reduced toxicity. *RSC Adv* 2013, **3**(25):9698 9704.

28. El Badawy AM, Silva RG, Morris B, Scheckel KG, Suidan MT, Tolaymat TM: Surface charge-dependent toxicity of silver nanoparticles. *Environ Sci Technol* 2010, **45**(1):283 287.

29. Barondes S, Rosen S: Cell surface carbohydrate-binding proteins: role in cell recognition. *Neuronal Recognit Springer US* 1976, **11**:331 335.

30. Petri-Fink A, Steitz B, Finka A, Salaklang J, Hofmann H: Effect of cell media on polymer coated superparamagnetic iron oxide nanoparticles (SPIONs): colloidal stability, cytotoxicity, and cellular uptake studies. *Eur J Pharm Biopharm* 2008, **68**(1):129 137.

31. Bajaj A, Samanta B, Yan H, Jerry DJ, Rotello VM: Stability, toxicity and differential cellular uptake of protein passivated-Fe3O4nanoparticles. *J Mater Chem* 2009, **19**(35):6328 6331.

32. Lesniak A, Fenaroli F, Monopoli MP, berg C, Dawson KA, Salvati A: Effects of the presence or absence of a protein corona on silica nanoparticle uptake and impact on cells. *ACS Nano* 2012, **6**(7):5845 5857.

33. Dorota Walczyk FBB, Monopoli MP, Lynch I, Dawson KA: What the cell sees in bionanoscience. *J Am Chem Soc* 2010, **132**(16):5761 5768.

34. Lynch I, Cedervall T, Lundqvist M, Cabaleiro-Lago C, Linse S, Dawson KA: The nanoparticle-protein complex as a biological entity; a complex fluids and surface science challenge for the 21st century. *Adv Colloid Interf Sci* 2007, **134 135**:167 174.

35. Monopoli M, Walczyk D, Campbell A, Elia G, Lynch I, Baldelli Bombelli F, Dawson K: Physical-chemical aspects of protein corona: relevance to in vitro and in vivo biological impacts of nanoparticles. *J Am Chem Soc* 2011, **133**:2525 2534.

9

Optimization of the magnetic labeling of human neural stem cells and MRI visualization in the hemiparkinsonian rat brain

Milagros Ramos-Gómez[1,2]*, Emma G Seiz[3] and Alberto Martínez-Serrano[3]*

Abstract

Background: Magnetic resonance imaging is the ideal modality for non-invasive *in vivo* cell tracking allowing for longitudinal studies over time. Cells labeled with superparamagnetic iron oxide nanoparticles have been shown to induce sufficient contrast for *in vivo* magnetic resonance imaging enabling the in vivo analysis of the final location of the transplanted cells. For magnetic nanoparticles to be useful, a high internalization efficiency of the particles is required without compromising cell function, as well as validation of the magnetic nanoparticles behaviour inside the cells.

Results: In this work, we report the development, optimization and validation of an efficient procedure to label human neural stem cells with commercial nanoparticles in the absence of transfection agents. Magnetic nanoparticles used here do not affect cell viability, cell morphology, cell differentiation or cell cycle dynamics. Moreover, human neural stem cells progeny labeled with magnetic nanoparticles are easily and non-invasively detected long time after transplantation in a rat model of Parkinson's disease (up to 5 months post-grafting) by magnetic resonance imaging.

Conclusions: These findings support the use of commercial MNPs to track cells for short- and mid-term periods after transplantation for studies of brain cell replacement therapy. Nevertheless, long-term MR images should be interpreted with caution due to the possibility that some MNPs may be expelled from the transplanted cells and internalized by host microglial cells.

Keywords: Magnetic nanoparticles, Neural stem cell, Cell tracking, Magnetic resonance imaging

Background

Neural stem cells (NSCs) represent a source of cells for regenerative medicine, in particular for cell replacement therapies both in the clinical and pre-clinical experimental settings. One of the goals of stem cell research is the *in vitro* and *in vivo* generation of neurons which could turn to be optimal candidates to replace specific lost neurons, for instance in Parkinson's disease (PD), in which the A9 subtype of dopaminergic neurons (DAn) in the Substantia nigra (SN) are lost [1]. Previous clinical studies of cell replacement in PD were based on the transplantation of fresh human fetal ventral mesencephalic (VM) tissue into the caudate and putamen of PD patients [1,2]. These initial experiments showed practical and ethical issues such as the need to obtain tissue from six to seven human fetuses to provide enough cells for one patient's transplantation, the lack of reproducibility between centers, poor survival in some cases, and the appearance of serious adverse side-effects in some patients. Recent work has thus aimed to obtain suitable sources of human NSCs (hNSCs) with the capacity to differentiate into DAn endowed with the required, genuine properties of Substantia Nigra pars compacta neurons (SNpc) lost in PD [3,4].

Recent pre-clinical research has demonstrated that immortalized human NSCs, derived from VM (hVM1 cell line) and modified for the elevated expression of Bcl-XL (hVM1-highBcl-XL cells), have the potential to differentiate

* Correspondence: milagros.ramos@ctb.upm.es; amserrano@cbm.csic.es
[1]Centre for Biomedical Technology, Polytechnic University of Madrid, 28223 Madrid, Spain
[3]Department of Molecular Biology and Center of Molecular Biology "Severo Ochoa", Autonomous University of Madrid-C.S.I.C, 28049 Madrid, Spain
Full list of author information is available at the end of the article

into DAn *in vivo* at a high rate [5-9]. After transplantation in hemiparkinsonian rats, these hVM cells survive, integrate, and differentiate into DAn, alleviating behavioral motor asymmetry and skilled paw use [5,9,10]. Thus, hVM1 cells and their derivatives represent a helpful tool for the development of cell therapies for neurodegenerative diseases, Parkinson disease in particular.

Tracking noninvasively the long-term spatial destination and final residence of transplanted cells *in vivo*, to monitor their survival, migration, differentiation and regenerative impact, has become a critical methodology in evaluating the efficacy of stem cell therapy procedures. Until now, it was only the behavioral testing or the DA determination by *in vivo* HPLC and the subsequent histological analysis the available methods used to evaluate grafting outcome, viability and differentiation of transplanted cells in hemiparkinsonian animal models. But, optimally, research in cell replacement therapy requires of non-invasive and sensitive imaging techniques to track the fate of transplanted cells; these techniques would increase reliability and reduce the total number of animals used in these experiments.

Labeling cells with magnetic nanoparticles (MNPs) has been shown to induce sufficient contrast for magnetic resonance imaging (MRI) of cells in the brain [11-15]. Therefore, MRI, in combination with other *in vivo* molecular imaging techniques, like PET, can provide insights into different cellular processes, including localization and migration of the cells, cell survival and proliferation kinetics, and cell differentiation patterns, which can aid clinical implementation of cell therapy [16].

Most labeling techniques currently take advantage of either the attachment of MNPs to the stem cell surface or the internalization of MNPs by endocytosis. Surface labeling normally results in lower iron content per cell and promotes a rapid reticulo-endothelial recognition and clearance of labeled cells [17,18]. Therefore, endocytosis of MNPs during *in vitro* cell cultivation stands as the preferred labeling method.

The most commonly used MNPs to label cells, dextran coated superparamagnetic iron oxide (SPIO) nanoparticles, as the ones used in the present study, do not efficiently label either nonphagocytic or non–rapidly dividing mammalian cells in vitro [19]. Consequently, these contrast agents are not used as isolated reagents to label hNSCs or other mammalian cells [20-22]. In most cases, internalization of nanoparticles by hNSCs requires the use of transfection agents (TAs), like protamine sulfate (PS) or poly-L-lysine (PL) to achieve an efficient labeling of the stem cells. TAs coat MNPs by means of electrostatic interaction with dextran-coated nanoparticles and help internalizing them into cells. PS, conventionally used to reverse heparin anticoagulation, has been used as a cationic TA to label human mesenchymal stem cells and hematopoietic

stem cells with SPIO nanoparticles [19]. The use of PL complexed with MNPs also reported a high labeling efficiency of NSCs, 80% [23]. However, the use of TAs to label cells might have a harmful side effect decreasing cell viability, since most TAs are toxic to cells when used alone and not complexed to DNA [19]. In addition, the use of relative high concentrations of magnetic nanoparticles to label hNSCs might be toxic or affect some of their functional properties, causing alterations in their differentiation processes. Thus, an extensive study of the properties of NSCs labeled with MNPs must be carried out to identify the effects of MNPs on hNSCs biology.

The aim of the present study was to efficiently label cells using magnetic nanoparticles detectable by MRI and to determine the effects of such particles on morphology, cell cycle and differentiation capacity of hVM cells. Here we report the development and validation of an efficient protocol to label hNSCs using several commercial magnetic nanoparticles as contrast agents for MRI. We optimized the incubation times and the concentration of MNPs to label hNSCs in the absence of any transfection agents that could damage the cellular integrity, in order to prevent impairment of the cells' functional properties.

Our results demonstrate that the use of MNPs to label hNSCs is feasible, efficient and safe for MRI tracking following grafting of hNSCs into hemiparkinsonian rat brain. The fate of MNPs-labeled hNSCs grafted into hemiparkinsonian rats can be successfully visualized using MRI at different time points, up to 5 months after transplantation.

Materials and methods
Cell cultures
Cell isolation and immortalization were described previously by Villa [8] and Courtois [9]. Briefly, human ventral mesencephalic cells were isolated from a 10-week-old aborted fetus (Lund University Hospital). Tissue procurement was in accordance with the Declaration of Helsinki and in agreement with the ethical guidelines of the European Network of Transplantation. Immortalization was carried out by infection with a retroviral vector coding for v-myc (LTR-vmyc-SV40p-Neo-LTR) generating the hVM1 cell line. The hVM1 polyclonal cell line was infected at passage 6 with a Bcl-XL coding (LTR-Bcl-XL-IRES-rhGFP-LTR) retroviral vector, to enhance their neurogenic capacity [24]. After infection, the cells were selected by fluorescence-activated cell sorting (FACS) generating the hVM1 Bcl-XL cell line (referred here as hVM to abbreviate). Both cell lines were routinely cultured under standard conditions as described before [8,9]. Briefly, cells were cultured on 10 µg/ml polylysine-pretreated plasticware in epidermalgrowth factor (EGF) and basic fibroblast growth factor (FGF) (20 ng/ml each; R&D Systems) supplemented chemically defined

Dulbecco's modified Eagle's medium/F-12 medium (Glutamax (Invitrogen), 1% Albumax (Invitrogen), 50 mM Hepes (Invitrogen), 0.6% glucose, N2 supplement (Invitrogen), 1x nonessential amino acids and penicillin/streptomycin,), referred to hereafter as "proliferation medium". To induce cell differentiation, cells were seeded at 10^5cells/cm^2 in proliferation medium on poly-L-lysine-treated glass coverslips. After 24 h, proliferation medium was replaced by differentiation medium without EGF or FGF, and containing 1 mM dibutyryl-cAMP (Sigma) and 2 ng/ml human recombinant glial cell-derived neurotrophic factor (GDNF) (Preprotech)). Differentiation medium was changed every second day until the end of the experiment. Cells were proliferated and differentiated at 37°C and 95% humidity in a low oxygen atmosphere (5% O_2, 5% CO_2, in a dual CO_2/O_2 incubator (Forma)).

Uptake of MNPs by hVM cells

Cell labeling with MNPs

hVM cells were seeded on glass coverslips at a density of 50,000 cells/cm^2. After 24 h with proliferation medium, different types of SPIO nanoparticles were added while varying several parameters such as concentration, precoating with transfection agents and the time of incubation. For these studies the following conditions were used: i) different core diameter types (all dextran coated): 250 nm (G.Kisker), 50 nm (G.Kisker), 100 nm (Endorem) and 100 nm-Cy3conjugated (Chemicell); ii) concentration: 50, 100 or 300 µgFe/ml of culture medium; iii) incubation time: 3, 6, 12, 24 or 72 h; iv) pre-treatment of MNPs with transfection agents: untreated (control) and treated with poly-lysine (0.03 µgPL/µgFe,Sigma) or protamine sulfate (0.5 µgPS/g Fe, Sigma). For pre-treatment MNPs were maintained in proliferation medium with constant stirring for 24 h at room temperature before being added to cell cultures.

Following the desired incubation time, cells were rinsed with culture medium and PBS to remove unincorporated MNPs. Subsequently, the cells were differentiated and fixed with 4% of paraformaldehyde (Merck) in 0.1 M phosphate buffer at pH 7.4 for 15 minutes at room temperature. Fixed cells were blocked for 1 h in PBS containing 10% normal horse serum, 0.25% Triton X-100 and incubated overnight at 4°C with a monoclonal antibody to mark the dextran coating of the MNPs (except in the case of 100 nm-Cy3) using an antibody anti-dextran (1:500, Stem Cell Technologies). Afterwards, cells were rinsed and incubated with an anti-mouse Cy3-conjugated secondary antibody (1:500, Jackson Immunoresearch). Last, the cells were incubated with ToPro-3 (1:750, Invitrogen) for nucleic acid staining and phalloidin A488 to mark F-actin filaments and define cell shape. The magnitude of MNPs uptake was determined as the proportion of cells with green fluorescent cytosolic dots corresponding to MNPs compared to the total number of cells identified by the nuclear counterstaining with ToPro-3 using a confocal microscope coupled to a LSM510 Axiovert200 inverted microscope (Zeiss) .

Cell viability assays

Cell viability in unlabeled (control) cultures and in those labeled with MNPs was evaluated by the MTT assay at day 0 and 7 of differentiation. Cells were seeded at a density of 50,000 cells/cm^2 in 0.5 ml culture medium. After labeling cells with MNPs for 72 h, 125 µl of 5 mg/ml MTT (Sigma M-2128) were added and the incubation was left to proceed for 60 min at 37°C. Then 1 ml of DMSO was added per well to extract the formazan and absorbance at 570 nm was determined.

Cell cycle analysis

To analyze the effect of MNPs on cell cycle, we performed a cell cycle analysis in unlabeled and MNP-labeled cells (100 nm, 50µgFe/ml) by propidium iodide staining and flow cytometry using the technique of Nicoletti [25]. The cells were seeded at a density of 30,000 cells/ cm^2. After 24 and 48 h of incubation with the MNPs, the cells were trypsinized and washed in PBS (without Ca/Mg^{2+}). Cells were centrifuged 10 min at 1000 rpm at 4°C and the cell pellets were mixed with 1 ml of cold 70% ethanol using the vortex. After18h fixation in ethanol at -20°C the samples were resuspended carefully and centrifuged 5 min at 1500 rpm. The supernatant was discarded and the pellet was resuspended in 0.5 ml of buffer cycle (50 µg/ml propidium iodide, 0.1% sodium citrate, 50 µg/ml ribonuclease A in PBS without Ca^{2+} or Mg^{2+}), and incubated 30 min at room temperature for staining. The cells were then analyzed by flow cytometry (flow cytometer FACSCalibur, Becton Dickinson) using the 488 nm argon laser for excitation and filter 585/42 nm for the collection of emission (channel FL-2). The intensity of fluorescence represented in linear scale (cell cycle distribution) and data were analyzed for quantification of the regions sub-G0-G1 (less than 2n of DNA and fragmented DNA), G0-G1 (2n DNA) and S-G2-M (mitotic phase active, i.e. an amount of DNA between 2n and 4n). A total of 10.000 events were acquired and analyzed using FloJo7 software. Samples were run in biological triplicates.

Immunocytochemistry

Unlabeled cells and cells labeled with MNPs (50 and 100 nm size; 50µgFe/ml) for 72 h, were grown on glass coverslips in differentiation medium for the specified time (4 and 7 days) and fixed with paraformaldehyde (PFA, Merck) at 4% in 0.1 M phosphate buffer at pH 7.4, for 15 minutes at room temperature. After 3 washes in PBS, samples were blocked for 1 h with a solution of

PBS containing 10% horse serum (HS) (Gibco/Life Tecnhologies) or goat serum (Gibco/Life Technologies) and 0.25% nonionic detergent Triton X-100 (Merck). Subsequently, cells were incubated overnight at 4°C with primary antibodies dissolved in a PBS solution with 0.25% triton and 1% horse-goat serum. To evaluate the possible effect of the MNPs in cell differentiation the following antibodies were used: nestin (day 0, 1:1000, Abcam; 1:500 BD Bioscience) and β-III-tubulin (day 7, 1:1000, Sigma), TH (day 7, 1:1000sigma) and GFAP (day 7, 1:1000 Sigma). To study the intracellular localization of MNPs the following antibodies were used: anti-manosidase II (day 7, 1:100, Millipore), anti-EEA1 (day 7, 1:200, BD Transduction) and anti-CD63 (day 7, 1:100, DSHB). After removing the solution with primary antibody and rinsing, the samples were incubated with secondary antibody conjugated with different fluorophores in PBS for 30 min at room temperature (1:500, Cy3, Cy5 or Alexa488, Jackson Inmunoresearch). Finally, nuclei were counterstained with Hoechst 33258 (0.2 μg/ml in PBS, Molecular Probes) or To-Pro-3 (1:500, Invitrogen). After the labeling, the coverslips were washed in PBS and distilled H_2O, allowed to dry and mounted with Mowiol (Calbiochem).

MNP labeling decay with passages in culture

Cells were labeled with MNPs (50 and 100 nm-Cy3 50μgFe/ml) for 72 hours in p60 plates ($3x10^4$ cells/cm^2). Subsequently, the cultures were trypsinized and seeded on glass coverslips (passage 0) and split to new p60 plates. After two days, cells of these plates were trypsinized and seeded on glass coverslips (passage 1), remaining cells were split to new p60 plates after 48 h in culture and seeded at the same initial density ($3x10^4$ cells/cm^2). We proceeded up to the passage 4. Cells were fixed and immuno-stained to detect dextran, Phalloidin A488 and To-Pro-3. Then, the percentage of cells containing MNPs at each passage was determined using a LSM510 laser confocal Microscope coupled to an Axiovert200 (Zeiss) inverted microscope.

Animal Experimentation
Lesion and Transplantation Procedures
Experiments were carried out according to the guidelines of the European Community (Directive 86/609/ECC, Directive 2010/63/EU) observing the 3Rs principles, and in accordance with the Society for Neuroscience recommendations. Animals used in this study were 3-month-old female Sprague-Dawley rats (Harlan), weighing 200–250 g at the beginning of the experiment, housed in a temperature- and humidity-controlled room, under 12-h light/dark cycles, with *ad libitum* access to food and water. Cells (in proliferative state) for transplantation in intact brains were dispersed and resuspended in Hanks' balanced salt solution

(Invitrogen) at a density of 10^5 cells/μl. Cell suspensions ($3x10^5$ cells in 3 μl) were injected into the left striatum (control unlabeled cells) and in the right striatum (cells labeled with 100 nm MNPs-Cy3 at 50 μg/ml for 72 h) at the following coordinates (in mm): anteroposterior -1; mediolateral +/-3; dorsoventral-4.5 (from dura), with the tooth bar set at -2.3. Hemiparkinsonian rats received a 6-hydroxydopamine (6-OH-DA) injection (9 μg/3 μl dissolved in 0.9% saline containing 0.2 mg/ml ascorbic acid; Sigma) in the right median forebrain bundle at the following stereotaxic coordinates (tooth bar set at -3.3 mm): anteroposterior -3.7 mm; mediolateral -1.6 mm (both from bregma); dorsoventral -8.8 mm from dura. The injection rate was 1 μl/min, and the syringe was kept in place for an additional 5 min before being slowly retracted. Four weeks after the lesion, the rats were tested for rotational behavior in automated rotameter bowls (Panlab) following an injection of apomorphine (0.2 mg/ml; Sigma) and 1 week later with D-amphetamine sulfate (5 mg/kg, intraperitoneally (Sigma); Rotational scores were collected every 2 min for 60 min for D-amphetamine test and 40 min for apomorphine test in a computer-assisted rotameter system (Panlab). Only rats exhibiting 5 or more ipsilateral rotations/min after D-amphetamine injection, and at least 4 contralateral rotations/min in response to apomorphine injection were selected for further unlabeled and MNPs-labeled hVM cell transplantation studies, performed under the same conditions as described above for intact rat brains, but transplanting cells into the right 6-OHDA lesioned striatum . The animals were immunosuppressed with daily intraperitoneal cyclosporin A injection (15 mg/kg; Novartis), starting 2 days before transplantation and throughout the experiment.

For *in vivo* studies, we analyzed the animals at 48 h and from 2 weeks to 5 months following transplantation.

Magnetic Resonance Imaging (MRI)
Magnetic resonance imaging (MRI) was performed in the C.A.I. Nuclear Magnetic Resonance and Electron Spin Centre at Complutense University of Madrid. We used the Biospec BMT 47/40 (Bruker, Ettlingen, Germany), which operates at 4.7 Tesla and is equipped with a gradient shield system active of 12 cm. Rats were anesthetized with a mixture of oxygen and isoflurane. Once anesthesized the animals were placed in prone position on a plate of 7 cm in length, head immobilized and connected to a radiofrequency probe to monitor their cardiac and respiratory frequency, and temperature of the animal during image acquisition. First global parameters were determined for centering and optimal collection of images in T2 *. Then, three images of spin echo were run in axial, sagittal, and coronal orientation (TR / TE = 200/10 ms, matrix = 128x128). Transplanted cells labeled with MNPs were visualized performing a gradient-echo sequence using the

following parameters: TR = 250 ms, TE = 10 ms, rotation angle = 30 °, thickness of section = 1 mm, number of slices = 8, mean number = 6, FOV = 3x3 cm 2, matrix = 256x192. The reconstructed matrix size was 256 x 256. The acquisition time for these experiments was 4 minutes and 48 seconds. The images were subsequently analyzed with the ImageJ program.

Immunohistochemistry

At the end of the experiments, the animals were anesthetized with an overdose of chloral hydrate and intracardially perfused with freshly prepared, buffered 4% paraformaldehyde (in 0.1 M phosphate buffer, pH 7.4). Brains were removed, postfixed for 12 h in the same fixative at 4°C, and dehydrated in 30% sucrose solution at 4°C until sunk. Eleven 30 μm-thick coronal sections were collected using a freezing microtome. Serial sections were used for immunohistochemistry with polyclonal antibodies against human Nestin (1:1000; Abcam), Ki67 (1:500, Neomarkers), Doublecortin (Dcx) (1:1000; Santa Cruz Biotechnology) and monoclonal antibodies against human GFAP (1:1000; Sternberger), human nuclei (hNu) (1:100; Chemicon). Briefly, free-floating sections were incubated overnight at 4°C with the primary antibodies diluted in PBS with 2% nonspecific serum. Sections were rinsed three times in PBS for a total time of 1 h and then incubated for 2 h with the secondary antibodies in PBS (Cy5-conjugated Ab (anti-mouse or anti-rabbit (1:500)), Cy3-conjugated Ab (antimouse or anti-rabbit (1:200), all from Jackson Immunoresearch), and mounted onto polylysine pre-treated glass slides (Menzel-Glaser). The slides were dried and coverslipped with Mowiol.

For staining with DAB, first endogenous tissue peroxidase activity was quenched using a 10% solution of methanol and 3% H_2O_2 in PBS for 20 minutes. Subsequently, brain sections were washed with PBS and incubated with blocking solution for 1 h. The sections were then incubated with primary antibody anti OX42 (1:1000, monoclonal, Chemicon) dissolved in a solution of PBS, 0.25% Triton X-100 and 1% horse serum overnight at 4°C. After rinsing, sections were incubated for 2 h with biotinylated secondary antibody (anti-mouse BA2001, dissolved 1:200, Vector) at room temperature. Then sections were incubated with a complex of avidin-biotin-peroxidase (ABC, Vector) and developed with the chromogen 1,3 - diaminobenzidine (DAB, dissolved in 0.05% PBS, 8% NiCl 2 and 0.03% H_2O_2 Sigma). Sections were washed in PBS and distilled H_2O and mounted on slides treated with polylysine, dried at room temperature over night, dehydrated by an increasing gradient of ethanol (50% -90% -100%) and delipidatted with xylene. Finally, slides were coverslipped with DPX (BDH).

Analyses and photography of fluorescent or DAB stained samples were carried out in an inverted Zeiss Axiovert 135 (Oberkochen, Germany), LeicaDMIRB microscope equipped with a digital camera Leica DC100 (Nussloch, Germany) or LSM510 laser confocal Microscope coupled to a Axiovert200 (Zeiss) inverted microscope.

Results

Internalization of MNPs by VM hNSCs: cell viability, and MNPs size and concentration effects

Different MNPs were compared to select the optimal one to magnetically label VM hNSCs. We tested iron MNPs (Fe_3O_4) with different core diameters: 50, 100 and 250 nm, all of them coated with dextran. The influence of different concentrations of MNPs on cell viability was studied at incubation times of 72 h. We also analyzed the effect of the transfection agents, PS and PL, in the case of the lowest concentration of 50 nm MNPs (50 μg/ml), to determine whether the transfection agents could avoid the use of higher doses of NPMs.

The highest concentrations of nanoparticles tested (100 and 300 μg/ml) produced a significant decrease in cell viability, as evaluated by MTT assay (Figure 1A). Moreover, 50 nm MNPs at 50 μg/ml in the presence of PL produced a significant decrease in cell viability (Figure 1B). In addition to the toxicity observed in the MTT assay, the highest concentrations of particles tested (100 and 300 μg/ml) induced a decrease in the number of cells per field (phalloidin staining), and the presence of numerous apoptotic nuclei stained with Hoechst (Figure 1C).

Thus, labeling of the cells at doses of 50 μg/ml and incubation times of 72 hours did not affect cell survival, evaluated by MTT assay, and resulted in 80 to 90% labeling efficiency (see below). This is valid for all sizes of nanoparticles tested (Figure 1A). These results led us to conclude that labeling hNSCs with MNPs did not significantly affect cell viability, using concentrations of 50 μg/ml for 72 h. Based on these results, we decided to use 50 μg/ml of MNPs for subsequent studies.

Effects of the incubation time and transfection agents used on the internalization of MNPs by VM hNSCs.

It has been previously reported that cationic transfection agents might improve the efficiency of MNPs intracellular uptake [19,26-29], especially in non-phagocytic cells. Thus, the ability of cationic compounds such as PL and PS might increase the capacity of VM hNSCs to internalize MNPs.

MNPs uptake was assessed by quantifying the number of cells that displayed fluorescent intracellular labeling of MNPs, as described in the methods section (see also Figure 1C). hVM cells were incubated for different times, ranging from 3 to 72 h, with 50 μg/ml of 50, 100 and 250 nm MNPs, since we had previously demonstrated that this dose did not affect cell viability in most

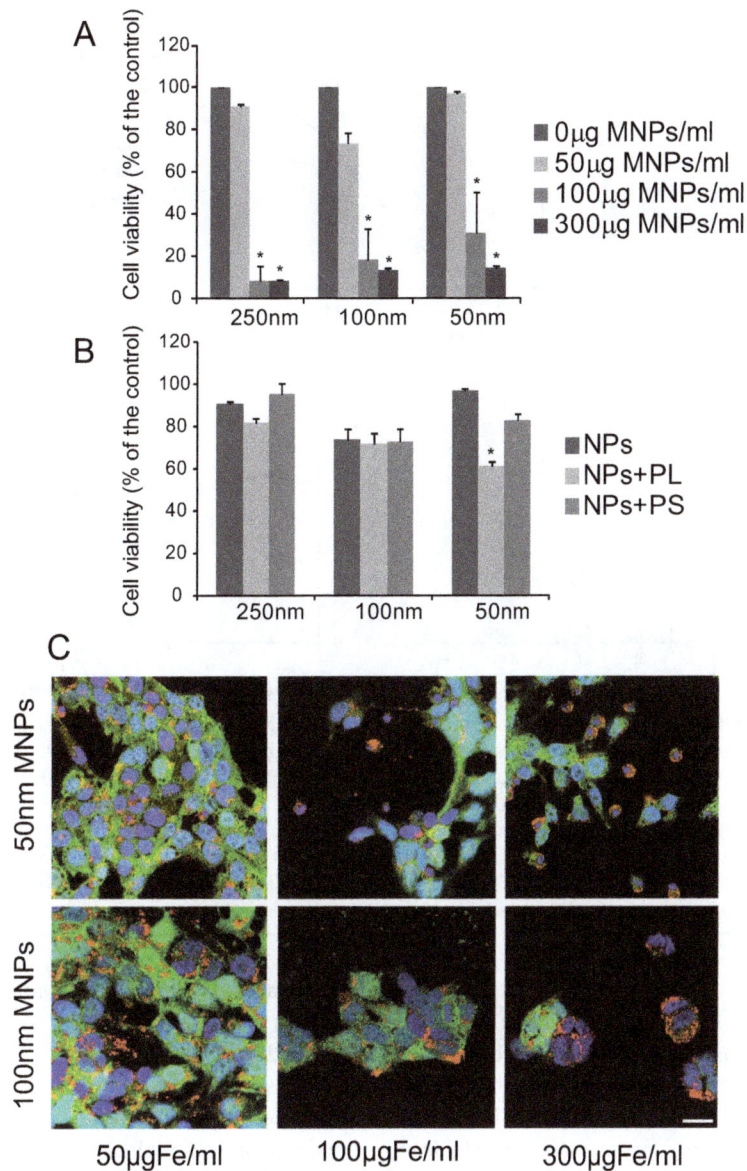

Figure 1 Cell viability in the presence of MNPs. A. hVM cells were incubated in the presence of 250, 100 and 50 nm in diameter MNPs for 72 h at increasing concentrations, ranging from 50 to 300 µg/ml. Doses up to 50 µg/ml did not significantly affect cell viability, as evaluated by MTT assay. **B** hVM cell viability (by MTT assay), after 72 h in the presence of 250, 100 and 50 nm MNPs at 50 µg/ml using poly-L-Lysine (PL) and protamine sulfate (PS) as transfection agents. Only PL produced a significant decrease in cell viability. **C** hVM-MNPs-Cy3 treated cells were fixed after 72 h and stained with Phalloidin A488. Nuclei were counterstained with Hoechst. Scale bar 20µm. Data represent mean +/- S.E. ($n = 4$). (*p <0.05, ANOVA, post hoc Tukey's test; * *versus* hVM cells treated with 0 µg/ml MNPs).

conditions (Figure 1A). The results in Figure 2A show a positive correlation between the uptake of MNPs and the incubation times for all types of MNPs tested. In the case of 50 nm MNPs coated with PL, uptake was maximum at all incubation times studied, indicating that both the size of the particle and the presence of the policationic agent PL significantly influence the time needed to label nearly 100% of the cells. PS had no effect for the 50 nm particles. The largest MNPs tested, 250 nm, were the ones uptaken most slowly by hVM cells, even when

PS or PL were added as transfection agents (Figure 2A). To demonstrate an efficient labeling of hVM cells with MNPs, a semi-quantitative assessment (Image J, NIH) of 100 nm MNPs fluorescence signal was performed in hVM cells comparing this signal to a well established SPIO uptake by a common cell line (COS cells) [30] indicating that the uptake of MNPs by hVM cells was similar to that observed in COS cells (Figure 2B).

MNPs with a diameter of 50 and 100 nm were the most appropriate MNPs to label hVM cells at short

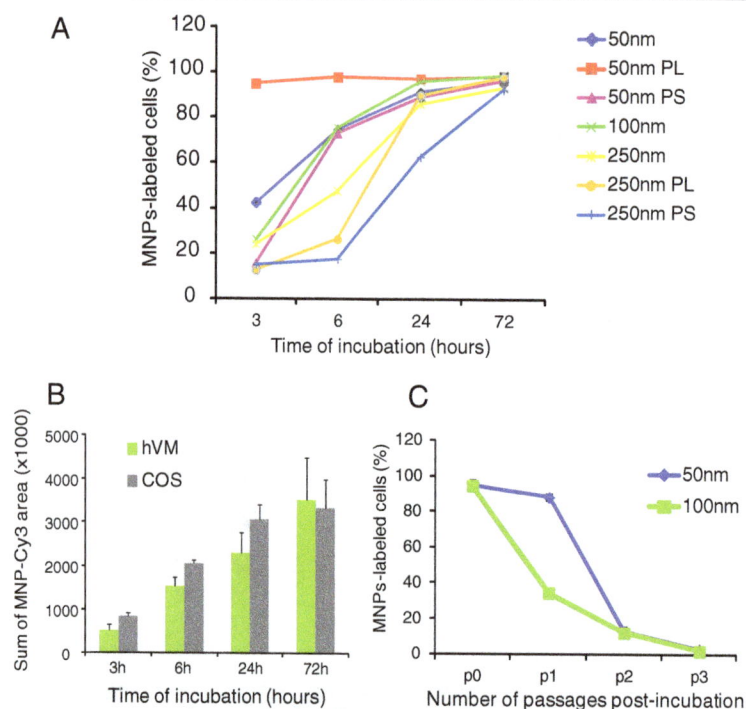

Figure 2 **Incorporation of MNPs by hVM cells and persistence of the MNPs load over time. A.** The incorporation of 50 μg/ml MNPs-Cy3 of 50, 100 and 250 nm in diameter by hVM cells was evaluated as percentage of Cy-3 labeled cells at 3, 6, 24 and 72 h. The effect of poly-L-Lysine (PL) and protamine sulfate (PS) was also evaluated in the uptake of 50 nm and 250 nm MNPs. Incubation times up to 72 h resulted in nearly 100% labeling efficiency even in the absence of PL or PS. **B.** hVM and COS cells were incubated with 100 nm MNP-Cy3 at 50 μg/ml for 3, 6, 24 and 72 hours. The sum of MNP-Cy3 area represents the total area occupied by nanoparticles-Cy3, measured by Image J (NIH) in units of pixels, corrected by the number of cells present in each field. **C.** hVM cells were labeled with 50 and 100 nm MNPs (50 μg/ml) for 72 h and serially subcultured to test the persistence of MNPs labeling after passages. The decreased numbers of MNP-labeled cells observed at progressive subculturing passages indicated that both 50 and 100 nm MNPs are progressively diluted over time.

times. Nevertheless, after a 72 h incubation period, more than 90% of the cells were labeled with MNPs in all conditions tested. The increase in the percentage of labeled cells provided by transfection agents can also be achieved increasing the incubation time with the MNPs. Thus, since these chemical agents induce a decrease in cell viability, especially PL in the case of 50 nm MNPs (Figure 1A), toxicity may be avoided by lengthening the incubation times with the MNPs.

Persistance of the MNPs load over time

To determine whether the MNPs were retained by hVM cells after passages, cells were incubated with 50 μg/ml of MNPs for 3 days (p0) and then passaged. At each subculturing passage, a fraction of the cells was plated separately and evaluated for the percentage of cells retaining MNP labeling. Dextran staining confirmed a labeling efficiency over 95% at passage 0 (Figure 2C). A progressive drop in the number of dextran-labeled cells with passages was observed, indicating that intracellular MNPs (both 50 and 100 nm sizes), were progressively diluted over time, likely due to cell proliferation. The labeling rate was similar for both nanoparticles 50 and 100 nm after the 2nd and

3rd passages. However, the 50 nm MNPS were retained by the cells much better than the 100nM ones after passage 1 (Figure 2C).

Effects of MNPs on cell cycle

Analysis of cell replication was determined by fluorescence labeling of the nuclei of the cells in suspension using propidium iodide, and then analyzing the fluorescence properties of each cell in the population by flow cytometry [25]. This study was conducted to determine whether the presence of MNPs was compatible with a normal cell cycle progression in hVM cells. To this end, the cells were exposed to MNPs for 24 and 48 h, and the distribution of cells in the different cell cycle phases was determined.

The percentage of cells in G0-G1, S and G2-M phases was assessed in both control and labeled cells (incubated with 50 μg/ml of 100 nm MNPs). Analysis of the DNA content between samples showed an identical distribution of cells in the different cell cycle phases at both time points (Figure 3). Therefore, MNP labeling of VM hNSCs does not alter cell cycle dynamics. The sub G0-G1 fraction (fragmented DNA) was also quantified to

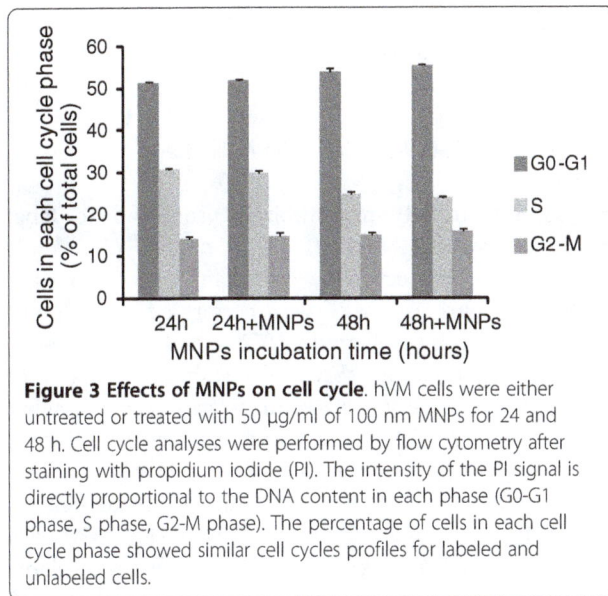

Figure 3 Effects of MNPs on cell cycle. hVM cells were either untreated or treated with 50 µg/ml of 100 nm MNPs for 24 and 48 h. Cell cycle analyses were performed by flow cytometry after staining with propidium iodide (PI). The intensity of the PI signal is directly proportional to the DNA content in each phase (G0-G1 phase, S phase, G2-M phase). The percentage of cells in each cell cycle phase showed similar cell cycles profiles for labeled and unlabeled cells.

assess cell death by flow cytometry. No cells were detected in subG0 phase in both conditions indicating that the addition of MNPs did not induce cell death, measured as DNA fragmentation (data not shown), consistent with the results of the MTT assay (Figure 1).

Effects of MNPs on cell stemness and differentiation

The expression of nestin, a typical marker of neural stem cells, was analyzed in hVM cells labeled with MNPs of different sizes (50 and 100 nm) at 50 µg/ml for 72 h. The percentage of nestin-positive cells in MNPs-labeled hVM cells and in control cells was quantified, and found to be the same in all the studied cultures (Figure 4A). Therefore, MNPs do not induce differentiation themselves.

To assess the differentiation potential of MNPs-labeled hVM cells, after labeling the cultures were subsequently differentiated for 7 days and stained for β-III-tubulin, tyrosine hydroxylase (TH) and glial fibrillary acid protein (GFAP), and counterstained with Hoechst (DNA stain). The percentage of total cells expressing markers characteristic of differentiated VM hNSCs, like β-III-tubulin, GFAP and TH, was compared among samples (Figure 4B, C). No significant differences between control and MNPs-labeled cells were observed, demonstrating that MNPs do not affect the ability of hVM cells to differentiate into glial cells, neurons and, specifically, dopaminergic neurons, obtaining similar results for TH- and β-III-tubulin-positive cells (17% and 25% respectively) to those previously described in [7]. Thus, these results demonstrate that labeling VM hNSCs cells with optimal doses of MNPs does not affect their stemness and differentiation potential.

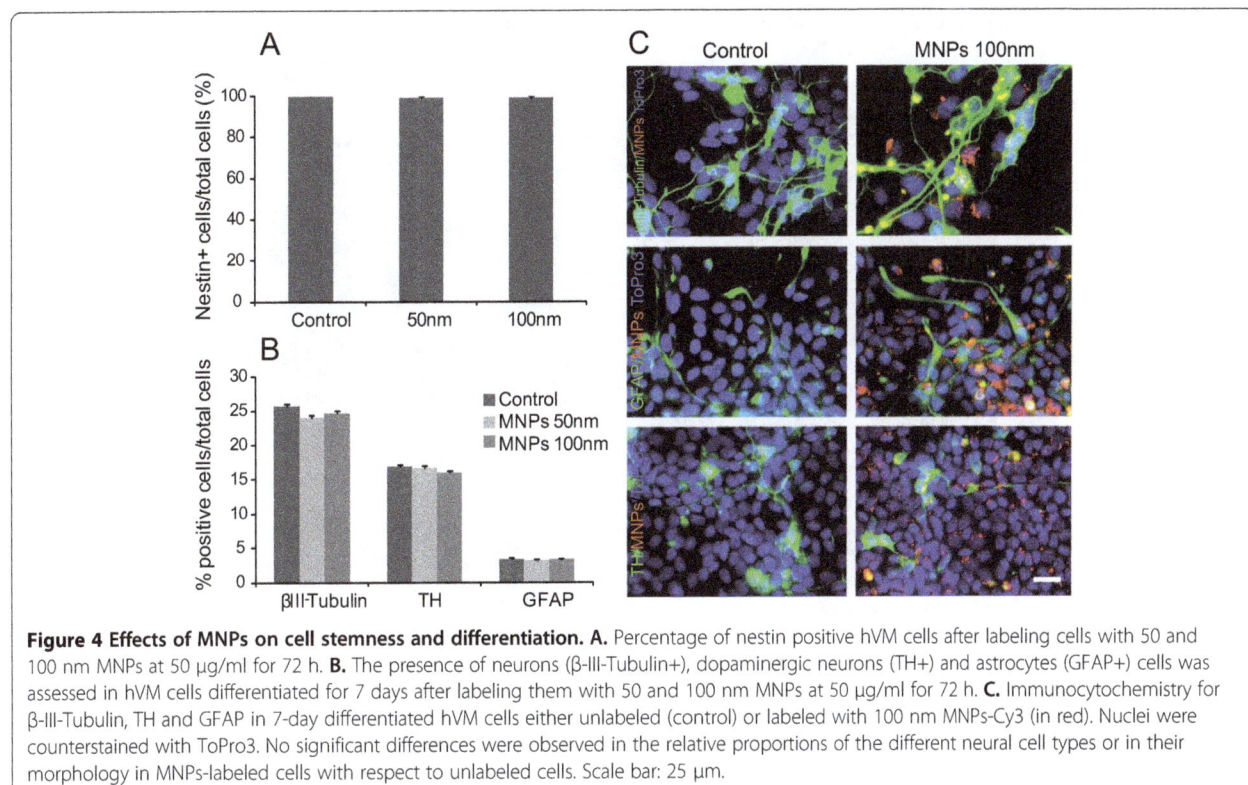

Figure 4 Effects of MNPs on cell stemness and differentiation. A. Percentage of nestin positive hVM cells after labeling cells with 50 and 100 nm MNPs at 50 µg/ml for 72 h. **B.** The presence of neurons (β-III-Tubulin+), dopaminergic neurons (TH+) and astrocytes (GFAP+) cells was assessed in hVM cells differentiated for 7 days after labeling them with 50 and 100 nm MNPs at 50 µg/ml for 72 h. **C.** Immunocytochemistry for β-III-Tubulin, TH and GFAP in 7-day differentiated hVM cells either unlabeled (control) or labeled with 100 nm MNPs-Cy3 (in red). Nuclei were counterstained with ToPro3. No significant differences were observed in the relative proportions of the different neural cell types or in their morphology in MNPs-labeled cells with respect to unlabeled cells. Scale bar: 25 µm.

Intracellular localization of MNPs in VM hNSCs.

To study the subcellular distribution of MNPs uptaken by VM hNSCs, the cells were incubated with 50 and 100 nm MNPs at 50 µg/ml for 72 h, and differentiated afterwards. After that time, proliferation medium was changed to differentiation medium for 4 days. Staining with an anti-dextran antibody or fluorophores directly coupled to MNPs, allowed us to track the intracellular location of the MNPs. To unambiguously determine if the detected NMPs were adhered to the plasma membrane of they were internalized/compartimentalized, series of images were taken along the XZ and XY planes under a confocal microscopy. The orthogonal projections in Figure 5A-B demonstrate that 50 and 100 nm MNPs did not remain superficially attached to cell membranes; on the contrary they were endocytosed by hVM cells.

To determine the final intracellular localization of the MNPs, different fluorescent antibodies, specific for early endosomes (antiEEA1), Golgi apparatus (manoxidase) and lysosomes (CD63) were used. Staining of the endosomes demonstrated that the green signals, labeling endosomes, and the MNPs, labeled in red, are located within the cytosol but do not co-localize (Figure 5C). Golgi apparatus (Figure 5D) and lysosomes (Figure 5E) both stained in green, do not co-localize with MNPs-Cy3 either.

These results indicate that MNPs are internalized by hVM cells labeling the cytosolic compartment, without being sequestered by endosomes, lysosomes or other subcellular organelles, as identified with the markers used in the present assays. In conclusion, the intracellular localization of MNPs in VM hNSCs makes them well suited for MRI analyses.

MRI analysis of MNP-labeled hVM cells after transplantation

Having established an optimized procedure to label hVM cells with MNPs (50 µg/ml of 100 nm MNPs for 72 h) we proceeded to transplant them into right striatum (Figure 6). MRI was performed at different times after cell transplantation (48 h, 2, 4 and 8 weeks). The same number of hVM cells without MNPs were transplanted in the left striatum as an internal imaging control. Strong MRI contrast was observed in the right striatum. No MRI signal was detected when similar numbers of unlabeled cells were injected in the contra-lateral striatum (Figure 6). The MNPs-labeled hVM cells can be detected from 48 h to 8 weeks after transplant-ation (Figure 6). 48 h after transplantation, labeled hVM cells are clearly visualized by MRI as a large hypointense signal in the area of the transplant. Although the inten-sity and size of the magnetic resonance signal decreases slightly during the time periods studied, it is still clearly visible even at 8 weeks after transplantation (Figure 6D). The slight decrease in MRI signal over time might be due to the migration of a small percentage of hVM cells through the corpus callosum that reach this area follow-ing the injection tract (arrow in Figure 6C). This observ-able fact demonstrates that the technique is also valid for tracking small migrating cell populations.

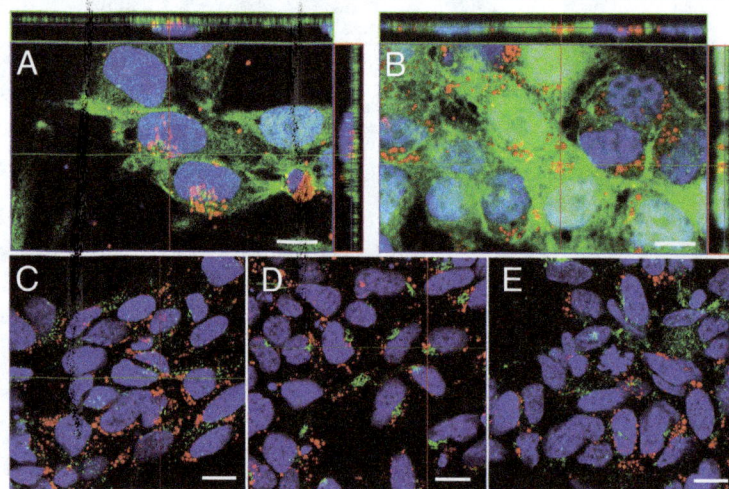

Figure 5 Intracellular location of MNPs on hVM cells. A-B. hVM cells were incubated with 50 nm **(A)** and 100 nm MNPs-Cy3 **(B)** at 50 µg/ml for 72 h. After fixation, cells were stained with Phalloidin A488 (green). MNPs in A were detected by staining with an anti-dextran antibody (red). Orthogonal projections are shown in A and B, confirming the colocalization of MNPs and phalloidin. **C-E.** To determine the final intracellular localization of MNPs-Cy3, hVM cells were incubated with 100 nm MNPs-Cy3 at 50 µg/ml for 72 h, antibodies anti-EEA1 specific for early endosomes **(C)**, anti-mannosidase II specific for Golgi apparatus **(D)** and anti-CD63 specific for lysosomes **(E)**, were used. Nuclei were counterstained with ToPro3. Scale bars: 10 µm in A-B and 15 µm in C-E.

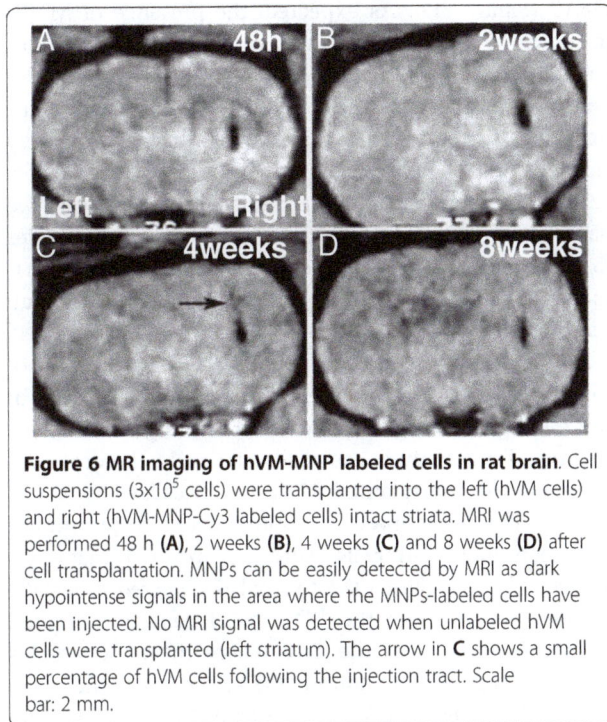

Figure 6 MR imaging of hVM-MNP labeled cells in rat brain. Cell suspensions (3×10^5 cells) were transplanted into the left (hVM cells) and right (hVM-MNP-Cy3 labeled cells) intact striata. MRI was performed 48 h (**A**), 2 weeks (**B**), 4 weeks (**C**) and 8 weeks (**D**) after cell transplantation. MNPs can be easily detected by MRI as dark hypointense signals in the area where the MNPs-labeled cells have been injected. No MRI signal was detected when unlabeled hVM cells were transplanted (left striatum). The arrow in **C** shows a small percentage of hVM cells following the injection tract. Scale bar: 2 mm.

Thus, MRI analysis presents a high spatial resolution and the advantage of visualizing transplanted cells within their anatomical surroundings, which is crucial to check the correct position of the cells after transplantation and the visualization of the cell migration processes.

Histological analysis of MNPs-hVM cells after transplantation

The localization of MNPs inside the cells after transplantation was performed using confocal fluorescence microscopy. One week after MNP-hVM cells transplantation in hemiparkinsonian rats, MNPs were identified by MRI scans previous to sacrifice (Figure 7A). Red fluorescent MNPs were evenly distributed throughout the transplant region (Figure 7B). Immunocytofluorescence to detect hVM cells was carried out using an antibody anti-human nucleus (Figure 7C). One week after transplantation, hVM cells present an immature phenotype, as evidenced by the fact that most of these cells are still nestin-positive (Figure 7D). Furthermore, a small percentage of hVM cells are Ki67-positive (Figure 7C), indicating that one week after transplantation some hVM cells are still dividing.

As it has been previously described, two months after transplantation the location of MRI signal is the same to that obtained few hours after transplantation, although there is a slight decrease in signal size and intensity (Figure 6A and D). Analyses by fluorescence microscopy, showed that 2 months after transplantation, the

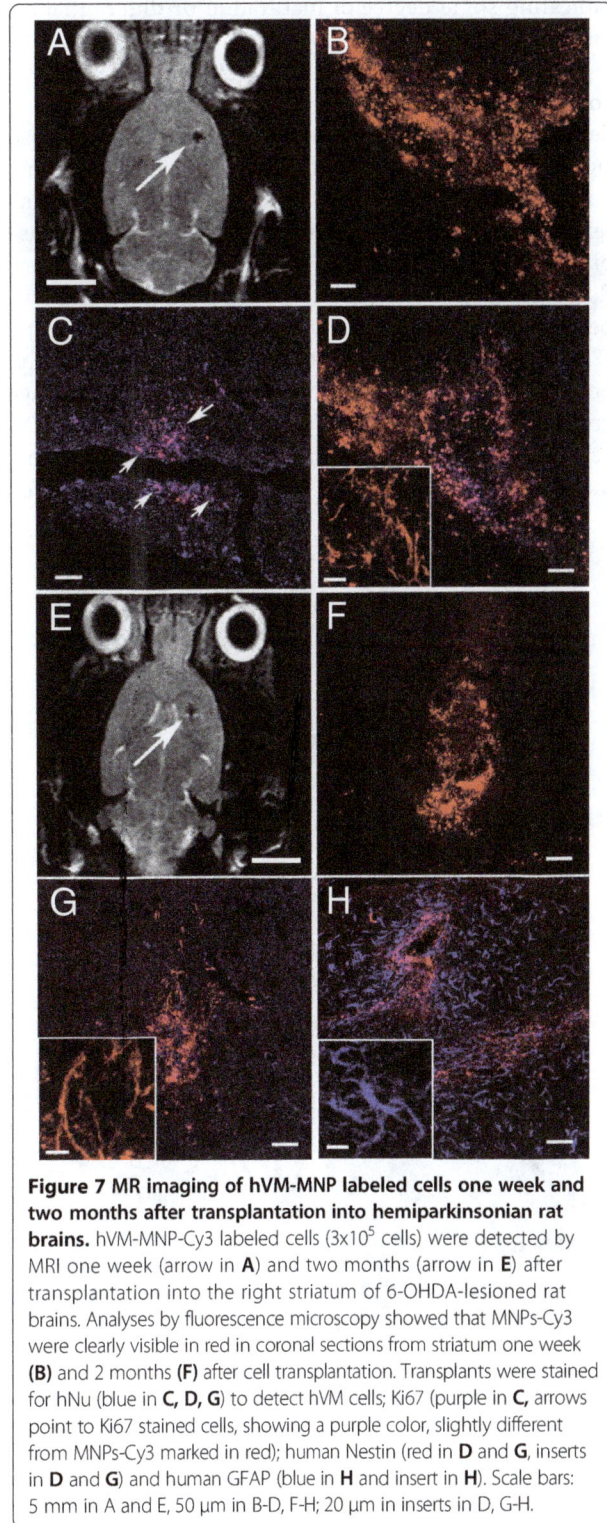

Figure 7 MR imaging of hVM-MNP labeled cells one week and two months after transplantation into hemiparkinsonian rat brains. hVM-MNP-Cy3 labeled cells (3×10^5 cells) were detected by MRI one week (arrow in **A**) and two months (arrow in **E**) after transplantation into the right striatum of 6-OHDA-lesioned rat brains. Analyses by fluorescence microscopy showed that MNPs-Cy3 were clearly visible in red in coronal sections from striatum one week (**B**) and 2 months (**F**) after cell transplantation. Transplants were stained for hNu (blue in **C, D, G**) to detect hVM cells; Ki67 (purple in **C,** arrows point to Ki67 stained cells, showing a purple color, slightly different from MNPs-Cy3 marked in red); human Nestin (red in **D** and **G**, inserts in **D** and **G**) and human GFAP (blue in **H** and insert in **H**). Scale bars: 5 mm in A and E, 50 μm in B-D, F-H; 20 μm in inserts in D, G-H.

MNPs were aggregated around the limits of the transplant zone, with less intense signal in the graft core (Figure 7F). At this time some hVM cells are still nestin positive (Figure 7G), but some GFAP positive cells are present in the area of the transplant (Figure 7H), indicating some degree of maturation or differentiation in transplanted hVM cells.

Long term MRI analysis

5 months after cell transplantation, a clear hypointense signal at the site of the transplantion is still visible (arrow in Figure 8 A). This signal is quite similar in intensity and size, to that obtained at 2 months after transplantation (Figures 6D and 7E). MNPs maintained their fluorescence five months after transplantation and can be detected in brain sections by fluorescence microscopy (Figure 8B). However, MNPs were found mainly in the boundaries of the transplantation area forming large aggregates outside the cells (arrows in Figure 8B). hVM cells identified by hNu immunoreactivity were predominantly positive for nestin (Figure 8C and insert) and for

GFAP (Figure 8D). As expected, the presence of MNPs inside hVM cells was detected, in both nestin (data not shown) and GFAP (Figure 8E). Neurons were not detected in the present experiment. It is important to highlight at this point that human neurons mature very slowly, even when fresh VM tissue is used (over half a year) [31-33]. In fact in other studies using hVM cells we have only observed immature neuronal morphologies [5]. Currently, we are conducting a one-year long experiment to allow the transplant enough time for full maturation.

In the striatum, and especially surrounding the area of the transplant, we can observe the presence of cells of the microglial lineage stained by OX42 (Figure 8F). This microglial reactivity is especially strong in the area surrounding the transplanted MNPs-hVM cells. This OX42 labeling is not observed in brain sections from animals in which unlabeled cells were transplanted (data not shown).The strongest OX42 signal is detected especially in the boundaries of the MNP-hVM cells transplanted area, coinciding with the area where the nanoparticles are more abundantly found. OX42 positive microglial cells co-localized with the signal from MNPs, suggesting that MNPs are removed from the brain parenchyma by microglial cells.

Discussion

The efficient application of stem cells for the treatment of neurodegenerative diseases requires safe cell tracking to follow stem cell fate over time in the host after transplantation. The present study illustrates the use of commercial nanoparticles to label human neural stem cells without using any transfection agents, for long term tracking by MRI after their transplantation.

In general, most nonphagocytic cells do not take up MNPs efficiently or require cells to be exposed to high amounts of iron in culture [21,22,34-36]. Most magnetic nanoparticles require transfection agents for adequate internalization by stem cells, although it has been reported that Feridex complexed with PL blocked the differentiation of human MSCs into chondrocytes [37]. Only very small superparamagnetic iron oxide particles with a core diameter of 5 nm have been used to label human mesencephalic neural precursor cells [14]. Our data show that nearly 100% of hVM cells were labeled with MNPs of different sizes (50 and 100 nm) in the absence of any transfection agents. These results are similar to those obtained in human NSCs using PS [38] or PL [23,38] as transfecting agents. The present method showed very clean labeling with minimum extracellular iron nanoparticles. These results indicate that with relatively low concentrations of iron (ie, 50 μg of iron per milliliter) in culture media, an almost complete cellular labeling with MNPs can be achieved.

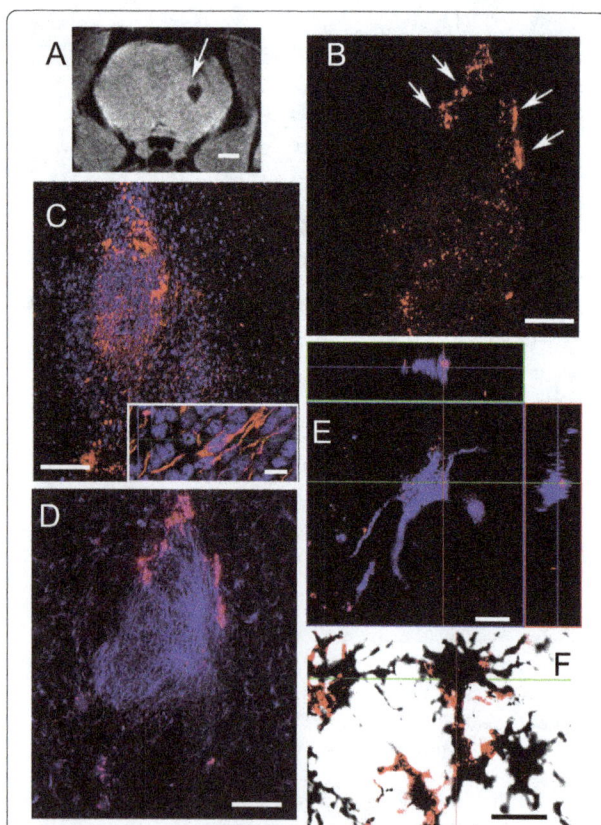

Figure 8 MR imaging of hVM-MNP labeled cells 5 months after transplantation into hemiparkinsonian rat brains. hVM-MNP-Cy3 labeled cells (3x10^5 cells) were detected by MRI 5 months (arrow in **A**) after transplantation into the right striatum of 6OHDA-lesioned rat brains. Analyses by fluorescence microscopy showed that MNPs-Cy3 were clearly visible in red in coronal sections from striatum 5 months after cell transplantation (**B, D**). Transplants were stained for hNu (blue in **C**) to detect hVM cells; Nestin (red in **C** and insert in **C**); human GFAP (blue in **D** and **E**, orthogonal projections are shown in **E**, confirming the colocalization of GFAP and MNPs-Cy3) and OX42 (stained with DAB) combined with red fluorescence to visualize MNPs-Cy3 (**F**). Scale bars: 2 mm in A, 100 μm in B-D; 10 μm in insert in C, E-F.

The distribution of cells in the different phases of the cell cycle of MNPs-labeled hVM cells was unaffected, coinciding with findings from other studies, in which MNPs were used to label embryonic stem cells [39], hematopoietic progenitor cells [19] and bone marrow stromal cells [40]. Thus, this procedure that avoids the use of transfecting agents can be considered safe for cell labeling.

We demonstrate the presence of MNPs in the hVM cytosol indicating that MNPs are incorporated by hVM cells via endocytosis. Uptake of 50 nm, 100 nm and 250 nm MNPs at 50 µg/ml occurred in almost 100% of hVM cells, after an incubation period of 72 h without affecting hVM cells survival rate, as evaluated by MTT assay. When using higher concentrations of MNPs (100-300 µg/ml) a significant decrease in cell viability was observed. These results are in agreement with other observations on MNPs-labeling of human NPCs that had no adverse effects on cell survival [23]. However, functional tests in rats transplanted with MNPs-labeled hVM cells must be performed to exclude adverse effect of the MNP labeling on functional outcome of the transplanted rats. Also, before thinking in a future clinical translation, oxidative damage as the one described in mesenchymal stem cells by MNPs [41] should be evaluated, even in the absence of effects of cell proliferation, cell viability and cell death as demonstrated in the present study with hNSCs.

Loss of MNPs over time

Intracellular MNPs were progressively diluted and almost completely undetectable by immunocytochemistry after 3 subculturing passages. Predicted dilution of MNPs inside hVM cells was most likely due to cell division. Loss of MNPs due to cell proliferation or differentiation has previously been reported for different cell types labeled with different types of MNPs [42-45]. This fact may hinder the reliable long term cell tracking by MRI after cell transplantation [46], in case the cells would have a large replication potential *in vivo*.

MNPs-labeled hVM cells maintain their stemness and differentiation potential

The feasibility of cell-based therapies depends upon the ability of transplanted cells to maintain their proposed therapeutic functions *in vivo*. Previous studies have revealed that hVM cells differentiate into a high percentage of dopaminergic neurons [8,9]; this characteristic makes them particularly attractive for cell replacement therapy in Parkinson's disease. It has been also demonstrated that most stem cells labeled with MNPs retain their regenerative and therapeutic potential in vivo [38,47]. In order to assess that the optimized labeling protocols did not affect the stemness or differentiation potential of hVM cells, we analyzed the presence of nestin, a typical marker of neural stem cells, in cells treated with MNPs compared with control unlabeled cells. As described in the Results section, the labeling of hVM cells with MNPs did not affect lineage commitment during the differentiation period of hVM cells. We have analyzed the percentage of TH$^\pm$ neurons generated after 7 days of differentiation. Labeling hVM cells with MNPs did not affect the percentage of TH$^+$ cells obtained during the differentiation process. We did not observe any significant changes in the percentage of nestin, GFAP or β-III-tubulin hVM cells compared to MNPs-hVM cells, after 7 days of differentiation. These results confirm previous data reported in human [14,23,48] and mouse [12] neural precursor cells labeled with ferumoxides. However, it has been observed that MNPs could cause morphological changes and alterations in the differentiation pattern of human adult mesenchymal stem cells [37,49] but not in mouse mesenchymal stem cells [50]. In the present study with hNSCs, we did not observe any morphological alteration of the labeled and differentiated cells in vitro. On the contrary to these findings, others have demonstrated an increase in GFAP expression and changes in cell morphology when treating hNPCs with Au- and Ag-nanoparticles [51].

Using confocal microscopy, the nanoparticles did not co-localize with any of the markers of the subcellular organelles analyzed (endosomes, lysosomes, endoplasmic reticulum or Golgi apparatus). This suggests that these nanoparticles can be freely distributed in the cytoplasm and eventually enter the normal iron metabolic pathways [26]. However, further electron microscopy analysis must be carried out to unequivocally determine the exact subcellular localization of the nanoparticles.

MRI and hVM cells

Previous studies have shown that magnetically labeled cells maintained their contrast for 6 weeks [44], 1 month [47] and even 3 months after transplantation [14], suggesting that labeling cells with MNPs can be a suitable method for cell tracking by MRI over time. However, long term analysis after cell transplantation might be deceiving, especially if transplanted cells continue dividing in the host brain [52]. As shown here, MNPs-labeled cells show a gradual reduction in intracellular iron particles after 2 or 3 passages in cell cultures. Therefore, dilution but also the loss of MNPs from transplanted cells to the host brain parenchyma might occur during cell division. MNPs released to the host parenchyma during cell division, exocytosis processes or even after cell death, can contribute to the MR signal.

In short term MRI analyses (1 week and 2 month after cell transplantation) we observed an excellent agreement between the areas of MR contrast enhancement and histological staining for nanoparticles. Therefore, we can conclude that this technique could also be valid to assess the precise position of the grafted cells shortly after transplantation.

However, long term MRI analyses are needed to perform because transplanted cells are expected to slowly mature into the correct phenotypes and participate to some extent in tissue reconstruction (over half a year). Our results showed that there were still some nestin positive hVM cells at 5 months post-transplantation. Although at that time we found numerous GFAP positive cells, grafted cells mainly express progenitor markers at short times after transplantation [45]. The ability of human embryonic stem cell neurons to provide extensive reinnervation of the host striatum and begin to restore DAergic neurotransmission have been described at least 6 months post-transplantation [53]. This is in line with previous clinical observations using human fetal VM grafted to patients, where progressive recovery of DA neurotransmission starts at 6 months, reflecting the gradual maturation of the transplanted cells [54]. Similar results have been recently described when using human pluripotent stem cells (hPSCs). Functional maturation of hPSC-derived neurons requires an extended timeline of up to seven months, mimicking endogenous human neural development [55].

Although some studies found no co-localization of mouse macrophages and iron-containing areas of brain tissue [56] this event has been described previously in a spinal cord injury model transplanted with Endorem-labeled mesenchymal stem cells [57]. In our case, long term MRI analyses followed by histological analyses (up to 5 months after cell transplantation) showed that MNPs can be found in the brain parenchyma co-localizing with areas rich in markers of reactive microglia, like OX42. This fact is a cause of concern because of the possible false positive interpretation of the MRI signal, which may be produced by macrophages that have engulfed some non-viable labeled NSCs or freely-dispersed iron nanoparticles in the brain tissue [58].

Conclusion

NSCs are one of the most attractive cells sources for stem cell therapies of neurodegenerative diseases. For therapeutic application, transplanted cells need to be tracked both spatially and temporally in the living brain, in order to assess their migration and survival in the host tissue. Therefore, for the application of these techniques *in vivo*, additional live imaging tools such as MRI are necessary. The procedures described here are suitable for studying the *in vivo* localization and migration of grafted human neural stem cells in longitudinal studies. We have shown that labeling of hVM cells with MNPs does not affect their overall biology, survival rate

and *in vitro* differentiation. These findings support the use of commercial MNPs to track cells for short- and mid-term periods after transplantation for studies of brain cell replacement therapy. Nevertheless, long-term MR images should be interpreted with caution due to the possibility that some MNPs may be expelled from the transplanted cells and internalized by host microglial cells.

Competing interests
The authors declare that they have no competing interests.

Authors' contributions
MRG and AMS conceived of the study, participated in its design and coordination and drafted the manuscript. EGS carried out all experimental studies and helped to draft the manuscript. All authors read and approved the final manuscript.

Acknowledgements
The authors would like to thank the expert technical assistance of Beatriz Moreno Moreno and Beatriz García Martínez at the CBMSO lab. This work was supported by grants from (to AM-S): Spanish Ministry of Economy and Competitiveness (SAF2010-17167), Comunidad Autónoma Madrid (S2011-BMD-2336), Instituto Salud Carlos III (RETICS TerCel, RD12/0019/0013). This work was also supported by an institutional grant from Fundación Ramón Areces to the Center of Molecular Biology Severo Ochoa. The authors gratefully acknowledge the financial support of the Reina Sofia Foundation and Comunidad Autónoma Madrid (S2010-BMD-2460) to MR-G.

Author details
[1]Centre for Biomedical Technology, Polytechnic University of Madrid, 28223 Madrid, Spain. [2]Biomedical Research Networking Center in Bioengineering Biomaterials and Nanomedicine (CIBER-BBN), Madrid, Spain. [3]Department of Molecular Biology and Center of Molecular Biology "Severo Ochoa", Autonomous University of Madrid-C.S.I.C, 28049 Madrid, Spain.

References
1. Lindvall O, Bjorklund A. Cell therapeutics in Parkinson's disease. Neurotherap. 2011;8:539–48.
2. Lindvall O, Kokaia Z. Prospects of stem cell therapy for replacing dopamine neurons in Parkinson's disease. Trends Pharmacol Sci. 2009;30:260–7.
3. Politis M, Lindvall O. Clinical application of stem cell therapy in Parkinson's disease. BMC Med. 2012;4:10–1.
4. Lindvall O, Kokaia Z. Stem cells in human neurodegenerative disorders–time for clinical translation? J Clin Invest. 2010;120:29–40.
5. Ramos-Moreno T, Castillo CG, Martínez-Serrano A. Long term behavioral effects of functional dopaminergic neurons generated from human neural stem cells in the rat 6-OH-DA Parkinson's disease model. Effects of the forced expression of BCL-X(L). Behav Brain Res. 2012;232:225–32.
6. Krabbe C, Bak ST, Jensen P, von Linstow C, Martínez-Serrano A, Hansen C, et al. Influence of oxygen tension on dopaminergic differentiation of human fetal stem cells of midbrain and forebrain origin. PLoS One. 2014;9:e96465.
7. Seiz EG, Ramos-Gómez M, Courtois ET, Tønnesen J, Kokaia M, Liste Noya I, et al. A Human midbrain precursors activate the expected developmental genetic program and differentiate long-term to functional A9 dopamine neurons in vitro. Enhancement by Bcl-X(L). Exp Cell Res. 2012;318:2446–59.
8. Villa A, Liste I, Courtois ET, Seiz EG, Ramos M, Meyer M, et al. Generation and properties of a new human ventral mesencephalic neural stem cell line. Exp Cell Res. 2009;315:1860–74.
9. Courtois ET, Castillo CG, Seiz EG, Ramos M, Bueno C, Liste I, et al. In vitro and in vivo enhanced generation of human A9 dopamine neurons from neural stem cells by Bcl-XL. J Biol Chem. 2010;285:9881–97.
10. Ramos-Moreno T, Lendinez JG, Pino-Barrio MJ, Del Arco A, Martinez-Serrano A. Clonal human fetal ventral mesencephalic dopaminergic neuron precursors for cell therapy research. PLoS One. 2012;7:e52714.

11. Chen A, Siow B, Blamire AM, Lako M, Clowry GJ. Transplantation of magnetically labeled mesenchymal stem cells in a model of perinatal brain injury. Stem Cell Res. 2010;5:255–66.

12. Cohen ME, Muja N, Fainstein N, Bulte JW, Ben-Hur T. Conserved fate and function of ferumoxides-labeled neural precursor cells in vitro and in vivo. J Neurosci Res. 2010;88:936–44.

13. Berman SC, Galpoththawela C, Gilad AA, Bulte JW, Walczak P. Long-term MR cell tracking of neural stem cells grafted in immunocompetent versus immunodeficient mice reveals distinct differences in contrast between live and dead cells. Magn Reson Med. 2011;65:564–74.

14. Focke A, Schwarz S, Foerschler A, Scheibe J, Milosevic J, Zimmer C, et al. Labeling of human neural precursor cells using ferromagnetic nanoparticles. Magn Reson Med. 2008;60:1321–8.

15. Muja N, Cohen ME, Zhang J, Kim H, Gilad AA, Walczak P, et al. Neural precursors exhibit distinctly different patterns of cell migration upon transplantation during either the acute or chronic phase of EAE: a serial MR imaging study. Magn Reson Med. 2011;65:1738–49.

16. Swijnenburg R, van der Bogt KE, Sheikh AY, Cao F, Wu JC. Clinical hurdles for the transplantation of cardiomyocytes derived from human embryomic stem cells: role of molecular imaging. Curr Opin Biotech. 2007;18:38–45.

17. Lewin M, Carlesso N, Tung CH, Tang XW, Cory D, Scadden DT, et al. Tat peptide-derivatized magnetic nanoparticles allow in vivo tracking and recovery of progenitor cells. Nat Biotechnol. 2000;18:410–4.

18. Syková E, Jendelová P. Magnetic resonance tracking of implanted adult and embryonic stem cells in injured brain and spinal cord. Ann NY Acad Sci. 2005;1049:146–60.

19. Arbab AS, Yocum GT, Kalish H, Jordan EK, Anderson SA, Khakoo AY, et al. Efficient magnetic cell labeling with protamine sulfate complexed to ferumoxides for cellular MRI. Blood. 2004;104:1217–23.

20. Sipe JC, Filippi M, Martino G, Furlan R, Rocca MA, Rovaris M, et al. Method for intracellular magnetic labeling of human mononuclear cells using approved iron contrast agents. Magn Reson Imaging. 1999;17:1521–3.

21. Bulte JW, Laughlin PG, Jordan EK, Tran VA, Vymazal J, Frank JA. Tagging of T cells with superparamagnetic iron oxide: uptake kinetics and relaxometry. Acad Radiol. 1996;3 Suppl 2:301–3.

22. Dodd SJ, Williams M, Suhan JP, Williams DS, Koretsky AP, Ho C. Detection of single mammalian cells by high-resolution magnetic resonance imaging. Biophys J. 1999;76:103–9.

23. Neri M, Maderna C, Cavazzin C, Deidda-Vigoriti V, Politi LS. Efficient in vitro labeling of human neural precursor cells with superparamagnetic iron oxide particles: relevance for in vivo cell tracking. Stem Cells. 2008;26:505–16.

24. Martínez-Serrano A, Castillo CG, Courtois ET, García-García E, Liste I. Modulation of the generation of dopaminergic neurons from human neural stem cells by Bcl-X(L): mechanisms of action. Vitam Horm. 2011;87:175–205.

25. Nicoletti I, Migliorati G, Pagliacci MC, Grignani F, Riccardi C. A rapid and simple method for measuring thymocyte apoptosis by propidium iodide staining and flow cytometry. J Immunol Methods. 1991;139:271–9.

26. Arbab AS, Yocum GT, Rad AM, Khakoo AY, Fellowes V, Read EJ, et al. Labeling of cells with ferumoxides-protamine sulfate complexes does not inhibit function or differentiation capacity of hematopoietic or mesenchymal stem cells. NMR Biomed. 2005;18:553–9.

27. Pawelczyk E, Arbab AS, Pandit S, Hu E, Frank JA. Expression of transferrin receptor and ferritin following ferumoxides-protamine sulfate labeling of cells: implications for cellular magnetic resonance imaging. NMR Biomed. 2006;19:581–92.

28. Panizzo RA, Kyrtatos PG, Price AN, Gadian DG, Ferretti P, Lythgoe MF. In vivo magnetic resonance imaging of endogenous neuroblasts labeled with a ferumoxide-polycation complex. Neuroimage. 2009;44:1239–46.

29. Kedziorek DA, Muja N, Walczak P, Ruiz-Cabello J, Gilad AA, Jie CC, et al. Gene expression profiling reveals early cellular responses to intracellular magnetic labeling with superparamagnetic iron oxide nanoparticles. Magn Reson Med. 2010;63:1031–43.

30. Gao J, Gu H, Xu B. Multifunctional Magnetic Nanoparticles: Design, Synthesis, and Biomedical Applications. Acc Chem Res. 2009;42:1097–107.

31. Winkler C, Kirik D, Bjorklund A. Cell transplantation in Parkinson's disease: how can we make it work. Trends Neurosci. 2005;28:86–92.

32. Brundin BP, Strecker RE, Widner H, Clarke DJ, Nilsson OG, Astedt B, et al. Human fetal dopamine neurons grafted in a rat model of Parkinson's disease: immunological aspects, spontaneous and drug-induced behaviour and dopamine release. Exp Brain Res. 1988;70:192–208.

33. Geeta R, Ramnath RL, Rao HS, Chandra V. One year survival and significant reversal of motor deficits in parkinsonian rats transplanted with hESC derived dopaminergic neurons. Biochem Biophys Res Commun. 2008;373:258–64.

34. Weissleder R, Cheng HC, Bogdanova A, Bogdanov Jr A. Magnetically labeled cells can be detected by MR imaging. J Magn Reson Imaging. 1997;7:258–63.

35. Cromer Berman SM, Walczak P, Bulte JW. Tracking stem cells using magnetic nanoparticles. Rev Nanomed Nanobiotechnol. 2011;3:343–55.

36. Jasmin, Torres AL, Jelicks L, de Carvalho AC, Spray DC, Mendez-Otero R. Labeling stem cells with superparamagnetic iron oxide nanoparticles: analysis of the labeling efficacy by microscopy and magnetic resonance imaging. Methods Mol Biol. 2012;906:239–52.

37. Kostura L, Kraitchman DL, Mackay AM, Pittenger MF, Bulte JW. Feridex labeling of mesenchymal stem cells inhibits chondrogenesis but not adipogenesis or osteogenesis. NMR Biomed. 2004;17:513–7.

38. Thu MS, Najbauer J, Kendall Q, Harutyunyan I, Sangalang N, Gutova M, et al. Iron labeling and pre-clinical MRI visualization of therapeutic human neural stem cells in a murine glioma model. PLoS One. 2009;4:e7218.

39. Hoehn M, Kustermann E, Blunk J, Wiedermann D, Trapp T, Wecker S. Monitoring of implanted stem cell migration in vivo: a highly resolved in vivo magnetic resonance imaging investigation of experimental sroke in rat. Proc Natl Acad Sci U S A. 2002;99:16267–72.

40. Yano S, Kuroda S, Shichinohe H, Hida K, Iwasaki Y. Do bone marrow stromal cells proliferate after transplantation into mice cerebral infarct?—a double labeling study. Brain Res. 2005;1065:60–7.

41. Novotna B, Jendelova P, Kapcalova M, Rossner Jr P, Turnovcova K, Bagryantseva Y, et al. Oxidative damage to biological macromolecules in human bone marrow mesenchymal stromal cells labeled with various types of iron oxide nanoparticles. Toxicol Lett. 2012;210:53–63.

42. Sun R, Dittrich J, Le-Huu M. Physical and biological characterization of superparamagnetic iron oxide- and ultrasmall superparamagnetic iron oxide-labeled cells: A comparison. Invest Radiol. 2005;40:504–13.

43. Arbab AS, Bashaw LA, Miller BR, Jordan EK, Bulte JW, Frank JA. Intracytoplasmic tagging of cells with ferumoxides and transfection agent for cellular magnetic resonance imaging after cell transplantation: methods and techniques. Transplantation. 2003;76:1123–30.

44. Bulte JW, Douglas T, Witwer B, Zhang SC, Strable E, Miller BP, et al. Magnetodendrimers allow endosomal magnetic labeling and in vivo tracking of stem cells. Nat Biotechnol. 2001;19:1141–7.

45. Amemori T, Romanyuk N, Jendelova P, Herynek V, Turnovcova K, Prochazka P, et al. Human conditionally immortalized neural stem cells improve locomotor function after spinal cord injury in the rat. Stem Cell Res Ther. 2013;4:68–72.

46. Park KI, Liu S, Flax JD, Nissim S, Stieg PE, Snyder EY. Transplantation of neural progenitor and stem cells: Developmental insights may suggest new therapies for spinal cord and other CNS dysfunction. J Neurotrauma. 1999;16:675–87.

47. Guzman R, Uchida N, Bliss TM, He D, Christopherson KK, Stellwagen D, et al. Long-term monitoring of transplanted human neural stem cells in developmental and pathological contexts with MRI. Proc Natl Acad Sci U S A. 2007;104:10211–6.

48. Thu MS, Bryant LH, Coppola T, Jordan EK, Budde MD, Lewis BK, et al. Self-assembling nanocomplexes by combining ferumoxytol, heparin and protamine for cell tracking by magnetic resonance imaging. Nat Med. 2012;18:463–7.

49. Farrell E, Wielopolski P, Pavljasevic P, van Tiel S, Jahr H, Verhaar J, et al. Effects of iron oxide incorporation for long term cell tracking on MSC differentiation in vitro and in vivo. Biochem Biophys Res Commun. 2008;369:1076–81.

50. Crabbe A, Vandeputte C, Dresselaers T, Sacido AA, Verdugo JM, Eyckmans J, et al. Effects of MRI contrast agents on the stem cell phenotype. Cell Transplant. 2010;19:919–36.

51. Söderstjerna E, Johansson F, Klefbohm B, Englund Johansson U. Gold- and silver nanoparticles affect the growth characteristics of human embryonic neural precursor cells. PLoS One. 2013;8:e58211.

52. Rubio FJ, Bueno C, Villa A, Navarro B, Martínez-Serrano A. Genetically perpetuated human neural stem cells engraft and differentiate into the adult mammalian brain. Mol Cell Neurosci. 2000;16:1–13.

53. Grealish S, Diguet E, Kirkeby A, Mattsson B, Heuer A, Bramoulle Y, et al. Human ESC-Derived Dopamine Neurons Show Similar Preclinical Efficacy and Potency to Fetal Neurons when Grafted in a Rat Model of Parkinson's Disease. Cell Stem Cell. 2014;15:653–65.

54. Piccini P, Lindvall O, Bjorklund A, Brundin P, Hagell P, Ceravolo R, et al. Delayed recovery of movement-related cortical function in Parkinson's disease after striatal dopaminergic grafts. Ann Neurol. 2000;48:689–95.

55. Nicholas CR, Chen J, Tang Y, Southwell DG, Chalmers N, Vogt D, et al. Functional maturation of hPSC-derived forebrain interneurons requires an extended timeline and mimics human neural development. Cell Stem Cell. 2013;12:573–86.

56. Arbab AS, Pandit SD, Anderson SA, Yocum GT, Bur M, Frenkel V, et al. Magnetic resonance imaging and confocal microscopy studies of magnetically labeled endothelial progenitor cells trafficking to sites of tumor angiogenesis. Stem Cells. 2006;24:671–8.

57. Urdzikova L, Jendelova P, Glogarova K, Burian M, Hajek M, Sykova E. Transplantation of bone marrow stem cells as well as mobilization by granulocyte-colony stimulating factor promotes recovery after spinal cord injury in rats. J Neurotrauma. 2006;23:1379–91.

58. Pawelczyk E, Arbab AS, Chaudhry A, Balakumaran A, Robey PG. In vitro model of bromodeoxyuridine or iron oxide nanoparticle uptake by activated macrophages from labeled stem cells: implications for cellular therapy. Stem Cells. 2008;26:1366–75.

10

The oxidative potential of differently charged silver and gold nanoparticles on three human lung epithelial cell types

Paul Schlinkert[1], Eudald Casals[2], Matthew Boyles[1], Ulrike Tischler[1], Eva Hornig[1], Ngoc Tran[2], Jiayuan Zhao[3], Martin Himly[1], Michael Riediker[3,4], Gertie Janneke Oostingh[1,5], Victor Puntes[2] and Albert Duschl[1*]

Abstract

Background: Nanoparticle (NPs) functionalization has been shown to affect their cellular toxicity. To study this, differently functionalized silver (Ag) and gold (Au) NPs were synthesised, characterised and tested using lung epithelial cell systems.

Methods: Monodispersed Ag and Au NPs with a size range of 7 to 10 nm were coated with either sodium citrate or chitosan resulting in surface charges from −50 mV to +70 mV. NP-induced cytotoxicity and oxidative stress were determined using A549 cells, BEAS-2B cells and primary lung epithelial cells (NHBE cells). TEER measurements and immunofluorescence staining of tight junctions were performed to test the growth characteristics of the cells. Cytotoxicity was measured by means of the CellTiter-Blue ® and the lactate dehydrogenase assay and cellular and cell-free reactive oxygen species (ROS) production was measured using the DCFH-DA assay.

Results: Different growth characteristics were shown in the three cell types used. A549 cells grew into a confluent mono-layer, BEAS-2B cells grew into a multilayer and NHBE cells did not form a confluent layer. A549 cells were least susceptible towards NPs, irrespective of the NP functionalization. Cytotoxicity in BEAS-2B cells increased when exposed to high positive charged (+65-75 mV) Au NPs. The greatest cytotoxicity was observed in NHBE cells, where both Ag and Au NPs with a charge above +40 mV induced cytotoxicity. ROS production was most prominent in A549 cells where Au NPs (+65-75 mV) induced the highest amount of ROS. In addition, cell-free ROS measurements showed a significant increase in ROS production with an increase in chitosan coating.

Conclusions: Chitosan functionalization of NPs, with resultant high surface charges plays an important role in NP-toxicity. Au NPs, which have been shown to be inert and often non-cytotoxic, can become toxic upon coating with certain charged molecules. Notably, these effects are dependent on the core material of the particle, the cell type used for testing and the growth characteristics of these cell culture model systems.

Keywords: Human lung epithelial cells, Nanoparticles, Cytotoxicity, ROS production, Surface charge

Background

Various characteristics of nanoparticles (NPs) can influence their toxicity, such as size [1], shape [2] or surface coating [3]. In the present study, silver (Ag) and gold (Au) NPs of similar size were synthesised with different coatings to provide 4 classes of surface charge ranging from −50 mV to +70 mV. One class of negatively charged NPs, coated with sodium citrate, (Ag/Au-SC) and three classes (low, medium, high) of positively charged NPs, coated with chitosan, (Ag/Au-CHIT-L/M/H) were synthesised. These metal NPs were chosen, as they are in widespread use, Au NPs are particularly used in medical applications [4]. In contrast, Ag NPs were reported to have anti-microbial properties [5]. Furthermore, Ag NPs have been described in many studies to be cytotoxic for human cells [6-9], whereas Au NPs were mainly found to be inert and only few studies report cytotoxicity of Au NPs [10-12]. The surface coating used to achieve the positive charge is

* Correspondence: albert.duschl@sbg.ac.at
[1]Department of Molecular Biology, Paris Lodron-University of Salzburg, Hellbrunnerstr. 34A-5020 Salzburg, Austria
Full list of author information is available at the end of the article

also of special interest, as chitosan-coated NPs are increasingly used in the field of nanobiotechnology. These type of functionalized NPs are very promising drug delivery systems [13,14], due to their low toxicity, high stability and biocompatibility [15]. The positive surface charge of these NPs renders them more suitable for an intravenous injection, as it has been reported that positively charged NPs remain in the blood stream longer than negatively charged NPs [16], which is a common route for the administration of anticancer agents [15]. Furthermore, chitosan NPs are also suitable to be administered orally and this administration route has been used for the delivery of drugs [17] and genes [18]. Finally, the inhalation of NPs appears to be a promising method for the delivery of drugs to the lung [19,20]

In a number of studies metal NPs were reported to induce oxidative stress (OS) [21-24]. However, it remains unclear which properties of NPs contribute to the induction of OS. Oxidative stress is a direct result of an imbalance of the cell's redox potential, where reactive oxygen species (ROS) are produced at a rate that the cell's antioxidant mechanisms are unable to detoxify [24]. The formation of ROS has also been linked to inflammation and apoptosis. Even though OS can also be induced by the production of reactive nitrogen species [25], this study focuses on the production of ROS. A cellular response to NPs may be dependent on the proteins from their surrounding biofluids which quickly adsorb to the NPs surface [26], first forming a weakly bound soft corona, which can be replaced by a hard corona over time [27]. This process is influenced by both the NPs properties and the composition of the solution [28]. Therefore, this study has examined whether NP functionalization, presenting different surface charges, can influence the interaction of NPs with cell culture medium components and in turn influence ROS production. The potential for some NPs to produce ROS directly on their surface is well recognized [29,30], however, it is important to determine whether this occurs under conditions as they exist in cell culture, which was performed in this study.

In vitro cell exposures provide a vital tool to assess potential risks to humans as these techniques reduce the need for animal studies and provide the opportunity to use human cells; so despite being artificial, a good understanding of in vitro cell systems is necessary. Regarding accidental exposure, inhalation is the most likely route [31]. Deposition within the lung is either due to interception, impaction, sedimentation or diffusion of particles, which is dependent on the size of the particles [32]. Large micron sized particles mainly deposit in the nasopharyngeal region (5–30 μm) as a result of impaction and interception and are then subject to mucociliary clearance [33]. Sedimentation of NPs commonly only occurs with particles with a diameter above 0.5 μm,

whereas the deposition of NPs within the lung is mostly due to diffusion [32]. In contrast to larger particles, NPs have been shown to travel deeper into the lung [32]. Several studies report deposition of NPs within the tracheobronchial region [32,34,35], but also within the deepest region of the lung, the alveolar region [36,37]. Once deposited in the alveolar region, clearance of small NPs has been proven difficult [38] and exposure to the NPs is thereby prolonged [39]. The prolonged exposure will allow the NPs to directly interact with the epithelial layer of the alveolar region, which can in turn lead to translocation of NPs into the blood stream and the subsequent deposition in other organs [40]. However, it has also been shown that insoluble NPs can remain in the lung indefinitely [41], thereby significantly increasing the risk of adverse effects.

The NPs used in this study are in a size range where they can deposit in the tracheobronchial as well as within in the alveolar region [35]. We therefore chose three different human lung epithelial cell types to assess the effects of NP exposure to the human lung for this study, which represent both the tracheobronchial and alveolar regions.

Two stable cell lines were used: the human alveolar adenocarcinoma cell line (A549) [42] and the human bronchial epithelial cell line (BEAS-2B) [43]. In addition, primary human bronchial epithelial cells (NHBE) [44], derived from healthy donors, were used since they represent the in vivo system more closely than the cell lines. These cell types are derived from different parts of the lung and have different properties. A549 cells are of interest since they originate from type II alveolar epithelial cells and not from bronchia, while the other two cell types do [45]. Even though alveolar epithelial cells are not covered by a mucosal layer, they produce a surfactant layer in vivo, which provides additional protection [46]. A549 cells are an important, well-established cell line and frequently used as a model in the assessment of NPs induced lung cytotoxicity, which is illustrated by the high number of publications that mention A549 cells and nanoparticles. In addition, A549 cells rapidly grow under submerged cell culture conditions that allows them to be used in high throughput screenings. The advantage of NHBE cells is that they are primary cells derived from healthy lung tissue, while BEAS-2B cells have the advantage of readily forming a tight epithelium. All three cell types require different culture media, bringing in further deviations to the in vivo situation. In light of their respective benefits and drawbacks it is likely that no single cell type will emerge as universal model in nanosafety research. The three cell types were used since they have all been used for studies on the nanosafety of inhaled NPs [47,48]. A comparison between them is especially useful as NPs that enter the respiratory system may deposit throughout the airways and lung sections, therefore contact with different types of lung cells is relevant.

Results

Cell development

Understanding the growth characteristics of the cell types used in this study is important in order to fully comprehend the observed responses to NPs insult. Epithelial cells grow in monolayers *in vivo* and therefore a tightly formed and well-functioning monolayer is preferred for *in vitro* experiments to increase the similarity to lung epithelia *in vivo*. TEER measurements and fluorescence microscopy using immunodetection of claudin-1 as a marker for tight junction proteins were performed to follow the formation of a tight cell layer. Figure 1a shows that A549 cells, when grown from a seeding density of 1×10^5 cells/ml, formed an intact monolayer after four days and remained stable for several days as a monocellular layer. The first successful staining of tight junction

proteins was also achieved at day 4 (Figure 1b). In contrast, BEAS-2B cells plated at the same cell density did not form a confluent monolayer until day 7 (Figure 1c). A different growth pattern was observed for BEAS-2B cells, which were shown to grow on top of each other and formed multilayers which also contained functional tight junctions (Figure 1d). This resulted in a tight epithelial cell layer, but the multiple cell layer phenotype does not correspond to *in vivo* situations. NHBE cells did not grow into a monolayer under our culture conditions, as maximum TEER values of only 12 Ω*cm^2 were determined (Figure 1e), while values of 67 Ω*cm^2 and 75 Ω*cm^2 were determined for A549 and BEAS-2B cells respectively (Figure 1a, c). NHBE cells did, however, synthesise the proteins necessary for the formation of tight junctions. Yet, the proteins were only found in the centre of

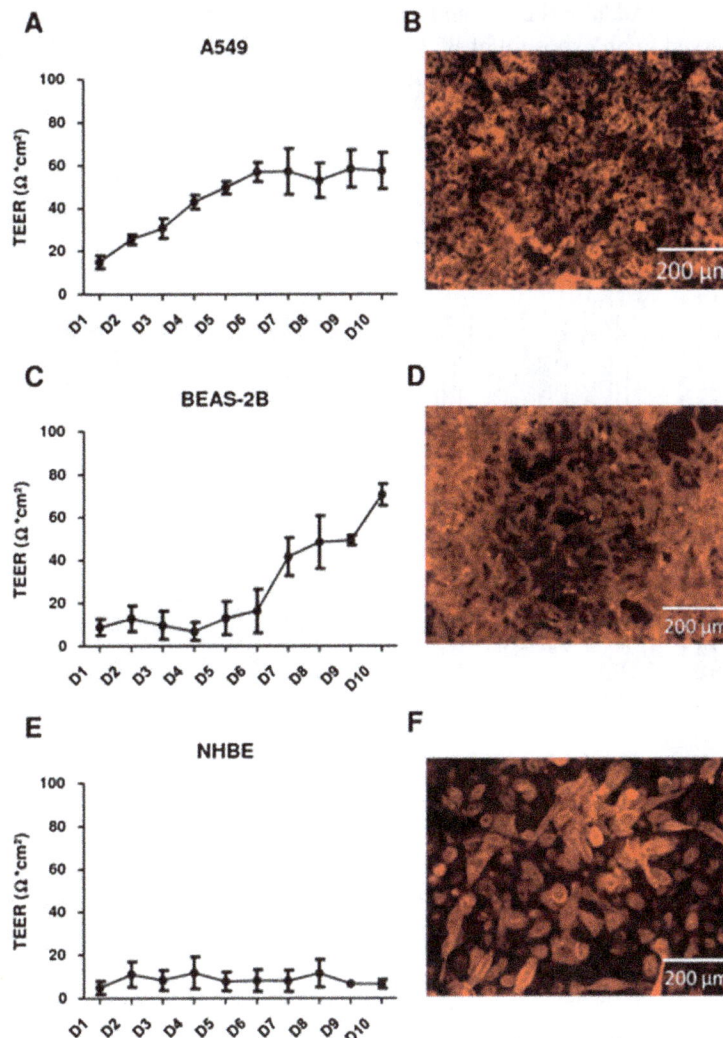

Figure 1 Development of the epithelial layer in (A-B) A549 cells, (C-D) BEAS-2B cells and (E-F) NHBE cells. TEER measurements (A, C and E) show the means ± SD of a minimum of 3 experiments. Staining of tight junction proteins: Claudin-1 staining (B) in A549 cells at day 4, (D) in BEAS-2B cells at day 7 and (F) in NHBE cells at day 7. All pictures were taken with a 10x magnification.

the cell and failed to move to the cell membrane where they would be needed for the formation of tight junctions (Figure 1f). This difference between cell lines of similar origin is also evident in other cell types as well and should be carefully monitored before performing a study [49]. All three cell types used here represent certain aspects of epithelia in the lung, but clearly display different properties.

Cytotoxicity

Effects of functionalized NPs on the cell membrane integrity

When A549 cells were exposed to increasing concentrations of differently functionalized Ag or Au NPs for 24 hours, no increase in LDH release was observed (Figure 2a). Only exposure to the Au NPs with the highest amount of chitosan (Au-CHIT-H) induced a small increase in LDH release, which was statistically not significant (Figure 2b). The same findings were observed at exposure periods of 4 and 48 hours (Additional files 1 and 2).

Similar results were observed upon exposure of BEAS-2B cells for 24 hours. Here, only the highest charged Au NPs (Au-CHIT-H) at a high concentration resulted in membrane impairment (Figure 1c, d). This increase in LDH release was also observed after 4 and 48 hour exposures (Additional files 1 and 2).

In contrast, NHBE cells were more susceptible towards both Ag and Au NPs. An increase in LDH release was observed at high concentrations (0.4 and 0.8 $\mu g/cm^2$) of chitosan-coated Ag NPs (Ag-CHIT-M/H). The two Au NP-preparations coated with the largest amount of chitosan (Au-CHIT-M/H) were shown to increase LDH release, without reaching statistical significance. A shorter exposure of 4 hours did not induce any membrane leakage (Additional file 1), yet a longer period of 48 hours showed the same trends as found after 24 hours (Additional file 2).

In addition, the solvents, sodium citrate and chitosan, were also tested with all cells and for each time point and no effects on membrane integrity or cell viability were observed. Furthermore, no interference of the NPs with the assay, such as binding of the end product, was observed (Additional file 3).

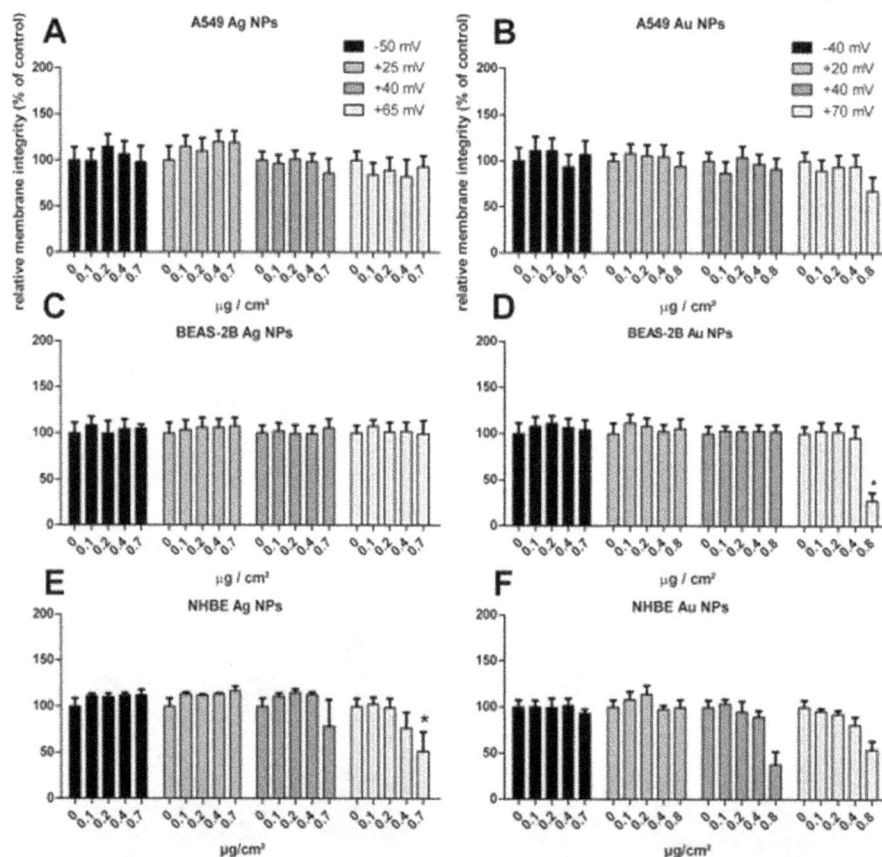

Figure 2 Cell membrane integrity, as measured by an increase in LDH-release, following a 24 h exposure of the different cell lines to Ag and Au NPs. An increase in LDH-release is indicated by a decrease in the membrane integrity. A549 cells (**A, B**; means ± SEM of n = 6), BEAS-2B (**C, D**; means ± SEM of n = 3) and NHBE cells (**E, F**; means ± SEM of n = 3). P-value * < 0.05. Cells treated with medium only were used as negative control (=100%).

Effects of functionalized NPs on cell viability

During a 24-hour exposure period, neither Ag nor Au NPs were shown to reduce the viability of A549 cells (Figure 3a,b). The same was observed after 4 and 48 hour exposures (Additional files 4 and 5).

Differently functionalized Ag NPs did not induce a significant decrease in cell viability in BEAS-2B cells after 24 hours (Figure 3c). Most of the chitosan-coated Au NPs showed no effect, however, a concentration of $0.8 \ \mu g/cm^2$ of Au NPs with the highest amount of chitosan (Au-CHIT-H) induced a significant decrease in cell viability (Figure 3d). The same trends were found at other time points (Additional files 4 and 5). These data are in agreement with those found for the LDH assay (Additional files 1 and 2).

The highest responses to functionalized Ag and Au NPs were observed for NHBE cells. Here, a significant decrease in cell viability was observed when NHBE cells were exposed to high concentrations of chitosan-coated Ag (Ag-CHIT-M/H) and Au NPs (Au-CHIT-H) for 24 hours (Figure 3e, f). These findings are in line with those of the LDH assay. When the cells were exposed

for only 4 hours, no decrease in cell viability was observed (Additional file 4). In contrast, increasing the exposure period to 48 hours induced a decrease in viability by the same NPs as after 24 hours (Additional file 5).

In contrast to the LDH assay, where no interference was found, there is a small interference of the light emitted during the endpoint measurement of the CTB assay, by both Ag and Au NPs coated with a high amount of chitosan (Additional file 6). There is an increase in the fluorescence signal when these particles are present during the measurement, which may be incorrectly interpreted as an increase in cell viability. This effect may have caused a small underestimation of the NP-induced reduction in cellular viability.

Oxidative stress
Induction of intracellular ROS production by functionalized NPs

To analyse the oxidative stress induced by NP exposure, shorter exposure periods compared to those of the cytotoxicity assays were chosen, since cell-mediated ROS production by NPs is a rapid process, which might be lost at later time points.

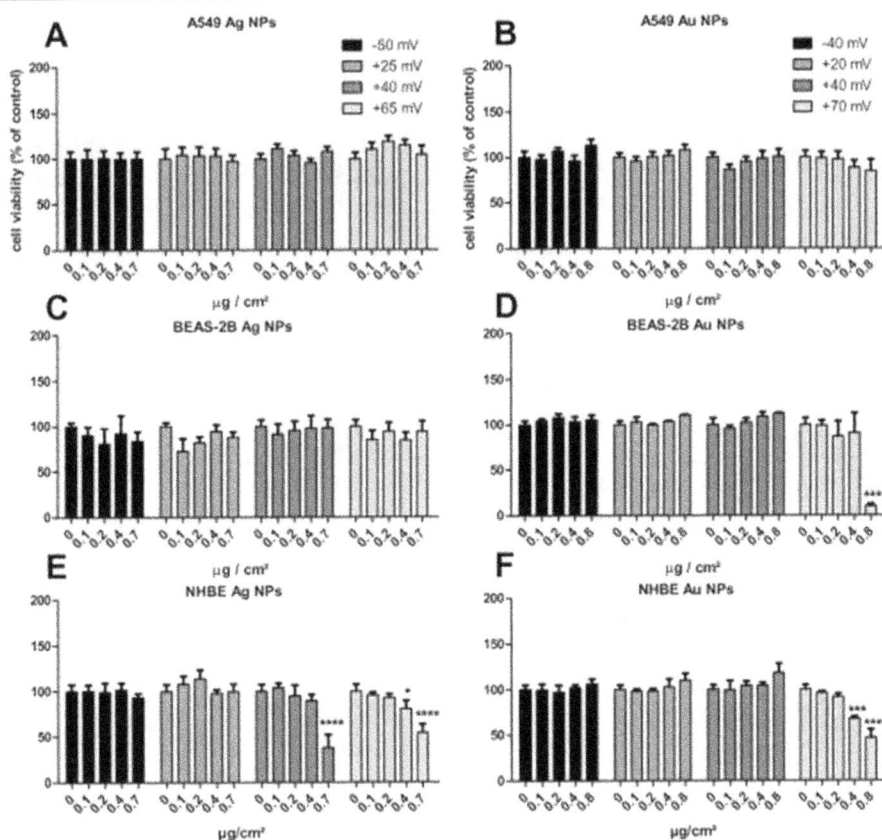

Figure 3 Cell viability following a 24 h exposure to functionalized Ag and Au NPs of A549 cells (A, B; means ± SEM of n = 6), BEAS-2B cells (C, D; means ± SEM of n = 3) and NHBE cells (E, F; means ± SEM of n = 3). P-values * < 0.05, *** < 0.005, **** < 0.001. Cells treated with medium only were used as control (100%).

Exposing A549 cells to chitosan coated Ag NPs (Ag-CHIT-L/M/H) for 4 hours induced low levels of ROS production, with the highest levels observed for the highest amount of chitosan (Ag-CHIT-H) (Figure 4a). In contrast, a significant concentration-dependent increase in ROS production, as determined by Spearman's rank correlation, could be observed following the exposure Au NPs with the highest amount of chitosan on their surface (Au-CHIT-H), whereas exposure to Au NPs with less amount of chitosan (Au-CHIT-L/M) only resulted in a small increase in ROS production (Figure 4b). The same observations were made when cells were exposed for 1 hour (Additional file 7).

Interestingly, both Ag and Au NPs induced some degree of ROS production after a 4 hour exposure (Figure 4c,d) in BEAS-2B cells, however, this was less prominent when compared to a 1 hour exposure (Additional file 7). Furthermore, ROS production induced by Au NPs with a large amount of chitosan (Au-CHIT-H) was not as high as that seen in A549 cells, yet a significant

concentration dependency was still determined by Spearman's rank correlation.

NPs induced the least ROS production in NHBE cells. Only small amounts of ROS were induced by Ag NPs after 4 hours (Figure 4e) and only slightly higher in response to Au NPs. However, a NPs concentration dependency was observed following the exposure to sodium citrate-coated Au NPs (Au-SC) (Figure 4f). Similar amounts of ROS were produced after 1 hour, yet here a concentration dependency could be observed for all of the NPs studied (Additional file 7).

Additionally, Spearman's rank coefficients were determined to assess if NP-induced ROS production was dependent on functionalization. The assessment can be viewed in Tables 1 and 2. The ROS production induced by both Ag and Au NPs in A549 cells appeared to be functionalization-dependent, as statistically significant Spearman's rank coefficients were found with a change from negatively to positively charged surface coatings and with further increases in positive charge, evident in

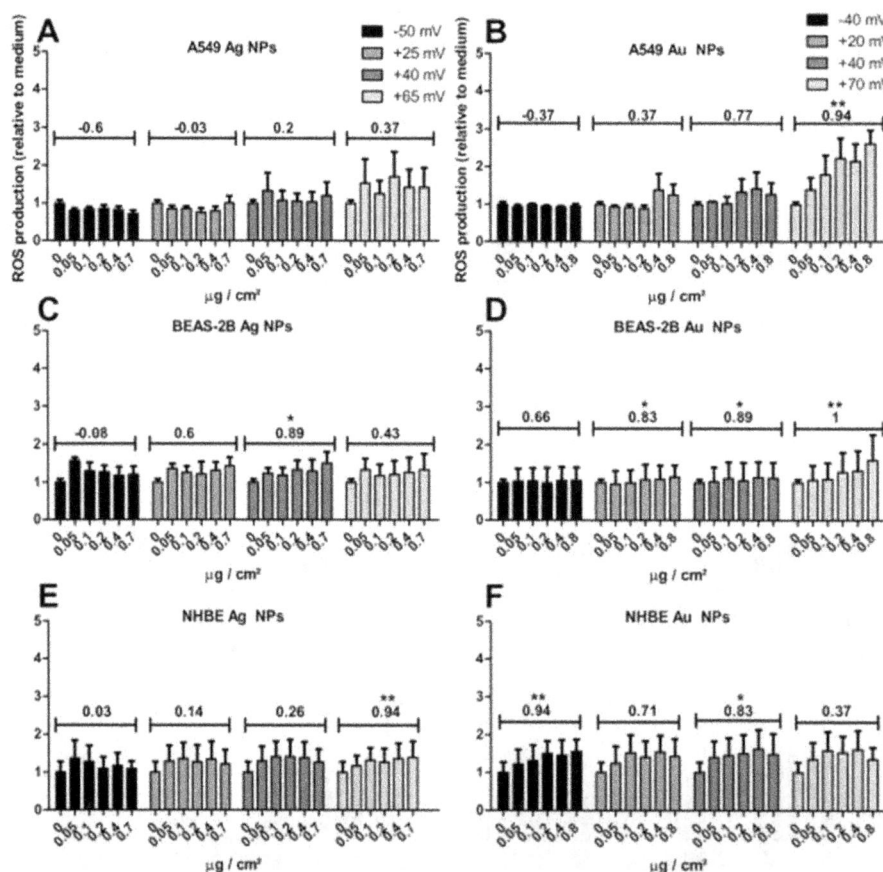

Figure 4 ROS production measured using the DCFH-DA assay following a 4 h exposure of A549 cells (A, B, means ± SEM of n = 3), BEAS-2B cells (C, D, means ± SEM of n = 3) and NHBE cells (E, F means ± SEM of n = 4) to charged Ag and Au NPs. Spearman's rank coefficients were calculated for each NP to assess possible charge dependent increases in ROS production. In addition, the coefficients of the highest concentrations of each NP surface charge were calculated to determine if ROS production was charge dependent. P-values * < 0.05, ** < 0.01.

Table 1 Overview of Spearman's rank coefficients to assess charge-dependent increase in ROS production following a 4 hour exposure to differently charged Ag NPs

NPs (µg/cm²)	A549	BEAS-2B	NHBE
0	0	0	0
0.05	1**	−0.20	−0.2
0.1	1**	−0.33	1
0.2	0.87*	0.07	0.60
0.4	0.87*	0.47	0.87*
0.8	1**	0.60	1*

P-values * < 0.05, ** < 0.01.

all NPs concentrations. This is in contrast to BEAS-2B, where no correlation between Ag NPs induced ROS production was found and only in response to higher concentrations of Au NPs. In NHBE cells, where ROS production could only be correlated to Ag NPs functionalization at higher NPs concentrations, while Au NPs did induce ROS production in a functionalization-dependent manner at all concentrations except the highest.

ROS production in a cell-free system

Since functionalized NPs can produce ROS via interactions on their reactive surface, the production of ROS in the presence of the NPs studied was measured in a cell-free system. As seen in Figure 5, Ag NPs produced only low levels of ROS, but in a functionalization-dependent fashion.

In contrast, all of the Au NPs produced relatively high levels of ROS. The particles with negatively charged functional groups produced the least ROS, whereas an increase in the amount of positively charged surface groups was shown to correlate with an increase in ROS production, with the Au NPs coated with the greatest amount of chitosan (Au-CHIT-H) being the most reactive (Figure 5).

In addition, ROS production of the respective solvents, sodium citrate and three concentrations of chitosan, was determined (Additional file 8). Interestingly, ROS production of the tested solvents appeared to be slightly

Table 2 Overview of Spearman's rank coefficients to assess charge-dependent increase in ROS production following a 4 hour exposure to differently charged Au NPs

NPs (µg/cm²)	A549	BEAS-2B	NHBE
0	0	0	0
0.05	1**	0.60	0.87*
0.1	0.87*	0.87*	0.87*
0.2	0.87*	0.60	0.87**
0.4	1**	1**	0.87*
0.8	1**	0.87*	−0.2

P-values * < 0.05, ** < 0.01.

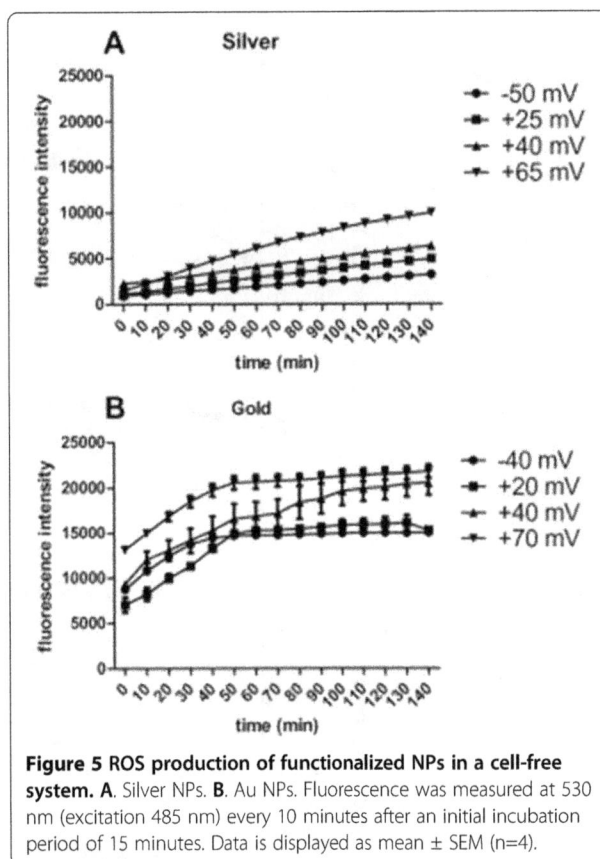

Figure 5 ROS production of functionalized NPs in a cell-free system. **A**. Silver NPs. **B**. Au NPs. Fluorescence was measured at 530 nm (excitation 485 nm) every 10 minutes after an initial incubation period of 15 minutes. Data is displayed as mean ± SEM (n=4).

increased in comparison to that of Ag NPs, yet much smaller compared to that of Au NPs. The trend observed with Au NPs, where an increase in positive charge created by an increase in chitosan concentration resulted in an increase in ROS production, could not be observed. The amount of ROS produced by the solvents was very similar and no increase in ROS production was observed with an increase in chitosan concentration.

Effects of biological solutions on ROS production in a cell-free system

As previously mentioned, the protein corona surrounding NPs may play a crucial role in cellular responses. It is therefore critical to study the NPs in the presence of cell culture media. Differently functionalized Ag and Au NPs were incubated in cell culture media corresponding to the three different cell type-specific media used in this study for different periods of time (0.5, 4 and 24 hours). Au-CHIT-H was chosen as representative class in Figure 6. The most prominent response was seen after NP-incubation with A549 medium, which contains 10% foetal calf serum (FCS) (Figure 6a), where ROS production by particles dropped to a minimal level. This effect was observed across all particles and functionalizations/charges. To further correlate the loss of ROS production with the presence of FCS in the media, NPs were incubated with

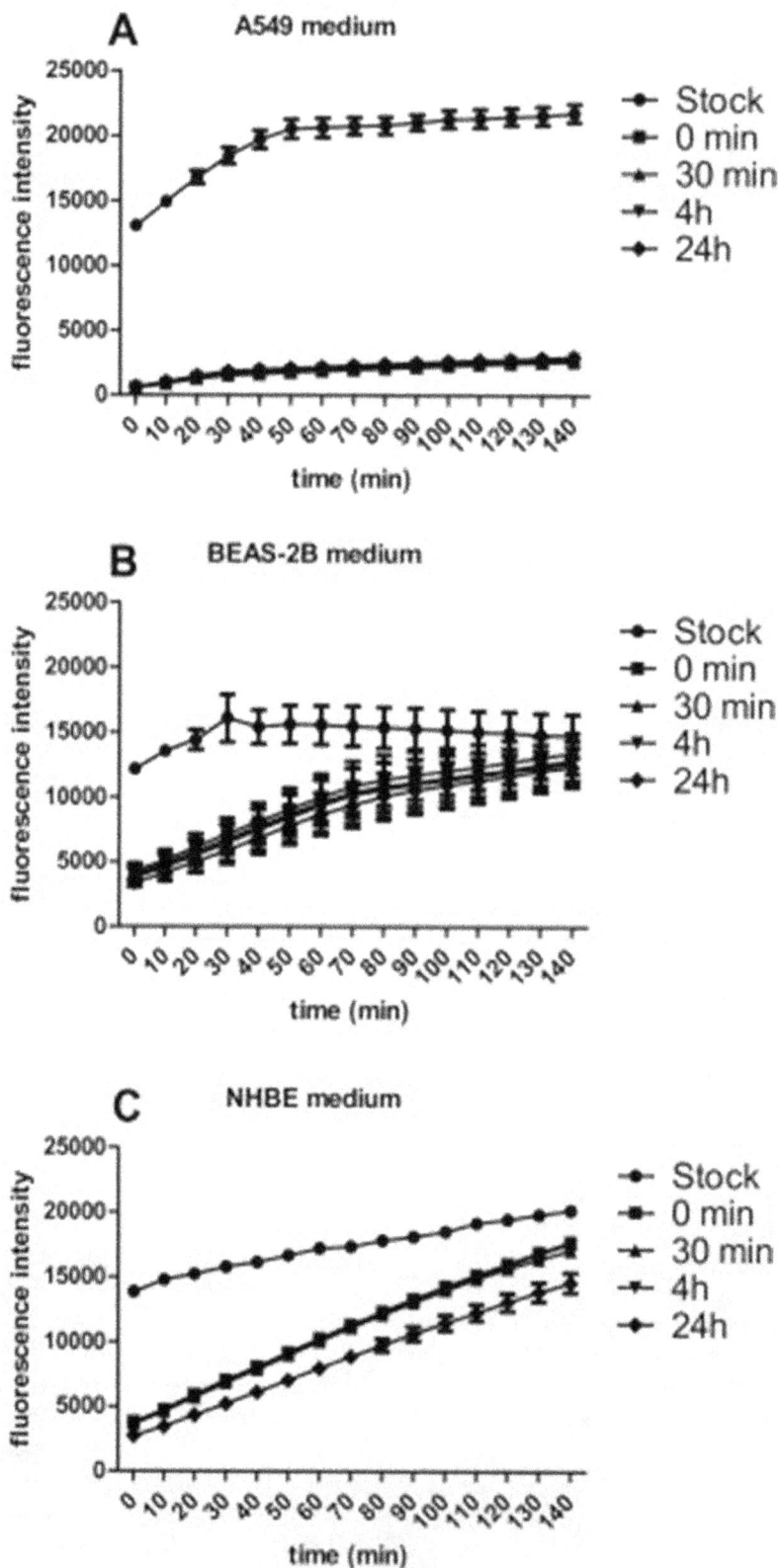

Figure 6 ROS production of chitosan-coated Au NPs (Au-CHIT-H) in a cell-free system after incubation in cell culture medium. **A**. A549 medium. **B**. BEAS-2B medium. **C**. NHBE medium. Fluorescence was measured at 530 nm (excitation 485 nm) every 10 minutes after an initial incubation period of 15 minutes. Means ± SEM of n=3 are shown.

A549 medium containing no FCS or with PBS containing FCS (Figure 7). PBS containing FCS and not full A549 medium was used to ensure that any observed effects are due to the FCS present in A549 medium and not to any other constituents of the medium. When the particles were incubated with medium containing no FCS (Figure 7a), ROS production did not change. However, when the particles were incubated with PBS containing FCS, ROS production by the particles decreased significantly (Figure 7b). Both, BEAS-2B and NHBE media do not contain serum. However, both media are supplemented with pituitary gland extract. The ROS production of Au-CHIT-H dropped by more than 50% after the incubation in both media (Figure 6b,c), yet was still found to be higher than the ROS production in A549 medium. The data show that ROS production observed in cell cultures derives essentially from the cells used, since the proteins contained in all the growth media effectively abrogate ROS production directly at particle surfaces.

Effects of cell culture components on the NPs surface charge

As depicted in Table 3, the surface charge of the NPs used in this study is altered dramatically after incubating the particles in cell culture medium. A 24-hour incubation in A549 cell culture medium resulted in a complete loss of the positive surface charge. This effect was observed in all positively charged Ag and Au NPs. Interestingly, when positively charged NPs where incubated in BEAS-2B medium, the surface charge also dropped significantly, but the positive charge of both Au and Ag NPs were maintained. In contrast, Ag and Au NPs incubated in NHBE medium all carried a negative surface charge after a 24 hour incubation. This effect was also observed when particles where incubated with cell culture media for 4 hours, yet it was less profound (Additional file 9).

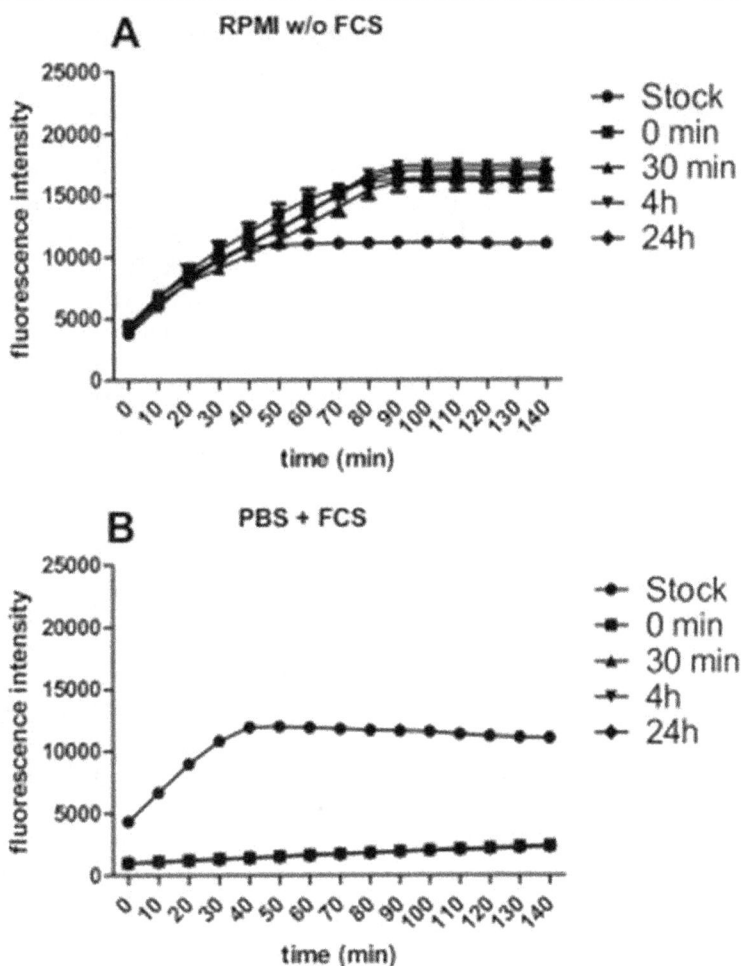

Figure 7 Effect of FCS on the ROS production of chitosan-coated Au NPs (Au-CHIT-H) in a cell-free system. A. Incubated in RPMI w/o FCS. **B**. Incubated in PBS + FCS. Fluorescence was measured at 530 nm (excitation 485 nm) every 10 minutes after an initial incubation period of 15 minutes. Means ± SEM of n=3 are shown.

Table 3 Surface charge (N = 3, mV ± SEM) of NPs as synthesised and after a 24 hour incubation in different cell culture media

NPs name	As synth	In A549 medium	In BEAS-2B medium	In NHBE medium
Ag-SC	−50 ± 1.4	−27 ± 0.5	−26.2 ± 0.2	−37 ± 1.2
Ag-CHIT-L	+25 ± 0.9	−35 ± 0.9	+10.3 ± 0.4	−26.6 ± 0.6
Ag-CHIT-M	+43 ± 0.4	−25 ± 0.7	+9.8 ± 0.2	−34 ± 0.9
Ag-CHIT-H	+60 ± 0.9	−14.4 ± 0.4	+10.2 ± 0.6	−25.7 ± 2.8
Au-SC	−45 ± 0.2	−32 ± 0.5	−24.4 ± 0.2	−25 ± 0.2
Au-CHIT-L	+23 ± 1.0	−33 ± 0.7	+6.8 ± 0.2	−31.9 ± 0.1
Au-CHIT-M	+40 ± 0.6	−33 ± 0.8	+7.9 ± 0.9	−35 ± 0.4
Au-CHIT-H	+65 ± 1.0	−23.3 ± 0.5	+9.0 ± 0.1	−28.0 ± 1.3

Here, NPs incubated in BEAS-2B medium carried a much higher positive charge and those incubated in NHBE medium ended up with a very low negative charge as compared to those incubated for 24 hours.

Discussion

The results of the TEER measurements and the tight junction staining demonstrated that the cell types tested had different growth capacities under the settings used in this study. These growth patterns corresponded well to the cytotoxic responses. A549 cells growing in a monolayer, and with well-developed tight junctions, as would be found *in vivo*, were not responsive to any of the particles tested. In contrast, NP exposure did not induce cytotoxicity in NHBE cells, which did not form a monolayer and no tight junctions. BEAS-2B cells did form a multilayer with tight junctions, with cytotoxicity observed in response to high levels of chitosan on functionalized Au NPs.

Another reason for the divergent cytotoxicity data found might be the difference between primary cells and cell lines. Both A549 and BEAS-2B cells are immortalized and might therefore be differently susceptible towards external stimuli. However, these two cell lines are derived using different methods, A549 were created from cancerous cells [42] and BEAS-2B were immortalized using an adenovirus [43]. One might describe the monolayer growth of A549 cells as more natural compared to that of BEAS-2B cells, where cells continue to grow which results in the formation of multilayers. The rapid growth of a cellular monolayer by A549 cells is one reason, why they are most frequently used in lung cytotoxicity studies. Unlike the two cell lines which originate from a single donor before immortalization primary NHBE cells were extracted from different healthy donors [44] by the supplier, so a donor-to-donor variation between cell charges is always given. It has been shown that NHBE cells grown at an air-liquid interface can form a full monolayer with working tight junctions after 8 days [50]. In the present study a confluent monolayer was not obtained; this, and the lower NHBE cell

number upon particle exposures, may be the reason for the observed cytotoxicity in these cells that was not observed in the other cell treatments.

It has been reported that NPs composed of chitosan are internalized by cells, e.g. A549 cells [51-54], yet it is also known, that the uptake of NPs by cells will not automatically result in a cellular response. In fact, chitosan-coated Au NPs have been shown to be taken up by A549 cells [55], and have been reported to be biocompatible [56,57] and for that reason are increasingly used as carriers in drug or gene delivery systems [58,59]. However, Choi *et al.* reported that chitosan-coated AuNPs internalised by A549 cells provoke cell damage through both apoptopic and necrotic pathways [55]. The NPs used by Choi *et al.* are of similar nature to those used in the study here, since they are also chitosan-coated Au NPs, yet are slightly larger, 17 nm compared to 7 nm in our study. The positive surface charge of their NPs, as determined by zeta-potential measurements, is comparable to that of the Au-CHIT-M presented here (~40 mV). The size difference of the particles used by Choi *et al.* does not allow to presume that our NPs would also be internalised, but the similarities in surface charge does infer that similar cell interactions would have occurred. As cell death was not observed in response to the comparable AuNPs (Au-CHIT-M) it can be assumed that the cytotoxicity of the AuNPs used by Choi *et al.* was due to the increase in NP size, or more likely due to the increased concentration which cells were exposed to.

Other studies have shown that chitosan can induce cytotoxicity in other cell types [51,52], but high levels of chitosan were used in these studies resulting in the observed cytotoxicity. Since the core of both Ag and Au NPs will stabilize the chitosan on the surface during synthesis, thereby forming robust conjugates [53], the amount of free chitosan used in our study will presumably be lower compared to those amounts in the studies mentioned above as no additional chitosan will be released by the NPs [54]. Thus, the observed cellular effects are more likely to be due to the chitosan-coated NPs and not due to the chitosan itself [55].

Possible differences in the cellular uptake of the NPs by the three cell types used in this study might provide an additional explanation to the divergent cellular responses [56], which has been previously compared [57]. Cellular uptake studies will therefore be performed as part of a future study.

Differently charged Ag and Au NPs were chosen for this study on purpose, as it is becoming increasingly evident that the surface charge of NPs is a crucial characteristic of NPs. Au NPs of similar sizes and surface charges (+45 mV, −41.5 mV) ,yet with a different surface coating (AUT, peptidic biomolecules) to those presented here were analysed by Ojea-Jiménez et al. [60]. They reported an increase in cellular uptake of positively charged NPs in comparison to negatively charged NPs. They also reported that once taken up by cells, a large fraction of the positively charged NPs migrated towards the vicinity of the nucleus. Thus, positively charged Au NPs might be more feasible for gene therapies than negatively charged NPs [60]. Similarly, Oh et al. reported an increase in cellular uptake with positively charged (+25 mV. +29 mV, +42 mV and + 55 mV) Au NPs in macrophages in comparison to negatively charged Au NPs (−15 mV, −30 mV, − 35 mV and −38 mV) [61]. Furthermore, the surface charge of NPs has been shown to affect the cytotoxicity of NPs. Schaeublin et al. reported an increase in toxicity in the human keratinocyte cell line HaCaT following the exposure to positively and negatively charged Au NPs, whereas neutral Au NPs were less harmful [62]. Others come to the conclusion that cationic can be considered to be more toxic than anionic NPs in red blood cells and COS-1 kidney cells [63]. Unfortunately, the latter study does not specify the surface charge of their NPs.

Our data demonstrated that an increase in chitosan coating affected the response to Au NPs more than to Ag NPs. The lower stability of the chitosan-Ag NPs conjugate as compared to the Au NPs conjugate could explain these findings. The chitosan molecules are not only attached more stably to the surface of the Au NPs, but also more densely [64,65]. Thus, the positive surface charge of Ag NPs decreased faster under cell culture conditions compared to Au NPs (observed during the four incubation of both NPs), as they will be subject to a greater amount of oxidation at their surface. This oxidation will weaken the interaction between the chitosan layer and the NP surface thereby reducing the compactness and robustness of it during exposure. This reduction in positive surface charge of Ag NPs may result in a less dense interaction of Ag NPs with cells compared to Au NPs, which may provide an explanation as to why Au NPs were found to induce greater membrane impairment than Ag NPs.

It is known that the protein corona of NPs may greatly affect their influence on cells [27]. Each of the cell types used in this study was cultured in a different medium. A549 cells were cultured in RPMI containing 10% FCS. Both BEAS-2B and NHBE cells were cultured in serum-free medium, however, other proteins, such as pituitary gland extract, were present in these media. It has previously been reported that the surface charge of NPs affects the composition of the protein corona formed [58] and that serum proteins present in solution will in turn affect the resulting surface charge of the NP-protein complex [59]. The present study shows that pre-incubation of NPs in the above-mentioned cell culture media affected the surface charge of the NPs, which influenced the cellular responses.

All NPs used in this study, irrespective of their core or surface functionalization, lost their positive surface charge following a 24 hour incubation in A549 medium. Our data suggests that the NPs quickly formed a protein corona from the FCS contained in the A549 medium, as even short incubation periods resulted in a negative charge. The new surface charge of the NPs can be considered to be the average of the surface charge of the FCS proteins that adsorbed onto the surface of the NPs. The adsorption process of the proteins onto the NPs is highly dependent on the affinity of the proteins towards the NPs surface [66]. The newly formed protein layer upon the NPs thus covers the original surface coating, thereby masking the original surface charge. We have previously shown that a nearly complete hard corona will surround the NPs after 24 hours [67], thus any observed effect suggests an impact of the protein corona. In contrast, all NPs incubated in medium without FCS, such as the BEAS-2B medium, retained their positive charge after 24 hours. They still adsorbed negatively charged proteins from the medium, as can be seen by the drop in surface charge, indicating the formation of a protein corona. However, the change in surface charge was not as big as the one observed when incubating in A549 medium, since lower amounts of protein are present. NPs incubated in NHBE medium behaved very similar to those incubated in BEAS-2B medium. BEAS-2B and NHBE medium are very similar in their composition, Amphotericin-B is found only in the NHBE medium. Although the exact concentrations of the supplements were not disclosed by the supplier, one can postulate that the difference in surface charge after incubation was due to concentration differences between the media. Notably, the NPs coated with the highest amount of chitosan ended up with the lowest negative charge of all NPs used.

All of the cells reacted to exposures of Au NPs coated with a large amount of chitosan, thereby carrying a high positive surface charge after synthesis. Since these were the NPs that maintained the least negative charge in A549 and NHBE medium and even remained positive in

BEAS-2B medium, it is suggested that their initial high positive charge is directly affecting the observed cytotoxicity. Different surface charges may affect various parameters, including amount and type of proteins and other biological compounds, corona hardening, and intensity as well as route of uptake into cells. Dissecting these interesting mechanistic aspects was beyond the scope of the present study.

Several studies have investigated NP-induced oxidative stress in cells [68-70]. Normally, ROS is only generated to a low extent in healthy cells and is quickly detoxified by the cells antioxidant defence mechanisms (mainly glutathione and antioxidant enzymes). However, an imbalance between ROS and defence mechanisms results in oxidative stress [22]. In the present study, functionalized NPs carrying a positive surface charge appear to induce a higher amount of ROS within the cells. The highest production of ROS following exposure to NPs was found in A549 cells. ROS production was increased in both BEAS-2B and NHBE cells, yet not to the same extent. These findings are in line with a study by Ekstrand-Hammarström et al. who analysed the effects of titanium dioxide NPs on A549, NHBE and BEAS-2B cells [57]. Similar to the results presented here, the group finds the least amount of NPs induced ROS in NHBE cells compared to the other two cell types, which might be due to different uptakes rates of NPs in the three cell types. Since uptake studies were beyond the scope of the study presented here, one can only postulate that the differences in ROS production are a result of the differences in cellular uptake, which will be the subject of a future study.

In order to understand the production of ROS with the characteristics of the NPs used in this study and to obtain information on the oxidative capacity of the NPs, ROS measurements in a cell-free system were performed. In this system, Ag NPs, irrespective of the surface charge, only produced very small amounts of ROS. In contrast, all of the Au NPs produced large amount of ROS, where the amount of ROS produced increased in correspondence with the surface charge. It is interesting that the oxidative capacity of, presumably neutral, Au was in this study found to be more detrimental than Ag, even though the latter is widely used as bacteriostatic agent due to its toxic potential [60].

Even though results on both cell-mediated and cell-free ROS production of the NPs used in this study were compared, no direct link between these two parameters could be detected. However, the results allow a better understanding of particle-mediated ROS production, in cell-free and cellular systems. It has previously been shown that the oxidative capacity of NPs can influence their ability to induce oxidative stress, but recently published data show that this might not always be the case. For example, a study performed by Weissenberg et al.

showed that intracellular ROS production still occurred after extracellular ROS produced by NPs was blocked using an inhibitor [71]. Furthermore, in the study presented here, extracellular ROS production was inhibited through binding of cell culture medium proteins to the NPs, while intracellular ROS production in response to the same particles was observed within cellular exposures.

Further experiments have shown that proteins adsorbed to the NPs surface reduced the amount of ROS being produced. During cell culture experiments, the NPs will encounter a vast amount of proteins. The amount and type of proteins is dependent on the cell culture medium used. In this study, we demonstrated that the widely used RPMI medium supplemented with 10% FCS greatly affects the ability of NPs to produce ROS. Even short incubation times in the medium resulted in a complete loss of ROS production by the particles in a cell-free system. Some ROS production was still observed in A549 cells, even though the cells presumably only interacted with cells coated with FCS. The loss of ROS production might be explained with the formation of a protein corona consisting of serum albumin. Albumin has been reported to have anti-oxidant properties [61]. Izak-Nau et al. used MALDI-TOF to confirm that the main component of the protein corona surrounding charged NPs is bovine serum albumin [59]. It is therefore likely, that ROS production by NPs is blocked once a protein corona consisting of albumin is formed. Further studies on protein-NPs interaction are needed to provide additional evidence for this.

Additional research will be needed to fully understand how the NPs affect the cells with respect to the protein corona of differently functionalized NPs in different biological media, how this affects cellular uptake of NPs and how intracellular ROS production is linked to the oxidative capacity of NPs.

Conclusion

Several conclusions can be drawn from the results of this study.

First, the type of lung epithelial cells used to analyse the effects of NPs greatly affects the results. The properties of the cell type used need to be considered for correct interpretation of data. By understanding the growth characteristics of different cell types and how a particle effect can be different between these cell types, an improved design of in vitro systems can be supported.

A surface charge of +30 to +40 mV might be considered as harmless under cell culture conditions, while a surface charge above +60 mV has to be designated as problematic and it can be postulated that inhalation of highly charged wet nanoaerosols, as may be stably produced in nature (water falls) or during nanomedical approaches by nebulization/atomization from NP suspensions [62,63], may carry serious risks.

Finally, this study supplies evidence towards the impact of the components of cell culture medium, mainly FCS, on the characteristics of NPs. FCS will not only reduce the surface charge of NPs, but will also affect their ability to produce ROS.

While the data allow conclusions about safety studies *in vitro* with widely used cell culture models, the behaviour of lung tissue under physiological conditions needs to be verified with other approaches. It can be predicted that the biological compounds present will play a major role in defining which types of NPs elicit cell damage through oxidative stress and that, based on cell culture models, high chitosan coating conferring strong positive surface charge may be a risk factor.

Materials and methods
Cells

The adenocarcinomic human alveolar basal epithelial cell line (A549) was maintained in RPMI 1640 medium (PAA Laboratories GMBH, Pasching, Austria) containing 10% FCS, 5% penicillin/streptomycin and 5% L-glutamine. The human bronchial epithelial cell line BEAS-2B, originally isolated from a non-cancerous patient and immortalised by an adenovirus 12-SV40 hybrid, was grown in LHC-9 basal medium containing supplements. Normal human bronchial epithelial cells NHBE (CC-2540, Lonza, Basel, Switzerland), isolated from healthy donors, were cultured in bronchial epithelial growth medium (BEGM) supplemented with BEGM® Single quots® (Lonza, Basel, Switzerland). All cells were sub-cultured by removing the cell culture medium from the cell culture flasks and washing the cells with 2 ml PBS. After removing the PBS, 2 ml of pre-warmed trypsin was added and the cells were incubated at 37°C for 2–3 minutes until all cells detached from the flask, observed via light microscopy. To stop the trypsin reaction, 8 ml pre-warmed cell culture medium was added to the flask. The number of living cells was determined via trypan blue exclusion and counted within a haemocytometer. The cells were plated on either 24-well or 96-well plates (Corning Inc., city) at a density of 1×10^5 cells/ml, where 1 ml of medium was added to a well of a 24-well plate and 100 µl to a well of a 96-well plate. A549 cells were grown for 4 days, BEAS-2B and NHBE cells were grown for 7 days, ensuring that the cells reached a confluent monolayer, before they were exposed to the NPs.

Cells were stored in liquid nitrogen and, after thawing, maintained at 37°C and 5% CO_2. Cells were never cultured for more than one month and the NHBE cells were only frozen and thawed once. The presence of mycoplasma was determined once a week using the MycoAlert™ kit (Lonza, Basel, Switzerland) and infected cultures were disposed of immediately.

Transepithelial electrical resistance measurements

In order to assess the integrity of the epithelial layer of A549, BEAS-2B and NHBE cells, transepithelial electrical resistance (TEER) was measured using a TEER electrode (WPI, Sarasotay, USA). Cells were grown on a 24-well plate containing well inserts (Millipore Corporation, Billerica, MA) with a pore size of 0.4 µM in diameter. TEER values were measured every 24 hours. Before each measurement, the medium was changed and the electrode was washed with RPMI medium between measurements. The TEER value was calculated for the dimension [Ohm*cm^2] by subtracting the medium only control from the measured value and then multiplying it by the surface area of the insert (0.33 cm).

Tight junction staining

A549, BEAS-2B or NHBE cells were grown on 24-well cell culture plates and tight junction staining was performed for each day after an initial growing phase of 24 hours. Cells were fixed with 3.7% formaldehyde in PBS for 10 minutes at RT and washed twice with 500 µl PBS. Following the fixation, 250 µl 1× saponin in PBS was added at RT for 5 minutes and again washed twice with 500 µl 1× PBS. The primary antibody (5 µl rabbit-anti-human claudin-1 in 250 µl saponin/PBS) was added and incubated for 45 minutes in the dark. The cells were washed with 1 ml saponin/PBS before the secondary antibody (2 µl goat-anti-rabbit IgG-PE in 250 µl saponin/PBS) was added. The plate was then incubated for 30 minutes in the dark, cells were washed with 500 µl 1× PBS. Fluorescence micrographs were taken using a fluorescence microscope (Olympus I × 70-S1F, Austria) using a 10× objective.

Nanoparticles
Synthesis

Nanoparticle preparation was performed following the most common synthesis recipes in water with some modifications to achieve the desired characteristics regarding size and surface charge. All reagents were purchased from Sigma-Aldrich (St. Quentin Fallavier, France) and used as received. All glass material was sterilized and depyrogenated in an oven prior to use.

Gold Nanoparticles (Au NPs)

Citrate-coated 10 nm Au NPs were obtained with a procedure based on Turkevich et al. [72] consisting of the fast injection of 1 mL of a solution of hydrogen tetrachloroaureate (HAuCl$_4$) 25 mM to a boiling solution containing trisodium citrate (SC) at 2.2 mM under vigorous stirring. After 3 minutes, when the suspension acquired the characteristic red colour of the colloidal gold, it was cooled down to room temperature (RT). NPs were loosely coated with the negatively charged citrate ions.

Chitosan-coated Au NPs of 7–10 nm mean diameter were obtained were obtained by a variation of a procedure based on the synthesis described in Jana *et al.* [73]. In brief, an aqueous solution containing as precursor HAuCl4 at 2.5×10^{-4} M and chitosan (0.001%, 0.01%, 0.1% (w/v) was prepared. To this solution, 1 ml of ice-cold 0.1 M NaBH4 was added under constant stirring. Increasing concentrations of chitosan were used as capping agent conferring a range of different positive surfaces charge to the Au NPs.

Silver Nanoparticles (Ag NPs)

Citrate-coated 10 nm Ag NPs. 5 mL of trisodium citrate 0.1 M were injected to a boiling solution of 50 mL of silver nitrate ($AgNO_3$) 1 mM and left under vigorous stirring for 5 minutes. The resulting solution was cooled down in another vial to avoid deposition of silver on the glass surface. Citrate ions were the coating agent as in the case of Au NPs. An overview of the NPs synthesized for this study can be viewed in Table 4.

Nanoparticle characterization

NPs were characterised as previously described by Casals *et al.* [27]. Briefly, the NPs sizes were determined by TEM measurements using a JEOL 1010 electron microscope (Jeol, Tokyo, Japan) at an accelerating voltage of 80 kV on carbon coated cooper TEM grids. A minimum of 1000 particles were computer-analysed and measured to obtain a size distribution (Figure 8). NPs surface charge was determined by zeta potential measurements using a Malvern ZetaSizer Nano ZS (Malvern Instruments, Malvern, UK) operating at a light source wavelength of 532 nm and a fixed scattering angle of 173°. Measurements were performed in the colloidal NPs solution after synthesis, with a pH of 7 for the negatively charged sodium citrate coated NPs and with a pH of 5 for the positively charged chitosan coated NPs.

Exposure conditions

NPs dispersed in the corresponding synthesis solvents were used in this study at different concentrations. Prior to the exposure of cells to the NPs, a serial dilution was performed to obtain the desired concentrations which ranged from 0.05 to 0.8 μg/cm². These concentrations were calculated as administered dose and not as delivered dose. This is the amount of NPs added to a cell culture well in correlation to the total surface area of the well, referred to as the administered dose. These administered concentrations expressed as μg/cm² convert to 6.25×10^{12}, 1.25×10^{12}, 2.5×10^{12} and 5×10^{12} NPs/ml. For the analysis of cytotoxicity (CTB, LDH), cells grown in 96-well plates were exposed to the NPs for 4, 24, 48 or 72 hours. ROS production was measured in 24-well plates. Here, the cells were exposed to NPs for up to 4 hours.

Cytotoxicity assays

Two different assays were performed to determine the viability of the cells. The viability of the cell is affected by various factors, one of which is cytotoxicity. One method to determine cytotoxicity is to measure the integrity of the cell membrane. When the membrane is disrupted and damaged, which can happen as a result of exposure to NPs, necrosis occurs and LDH is released into the supernatant. Measuring the amount of LDH is therefore a good parameter for determination of cell membrane integrity. To complement this, the CTB assay which measures the viability of the cells by their metabolism and also refers to the proliferation of the cells, was used as a second cytotoxicity test. Taken together, the results of these assays supply a clear picture of the cells well-being.

In house lactate dehydrogenase assay

For the determination of membrane integrity, a modified lactate dehydrogenase assay was used [64]. In brief, 50 μl of a solution containing 1 mg/ml NADH and 0.75 mM pyruvate was added to 10 μl test supernatant. The supernatants were incubated for 37°C for 30 minutes after

Table 4 Overview of the NPs used in this study

NPs name	Size (nm)	Surface coating	Z-potential (mV)	NPs/ml	mg/ml
Ag-SC	10 ± 2.0	Sodium citrate	−50 ± 1.4	1×10^{12}	0.027
Ag-CHIT-L	10 ± 4.0	Chitosan 0.001%	+25 ± 0.9	5×10^{12}	0.027
Ag-CHIT-M	10 ± 4.0	Chitosan 0.01%	+43 ± 0.4	5×10^{12}	0.027
Ag-CHIT-H	10 ± 4.0	Chitosan 0.1%	+60 ± 0.9	5×10^{12}	0.027
Au-SC	10 ± 1.5	Sodium citrate	−45 ± 0.2	3×10^{12}	0.032
Au-CHIT-L	7 ± 3.0	Chitosan 0.001%	+23 ± 1.0	5×10^{12}	0.025
Au-CHIT-M	7 ± 3.0	Chitosan 0.01%	+40 ± 0.6	5×10^{12}	0.025
Au-CHIT-H	7 ± 3.0	Chitosan 0.1%	+65 ± 1.0	5×10^{12}	0.025

The designation of NP name used in this table is used throughout the text. NPs size was determined by determining the size distribution of NPs in TEM images (N = 3, ±SEM). Zeta potential was measured in the colloidal solutions of the NPs after synthesis at a pH of 7 for the citrate coated NPs and a pH of 5 for the chitosan coated NPs.

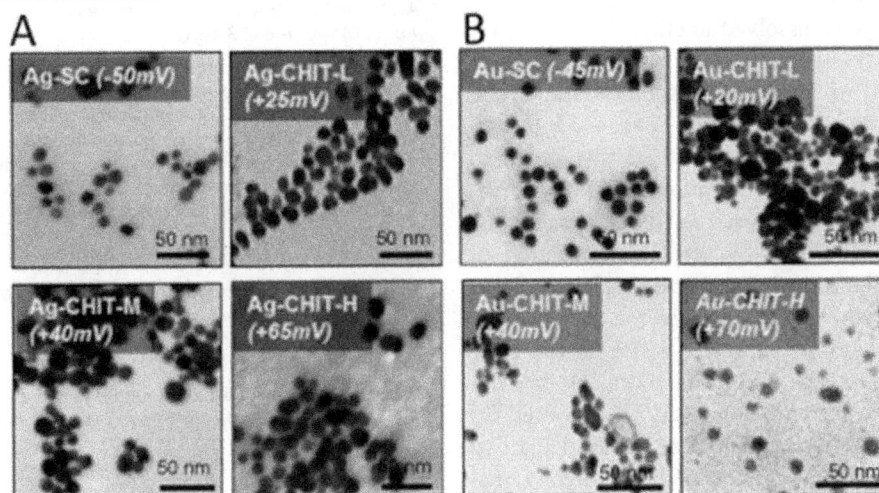

Figure 8 TEM images of sodium citrate or chitosan coated Ag (A) and Au (B) NPs after synthesis.

which 50 μl of 2,4-dinitrophenylhydrazine dissolved in 1 M HCl was added to all wells and the plate was incubated at room temperature (RT) for 20 minutes. Finally, 50 μl of 4 M sodium hydroxide (NaOH) was added and the absorbance at 540 nm was measured using a Tecan plate reader (Tecan infinity 200 pro, Tecan, Maennedorf, Switzerland) after leaving the plate to incubate for 5 minutes at RT.

CellTiter-Blue ® (CTB) assay

Cells were exposed to the particles as mentioned above. After exposure, cell viability was determined using the CTB cell viability kit (Promega, Madison, USA). For all tests performed, untreated cells were used as a negative and 0.1% Triton X-100 treated cells were used as a positive control. Fluorescence was measured at 590 nm upon excitation at 560 nm using a plate reader (Infinity 200 Pro, Tecan, Groedig, Austria).

Interference control for LDH and CTB assays

To control for interference of NPs within the detection of LDH released during exposure of A549 cells, cells were seeded in the same fashion as previously described. Au and Ag NPs were added at all the concentrations used throughout this study and left for 4 hours, after which Triton X-100 (0.1%) was added to all wells for 10 minutes to induce cell lysis. Control cells were treated with medium only. After this stimulation, the supernatant was removed and centrifuged at $25000 \times g$ to remove NPs, and the LDH assay was performed as previously described. This would enable detection of LDH bound to the NPs and therefore removed during centrifugation, which would result in false negative results. These control experiments were conducted with one biological replicate.

To determine if NPs interfere with either the optical readout of the CTB assay, or with any of the assay components, A549 cells were seeded at different cell densities (2×10^5, 4×10^5, 6×10^5, 8×10^5, 1×10^6), left to adhere for 4 hours and then exposed to Au and Ag NPs at all concentrations used in this study and to medium only. The CTB reagent was then added and the protocol followed as previously described. The use of different cell densities would allow the determination of whether present NPs interfere with the CTB reagent, or with the fluorescence readout, as there would be a deviation from each respective medium only control cell population. These control experiments were conducted with one biological replicate.

Detection of reactive oxygen species
DCFH-DA assay for cellular ROS production

Cell-mediated reactive oxygen species production was detected by a carboxy-dichlorofluorescein diacetate (carboxy-DCFH-DA) assay. This assay was carried out according to the manufacturer's instructions (Sigma-Aldrich, St. Louis, USA) [65]. Briefly, 5 μl carboxy-DCFH-DA (1 mM) was added to each well of a 24-well plate for 1 hour exposures. For longer exposures, 5 μl carboxy-DCFH-DA was added 60 minutes before the end of the exposure period. Cells exposed to cell culture medium only acted as negative controls, 500 μM H_2O_2 was used as a positive control. At the end of the exposure period, cells were washed twice with 500 μl phosphate buffer saline (PBS) and harvested using 75 μl Trypsin. 425 μl cell culture medium containing FCS was used to neutralize the trypsin reaction. Cells were then immediately analysed using a flow cytometer (FACS-Canto™, Becton Dickinson). All steps, including the flow cytometry assessment, were carried out with minimal light to avoid effects caused by the light sensitivity of the dye.

Cell-free ROS production

DCFH-DA powder was dissolved in ethyl alcohol to prepare a 1 mM stock. 10 ml 0.01 M sodium hydroxide (NaOH) was added to 2.5 ml stock solution and left in darkness for 30 min to deacetylate. Then 487.5 ml sodium phosphate buffer with a pH of 7.2 was added to the solution (final DCFH concentration of 5 μM). Horseradish peroxidase (HRP) was used as the catalyst for the oxidation reaction of DCFH-DA at a concentration of 0.5 units/ml. 200 μl of this dye and 100 μl NPs in solvent were added to the corresponding wells in a 96-well plate and left to incubate for 15 minutes at 37°C in the plate reader before fluorescence was measured using 485 nm excitation and 530 nm emission every 10 minutes for 4 hours. All steps were carried out in minimal light to avoid the analysis of artefacts caused by the light sensitivity of the dye.

Statistical analysis

Data are displayed as mean (±SEM) and were analysed using GraphPad Prism (GraphPad Software Inc., USA). A minimum of three replicates were performed for each method used. Statistical analysis was performed using a one-way analysis of variance (one-way Anova) and Tukey's test was used as a post-hoc analysis. *P-values* < 0.05 were regarded as statistically significant. Spearman's rank correlation was performed to analyse trends observed in the ROS production of cells.

Additional files

Additional file 1: Cell membrane integrity, as measured by an increase in LDH-release, following a 4 h exposure of the different cell lines to Ag and Au NPs. An increase in LDH-release is indicated by a decrease in the membrane integrity. A549 cells (**A**, **B**; means ± SEM of n = 6), BEAS-2B (**C**, **D**; means ± SEM of n = 3) and NHBE cells (**E**, **F**; means ± SEM of n = 3). P-value * < 0.05. Cells treated with medium only were used as negative control (=100%).

Additional file 2: Cell membrane integrity, as measured by an increase in LDH-release, following a 48 h exposure of the different cell lines to Ag and Au NPs. An increase in LDH-release is indicated by a decrease in the membrane integrity. A549 cells (**A**, **B**; means ± SEM of n = 6), BEAS-2B (**C**, **D**; means ± SEM of n = 3) and NHBE cells (**E**, **F**; means ± SEM of n = 3). P-value * < 0.05. Cells treated with medium only were used as negative control (=100%).

Additional file 3: Assessment of NP interference with the LDH assay. LDH release was induced using Triton-X-100 in A549 cells, prior to the measurement.

Additional file 4: Cell viability following a 4 h exposure to charged Ag and Au NPs of A549 cells (A, B; means ± SEM of n = 6), BEAS-2B cells (C, D; means ± SEM of n = 3) and NHBE cells (E, F; means ± SEM of n = 3). P-values * < 0.05, Cells treated with medium only were used as control (100%).

Additional file 5: Cell viability following a 48 h exposure to charged Ag and Au NPs of A549 cells (A, B; means ± SEM of n = 6), BEAS-2B cells (C, D; means ± SEM of n = 3) and NHBE cells (E, F; means ± SEM of n = 3). P-values * < 0.05, Cells treated with medium only were used as control (100%).

Additional file 6: Assessment of interference of Ag NPs (A, B) and Au (C, D) with the CTB assay.

Additional file 7: ROS production measured using the DCFH-DA assay following a 1 h exposure of A549 cells (A, B, means ± SEM of n = 3), BEAS-2B cells (C, D, means ± SEM of n = 3) and NHBE cells (E, F means ± SEM of n = 4) to functionalized Ag and Au NPs. Spearman's rank coefficients were calculated for each NP to assess possible charge dependent increases in ROS production. In addition, the coefficients of the highest concentrations of each NP surface charge were calculated to determine if ROS production was charge dependent. P-values * < 0.05, ** < 0.01.

Additional file 8: ROS production of the NPs solvents in a cell-free system. Fluorescence was measured at 530 nm (excitation 485 nm) every 10 minutes after an initial incubation period of 15 minutes. Means ± SEM of n = 3 are shown.

Additional file 9: Surface charge (mV) of NPs as synthesised and after a 4 hour incubation in different cell culture media.

Competing interests

The authors declare that they have no competing interests.

Authors' contributions

PS performed cytotoxicity measurements in A549 and BEAS-2B cells, ROS measurements in all cell types, cell-free ROS measurements and tight junction stainings/TEER measurements in A549 cells and wrote the manuscript. EH established and performed TEER measurements and tight junction stainings measurements in BEAS-2B and NHBE cells. MB performed cytotoxicity measurements in NHBE cells, as well as the interference control measurements and assisted in the preparation of the manuscript. UT was responsible for cell culturing. EC and NT synthesised and characterised the NPs and performed zeta potential measurement in cell media. JZ assisted in the cell-free ROS measurements. MR helped in the design and interpretation of the cell-free ROS measurements. GJO was involved in the design of the study as well as in the preparation of the manuscript. MR, MH and VP designed parts of the study and reviewed the manuscript. AD designed and supervised the study and assisted in the preparation of the manuscript. All authors read and approved the final manuscript.

Funding

This study was supported by the EU 7th framework programme, Marie Curie Actions, Network for Initial Training NanoTOES (PITN-GA-2010-264506).

Author details

[1]Department of Molecular Biology, Paris Lodron-University of Salzburg, Hellbrunnerstr. 34A-5020 Salzburg, Austria. [2]Institute Catalá de Nanotecnologia, Barcelona, Spain. [3]Institute for Work and Health, Lausanne, Switzerland. [4]Instiue for Occupational Medicine (IOM) Singapore, Downtown Core, Singapore. [5]Biomedical Sciences, Salzburg University of Applied Sciences, Puch, Salzburg, Austria.

References

1. Taylor U, Barchanski A, Garrels W, Klein S, Kues W, Barcikowski S, et al. Toxicity of gold nanoparticles on somatic and reproductive cells. Adv Exp Med Biol. 2012;733:125–33.
2. Stoehr LC, Gonzalez E, Stampfl A, Casals E, Duschl A, Puntes V, et al. Shape matters: effects of silver nanospheres and wires on human alveolar epithelial cells. Part Fibre Toxicol. 2011;8:36.
3. Suresh AK, Pelletier DA, Wang W, Morrell-Falvey JL, Gu B, Doktycz MJ. Cytotoxicity induced by engineered silver nanocrystallites is dependent on surface coatings and cell types. Langmuir. 2012;28:2727–35.
4. Gerber A, Bundschuh M, Klingelhofer D, Groneberg DA. Gold nanoparticles: recent aspects for human toxicology. J Occup Med Toxicol. 2013;8:32.
5. Sharma VK, Yngard RA, Lin Y. Silver nanoparticles: green synthesis and their antimicrobial activities. Adv Colloid Interf Sci. 2009;145:83–96.
6. Bartlomiejczyk T, Lankoff A, Kruszewski M, Szumiel I. Silver nanoparticles - allies or adversaries? Ann Agric Environ Med. 2013;20:48–54.

7. Kruszewski M, Brzoska K, Brunborg G, Asare N, Dobrzynska M-g, Dusinska M, et al. Toxicity of silver nanomaterials in higher eukaryotes. Adv Mol Toxicol. 2011;5:179–218.

8. Jena P, Mohanty S, Mallick R, Jacob B, Sonawane A. Toxicity and antibacterial assessment of chitosan-coated silver nanoparticles on human pathogens and macrophage cells. Int J Nanomedicine. 2012;7:1805–18.

9. Hackenberg S, Scherzed A, Kessler M, Hummel S, Technau A, Froelich K, et al. Silver nanoparticles: evaluation of DNA damage, toxicity and functional impairment in human mesenchymal stem cells. Toxicol Lett. 2011;201:27–33.

10. Ng CT, Li JJ, Gurung RL, Hande MP, Ong CN, Bay BH, et al. Toxicological profile of small airway epithelial cells exposed to gold nanoparticles. Exp Biol Med (Maywood). 2013;238:1355–61.

11. Pan Y, Wu Q, Liu R, Shao M, Pi J, Zhao X, et al. Inhibition effects of gold nanoparticles on proliferation and migration in hepatic carcinoma-conditioned HUVECs. Bioorg Med Chem Lett. 2014;24:679–84.

12. Tiedemann D, Taylor U, Rehbock C, Jakobi J, Klein S, Kues WA, et al. Reprotoxicity of gold, silver, and gold-silver alloy nanoparticles on mammalian gametes. Analyst. 2014;139:931–42.

13. Paul W, Shelma R, Sharma CP. Alginate Encapsulated Anacardic Acid-Chitosan Self Aggregated Nanoparticles for Intestinal Delivery of Protein Drugs. J Nanopharm Drug Deliv. 2013;1:82–91.

14. Limpeanchob N, Tiyaboonchai W, Lamlertthon S, Viyoch J, Jaipan S. Efficacy and Toxicity of Amphotericin B-Chitosan Nanoparticles in Mice with Induced Systemic Candidiasis. (Naresuan Univ J). 2013;14:27–34.

15. Tiyaboonchai W. Chitosan nanoparticles: a promising system for drug delivery. Naresuan Univ J. 2013;11:51–66.

16. Tabata Y, Ikada Y. Macrophage phagocytosis of biodegradable microspheres composed of lactic acid/glycolic acid homo-and copolymers. J Biomed Mater Res. 1988;22:837–58.

17. Ma Z, Lim TM, Lim L-Y. Pharmacological activity of peroral chitosan–insulin nanoparticles in diabetic rats. Int J Pharm. 2005;293:271–80.

18. Bowman K, Leong KW. Chitosan nanoparticles for oral drug and gene delivery. Int J Nanomedicine. 2006;1:117.

19. Garbuzenko OB, Winkler JS, Tomassone MS, Minko T. Biodegradable Janus Nanoparticles for Local Pulmonary Delivery of Hydrophilic and Hydrophobic Molecules to the Lungs. Langmuir. 2014;30(43):12941–9.

20. Guo X, Zhang X, Ye L, Zhang Y, Ding R, Hao Y, et al. Inhalable microspheres embedding chitosan-coated PLGA nanoparticles for 2-methoxyestradiol. J Drug Target. 2014;22:421–7.

21. Schrand AM, Rahman MF, Hussain SM, Schlager JJ, Smith DA, Ali SF. Metal-based nanoparticles and their toxicity assessment. Wires Nanomed Nanobi. 2010;2:544–68.

22. Nel A, Xia T, Madler L, Li N. Toxic potential of materials at the nanolevel. Science. 2006;311:622–7.

23. Foldbjerg R, Dang DA, Autrup H. Cytotoxicity and genotoxicity of silver nanoparticles in the human lung cancer cell line, A549. Arch Toxicol. 2011;85:743–50.

24. Limbach LK, Wick P, Manser P, Grass RN, Bruinink A, Stark WJ. Exposure of engineered nanoparticles to human lung epithelial cells: influence of chemical composition and catalytic activity on oxidative stress. Environ Sci Technol. 2007;41:4158–63.

25. Pourova J, Kottova M, Voprsalova M, Pour M. Reactive oxygen and nitrogen species in normal physiological processes. Acta Physiol. 2010;198:15–35.

26. Monopoli MP, Pitek AS, Lynch I, Dawson KA. Formation and characterization of the nanoparticle-protein corona. Methods Mol Biol. 2013;1025:137–55.

27. Casals EPT, Duschl A, Oostingh GJ, Puntes V. Time evolution of the nanoparticle protein corona. Am Chem Soc. 2010;4:10.

28. Saptarshi SR, Duschl A, Lopata AL. Interaction of nanoparticles with proteins: relation to bio-reactivity of the nanoparticle. J Nanobiotechnol. 2013;11:26.

29. Foucaud L, Wilson MR, Brown DM, Stone V. Measurement of reactive species production by nanoparticles prepared in biologically relevant media. Toxicol Lett. 2007;174:1–9.

30. Bouwmeester H, Lynch I, Marvin HJ, Dawson KA, Berges M, Braguer D, et al. Minimal analytical characterization of engineered nanomaterials needed for hazard assessment in biological matrices. Nanotoxicology. 2011;5:1–11.

31. Bachand GD, Allen A, Bachand M, Achyuthan KE, Seagrave JC, Brozik SM. Cytotoxicity and inflammation in human alveolar epithelial cells following exposure to occupational levels of gold and silver nanoparticles. J Nanopart Res. 2012;14:1–10.

32. Politis M, Pilinis C, Lekkas T. Ultrafine particles (UFP) and health effects. Dangerous. Like no other PM? Review and analysis. Global NEST J. 2008;10:439–52.

33. Bakand S, Hayes A, Dechsakulthorn F. Nanoparticles: a review of particle toxicology following inhalation exposure. Inhal Toxicol. 2012;24:125–35.

34. Saieg MA, Cury PM, Godleski JJ, Stearns R, Duarte LG, D'Agostino L, et al. Differential elemental distribution of retained particles along the respiratory tract. Inhal Toxicol. 2011;23:459–67.

35. Löndahl J, Möller W, Pagels JH, Kreyling WG, Swietlicki E, Schmid O. Measurement techniques for respiratory tract deposition of airborne nanoparticles: A critical review. J Aerosol Med Pulm Drug Deliv. 2013.

36. Witschi HR, Last JA. Toxic responses of the respiratory system. Casarett and Doull's Toxicology: The basic science of poisons. New York: McGraw-Hill; 2001. p. 515–34.

37. Siegmann K, Scherrer L, Siegmann H. Physical and chemical properties of airborne nanoscale particles and how to measure the impact on human health. J Mol Struct THEOCHEM. 1998;458:191–201.

38. Lundborg M, Johard U, Låstbom L, Gerde P, Camner P. Human alveolar macrophage phagocytic function is impaired by aggregates of ultrafine carbon particles. Environ Res. 2001;86:244–53.

39. Blank F, Gehr P, Rothen-Rutishauser B. In vitro human lung cell culture models to study the toxic potential of nanoparticles. Nanotoxicity: From in vitro, in vivo models to health risks. Chichester, England: Wiley; 2009. p. 379–95.

40. Oberdörster G, Oberdörster E, Oberdörster J. Nanotoxicology: An emerging discipline evolving from studies of ultrafine particles. Environ Health Perspect. 2005;113:823–39.

41. Muhlfeld C, Gehr P, Rothen-Rutishauser B. Translocation and cellular entering mechanisms of nanoparticles in the respiratory tract. Swiss Med Wkly. 2008;138:387.

42. Giard DJ, Aaronson SA, Todaro GJ, Arnstein P, Kersey JH, Dosik H, et al. In vitro cultivation of human tumors: establishment of cell lines derived from a series of solid tumors. J Natl Cancer Inst. 1973;51:1417–23.

43. Albright CD, Jones RT, Hudson EA, Fontana JA, Trump BF, Resau JH. Transformed human bronchial epithelial cells (BEAS-2B) alter the growth and morphology of normal human bronchial epithelial cells in vitro. Cell Biol Toxicol. 1990;6:379–98.

44. Lechner J, LaVeck M. A serum-free method for culturing normal human bronchial epithelial cells at clonal density. J Tissue Cult Methods. 1985;9:43–8.

45. Lieber M, Todaro G, Smith B, Szakal A, Nelson-Rees W. A continuous tumor-cell line from a human lung carcinoma with properties of type II alveolar epithelial cells. Int J Cancer. 1976;17:62–70.

46. Smith B. Cell line A549: a model system for the study of alveolar type II cell function. Am Rev Respir Dis. 1977;115:285–93.

47. AshaRani P, Low Kah Mun G, Hande MP, Valiyaveettil S. Cytotoxicity and genotoxicity of silver nanoparticles in human cells. ACS Nano. 2008;3:279–90.

48. Zhang H, Xia T, Meng H, Xue M, George S, Ji Z, et al. Differential expression of syndecan-1 mediates cationic nanoparticle toxicity in undifferentiated versus differentiated normal human bronchial epithelial cells. ACS Nano. 2011;5:2756–69.

49. Oostingh GJ, Schlickum S, Friedl P, Schon MP. Impaired induction of adhesion molecule expression in immortalized endothelial cells leads to functional defects in dynamic interactions with lymphocytes. J Investig Dermatol. 2007;127:2253–8.

50. Lin H, Li H, Cho HJ, Bian S, Roh HJ, Lee MK, et al. Air-liquid interface (ALI) culture of human bronchial epithelial cell monolayers as an in vitro model for airway drug transport studies. J Pharm Sci. 2007;96:341–50.

51. Qi L-F, Xu Z-R, Li Y, Jiang X, Han X-Y. In vitro effects of chitosan nanoparticles on proliferation of human gastric carcinoma cell line MGC803 cells. World J Gastroenterol. 2005;11:5136.

52. Hu Y-L, Qi W, Han F, Shao J-Z, Gao J-Q. Toxicity evaluation of biodegradable chitosan nanoparticles using a zebrafish embryo model. Int J Nanomedicine. 2011;6:3351–9.

53. Chen Z, Wang Z, Chen X, Xu H, Liu J. Chitosan-capped gold nanoparticles for selective and colorimetric sensing of heparin. J Nanopart Res. 2013;15:1–9.

54. Wei D, Qian W. Facile synthesis of Ag and Au nanoparticles utilizing chitosan as a mediator agent. Colloids Surf B: Biointerfaces. 2008;62:136–42.

55. Amidi M, Hennink WE. Chitosan-based formulations of drugs, imaging agents and biotherapeutics. Adv Drug Deliv Rev. 2010;62:1–2.

56. Ojea-Jiménez I, García-Fernández L, Lorenzo J, Puntes VF. Facile preparation of cationic gold nanoparticle-bioconjugates for cell penetration and nuclear targeting. ACS Nano. 2012;6:7692–702.

57. Ekstrand-Hammarstrom B, Akfur CM, Andersson PO, Lejon C, Osterlund L, Bucht A. Human primary bronchial epithelial cells respond differently to titanium dioxide nanoparticles than the lung epithelial cell lines A549 and BEAS-2B. Nanotoxicology. 2012;6:623–34.

58. Fleischer CC, Payne CK. Nanoparticle surface charge mediates the cellular receptors used by protein-nanoparticle complexes. J Phys Chem B. 2012;116:8901–7.

59. Izak-Nau E, Voetz M, Eiden S, Duschl A, Puntes VF. Altered characteristics of silica nanoparticles in bovine and human serum: the importance of nanomaterial characterization prior to its toxicological evaluation. Part Fibre Toxicol. 2013;10:56.

60. Marambio-Jones C, Hoek EM. A review of the antibacterial effects of silver nanomaterials and potential implications for human health and the environment. J Nanopart Res. 2010;12:1531–51.

61. Roche M, Rondeau P, Singh NR, Tarnus E, Bourdon E. The antioxidant properties of serum albumin. FEBS Lett. 2008;582:1783–7.

62. Gaisberger M, Šanovic R, Dobias H, Kolarz P, Moder A, Thalhamer J, et al. Effects of ionized waterfall aerosol on pediatric allergic asthma. J Asthma. 2012;49:830–8.

63. Lenz A-G, Stoeger T, Cei D, Schmidmeir M, Pfister N, Burgstaller G, Lentner B, Eickelberg O, Meiners S, Schmid O: Efficient bioactive delivery of aerosolized drugs to human pulmonary epithelial cells cultured at air-liquid interface conditions. Am J Respir Cell Mol Biol 2014 526-535

64. Brown DM, Stone V, Findlay P, MacNee W, Donaldson K. Increased inflammation and intracellular calcium caused by ultrafine carbon black is independent of transition metals or other soluble components. Occup Environ Med. 2000;57:685–91.

65. Robinson JP, Bruner LH, Bassoe CF, Hudson JL, Ward PA, Phan SH. Measurement of intracellular fluorescence of human monocytes relative to oxidative metabolism. J Leukoc Biol. 1988;43:304–10.

66. Monopoli MP, Walczyk D, Campbell A, Elia G, Lynch I, Baldelli Bombelli F, et al. Physical – chemical aspects of protein corona: relevance to in vitro and in vivo biological impacts of nanoparticles. J Am Chem Soc. 2011;133:2525–34.

67. Casals E, Pfaller T, Duschl A, Oostingh GJ, Puntes VF. Hardening of the Nanoparticle-Protein Corona in Metal (Au, Ag) and Oxide (Fe3O4, CoO, and CeO2) Nanoparticles. Small. 2011;7:3479–86.

68. Aschberger K, Johnston HJ, Stone V, Aitken RJ, Hankin SM, Peters SA, et al. Review of carbon nanotubes toxicity and exposure–appraisal of human health risk assessment based on open literature. Crit Rev Toxicol. 2010;40:759–90.

69. Li N, Xia T, Nel AE. The role of oxidative stress in ambient particulate matter-induced lung diseases and its implications in the toxicity of engineered nanoparticles. Free Radic Biol Med. 2008;44:1689–99.

70. Stone V, Donaldson K. Nanotoxicology: signs of stress. Nat Nanotechnol. 2006;1:23–4.

71. Weissenberg A, Sydlik U, Peuschel H, Schroeder P, Schneider M, Schins RP, et al. Reactive oxygen species as mediators of membrane-dependent signaling induced by ultrafine particles. Free Radic Biol Med. 2010;49:597–605.

72. Turkevich J, Stevenson PC, Hillier J. A study of the nucleation and growth processes in the synthesis of colloidal gold. Discussions of the Faraday Society. 1952;11:55–75.

73. Jana NR, Gearheart L, Murphy CJ. Seeding Growth for Size Control of 5–40 nm Diameter Gold Nanoparticles. Langmuir. 2001;17:6782–6786.

11

Evaluating the effect of assay preparation on the uptake of gold nanoparticles by RAW264.7 cells

Simona Bancos and Katherine M Tyner[*]

Abstract

Background: Cell culture conditions can greatly influence the results of nanoparticle (NP) uptake assays. In this study, 10 nm gold nanoparticles (AuNPs) and RAW 264.7 macrophages were used as a model system, while instrumental neutron activation analysis (NAA) was used as the elemental analysis technique to determine AuNP levels produced by the various culturing conditions. Static plate-based and insert-based culture conditions were compared with a dynamic suspension culture to evaluate the conditions' effect on the rate and extent of AuNP uptake.

Results: The results indicate that a dynamic culturing condition allows for the greatest NP uptake (approximately 3-5 times over the adherent conditions), whereas the plate-based assays have the initial highest rate of NP incorporation.

Conclusions: These data highlight the importance of judiciously choosing the assay conditions prior to evaluating NP uptake.

Keywords: Gold nanoparticle, Nanoparticle uptake, *In vitro* assay, Nanoparticle dosimetry, Macrophage

Introduction

In vitro assays are commonly used during the drug discovery process to provide a rapid assessment of a wide variety of pharmacological endpoints including drug uptake, cell proliferation, cytotoxicity, etc. With the emergence of nanotechnology in medicine, using these *in vitro* assays for drugs containing nanomaterials was a logical continuation of the discovery pathway. However, due to multiple aspects of nanomaterials, including high surface area, enhanced surface activity, and the particulate nature of many drugs incorporating nanomaterials, the direct translation of these *in vitro* assays to nanomaterials has not been straightforward. There have been multiple reports throughout the literature of nanomaterials interfering with cell-based *in vitro* assays, producing false positives, false negatives, or nonsensical data [1,2].

There have been multiple studies evaluating nanoparticle (NP) dosimetry in *in vitro* systems. Teeguarden et al [3] reviewed the many ways different particle dosimetry metrics may impact particle uptake and analysis. In particular, sedimentation through gravitation and agglomeration were highlighted as was the appropriate calculation of delivered NP dose [3]. These considerations are now widespread throughout the field, although some groups have found that there is little effect of sedimentation with some NPs, eliminating many artifactual dosing concerns [4]. In order to demonstrate and/or alleviate the effect of NP sedimentation on apparent NP cellular uptake, several groups have attempted non-conventional culturing techniques. For example, Cho et al. [5] used gold NPs and either traditional plate based assays or inverted assays to monitor cellular uptake of NPs.

There are multiple methods that may be used to monitor NP uptake, each with their unique set of benefits and restrictions. For example, microscopy techniques have been used to determine cellular uptake of nanomaterials. For optical microscopy, sample preparation is often facile, however, NPs must either be labeled or intrinsically fluorescent/luminescent/reflectant, as well as be large enough to be discriminated in the image. Transmission electron microscopy (TEM) can also be used for a confirmation of cellular uptake, but it is rarely used as a direct or quantitative measure of uptake due to the extremely small sample size evaluated. For non-carbon NPs, quantitative assays

* Correspondence: Katherine.Tyner@fda.hhs.gov
Center for Drug Evaluation and Research, Food and Drug Administration, Building 51 Room 4159, 10903 New Hampshire Ave., Silver Spring, MD 20993, USA

had been developed to evaluate nanomaterial levels inside the cells such as chemiluminescence measurements, inductively coupled plasma mass spectrometry (ICP-MS), laser desorption/ionization mass spectrometry, and UV-Vis spectrometry [6,7]. Many of these techniques require dilution or digestion of cellular matrix and NP, introducing variability into the measurements, especially for small sample sizes and hard-to-digest materials.

Instrumental neutron activation analysis (NAA) is an elemental analysis technique that can detect gold down to the parts per billion (ppb) level. While not a widespread technique due, in part, to the requirement of a nuclear reactor in order to irradiate the samples, it has some benefits, particularly in the area of sample preparation. The most obvious benefit is the ability to interrogate samples "as is" (i.e. with no digestion or additional sample preparation steps). When interrogating hundreds of samples at a time, this is a non-trivial consideration. In addition, the lack of manipulation may reduce sample variability, due to the reduction of sample processing steps. In this study, NAA is used as the elemental analysis technique to evaluate the uptake of 10 nm AuNPs by the mouse macrophage-like cell line RAW264.7 under different assay conditions. Experimental parameters and culture conditions were varied to determine their impact on the cellular uptake.

Results and discussion
NPs characterization
DLS and TEM were used to monitor the size of the AuNPs. Figure 1A contains the summary of the size characterization of the AuNPs dispersed in both media and water before and after incubation at 37°C. In general, NPs dispersed in water were spherical and approximately 10 nm in diameter (Figure 1B), with some agglomeration noted in the intensity weighted DLS histogram as well as the TEM micrographs (Figure 1B-C). This agglomeration translates into a larger overall Zave seen in Figure 1A. Agglomeration of the NPs in water increased over the 72 hour incubation at 37°C as shown by both the increase in Zave and intensity weighted histogram (as indicated by the appearance and increase of additional peaks at larger diameters). Zeta potential decreased during this time period.

To better understand the AuNPs under experimental conditions (full media at 37°C over 72 hours), AuNPs were also dispersed in cell culture media. DLS data is confounded by the presence of serum proteins, which nominally have the same size as the AuNPs (Figure 1D). However, a large peak not present in the media control appears after the addition of the AuNPs. This peak shift and broadening, along with the increase in Zave and lessening of zeta potential is consistent with serum protein opsonization of the AuNPs [8]. Observations with TEM

also noted more agglomerates, although this is a qualitative observation (Figure 1E). Further agglomeration was also noted over time in both DLS and TEM. No change was observed in the media control parameters. AuNPs tested below detection levels for endotoxin as determined by *Limulus amoebocytes* lysate (LAL) gel clot assay.

In order to verify AuNPs dosing solution concentrations, culture media containing increasing concentrations of AuNPs (10 nm; 0.0005 g/L- 0.05 g/L) along with AuNPs stock concentration (4.56 g/L as determined by TGA) and controls were prepared and assayed by NAA (Figure 1F). PBS and media without AuNPs were used as negative controls. PBS and control media did not have any detectable levels of AuNPs. For the media dilutions as well as the stock solutions, the expected media concentrations determined by TGA corresponded to the actual concentrations as measured by NAA.

NAA limits of detection
NAA was used as an elemental analysis technique to eliminate the need for additional sample processing steps. In order to determine the utility of the technique for the study, the limits of detection within the cell line was first determined. NAA detected AuNPs levels in all AuNPs treated macrophage groups. The range was ~0.01 µg (in 0.2×10^4 cells exposed to 0.01 g/L AuNPs) to ~14.5 µg (in 20×10^4 cells exposed to 0.1 g/L). The background levels (cells exposed to media without AuNPs) were 0.001- 0.03 µg. By normalizing these values to cell number, the results indicate that NAA can detect AuNP concentrations in as few as 2000 RAW 264.7 cells. Previous work has shown that the majority of the AuNPs are internalized within the cells and not adhered to the outer membrane [9].

Time dependent uptake of AuNPs in different cell culture conditions
The time dependent uptake of AuNPs by RAW264.7 cells was evaluated using different culture set-ups. RAW 264.7 cells present an adherent phenotype and are typically cultured in plates in upright conditions. This "basic plate" set-up represents the typical cell assay configuration, where cells are grown in adherent conditions in a multi-well plate, and dosed by media containing the test compound covering the cells. However, in this set-up, there exists the possibility that NPs will settle onto the cell surface which can cause non-representative uptake [3]. To avoid this scenario, a second type of adherent conditions termed "insert" was employed. In this set-up, cells were cultured in a trans-well membrane and covered with media on the apical side. The basal side of the membrane contained the dosing solution (media with AuNPs). The insert set-up allows the RAW 264.7 cells to incorporate AuNPs that are only dispersed within the bulk media,

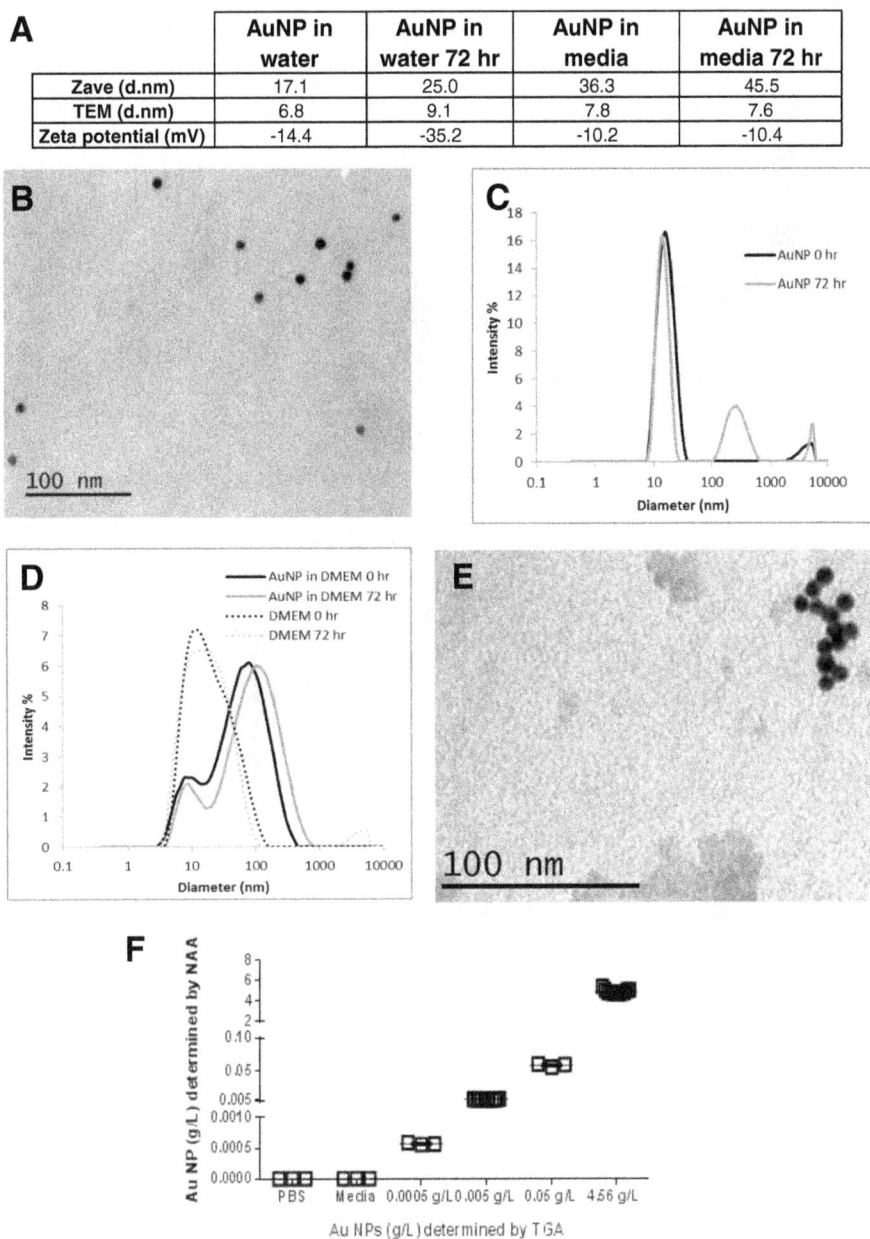

A	AuNP in water	AuNP in water 72 hr	AuNP in media	AuNP in media 72 hr
Zave (d.nm)	17.1	25.0	36.3	45.5
TEM (d.nm)	6.8	9.1	7.8	7.6
Zeta potential (mV)	-14.4	-35.2	-10.2	-10.4

Figure 1 AuNPs (10 nm) characterization by DLS, TEM and NAA. A. Summary of size and zeta potential results for AuNPs dispersed in water and media before and after 72 hours incubation at 37°C. **B**. TEM micrograph of 10 nm AuNP dispersed in water. **C**. Representative DLS histogram (intensity weighted) for 10 nm AuNPs before and after incubation at 37°C for 72 hours. **D**. DLS of AuNPs dispersed in media and media control before and after incubation at 37°C for 72 hours. **E**. TEM micrograph of 10 nm AuNPs dispersed in full cell culture media. **F**. Determination of AuNPs concentration by NAA as compared to TGA.

thus avoiding artifacts of uptake due to NP sedimentation on the cell or plate surface.

Figure 2A shows the time course of AuNPs (10 nm) incorporation as monitored by pg AuNP/cell in adherent culture conditions (plate and insert set-up). In general, the uptake profiles are similar with an increase in AuNP concentration per cell over time, which peaks at ~12 hr. After 12 hours, the amount of AuNP per cell levels out

and then has decreased by 72 hours. Throughout the time course, AuNP incorporation is slightly higher in the plate set-up versus insert set-up and reaches a statistically significant difference at 48 h. It is important to mention that the level of AuNP was measured by NAA in a cell batch with a known number of cells, after which the value of AuNPs obtained in that batch was divided by the number of cells to obtain AuNP/cell. This calculation, however,

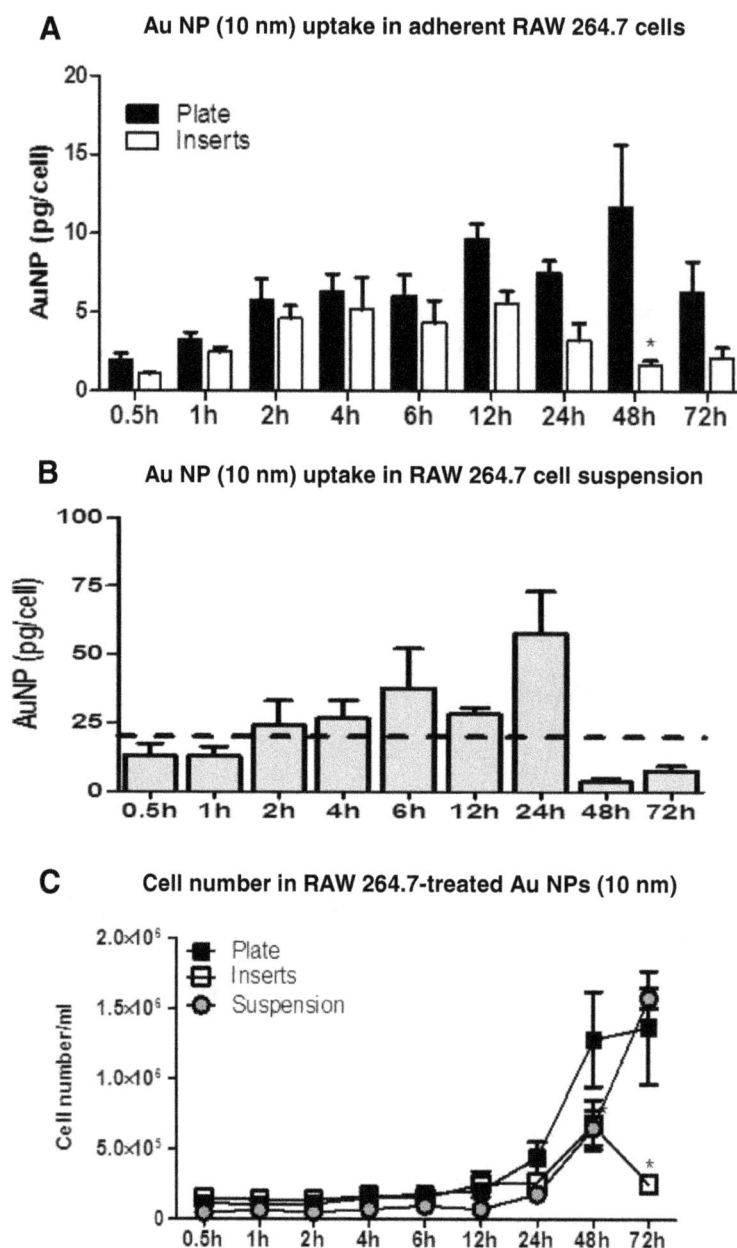

A Au NP (10 nm) uptake in adherent RAW 264.7 cells

B Au NP (10 nm) uptake in RAW 264.7 cell suspension

C Cell number in RAW 264.7-treated Au NPs (10 nm)

Figure 2 AuNP (10 nm) uptake in RAW 264.7 cells cultured in adherent (plate and insert) or suspension set-up and their proliferation pattern. **A**. AuNP (10 nm) uptake in RAW 264.7 cells show a time-dependent pattern with the highest concentrations determined at 12 h for both traditional and insert set-up. There was a significant difference between AuNPs measured in cells grown in plate versus AuNPs in cells cultured in insert set-up. **B**. AuNP uptake in RAW 264.7 cells cultured in suspension. The peak in AuNP (10 nm) uptake (pg/cell) was detected at 24 h. Except for the 24 h time-point all the time-points were close to the background levels. **C**. RAW 264.7 cells cultured in plates and in suspension proliferate the most when exposed to AuNPs (10 nm) compared to RAW 264.7 cells grown in inserts. Graph shows mean and SEM (N = 6; 2 experiments for Figure 4A and N = 3, 1 experiment for Figure 4B) *, p < 0.05 as determined by two-way ANOVA when comparing traditional to insert set-up.

does not take in consideration the heterogeneous uptake of RAW267.4 cells, but shows an average for the cell population.

RAW264.7 cells can also be cultured in suspension when using Teflon inserts and gentle agitation to prevent cell adhesion. Figure 2B shows the AuNP uptake of cells

under suspension conditions. In this case, the peak AuNP/cell is reached by ~24 hours. The analysis of the uptake, however, is confounded by the assay configuration. In the suspension set-up, the cells are harvested via centrifugation. Due to opsonization and the agglomeration noted in the stability studies, however, many of the AuNPs not incorporated with the cells sediment and form a pellet along with the cell pellet. It is, therefore, difficult to separate the AuNPs in media from the AuNPs incorporated by cells. In this manner, NAA suffers from the same limitation as those in many bulk elemental analysis techniques (such as ICPMS) in that there is no easy way to differentiate between internalized NPs and ones that co-precipitated during the course of the assay. In order to determine the background values (caused by AuNPs in the media sedimenting with the cell pellet), "no cell" controls were performed. These controls were prepared in the same way as the experimental samples, with the exception that cells were not included. NAA of the controls indicates a high background of free AuNPs, which is indicated by the dashed line in Figure 2B. Figure 2B shows the amount of AuNP per cell as compared to the background levels at 24 hours (dashed line). In general, the uptake profile trend is similar to the adherent set-ups with an initial increase in AuNPs/cell, with a decrease after the 24 hour time point.

All tested cell culture set-ups show an initial increase in cell incorporation followed by a decrease by the last time point of 72 hours. In order to understand the uptake profile, the cell proliferation profile was evaluated. Figure 2C shows the number of cells that correspond to each time point for the different set-ups. RAW264.7 cells have a typical doubling time of approximately 12 hours for a standard plate set-up. RAW 264.7 cells exposed to AuNPs (10 nm) proliferated the most when cultured in plate set-up and in suspension. At 48 h and 72 h there are ~2-5 times more cells in plate and suspension versus culture insert. Cells grown on inserts, however, show a reduction in growth at 72 hours, most likely due to cell overgrowth conditions on the insert surface. In adherent conditions (plate and inserts) RAW 264.7 cells reached ~100% confluency at 72 h. It should be noted that the concentration of AuNPs used in this experiment have been previously shown not to cause toxicity in *in vitro* culture conditions [9].

Since, for all the culturing conditions, the amount of AuNPs available is a fixed amount during the time course of the experiment, by 24 hours, there are less AuNPs available per cell simply due to more cells being present in the wells at t =24 hours compared to t =0 hr. Thus, it is possible that the amount of gold per cell decreases over time. In addition, the concentration of AuNPs is further reduced by incorporation into the cells over time, decreasing the amount of free AuNPs available to cells. This dilution effect may be what is responsible for the lower AuNP incorporation into cells at 48 and 72 hours. Such an effect has been analyzed by Summers et al. [10]. Another possibility is the mechanism of exocytosis, which has been demonstrated to occur for AuNPs in macrophage cell lines [11].

In general, cell proliferation was similar between the different culture conditions up to 6 hours, with AuNP incorporation also increasing steadily among the culture conditions up to 4 hours. When examining the rate of uptake over this time period, the plate-based assay had the highest rate of uptake, with a slope of 6.9 (R^2 0.90). The insert and suspension uptake conditions had a lower rate of uptake, with slopes of 2.1 (R^2 0.96) and 2.5 (R^2 0.98), respectively. The enhanced uptake of the plate-based assay compared to the insert may be explained by the culturing conditions, where the cells cultured in the standard plate assay encounter particles both dispersed throughout the media, as well as any settled agglomerates. In contrast, the cells cultured on the inserts only have access to the AuNPs that remain dispersed in the media. Cho et al. [5] using an inverted set-up demonstrated that cells cultured in an inverted set-up incorporated less AuNPs versus cells cultured in an upright set-up. The difference between the set-ups was least apparent for their smallest AuNPs tested (15 nm). Although the suspension assay conditions have a similar rate of AuNP uptake as the insert set-up, the actual amount of AuNP incorporated per cell is much higher than either the plate or insert-based assays, even when accounting for the baseline AuNPs. Again, culture conditions may inform these results. For the suspension cells, the culturing conditions are not static, with constant movement of the cells and AuNPs (both single and agglomerated) dispersed within the media. These dynamic conditions provide additional interactions between the cells and AuNPs, allowing for enhanced AuNP uptake.

One of the drawbacks of adherent/plate assays is that the exact cell number at the time of experiment can only be approximated (due to the cell proliferation that is occurring during the plate incubation time, e.g. "plates were incubated overnight prior to treatments). The cell numbers can be known exactly when plating, and can be determined after the assay via cell counts, but the number of cells at the time of assay dosing is an approximation. This is another advantage of using cells in suspension. The cell number in our "adherent/plate set-up" conditions were determined in each experiment at the time of plating and at the end of the assay.

AuNPs distribution and recovery in culture conditions
in vitro

NPs can adhere to not only external cell membranes, but to the tissue culture plates, pippettor surfaces, etc., reducing the actual dosing concentration of the NPs [12]. In order to have a better understanding of the dosing conditions of the AuNPs within the cell culture set-up, a mass balance study was undertaken for all cell culture conditions. At each time point for the adherent set-ups (and at 24 hours for the suspension set up), media was removed from the cell monolayer, placed in NAA vials, and allowed to dry. PBS used to rinse the cell monolayers was also collected in a separate NAA vial, as was the cell monolayer itself (reserving a small aliquot to use for cell counting). All labware that the AuNPs had contact with (pipettor tips, cell scrapers, culture plate dishes, Teflon inserts) were collected and pooled into a final NAA vial. Figure 3 represents the results of each of the conditions tested in % AuNP recovered. Since in this experiment the aim was to obtain a mass balance (AuNP distribution in cells and culture media throughout the incubation time), AuNPs levels were determined in the harvested cell population and are shown as percentage of the starting AuNPs (0.005 g/L) rather than AuNPs/cell.

For the adherent set-ups (Figure 3A and B), recovery was close to 100% and no less than 90% for any of the time points evaluated. Figure 3A shows the recovery profile for the plate-based set up. At the beginning of the experiment, the majority of the AuNPs are in the media, and a small fraction of the AuNPs within the cells. As the incubation time increases, the fraction of AuNPs within the media decreases with a corresponding increase of the amount of AuNP in the cells, in agreement with the uptake of the AuNPs by the macrophages over time. The amount of AuNPs measured in the PBS rinse also increases over time, and could be a result of dead or dying cells that had incorporated AuNPs and were removed from the cell layer upon rinsing, or a result of an exocytosis mechanism. Only a small fraction of the AuNPs remained adhered to the labware.

As anticipated with the culture inserts set-up, the majority of AuNPs remain in the media portion of the set-up throughout the incubation time, with a minor fraction (up to 14%) measured in the cells (Figure 3B). Towards the end of culture period, the amount of AuNP detected in PBS washes increased to ~15-20% again, likely due to cell death and detachment and/or exocytosis. The amount of AuNPs recovered in the labware increases over time, with

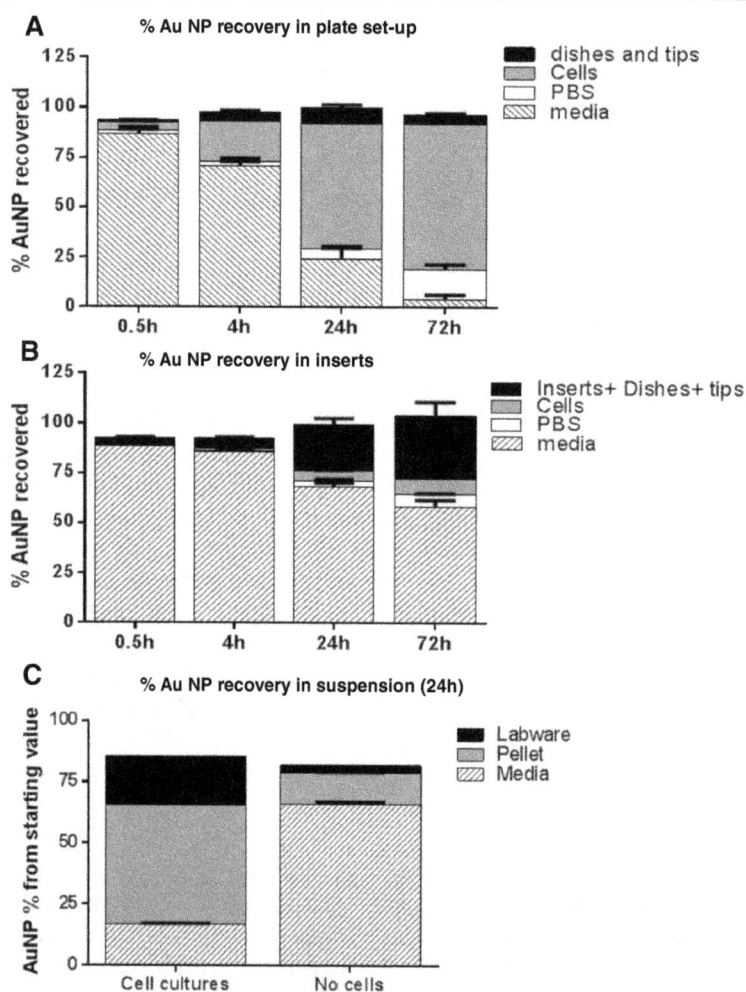

Figure 3 AuNPs recovery in RAW 264.7 cell cultures. There is a ~100% recovery of AuNPs in plate (**A**), insert (**B**) and suspension (**C**) set-up. Graph shows mean and SEM (N = 6; 2 independent experiments for A and B and N = 3 for C).

the largest percentage found at 72 hours, possibly indicating adhesion/interaction of the AuNPs with the insert membrane over time.

Figure 3C shows the mass balance for the 24 hour time point. Due to the s cost considerations of the Teflon inserts, only the 24 hour time point was evaluated. As there was a high AuNP baseline (as noted above), a media control was also performed for this measurement and is shown in the right side of the figure. At 24 hours, the majority of AuNPs are associated with the cells, with only a small fraction remaining with the media. This result, however, is cofounded by free AuNPs pelleting with the cells. Approximately 13% of the control media pellet contained AuNPs that is not incorporated by the cells. A large portion of the AuNPs also appears with the labware. This may be due to the multiple transfer steps that this set-up requires. These multiple transfers plus not measuring the Petri dishes where the inserts were housed (along with any media drips that occurred during the incubation) may account for the lower recovery of the AuNPs.

The mass balance distribution highlights the differences between the different adherent culture conditions. For the plate based conditions, there is strong correlation between the AuNPs concentration in the media (decreasing over time) and the incorporation of AuNPs within the cell monolayer (increasing over time). Whereas this correlation is also observed for the insert conditions, it is not as robust a response, and correlates with the rate of uptake that was observed in Figure 2, where the cells in plate-based conditions have more access to the AuNPs. Suspension conditions also show a large incorporation of AuNPs within the cells at 24 hours, which also agrees with the rate of uptake studies, where the dynamic culture conditions bring the cells and NPs into contact more frequently.

AuNP settling studies

Based upon the stability studies conducted in media (Figure 1), as well as common belief in the literature, it was assumed that the AuNPs would over time agglomerate and settle to the bottom of a plate-based set-up in significant quantities [13]. To test this hypothesis, a settling experiment was conducted where the plate-based set-up was evaluated for the amount of AuNPs in the top aliquot of media, the bottom layer of media, and settled onto the plate. Conditions were run both with and without cells (Figure 4). Figure 4A shows the mass balance results for the plate-set-up without cells. For this experiment, approximately the same amount of AuNPs is found in the top media layer, bottom media layer and on the culture plates throughout the 72 hours of incubation. No statistical difference was found between the top and bottom media layers. For the plates that contained cells (Figure 4B), the amount of AuNPs within the media

decreases over the time course of the experiment, while the percentage of AuNPs associated with the plates increases. However, due to the manipulation steps (plates transfer from the incubator to the sterile hood, pipetting, etc) it is difficult to accurately collect the top and bottom layers. Therefore, as shown in Figure 4B, some errors due to the handling of the culture plates occurred. Nevertheless, recovery of AuNPs in the plates containing cells is in line with the uptake experiments presented in Figures 2 and 3, with the cells layer incorporating most of the AuNPs from the media by 72 hours. Again, there was no difference in the gold concentrations in the different media layers.

This lack of significant settling is consistent with the experimental results demonstrated with Cho et al [5], who determined that the uptake of uncoated 15 nm AuNPs was essentially the same in inverted versus plate-based set-ups. Based upon their results, the authors hypothesized that diffusion of NPs within the cell culture media was the main mode of transport (as opposed to sedimentation) for these smaller NPs. Indeed our own DLS stability study, while indicating agglomeration over time, did not indicate significant settling, with no visible settling and no significant decrease of the count rate over the course of the experiment. The lack of significant sedimentation, however, confounds the analysis of the greater cellular uptake of AuNPs in the plate-based setting (compared to the insert culture conditions). Several factors could contribute to this observation. The first is that the membrane insert is impeding free diffusion of the AuNPs to the cell monolayer, thus artificially lowering the AuNP dose. Another possibility is that the control sedimentation conditions used in Figure 4A do not account for the extracellular matrix. This matrix could cause the AuNPs to agglomerate and stick near the macrophages, thus allowing for the observed increased uptake.

Basic culture conditions impact cellular uptake

Finally, other aspects of the assay set up may greatly influence the uptake of NPs by macrophages. Figure 5A shows the amount of AuNPs taken up by cells in a plate-based set-up and exposed to increasing amounts of gold. Not surprisingly, increasing the amount of AuNPs within the media increases not only the amount of NPs internalized in the cell, but the rate of uptake as well, with the higher AuNP concentrations showing a higher rate of uptake over the first 4 hours (0.0005 g/L, 1.4, R^2 0.92; 0.005 g/L, 6.9, R^2 0.90; 0.05 g/L, 78.0, R^2 0.96). The starting cell density (cells/ NP), on the other hand, did not significantly affect the cellular uptake profile, with the exception of when cells overgrew and died (data not shown).

To better understand NP-induced cytotoxicity *in vitro* several laboratories have developed platforms in which

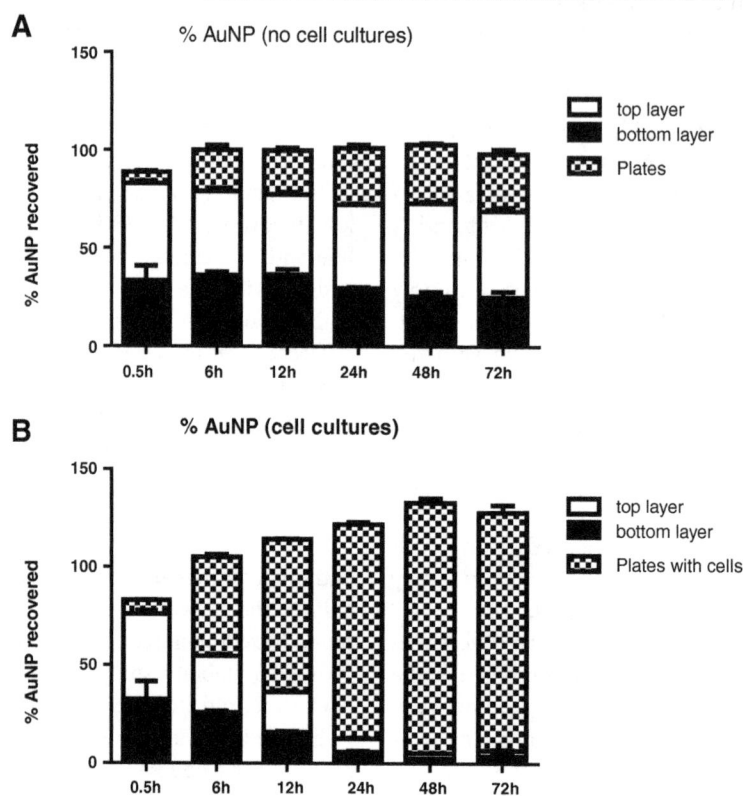

Figure 4 Settling studies for plate-based set-up with (A) no cells present and (B) cells present. No significant difference found between time points or between the top and bottom media layers. $P < 0.05$ two-way ANOVA.

several cell types are exposed to different NPs and their cytotoxicity is evaluated using a variety of techniques and assays. It was shown that 3 T3 fibroblasts are more resistant to NPs (TiO_2 NPs, SiO_2 NPs, MWCNT) while RAW 264.7 cells are more sensitive [14]. Using a panel of 10 different cell lines and 23 different NPs it was shown that there were different toxicity responses in the cells analyzed [15]. While this is a thorough approach, a clear conclusion cannot be drawn unless there is a quantitative determination of NPs level in every cell type. The difference in cytotoxcity responses among cell types could be caused by the incorporation of higher NPs levels in some of the cells studied. Our data show that RAW 264.7 cells incorporate more AuNPs versus rat hepatocytes (Figure 5B). This result is not overly surprising given that RAW264.7 cells are derived from a specialized phagocytyic cell line that has been noted for rapid uptake of NP [16]. The rH4IIR hepatocyte cell line, in contrast, has less phagocytic capabilities. The phagocyotic macrophages have almost triple the amount of AuNPs in them after 24 hours as compared to the hepatic cells. These data underscore the importance of using not only the correct assay conditions, but the cells most relevant to the desired endpoint (e.g. if studying toxicity for a

possibly hepatotoxicant, a hepatocyte might be more relevant than a phagocytic cell).

Conclusions

While there is overwhelming information regarding the factors that influence NP uptake *in vitro* (physical and chemical characteristics, dosimetry, cell lines used, etc.) there are few communications related to the actual amount of NPs incorporated inside the cells. This study looked at the effect of different plating conditions on the uptake of AuNPs. While this study only evaluates the uptake in three assay conditions, the techniques and evaluation could be applied toward other in vitro models, including flow-through and 3 dimensional models. The suspension culture, while exhibiting some interference with agglomerated AuNPs showed more uptake than the static systems and has the potential to be more representative *in vivo* (where sedimentation of the NPs onto tissues is less likely to occur). In contrast, the insert culture has the least AuNP uptake but mitigates settling effects as well as the pelleting issues observed for the suspension system. While none of the set-ups employed (plate, inserts and cell suspension) are ideal, the current study draws attention to the variability in NPs incorporation that can be induced

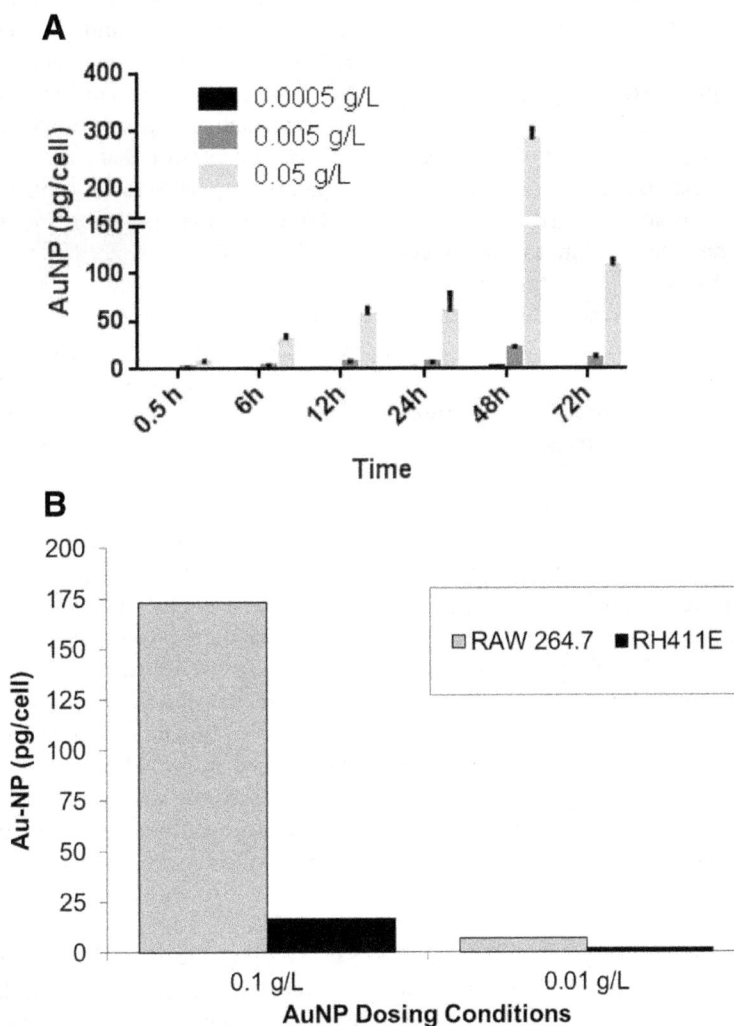

Figure 5 NP dose and type of cell line impacts the amount of AuNP uptake. (A) Uptake of AuNPs in RAW264.7 cells when exposed to increasing concentrations of AuNPs. **(B)** AuNP uptake in pg/cell for a hepatocyte cell line (rH4IIE) versus a macrophage cell line (RAW264.7).

by assay set-up as well as the necessity to rationally choose the most appropriate assay conditions.

Methods

Reagents

RAW 264.7 cells and DMEM culture media were purchased from ATCC; FBS, Penicillin/Streptomycin (p/s), PBS, propidium iodide (PI) were purchased from Invitrogen. AuNPs (10 nm) were purchased from Structure Probe Inc.

NPs characterization

AuNPs were concentrated as previously described [8]. Concentration was determined by TGA and later confirmed by NAA. For size and stability analysis, AuNPs were diluted in water or media with 10% FBS and 1% penicillin/streptomycin. Full media without AuNPs were

run as a control. AuNPs were dispersed into the water or media at a concentration of 0.005 g/L and then aliquoted into individual zeta cuvettes (Malvern) and stored at 37°C. At various time points, cuvettes were removed and size via dynamic light scattering (DLS) and zeta potential were measured on a Malvern Zetasizer. Parameters were adjusted for viscosity and refractive index for the media and water. Measurement position and attenuator were standardized across all runs. Three measurements of 12 runs each were taken for each time point after an incubation time of 120 sec. For zeta potential measurements, 50 runs were performed. Z average, intensity weighted histograms, and zeta potential were recorded. Lack of endotoxin contamination was confirmed by LAL gel clot formation assay. For TEM analysis an aliquot of AuNPs was triple dripped onto a holey carbon coated copper grid (quantfoil, EMS, PA USA) and allowed to air dry. Grids were analyzed on a JEOL 1400 TEM at 80 kV. At least 5 images were taken for each sample with 36-96 particles counted.

Dosing solutions were prepared based upon AuNPs concentrations determined by TGA. To compare the AuNPs concentration in media as measured by TGA and NAA, an aliquot of media containing AuNPs or the stock solution was placed into an NAA vial and allowed to dry. NAA vials were analyzed within the main cell experiments (see details below). N = 12 for 0.005 g/L and stock concentrations; N = 3 for rest of dilutions.

RAW 264.7 cell culture conditions

RAW 264.7 cells were purchased from ATCC, propagated and aliquots stored in liquid nitrogen. Cells were maintained as adherent cell cultures and passaged 3-25 times after which a new frozen aliquot was used. Cell treatments were performed in adherent or cell suspension conditions. Cell density was 10^5 cells/ml at time of AuNP treatment. Cell suspensions were exposed to AuNPs at the time of plating. In adherent cultures, cells were plated at a density of 0.8×10^4 cells/cm^2 in either 12-well plates (3.8 cm^2 surface area) or tissues culture Transwell inserts (0.9 cm^2). For adherent cultures, cells were allowed to propagate for 48 h (at which point they reached a density of ~ 10^5 cells/well or insert) and were exposed to AuNPs. Cells were harvested and processed depending on experiment requirements (described below).

Cellular uptake and mass balance studies

For adherent set-ups, RAW 264.7 cells cultured in 12 well multi-well plates were exposed to AuNPs (0.005 g/L) for varying time points, after which supernatants were harvested, cells washed with PBS, trypsinized, counted and transferred into NAA vials. 12 well plates were sectioned to isolate the individual wells for analysis. With the exception of the cells that were used to determine cell count, all cells were used for NAA. RAW 264.7 cells grown in trans-well culture inserts were exposed to theAuNPs only in the basal side of the membrane (media only was added to the top of inserts). When mass balance experiments were performed, cells, culture media, PBS washes, and the labware that came in contact with the AuNPs (pipettor tips, culture inserts and culture wells) were collected, transferred into individual NAA vials and measured for AuNP levels. Conditions were run in triplicate on two separate days for an N = 6.

For suspension cells, RAW 264.7 cells were plated in 60 mm Petri plates containing Teflon inserts in 5 mL medial containing 0.005 g/L AuNPs. Cells were placed on a shaker table within the incubator during the experiment. At variable time points, cells were removed from the inserts with a cell scraper and transferred to centrifuge tubes. Cells were centrifuge at 1500 rpm (~500 g) for

10 minutes at room temperature. The supernatant was saved in a NAA vial and the cell pellet resuspended in 500 μL PBS. Cells were counted and transferred to an NAA vial for analysis. For mass balance experiments, the experimental set up separated cells, culture media, PBS washes, and labware that came in contact with the AuNPs (pipettor tips, cell scraper, centrifuge tube, and Teflon inserts). Conditions were run in triplicate on two separate days for an N = 6.

NAA limits of detection

RAW 264.7 cells were plated at a density of $0.2 \times 10^4 - 20 \times 10^4$ cells/measurement. Cells were exposed to media with no AuNPs or media containing 0.01 g/L or 0.1 g/L AuNPs for 24 h and processed as above.

Settling studies

RAW 264.7 cells cultured in 12 well multi-well plates (as noted above) were exposed to AuNPs (0.005 g/L) for varying time points. 400 μL of media was carefully aspirated from the top of the well to avoid dispersing AuNPs and transferred to an NAA vial. The remaining media was then transferred to a NAA vial. Plates were then sectioned to isolate the individual wells and the wells were placed in an NAA vial. For the media control, the above procedure was repeated with the exception that no cells were cultured in the plates. The experiment was run in triplicate with N =3.

NAA

Sample vials were allowed to dry prior to capping the vials to prevent leakage during sample transfer and analysis. For the 12 well plates, plates were dried in a hood prior to sectioning. NAA analysis was performed at Becquerel Laboratory (Ontario, Canada) using their standard operation procedure SOP (BQ-NAA-4, Elemental Analysis via INAA) through a contract with Elemental Analysis Inc (Kentucky, USA). Gold values were reported as total gold in μg/vial.

Statistical analyses were performed using two-way ANOVA followed by Bonferroni's multiple comparison tests.

Competing interests
The authors declare that they have no competing interests.

Authors' contributions
SB designed and carried out the NAA studies, participated in the interpretation of the data, and helped draft the manuscript. KMT conceived of the study, participated in its design and interpretation of data, and helped to draft the manuscript. All authors read and approved the final manuscript.

Acknowledgments
The authors acknowledge the invaluable help of David Stevens CDER/FDA and Oleg Vesnovsky CDRH/FDA for assistance in cutting up the cell culture dishes. The authors acknowledge the Office of Science and Engineering Laboratories, FDA, for the use of their TGA. The authors would also like to acknowledge the FDA White Oak Nanotechnology Core Facility for instrument use, scientific and

technical assistance. This project was supported in part by an appointment to the Research Participation Program at the Center for Drug Evaluation and Research administered by the Oak Ridge Institute for Science and Education through an interagency agreement between the U.S. Department of Energy and the U.S. Food and Drug Administration.

Disclosure

The findings and conclusions in this article have not been formally disseminated by the Food and Drug Administration and should not be construed to represent any Agency determination or policy. The mention of commercial products, their sources, or their use in connection with material reported herein is not to be construed as either an actual or implied endorsement of such products by the Department of Health and Human Services.

References

1. Keene AM, Bancos S, Tyner KM: **Considerations for in vitro nanotoxicity testing.** In *Handbook of Nanotoxicology and Nanomedicine.* Edited by Sahu S, Casciano D. The Atrium, Southern Gate, Chichester, West Sussex, PO19 8SQ, United Kingdom: John Wiley and Sons; 2014.
2. Monteiro-Riviere NA, Inman AO, Zhang LW: **Limitations and relative utility of screening assays to assess engineered nanoparticle toxicity in a human cell line.** *Toxicol Appl Pharmacol* 2009, **234**(2):222–235.
3. Teeguarden JG, Hinderliter PM, Orr G, Thrall BD, Pounds JG: **Particokinetics in vitro: dosimetry considerations for in vitro nanoparticle toxicity assessments.** *Toxicol Sci* 2007, **95**:300–312.
4. Lisson D, Thomassen LCJ, Rabolli V, Gonzalez L, Napierska D, Seo JW, Kirsch-Volders M, Hoet P, Kirschhock CEA, Martens JA: **Nominal and effective dosimetry of silica nanoparticles in cytotoxicity assays.** *Toxicol Sci* 2008, **104**:155–162.
5. Cho EC, Zhang Q, Xia Y: **The effect of sedimentation and diffusion on cellular uptake of gold nanoparticles.** *Nat Nanotechnol* 2011, **6**:385–391.
6. Aggarwal P, Dobrovolskaia M: **Gold nanoparticle quantitation via fluorescence in solution and cell culture, characterization of nanoparticles intended for drug delivery.** In *Methods in Molecular Biology.* Edited by McNeil SE. New York: Humana Press; 2011:137–145.
7. Zhu ZJ, Ghosh PS, Miranda OR, Vache RW, Rotello VM: **Multiplexed screening of cellular uptake of gold nanoparticles using laser desorption/ionization mass spectrometry.** *J Am Chem Soc* 2008, **130**(43):14139–14143.
8. Keene AM, Tyner KM: **Analytical characterization of gold nanoparticles primary particles, aggregates, agglomerates, and agglomerated aggregates.** *J Nanopart Res* 2012, **13**:3465–3481.
9. Bancos S, Stevens D, Tyner KM: **The effect of silica and gold nanoparticles on macrophage proliferation, activation markers, cytokine production and phagocytosis *in vitro*.** *Int J Nanomed* 2014, accepted.
10. Summers HD, Rees P, Holton MD, Brown RM, Chappell SC, Smith PJ, Errington RJ: **Statistical analysis of nanoparticle dosing in a dynamic cellular system.** *Nat Nanotechnol* 2011, **6**:170–174.
11. Oh N, Park JH: **Surface chemistry of gold nanoparticles mediates their exocytosis in macrophages.** *ACS Nano* 2014, **8**(6):6232–6241.
12. Keene AM, Peters D, Rouse R, Stewart S, Rosen ET, Tyner KM: **Tissue and cellular distribution of gold nanoparticles varies based on aggregation/agglomeration status.** *Nanomedicine (Lond)* 2012, **7**(2):199–209.
13. Deloid G, Cohen JM, Darrah T, Derk R, Rojanasakul L, Pyrgiotakis F, Wohlleben W, Demokritou P: **Estimating the effective density of engineered nanomaterials for in vitro dosimietry.** *Nat Commun* 2014, **5**:3514. doi:10.1038/ncomms4514.
14. Sohaebuddin SK, Thevenot PT, Baker D, Eaton JW, Tang L: **Nanomaterial cytotoxicity is composition, size, and cell type dependent.** *Part Fibre Toxicol* 2010, **7**:22.
15. Kroll A, Dierker C, Rommel C, Hahn D, Wohlleben W, Schulze-Isfort C, Göbbert C, Voetz M, Hardinghaus F, Schnekenburger J: **Cytotoxicity screening of 23 engineered nanomaterials using a test matrix of ten cell lines and three different assays.** *Part Fibre Toxicol* 2011, **8**:9.
16. Dos Santos T, Varela J, Lynch I, Salvati A, Dawson KA: **Quantitative assessment of the comparative nanoparticle-uptake efficiency of a range of cell lines.** *Small* 2011, **7**:3341–3349.

12

Quantum dot assisted tracking of the intracellular protein Cyclin E in *Xenopus laevis* embryos

Yekaterina I Brandt[1], Therese Mitchell[1], Gennady A Smolyakov[2], Marek Osiński[2] and Rebecca S Hartley[1*]

Abstract

Background: Luminescent semiconductor nanocrystals, also known as quantum dots (QD), possess highly desirable optical properties that account for development of a variety of exciting biomedical techniques. These properties include long-term stability, brightness, narrow emission spectra, size tunable properties and resistance to photobleaching. QD have many promising applications in biology and the list is constantly growing. These applications include DNA or protein tagging for *in vitro* assays, deep-tissue imaging, fluorescence resonance energy transfer (FRET), and studying dynamics of cell surface receptors, among others. Here we explored the potential of QD-mediated labeling for the purpose of tracking an intracellular protein inside live cells.

Results: We manufactured dihydrolipoic acid (DHLA)-capped CdSe-ZnS core-shell QD, not available commercially, and coupled them to the cell cycle regulatory protein Cyclin E. We then utilized the QD fluorescence capabilities for visualization of Cyclin E trafficking within cells of *Xenopus laevis* embryos in real time.

Conclusions: These studies provide "proof-of-concept" for this approach by tracking QD-tagged Cyclin E within cells of developing embryos, before and during an important developmental period, the midblastula transition. Importantly, we show that the attachment of QD to Cyclin E did not disrupt its proper intracellular distribution prior to and during the midblastula transition. The fate of the QD after cyclin E degradation following the midblastula transition remains unknown.

Keywords: Quantum dots, Intracellular protein tracking, Cyclin E, *Xenopus laevis*

Background

Luminescent quantum dots (QD) are semiconductor nanocrystals with unique spectroscopic properties. QD have generated much interest in the past two decades, with current and projected applications including use as FRET donors, fluorescent labels for cellular labeling, intracellular sensors, deep-tissue and tumor imaging agents, sensitizers for photodynamic therapy, and vectors for studying nanoparticle-mediated drug delivery [1]. The advantages of QD over organic dyes and genetically engineered fluorescent proteins for tagging biomolecules are their broad excitation and narrow emission spectra, brightness, resistance to photobleaching, and the fact that they can be synthesized from a single material to emit a variety of wavelengths [2]. QD have size tunable fluorescent properties, meaning that changes in QD size result in different colors of emitted light. Larger QD emit redder (lower energy) light while smaller QD emit bluer (higher energy) light. Multiple molecules can be labeled with QD of various colors and simultaneously imaged after excitation with a single UV source, which prevents overheating of cells, a quality desirable for both *in vitro* and *in vivo* applications.

The colloidal nanocrystals most often used in fundamental or applied studies are spherical, with cores varying between 15 and 120 Å in diameter. CdSe nanocrystals are prepared by reacting organometallic precursors at high temperatures in a coordinating solvent mixture. This results in capping of the inorganic core with an organic layer of trioctylphosphine/trioctylphosphineoxide mixture (TOP/TOPO) [3]. Overcoating of the CdSe core with several layers (3–5) of wider band gap semiconducting material, such as ZnS or CdS permits passivation of the core surface and produces highly luminescent CdSe-ZnS or CdSe-CdS core-shell QD [4]. In addition, the shell protects the QD from oxidation and prevents

* Correspondence: rhartley@salud.unm.edu
[1]Department of Cell Biology and Physiology, University of New Mexico Health Sciences Center, Albuquerque, New Mexico 87131-0001, USA
Full list of author information is available at the end of the article

oozing of heavy metal core components into the environment, which is very important for *in vivo* applications [5]. For biological studies, QD also need to be soluble in water. Several methods have been used to render QD water soluble, such as exchange of native TOP/TOPO organic surface ligand for a water soluble surface ligand (cap exchange) [6], encapsulation within amphiphillic molecule-copolymer micelles [7], and coating with silica [8-11].

Despite the obvious optical advantages of QD-assisted visualization, most fluorescent QD labeling is being utilized either outside of cell boundaries or directly on the surface of the plasma membrane. Targeted labeling of biological molecules with QD is utilized for a variety *in vitro* applications, including using antibody-coupled QD for immunoassays, immunochromatography assays [12-16], and as biosensors that are based on QD as FRET donors [17,18]. The antibody mediated specificity approach of tagging proteins on the cell surface i.e., receptors or receptor ligands, allows for either studying their dynamics via single-particle tracking [19-21] or targeting tumor cells for detection and destruction [22-24]. QD have also been used for cell-tracking in *Xenopus laevis* embryos, either unconjugated [7], or conjugated to a nuclear localization signal [25] in order to track morphogenetic movement of cells during development.

As protein function is directly linked to its localization within the cell, accurate assessment of protein spatiotemporal dynamics is crucial. Moreover, protein localization to different compartments within a cell provides an important regulatory role. We set out to use QD to track the movement of Cyclin E, a cell cycle regulatory protein whose expression is restricted to G1/S phase of the mammalian cell cycle, where it mediates this cell cycle transition. Cyclin E has been shown to shuttle between the nucleus and cytoplasm in mammalian cells [26]. This shuttling makes sense as two of the best defined functions of Cyclin E, initiation of DNA replication and centrosome duplication, require its presence in the nucleus and cytoplasm, respectively. Cyclin E shuttling and its degradation

after the G1/S transition is regulated by phosphorylation and other post-translational modification. Malfunction of mechanisms controlling Cyclin E leads to its elevation in many types of human malignancy, where it is often present throughout the cell cycle [27]. Experimental evidence indicates that Cyclin E elevation is a cause rather than an effect of tumorigenesis [28,29].

In this study, we synthesized DHLA-capped CdSe-ZnS QD, as they are not available commercially, and use them to label polyhistidine (His_6)-tagged recombinant Cyclin E. We demonstrate feasibility of QD-assisted tracking inside cells by visualizing the intracellular localization of QD-labeled Cyclin E in live embryos in real time during early cell division cycles.

Results

CdSe-ZnS core-shell quantum dots with the photoluminescence emission peak of 564 nm were manufactured according to the procedure of *Clapp et al.* [4]. To make the QD soluble in water, we utilized the cap exchange strategy, exchanging TOP/TOPO with dihydrolipoic acid (DHLA). DHLA is a bifunctional ligand, containing a bidentate thiol moiety on one end, allowing its stable attachment to the inorganic QD surface, and an opposing hydrophilic end group, which permits its aqueous dispersion [4]. The DHLA-capped QD maintain their high photoluminescence and quantum yield. After synthesis and cap exchange, the morphology of DHLA coated CdSe-ZnS (QD_{564}) was assessed by transmission electron microscopy (TEM). As seen in Figure 1, the QD were homogeneous, of high crystallinity (Figure 1c), consistent size (4.1 ± 0.88 nm in diameter) and shape (spherical) and did not form aggregates, as shown in Additional file 1.

Next, we coupled the (His_6)-tagged cyclin E to QD (Figure 2) and confirmed the attachment by measuring photoluminescence (PL) and quantum yield (QY). The observed 30% increase in PL intensity and 0.4% increase in QY (from 2.2% to 2.6%) of conjugated versus unconjugated QD confirmed successful bonding of

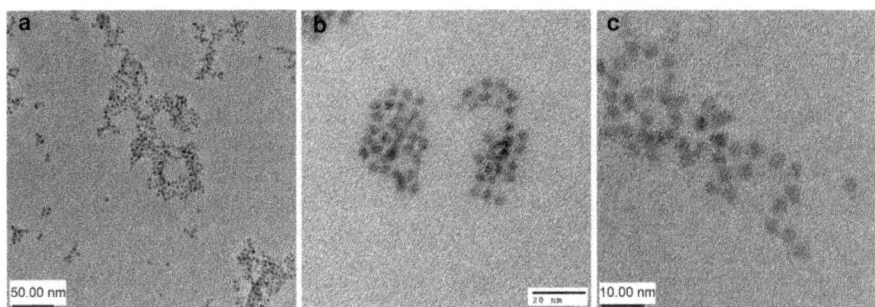

Figure 1 Transmission electron microphotographs of the synthesized CdSe-ZnS QD. **(a)** Scale bar is 50 nm. **(b)** Scale bar is 20 nm. **(c)** Scale bar is 10 nm.

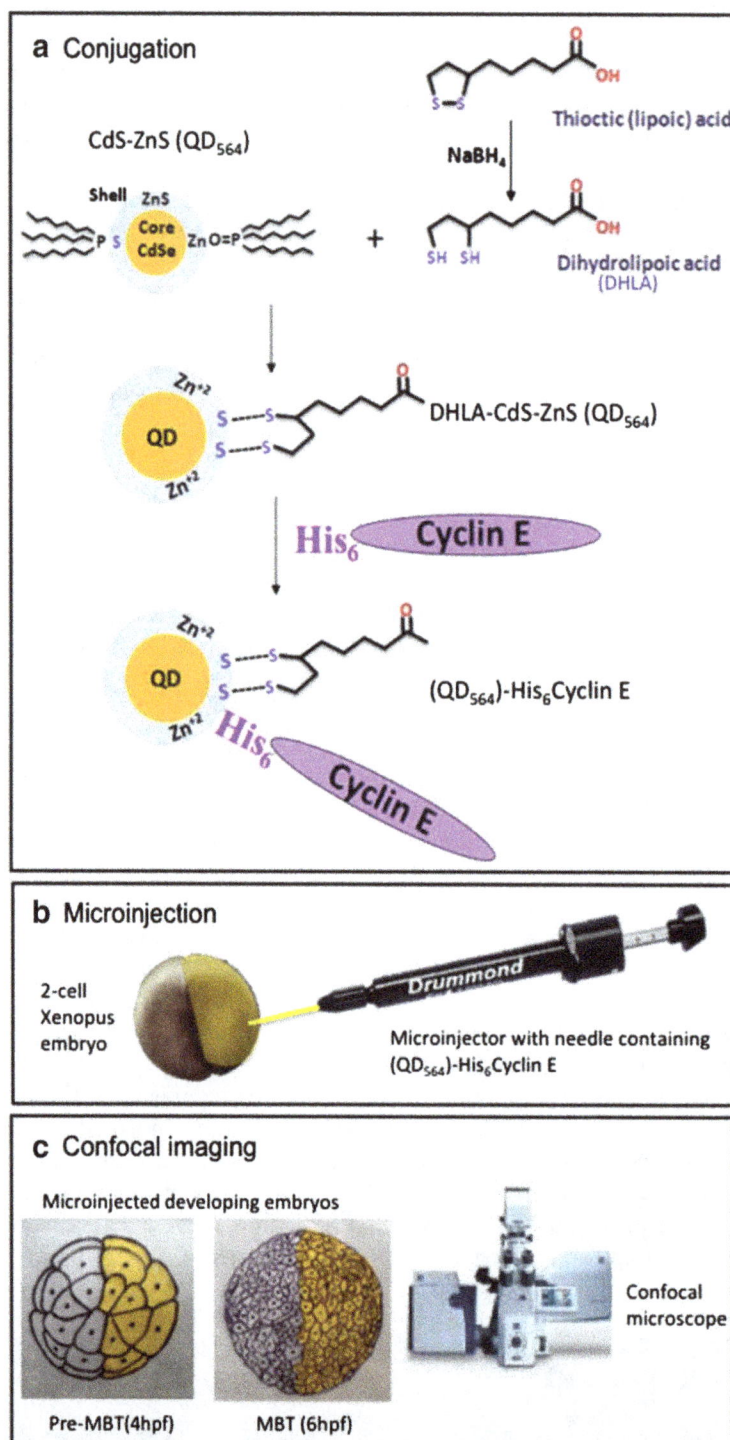

Figure 2 Schematic showing experimental design. **(a)** DHLA capping of CdSe-ZnS (QD₅₆₄) and subsequent conjugation of (His₆)-Cyclin E (modified from [5]). **(b)** Microinjection of (QD₅₆₄)-His₆Cyclin E into 2-cell Xenopus embryos and **(c)** confocal imaging of microinjected pre-MBT and MBT Xenopus embryos. In **(a)** the DHLA molecule contains a bidentate thiol moiety on one end, allowing its stable attachment to the inorganic QD surface. Coupling of (His₆)-Cyclin E to QD is achieved via strong metal affinity between the histidine tag of the protein and Zn^{+2} atoms on the QD surface. The schematic is not to scale.

protein moieties to the surfaces of the QD (Additional file 2).

(QD_{564})-His$_6$Cyclin E complexes were microinjected into 1 cell of a 2-cell *Xenopus laevis* embryo between 1.5-2 hours post fertilization (hpf) and visualized in live, developing embryos by confocal microscopy. Embryos were viewed between 4 to 7 hpf (stages $6^{1/2}$-8), during which time embryos undergo cell divisions 5–12. This developmental period encompasses the midblastula transition (MBT), an important transition that begins around 6 hpf (stage 8) when global transcription initiates, cell motility begins, and the rapid embryonic cell cycle is remodeled to the adult cell cycle. Remodeling adds gap (G1 and G2) phases to a cell cycle that consists of rapid oscillations between DNA synthesis (S phase) and mitosis (M phase) [30]. Cyclin E is known to be nuclear at the MBT [31] but during the time we were performing these experiments, its localization before the MBT had not been determined. Therefore, we assessed Cyclin E localization before and during the MBT.

Images of embryos taken at 4 hpf, about 2 hrs post-injection (Figure 3a and c) show that localization of (QD_{564})-His$_6$Cyclin E was cytoplasmic prior to the MBT, as evident from the diffuse staining seen within individual blastomeres (large cells). At the MBT (6 hpf), (QD_{564})-His$_6$Cyclin E began to accumulate in the nuclei of the cells (Figure 3d and f) (white arrows). The daughter cells that originated from the uninjected cell of the 2-cell embryo

lack luminescence and serve as a negative control for QD detection, as only cells derived from the injected cell should contain the QD. The cells derived from the uninjected blastomere can be clearly seen as they lack luminescence in Figure 3c and f, which show merged images of the UV and light channels. Gap junctions and cytoplasmic bridges couple cells of the early embryo during cleavage [32,33]. Gap junctions are too small for QD to cross (1–2 nm). Lack of QD luminescence in the sister cell at 2-cell or 4-cell stages (not shown), as well as in a very specific subset of cells at later stages (Figure 3), show that QD are not diffusing via gap junctions or cytoplasmic bridges (QD are injected after the first cleavage division is complete). It should also be noted that individual cells of the embryo become progressively smaller as embryos continue to divide without growing during this developmental timeframe (see Figure 2c for a graphical illustration).

Our tracking results of Cyclin E labeled with QD are in agreement with a previous study that reported nuclear localization of Cyclin E at the MBT [31]. They are also consistent with our published work showing the localization pattern of endogenous and exogenous Cyclin E [34]. Similarly to these published results, Figure 4 shows immunofluorescence staining of Myc$_6$-GFP-Cyclin E in embryos, produced from *in vitro* transcribed mRNA also injected at the 2-cell stage. The embryos were probed with an antibody to the Myc$_6$-tag followed by an Alexa Fluor488 conjugated secondary antibody. Exogenous Cyclin E was diffusely

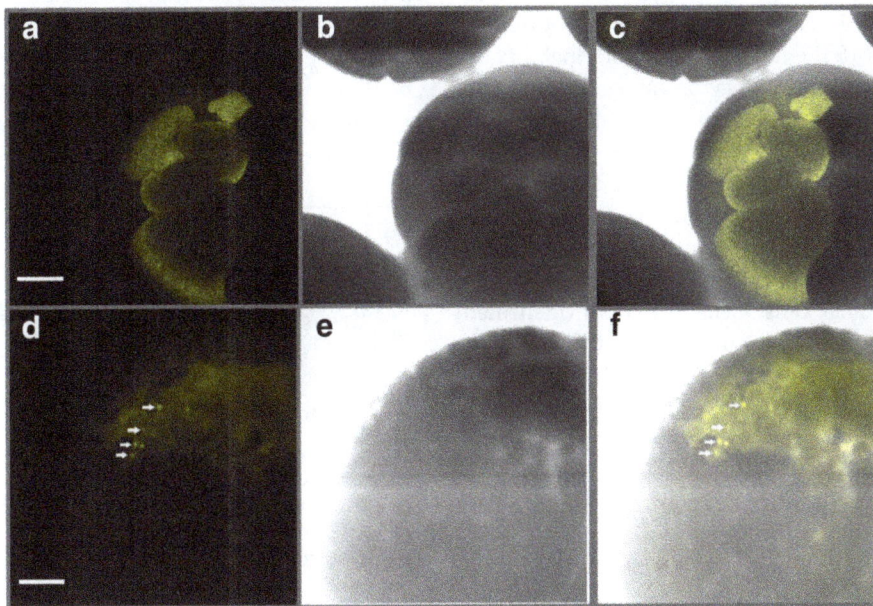

Figure 3 Localization of (QD_{564})-His$_6$Cyclin E in live pre-MBT (4 hpf, 64-cell embryo, **a-c**) and MBT (6 hpf, 2048-cell embryo, **d-f**) *Xenopus laevis* embryos. One cell of embryos at the 2-cell stage was microinjected with (QD_{564})-His$_6$Cyclin E and visualized using confocal microscopy. **(a, d)** fluorescence channel; **(b, e)** light channel; **(c, f)** merged fluorescence and light channels. Nuclei are marked with white arrowheads in panels b, f. Embryos were viewed with a 10X objective on a Zeiss LSM 510 confocal microscope equipped with a META detector, and analyzed using LSM510 Image Acquisition software. Scale bars are 100 μM. At least 20 embryos were injected and viewed in at least 3 separate experiments.

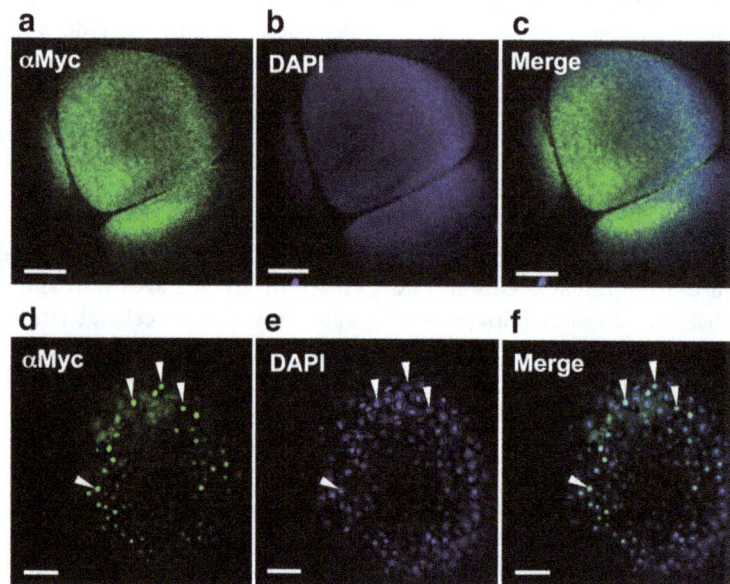

Figure 4 Localization of exogenous Cyclin E in pre-MBT and MBT *Xenopus laevis* embryos. One cell of 2-cell embryo was microinjected with *in vitro* transcribed Myc$_6$_GFP-Cyclin E RNA, collected at indicated time points, and the translated protein detected in fixed and stained embryos. For immunofluorescence analysis of Cyclin E localization, embryos were collected at 4 hpf, pre-MBT **(a-c)** or at 6 hpf, MBT **(d-f). (a, d)** Embryos were fixed and stained with an antibody against the Myc$_6$ tag (αMyc) followed by an Alexa488 conjugated secondary antibody. **(b, e)** Embryos were counterstained with DAPI to visualize the nuclei. **(c, f)**. Merged image of the Alexa488 and DAPI. White arrowheads in d-f indicate nuclei. Embryos were viewed with a 10X objective on a Zeiss LSM 510 confocal microscope equipped with a META detector, and analyzed using LSM510 Image Acquisition software. Scale bars are 100 μM. At least 20 embryos were injected in at least 3 separate experiments, with at least 5 embryos fixed per timepoint for analysis.

distributed in the cytoplasm pre-MBT (Figure 4a) and accumulated in the nuclei at the MBT in fixed embryos (Figure 4d). Figure 4b shows the diffuse distribution of DAPI in the pre-MBT embryos, due to its inability to diffuse though the lipid-rich cytoplasm even in fixed embryos. Figure 4c is a merged image of the Alexa Fluor488 and DAPI channels. Figure 4e shows DAPI stained nuclei in an MBT embryo (white arrowheads), while Figure 4f shows the coincidence of DAPI and exogenous cyclin E. DAPI easily stains the nuclei in MBT embryos as the ratio of cytoplasm to nucleus is greatly reduced in the smaller cells. Cells in mitosis (identified by the condensed chromatin and mitotic figures) have diffusely distributed cytoplasmic cyclin E (Figure 4d-f), also in agreement with previous studies [30].

Additional GFP imaging of fluorescent Myc$_6$-GFP-Cyclin E in live embryos confirmed its residence in the nucleus at the MBT in merged (Figure 5a) and fluorescence (Figure 5b) images (white arrowheads). Altogether, these results provide evidence that attachment of QD to Cyclin E leads to excellent tracking capabilities, without altering the ability of a protein to follow a proper and timely localization pattern intracellularly.

We also assessed potential cytotoxicity of the (QD$_{564}$)-His$_6$Cyclin E complexes by carefully monitoring embryo morphology and survival for up to 3 days after

microinjection. In a typical experiment, 20 embryos each were microinjected with either QD solution or solvent (water). Of these 20 embryos, 10–15 survived in both the QD-injected and Solvent-injected embryos. Of these, approximately 6–10 developed into free-swimming tadpoles. This number varied depending on the quality of the eggs and sperm. Germ cell quality

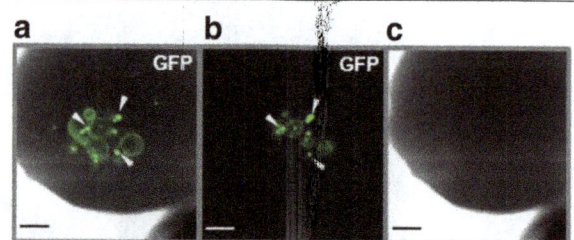

Figure 5 Cyclin E accumulates in the nucleus of live *Xenopus laevis* embryos at the MBT (6 hpf). One cell of the 2-cell embryo was microinjected with *in vitro* transcribed Myc$_6$-GFP-Cyclin E RNA and the translated protein visualized in live embryos using confocal microscopy in real time. **(a, b)** Fluorescence channel, Z stack images #5 and #8 from the top, respectively. Nuclei are marked with white arrowheads. **(c)** Light field image. A 3D image is shown. Scale bars are 100 μM. Embryos were viewed with a 10X objective on a Zeiss LSM 510 confocal microscope equipped with a META detector, and analyzed using LSM510 Image Acquisition software. At least 20 embryos were injected in at least 3 separate experiments.

varies with age of the female and male frogs, the number of times the female frog has been ovulated, as well as environmental factors such as temperature. Injected embryos progressed through the blastula stage, gastrulated normally, and became free-swimming tadpoles. We were able to visualize QD-mediated fluorescence in embryos microinjected with (QD_{564})-His_6Cyclin E complexes for up to three days after microinjection without any obvious deleterious effects. These results suggest that toxicity associated with the proposed application of QD-assisted tracking in Xenopus embryos remained low. This low toxicity differs from reports using commercial QD in Xenopus embryos, which showed variable cytotoxicity [25], likely due to variable overcoating. The low toxicity of our QD is probably due to consistent overcoating, which can be controlled when they are synthesized in lab.

Discussion

In this paper we conducted "proof-of-principle" experiments for an exciting application for engineered nano-scale products in the biomedical field, tracking of intracellular protein inside cells. We accomplished both objectives we set out to achieve. First, we successfully synthesized functional DHLA-capped CdSe-ZnS QD specifically tailored to our application of attachment of recombinant (His_6)-Cyclin E protein to its surface. Second, we followed our QD-tagged Cyclin E complexes inside the cells of developing Xenopus embryos in real time, which allowed us to discriminate between cytoplasmic Cyclin E localization before the MBT (4 hpf) and its accumulation in the nucleus at the MBT (6 hpf). Therefore we provide evidence for the feasibility of this approach.

For our experiments we chose ZnS overcoated CdSe quantum dots capped with DHLA. Cap exchange with DHLA not only renders QD water soluble [35], but also allows direct attachment of Histidine-tagged proteins. Alternatively, a more traditional conjugation approach involves the use of the EDC (1-ethyl-3-(3-dimethylaminopropyl) carbomide) crosslinking agent that reacts with carboxyl groups on the QD surface to the primary amine groups of the protein. The direct coupling strategy that we utilized has several unique advantages. First, it allows a unidirectional attachment of protein to the QD. Second, QD-protein complexes are less prone to aggregation in neutral and acidic buffers. Third, the attachment of His-tagged proteins to the DHLA capped quantum dots is mediated by a strong metal affinity between the His_6-tag and Zn^{2+} atoms of the QD shell [5] (Figure 2), with a dissociation constant (K_d) of 1×10^{-13}. This interaction is stronger than most antibody binding, which has a K_d of 1×10^{-6}-10^{-9} [36].

Unlike most mammalian cells, *Xenopus laevis* embryos provide an ideal model for a straightforward and efficient intracellular delivery of QD-protein complexes. Since QD-His_6-protein complexes are stable at physiological pH, they can be microinjected into the 1- or 2-cell (or even later), 1–1.5 mm diameter Xenopus embryo and imaged in real time *in vivo*. One advantage of self-synthesized QD is that the synthesis, overcoating and capping can be carefully controlled, resulting in low cytotoxicity after microinjection. Another advantage is that QD are too large to pass through the numerous gap junctions that couple cells in the early embryo [32,33], allowing tracking of only the progeny of the injected cell. One disadvantage of imaging live Xenopus embryos, is that it is very difficult to stain and image the nuclei. DAPI and other dyes cannot efficiently diffuse through the lipid rich cytoplasm, necessitating fixation and staining of the embryos. Visualizing the nuclei in living embryos can be accomplished by producing transgenic frogs expressing GFP-tagged nuclear membrane proteins [37] or by coupling QD to a nuclear localization signal [25], which in our case would have interfered with monitoring cyclin E intracellular localization.

QD were detected in the embryos past the time that endogenous and exogenous cyclin E are normally degraded [34]. Based on our prior studies, we assume that the cyclin E1 conjugated to the QD was likely degraded following the MBT but that the QD remain. It is possible that cyclin E conjugated to the QD is not phosphorylated and therefore not degraded at the developmentally appropriate time. As the time period of interest to us in this study was prior to the time that cyclin E is degraded physiologically, we did not ask if cyclin E was degraded with normal kinetics. This is an intriguing question that will be pursued in future studies.

Conclusions

Our approach provides an excellent alternative to the more commonly used traditional method of genetically encoded fluorophores, with its sometimes cumbersome and labor intensive cloning steps and also eliminates the risks of fusing the large protein moiety to the native protein (~30 kDa in case of GFP, YFP, and RFP) that can lead to protein misfolding, missorting, loss of fluorescent properties, or even changes in the host protein behavior and activity. Even though failures are seldom reported, many representative examples of altered fusion protein integrity are present in the literature [38-44]. Moreover, the QD-based approach is highly beneficial due to inherent fluorescent signal longevity of QD that allow deeper imaging during prolonged periods of time throughout embryo development with the additional capabilities of cell lineage tracing. It is widely applicable to other developmental systems such as zebrafish, Drosophila, *Caenorhabditis elegans*, and mouse blastocysts as well as to attached and suspended cells where cytoplasmic/ nuclear microinjections are either desirable or permissible [45]. With the future advances in methods for cargo delivery inside eukaryotic/mammalian cells across lipid bilayers and the commercial availability of various QD products, the application of this method will be vastly expanded.

Methods

CdSe-ZnS quantum dot preparation

QD were prepared using a stepwise approach consisting of core CdSe nanocrystal growth, overcoating with five layers of ZnS, size selective precipitation, and surface ligand exchange and purification [36]. Growth of nanocrystals was monitored by a change in the absorption spectrum by UV–VIS spectroscopy. QD were made water soluble through exchanging the native capping shell (trioctylphosphine (TOP)/trioctylphosphine oxide(-TOPO)/hexadecylamine) (Sigma Aldrich, St. Louis, MO) with freshly prepared DHLA [4] . DHLA was prepared by ring opening of the DHLA precursor, thioctic acid (Sigma Aldrich, St. Louis, MO) using NaBH$_4$ (Fisher, Waltham, MA) as a reducing agent in aqueous solution (Figure 2) [4,46] followed by DHLA distillation to remove impurities. The QD morphology was assessed by transmission electron microscopy.

Optical characterization

UV–VIS absorption spectra were obtained at room temperature using a Varian Carey 400 Spectrophotometer. Each aliquot was quenched directly in a UV cell containing toluene and the spectra were collected from 300 to 800 nm. QD photoluminescence (PL) and quantum yield (QY) were measured on a Horiba Jobin-Yvon Fluorolog-3 Spectrofluorometer with the excitation and emission monochromators both set at 1 nm. Additional file 2 shows results of this analysis. The optimal excitation wavelength of 450 nm was determined from PL excitation spectroscopy measurements. QY measurements were performed as described [47] using the Fluorolog-3 integrating sphere attachment and a liquid sample holder. QY determination involves five separate spectral measurements: three excitation scatter spectra taken for 1) an empty integrating sphere, 2) integrating sphere with the sample inside directly hit with the excitation light, 3) integrating sphere with the sample excited indirectly by the excitation light scattered by the integrating sphere; and two photoluminescence spectra taken for 4) the sample inside the integrating sphere directly hit by the excitation light, and 5) the sample inside the integrating sphere

excited indirectly by the excitation light scattered by the integrating sphere.

Transmission electron microscopy

Transmission electron microscopy (TEM) was used to verify size and quality of the cap-exchanged QD. An aliquot of the CdSe-ZnS dots was placed onto a carbon coated TEM grid (300 mesh) and allowed to dry for either 0.5 hour or overnight. The QD were imaged using either a Phillips CM 30 TEM at 300 kV accelerating voltage (Figure 1a and c) or a Hitachi H7500 equipped with an AMT XR60 camera (Figure 1b). Image J software was used to determine QD size, which was 4.1 ± 0.88 nm (average of 20 non overlapping QD with visible borders).

Recombinant protein purification and conjugation to QD

The full-length open reading frame of Cyclin E was subcloned into Pepex vector [31], expressed in *E. coli*, grown at 37°C to an OD600 of 0.7, then induced with a final concentration of 1 mM of isopropyl-β-D-thiogalactopyranoside (Promega, Madison, WI) and grown to an OD600 of 1. The Pepex vector contains a DNA sequence specifying a string of six histidine residues at the amino terminus of the inserted coding region for a protein of interest. The result is expression of a recombinant protein with a 6xHis tag fused to its amino-terminus. *E. coli* cells were lysed followed by centrifugation and the supernatant loaded onto BioRad (Hercules, CA) econo-pack columns filled with Ni-NTA agarose (Qiagen, Ventura, CA) for purification (according to the manufacturer's specifications). His$_6$-Cyclin E was eluted by addition of phosphate buffered saline containing imidazole (Sigma-Aldrich, St. Louis, MO). Protein quality and quantity were assessed by SDS-PAGE and visualization on Coomassie stained gels. (His$_6$)-Cyclin E attachment to DHLA capped QD was carried out by addition of an equal amount of (His$_6$)-Cyclin E (in phosphate buffered saline) to QD in an aqueous solution of 10 mM sodium tetraborate buffer (pH 9.5); the mixture was incubated for 15 min at room temperature. pH was tightly controlled to 9.5 during all steps.

QD intracellular delivery and microscopy

X. laevis embryos were obtained by inducing egg laying with hormones followed by in vitro fertilization using standard methods [48], except albino females were used for egg laying. Fertilized embryos were dejellied using 2% cysteine–HCl, pH 7.8, then maintained in 0.1X Marc's Modified Ringer's (0.1X MMR). Microinjections were performed in 4% Ficoll in 0.33X MMR [49]. Embryos were microinjected into the animal pole of one cell at the two-cell stage with 27.6 nl of (QD$_{564}$)-His$_6$Cyclin E. Embryos were staged according to Nieuwkoop and Faber [50] Twenty embryos were microinjected, placed in depression slides containing 0.1X MMR in 3% Ficoll and imaged live at the stated time post-fertilization. An empirically chosen constant exposure time was used for imaging on a Zeiss LSM510 confocal microscope equipped with a META detector and analyzed using LSM510 Image Acquisition software. The Z step size was 70.71 μm. The pixel size was 2.54 μm × 2.54 μm.

Immunofluorescence analysis

The Cyclin E coding sequence (Genbank accession no. Z13966) in the pCS2mt-GFP vector (Addgene, Cambridge, MA) was used to express Myc_6-GFP-tagged cyclin E (Myc_6-GFP-Cyclin E), in Xenopus embryos for live imaging (Figure 5) or for immunofluorescence analysis (Figure 4), as previously described [34]. pCS2mt-GFP-cyclin E plasmid was linearized with Not I and transcribed with SP6 RNA polymerase according to the manufacturer's instructions (Promega, Madison, WI). 1–2 μg of the in vitro transcribed RNA was analyzed on a formaldehyde gel to check quality. Capped RNA was injected into 1-cell of a 2-cell embryo (0.5 ng), between 1.5–2 hours post-fertilization. Five embryos were collected at the indicated time points and either imaged live or fixed in 3.7% formaldehyde/phosphate buffered saline (PBS) for 2 hours on a Nutator shaker. Embryos were then transferred into Dent's fixative (4 parts of MeOH and 1 part DMSO), and incubated at – 20°C for at least 48 hours. Embryos were rehydrated in a graded series of methanol and washed twice for 10 min in PBS. Rehydrated embryos were hemi-sectioned and all incubations and washes were performed at 4°C on a Nutator. As GFP fluorescence could not be easily detected in the fixed embryos (not shown), hemi-sectioned embryos were next incubated in anti-Myc (1:2,000; Cell Signaling Technology, Danvers, MA) in PBS containing 0.05% Tween20 (PBT) overnight in order to detect the Myc tag. The embryos were washed four times with PBT, and incubated in Alexa Fluor 488 conjugated goat-anti-mouse (Molecular Probes, Eugene, OR) in PBT (1:200) overnight. Embryos were washed 3 times with PBT, nuclei counterstained with 4',6'-diamidino-2-phenylindole (DAPI) in PBS (1:3,000 dilution of 1 mg/ml) for 30–60 min, washed extensively with PBS, and mounted in depression slides using Vectashield mounting media (Vector Laboratories, Burlingame, CA) and coverslipped. Empirically chosen acquisition parameters was used for imaging on a Zeiss LSM 510 confocal microscope equipped with a META detector, and analyzed using LSM510 Image Acquisition software. The Z step size was 70.71 μm. The pixel size was 2.54 μm × 2.54 μm.

Additional files

Additional file 1: Figure S1. DHLA coated CdSe-ZnS (QD_{564}) solution. (a) Photograph of 0.6 ml clear-walled PCR tubes containing solvent (water, left) or quantum dot solution (right), showing uniform quantum dot dispersal. (b) Stereomicroscopic image of droplets of solvent (water, left) or quantum dot solution (right) under bright field illumination. The dotted circles on the droplets are a reflection of the ring light on the stereomicroscope. Scale bar is 3 mm. (c) Stereomicroscopic image of droplets of solvent (water, left) or quantum dot solution (right) under UV illumination. Identical field is shown, except (b) is bright field and (c) UV illumination. Droplets were viewed using a Leica MZ FLIII fluorescence stereomicroscope equipped with a Leica GFP1 filter set (ex 425/60, em 480 long pass). Photographs were taken with a Photometrics Coolsnap ES camera.

Additional file 2: Figure S2. (a, b) Photoluminescence spectra of DHLA-CdSe-ZnS (QD_{564}) before (a) and after (b) conjugation to (His_6) Cyclin E. (c, d) Results of quantum yield measurements of DHLA-CdSe-ZnS (QD_{564}) before (c) and after (d) conjugation to (His_6)Cyclin E represented by excitation scatter spectra (upper panel) and the corresponding PL spectra (lower panel) of the samples. The excitation scatter spectra show the actual spectral content of the excitation source (Xenon lamp) scattered by the integrating sphere.

Abbreviations
QD: Quantum dot; DHLA: Dihydrolipoic acid; FRET: Fluorescence resonance energy transfer; DNA: Deoxyribonucleic acid; TOP: Trioctyl phosphine; TOPO: Trioctylphosphineoxide; PL: Photoluminescence; QY: Quantum yield; His_6: Six histidine; MBT: Mid blastula transition; hpf: Hours post fertilization; Myc_6: Six myc; GFP: Green fluorescent protein.

Competing interests
The authors do not have any financial or non-financial competing interests.

Authors' contributions
Y.B. designed and performed experiments, analyzed and interpreted the data and wrote the manuscript. T.M. assisted in embryo injections. G.S. contributed to optical characterization of the NPs by performing the PL and QY measurements M.O. contributed to conception and experimental design. R.H. contributed to conception, experimental design, editing and proofreading of the manuscript. All authors read and approved the final manuscript.

Acknowledgements
We thank Tamara Howard for sharing her expertise on immunofluorescence imaging and for helpful discussions of the work and manuscript. TEM data were generated in the UNM Electron Microscopy Shared Facility supported by the University of New Mexico Health Sciences Center and the University of New Mexico Cancer Center by Dr. Steve Jett, Facility Director, and by Krishnaprasad Sankar at the UNM Center for High Technology Materials. This work was supported by a National Cancer Institute-National Institutes of Health R01CA095898 to RSH and a National Science Foundation IGERT Fellowship on Integrating Nanotechnology with Cell Biology and Neuroscience, NSF DGE-0549500 (Dr. Marek Osiński, UNM-CHTM) to Y.B. Images were generated in the UNM Cancer Center Microscopy Facility, supported from NCRR 1 S10 RR14668, NSF MCB9982161, NCRR P20 RR11830, NCI P30 CA118100, NCRR S10 RR19287, NCRR S10 RR016918, the UNM HSC, and the UNM Cancer Center.

Author details
[1]Department of Cell Biology and Physiology, University of New Mexico Health Sciences Center, Albuquerque, New Mexico 87131-0001, USA. [2]Center for High Technology Materials, University of New Mexico, 1313 Goddard SE, Albuquerque, New Mexico 87106-4343, USA.

References

1. Delehanty JB, Mattoussi H, Medintz IL. Delivering quantum dots into cells: strategies, progress and remaining issues. Anal Bioanal Chem. 2009;393:1091–105.
2. Jaiswal JK, Goldman ER, Mattoussi H, Simon SM. Use of quantum dots for live cell imaging. Nat Methods. 2004;1:73–8.
3. Peng ZA, Peng X. Formation of high-quality CdTe, CdSe, and CdS nanocrystals using CdO as precursor. J Am Chem Soc. 2001;123:183–4.
4. Clapp AR, Goldman ER, Mattoussi H. Capping of CdSe-ZnS quantum dots with DHLA and subsequent conjugation with proteins. Nat Protoc. 2006;1:1258–66.
5. Medintz IL, Uyeda HT, Goldman ER, Mattoussi H. Quantum dot bioconjugates for imaging, labelling and sensing. Nat Mater. 2005;4:435–46.
6. Chan WC, Nie S. Quantum dot bioconjugates for ultrasensitive nonisotopic detection. Science. 1998;281:2016–8.
7. Dubertret B, Skourides P, Norris DJ, Noireaux V, Brivanlou AH, Libchaber A. In vivo imaging of quantum dots encapsulated in phospholipid micelles. Science. 2002;298:1759–62.
8. Bakalova R, Zhelev Z, Aoki I, Ohba H, Imai Y, Kanno I. Silica-shelled single quantum dot micelles as imaging probes with dual or multimodality. Anal Chem. 2006;78:5925–32.
9. Koole R, van Schooneveld MM, Hilhorst J, Castermans K, Cormode DP, Strijkers GJ, et al. Paramagnetic lipid-coated silica nanoparticles with a fluorescent quantum dot core: a new contrast agent platform for multimodality imaging. Bioconjug Chem. 2008;19:2471–9.
10. Yi DK, Selvan ST, Lee SS, Papaefthymiou GC, Kundaliya D, Ying JY. Silica-coated nanocomposites of magnetic nanoparticles and quantum dots. J Am Chem Soc. 2005;127:4990–1.
11. Zhelev Z, Ohba H, Bakalova R. Single quantum dot-micelles coated with silica shell as potentially non-cytotoxic fluorescent cell tracers. J Am Chem Soc. 2006;128:6324–5.
12. Goldman ER, Mattoussi H, Anderson GP, Medintz IL, Mauro JM. Fluoroimmunoassays using antibody-conjugated quantum dots. Methods Mol Biol. 2005;303:19–34.
13. Ji J, He L, Shen Y, Hu P, Li X, Jiang LP, et al. High-efficient energy funneling based on electrochemiluminescence resonance energy transfer in graded-gap quantum dots bilayers for immunoassay. Anal Chem. 2014;86:3284–90.
14. Lingerfelt BM, Mattoussi H, Goldman ER, Mauro JM, Anderson GP. Preparation of quantum dot-biotin conjugates and their use in immunochromatography assays. Anal Chem. 2003;75:4043–9.
15. Wang L, Zhang J, Bai H, Li X, Lv P, Guo A. Specific detection of Vibrio parahaemolyticus by fluorescence quenching immunoassay based on quantum dots. Appl Biochem Biotechnol. 2014;173:1073–82.
16. Liu F, Deng W, Zhang Y, Ge S, Yu J, Song X. Application of ZnO quantum dots dotted carbon nanotube for sensitive electrochemiluminescence immunoassay based on simply electrochemical reduced Pt/Au alloy and a disposable device. Anal Chim Acta. 2014;818:46–53.
17. Boeneman K, Mei BC, Dennis AM, Bao G, Deschamps JR, Mattoussi H, et al. Sensing caspase 3 activity with quantum dot-fluorescent protein assemblies. J Am Chem Soc. 2009;131:3828–9.
18. Medintz IL, Clapp AR, Mattoussi H, Goldman ER, Fisher B, Mauro JM. Self-assembled nanoscale biosensors based on quantum dot FRET donors. Nat Mater. 2003;2:630–8.
19. Dahan M, Levi S, Luccardini C, Rostaing P, Riveau B, Triller A. Diffusion dynamics of glycine receptors revealed by single-quantum dot tracking. Science. 2003;302:442–5.
20. Bouzigues C, Levi S, Triller A, Dahan M. Single quantum dot tracking of membrane receptors. Methods Mol Biol. 2007;374:81–91.
21. Andrews NL, Lidke KA, Pfeiffer JR, Burns AR, Wilson BS, Oliver JM, et al. Actin restricts FcepsilonRI diffusion and facilitates antigen-induced receptor immobilization. Nat Cell Biol. 2008;10:955–63.
22. Gao X, Cui Y, Levenson RM, Chung LW, Nie S. In vivo cancer targeting and imaging with semiconductor quantum dots. Nat Biotechnol. 2004;22:969–76.
23. Lee KH, Galloway JF, Park J, Dvoracek CM, Dallas M, Konstantopoulos K, et al. Quantitative molecular profiling of biomarkers for pancreatic cancer with functionalized quantum dots. Nanomedicine. 2012;8:1043–51.
24. Lu Y, Zhong Y, Wang J, Su Y, Peng F, Zhou Y, et al. Aqueous synthesized near-infrared-emitting quantum dots for RGD-based in vivo active tumour targeting. Nanotechnology. 2013;24:135101.
25. Stylianou P, Skourides PA. Imaging morphogenesis, in Xenopus with Quantum Dot nanocrystals. Mech Dev. 2009;126:828–41.
26. Jackman M, Kubota Y, den Elzen N, Hagting A, Pines J. Cyclin A- and cyclin E-Cdk complexes shuttle between the nucleus and the cytoplasm. Mol Biol Cell. 2002;13:1030–45.
27. Hwang HC, Clurman BE. Cyclin E in normal and neoplastic cell cycles. Oncogene. 2005;24:2776-86.
28. Keyomarsi K, O'Leary N, Molnar G, Lees E, Fingert HJ, Pardee AB. Cyclin E, a potential prognostic marker for breast cancer. Cancer Res. 1994;54:380–5.
29. Dutta A, Chandra R, Leiter LM, Lester S. Cyclins as markers of tumor proliferation: immunocytochemical studies in breast cancer. Proc Natl Acad Sci U S A. 1995;92:5386–90.
30. Newport JW, Kirschner MW. Regulation of the cell cycle during early Xenopus development. Cell. 1984;37:731–42.
31. Chevalier S, Couturier A, Chartrain I, Le Guellec R, Beckhelling C, Le Guellec K, et al. Xenopus cyclin E, a nuclear phosphoprotein, accumulates when oocytes gain the ability to initiate DNA replication. J Cell Sci. 1996;109(Pt 6):1173–84.
32. Guthrie S, Turin L, Warner A. Patterns of junctional communication during development of the early amphibian embryo. Development. 1988;103:769–83.
33. Weber PA, Chang HC, Spaeth KE, Nitsche JM, Nicholson BJ. The permeability of gap junction channels to probes of different size is dependent on connexin composition and permeant-pore affinities. Biophys J. 2004;87:958–73.
34. Brandt Y, Mitchell T, Wu Y, Hartley RS. Developmental downregulation of Xenopus cyclin E is phosphorylation and nuclear import dependent and is mediated by ubiquitination. Dev Biol. 2011;355:65–76.
35. Bunge SD, Krueger KM, Boyle TJ, Rodriguez MA, Headley TJ, Colvin VL. J Mater Chem. 2003;13:1705–9.
36. Hainfeld JF, Liu W, Halsey CM, Freimuth P, Powell RD. Ni-NTA-gold clusters target His-tagged proteins. J Struct Biol. 1999;127:185–98.
37. Takagi C, Sakamaki K, Morita H, Hara Y, Suzuki M, Kinoshita N, et al. Transgenic Xenopus laevis for live imaging in cell and developmental biology. Dev Growth Differ. 2013;55:422–33.
38. Feilmeier BJ, Iseminger G, Schroeder D, Webber H, Phillips GJ. Green fluorescent protein functions as a reporter for protein localization in Escherichia coli. J Bacteriol. 2000;182:4068–76.
39. Landgraf D, Okumus B, Chien P, Baker TA, Paulsson J. Segregation of molecules at cell division reveals native protein localization. Nat Methods. 2012;9:480–2.
40. Zhu M, Ni W, Dong Y, Wu ZY. EGFP tags affect cellular localization of ATP7B mutants. CNS Neurosci Ther. 2013;19:346–51.
41. Trombetta ES, Parodi AJ. Quality control and protein folding in the secretory pathway. Annu Rev Cell Dev Biol. 2003;19:649–76.
42. D'Alessio C, Trombetta ES, Parodi AJ. Nucleoside diphosphatase and glycosyltransferase activities can localize to different subcellular compartments in Schizosaccharomyces pombe. J Biol Chem. 2003;278:22379–87.
43. Vitale A, Pedrazzini E. Recombinant pharmaceuticals from plants: the plant endomembrane system as bioreactor. Mol Interv. 2005;5:216–25.
44. Foresti O, De Marchis F, de Virgilio M, Klein EM, Arcioni S, Bellucci M, et al. Protein domains involved in assembly in the endoplasmic reticulum promote vacuolar delivery when fused to secretory GFP, indicating a protein quality control pathway for degradation in the plant vacuole. Mol Plant. 2008;1:1067–76.
45. Zhang Y. Single-cell microinjection technologies. Methods Mol Biol. 2012;853:169–76.
46. Gunsalus IC, Barton LS, Gruber W. Biosynthesis and structure of lipoic acid derivatives. J Am Chem Soc. 1956;78:1763–6.
47. Mello JC, Wittmann HF, Friend RH. An improved experimental determination of external photoluminescence quantum efficiency. Adv Mater. 1997;9:230–2.
48. Sive HL, Grainger RM, Harland RM. Xenopus laevis In Vitro Fertilization and Natural Mating Methods. CSH Protoc. 2007;2007:pdb prot4737.
49. Sive HL, Grainger RM, Harland RM. Microinjection of Xenopus embryos. Cold Spring Harb Protoc. 2010;2010:pdb ip81.
50. Na F. Normal Table of Xenopus laevis (Daudin). New York: Garland Publishing Inc; 1994.

13

Effect of carbon black nanomaterial on biological membranes revealed by shape of human erythrocytes, platelets and phospholipid vesicles

Manca Pajnič[1], Barbara Drašler[2], Vid Šuštar[3], Judita Lea Krek[1], Roman Štukelj[1], Metka Šimundić[1], Veno Kononenko[2], Darko Makovec[4], Henry Hägerstrand[5], Damjana Drobne[2] and Veronika Kralj-Iglič[1*]

Abstract

Background: We studied the effect of carbon black (CB) agglomerated nanomaterial on biological membranes as revealed by shapes of human erythrocytes, platelets and giant phospholipid vesicles. Diluted human blood was incubated with CB nanomaterial and observed by different microscopic techniques. Giant unilamellar phospholipid vesicles (GUVs) created by electroformation were incubated with CB nanomaterial and observed by optical microscopy. Populations of erythrocytes and GUVs were analyzed: the effect of CB nanomaterial was assessed by the average number and distribution of erythrocyte shape types (discocytes, echinocytes, stomatocytes) and of vesicles in test suspensions, with respect to control suspensions. Ensembles of representative images were created and analyzed using computer aided image processing and statistical methods. In a population study, blood of 14 healthy human donors was incubated with CB nanomaterial. Blood cell parameters (concentration of different cell types, their volumes and distributions) were assessed.

Results: We found that CB nanomaterial formed micrometer-sized agglomerates in citrated and phosphate buffered saline, in diluted blood and in blood plasma. These agglomerates interacted with erythrocyte membranes but did not affect erythrocyte shape locally or globally. CB nanomaterial agglomerates were found to mediate attractive interaction between blood cells and to present seeds for formation of agglomerate - blood cells complexes. Distortion of disc shape of resting platelets due to incubation with CB nanomaterial was not observed. CB nanomaterial induced bursting of GUVs while the shape of the remaining vesicles was on the average more elongated than in control suspension, indicating indirect osmotic effects of CB nanomaterial.

Conclusions: CB nanomaterial interacts with membranes of blood cells but does not have a direct effect on local or global membrane shape in physiological *in vitro* conditions. Blood cells and GUVs are convenient and ethically acceptable methods for the study of effects of various substances on biological membranes and therefrom derived effects on organisms.

Keywords: Carbon black nanomaterial, Carbon black, Air pollution, Nanotoxicity, Cell membranes, Phospholipid bilayer, Model systems, Blood cells

* Correspondence: veronika.kralj-iglic@fe.uni-lj.si
[1]Laboratory of Clinical Biophysics, University of Ljubljana, Faculty of Health Sciences, Zdravstvena pot 5, Ljubljana SI-1000, Slovenia
Full list of author information is available at the end of the article

Background

Recent advances in development and industrial production of nanomaterials require assessment of their effects on human and animal health. Toxicity of nanomaterials has therefore become an issue regarding the question which methods are the most appropriate for determining health risk [1,2]. Nanomaterial introduces also effects that cannot be explained by composition of the material and chemical reactions, but require consideration of non-specific properties, such as size and shape of particles, their electromagnetic properties and interactions with biological material [3]. Since the underlying mechanisms are largely unknown, standard methods for research, testing and safety are not necessarily relevant for these materials. Standard methods for research and testing of various compounds include experiments on experimental animals. These methods were found to have poor predictability regarding other species, in particular human, as shown by carcinogeneity studies [4-6]. It was suggested that *in vitro* research may provide essential information pertaining to the human health risks posed by nanoparticle exposure [7]. With fast and extensive development of new nanomaterials and their use in medicine and industry there is urgent need to develop effective and low cost methods that will enable understanding basic processes in living organisms. The role of nonspecific biophysical mechanisms has hitherto been underestimated as potentially essential in revealing biological processes but should be considered in future paradigms of research and testing of nanomaterials.

Exogenously added substances first come in contact with cells by interacting with the membrane, so it is of great interest to study the interaction of these substances with biological membranes. Convenient systems for such studies are mammalian erythrocytes and giant unilamelar phospholipid vesicles (GUVs) composed of a closed bilayer membrane. These entities do not have internal structure (other than cortical membrane skeleton in case of erythrocytes) so their equilibrium shape is determined by the minimum of the membrane free energy [8]. Interaction of the added substance with the membrane causes changes in the membrane properties and in the constraints imposed upon the cell/vesicle (fixed area of the membrane, fixed difference between the areas of the two membrane layers, fixed cell/vesicle volume) which is reflected in the shape change [9]. Since mammalian erythrocytes and phospholipid vesicles (sized up to 100 micrometers) can be observed under the optical microscope, the effect of the exogenously added substances can be directly followed in real time.

Further, of special interest are processes caused by exogenously added substances that increase the risk for thromboembolic events [10]. Some parameters of these processes, e.g. activation of platelets and coalescence of

membranous structures due to the presence of nanomaterial have previously been studied [11].

Membrane – nanomaterial interactions could be considered as one of the basic elements in an effective strategy of research and testing. These interactions are subject to biophysical methods which provide theoretical background. The accompanying experiments on the microscopic level are performed *in vitro*, on the artificial membranes and cell membranes.

Carbon black (CB) is a material produced by the incomplete combustion of carbon. Commercially available CB is a well characterized carbonaceous core particle that has been used extensively as a model for Diesel exhaust particles, which showed proinflamatory and prothrombotic systemic effects [12]. CB nanoparticles that are produced in traffic are present in the air, so living beings are commonly exposed to them. CB nanoparticles thus enter the body by means of inhalation. As they are very small, they cross biological barriers, enter the circulation and are spread in tissues and fetal organs. In the body, they affect the cell function and due to intercellular interactions [13] affect the entire organism. CB nanomaterial was found to have a size-dependent effect on *in vitro* cultures [14]. The reported effects of CB on animals include changes in development, in the immune response, and in gene expression [15]. Epidemiological studies on human have shown that CB nanomaterial is connected with adverse effects on health [16], in particular, with respiratory diseases and lung cancer.

In this work we studied the effect of CB nanomaterial on biological membranes as revealed by shape changes of erythrocytes and GUVs. These model systems and their interactions with the added substances have previously been used in the study of the effects of various compounds [17-21] including nanoparticles [11,22-26] on cells and their membranes and were found to contribute to better understanding of the relevant mechanisms which are common in *ex vivo* and *in vivo* exposures.

Results

Characterization and imaging of carbon black nanomaterial suspensions

Transmission and scanning electron microscope (TEM and SEM, respectively) images of CB nanomaterial suspended in citrated and phosphate-buffered saline (PBS) showed agglomerates of globular nanoparticles of homogeneous size about 20 nm (Figure 1A and B, respectively) and a small amount of larger aggregates (Figure 1C). By using the optical microscope, we could inspect relatively large volume of the test sample. We observed even larger aggregates in PBS (Figure 1D1) and in platelet rich plasma (Figure 1D2). These large aggregates had different sizes, some of them extending over 10 μm (Figure 1D).

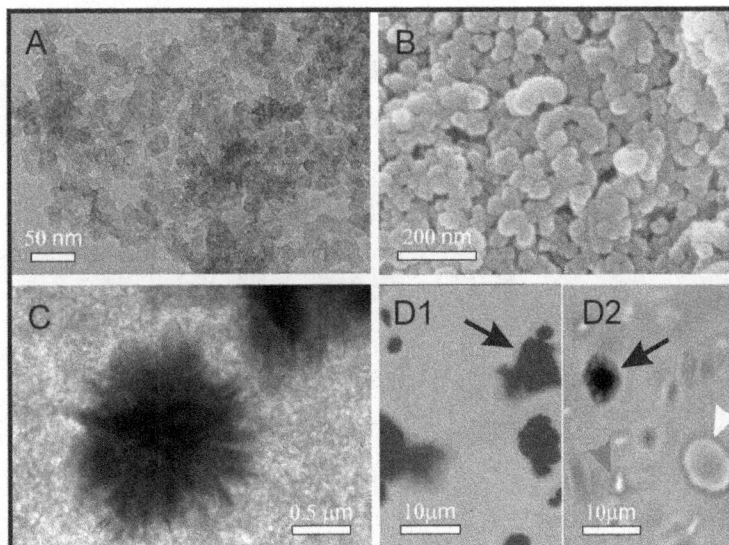

Figure 1 Micrographs of agglomerated carbon black nanomaterial suspended in citrated and phosphate–buffered saline. Agglomerated carbon black (CB) nanomaterial suspended in citrated and phosphate buffered saline (PBS) as observed by transmission electron microscope (TEM) **(A)** and scanning electron microscope (SEM) **(B)**. The primary particles are amorphous with homogeneous size about 20 nm. A small number of larger agglomerates were found in PBS, as observed by TEM **(C)** and by the optical microscope **(D1)**, and in platelet rich plasma **(D2)**. Black arrows point to agglomerates, white arrow points to an erythrocyte and gray arrow points to a platelet.

Dynamic light scattering (DLS) measurements of previously sonicated CB nanomaterial showed a wide distribution of the agglomerates with respect to size (ranging from 500 to 2500 nm). Zeta potential of the PBS suspension of CB nanomaterial at pH = 7.32 was -29 mV. We could not perform DLS measurements of CB nanomaterial suspensions in 0.3 M glucose solution due to strong agglomeration and fast sedimentation of CB nanomaterial.

Carbon black nanomaterial interactions and effects on erythrocyte shape

Figure 2 shows the effect of CB nanomaterial on the shape of washed human erythrocytes as observed by SEM. Panels A, D and G show the control samples (with added PBS) and panels B, E and H show the samples with added PBS-suspended CB nanomaterial after 1, 3 and 24 hours of incubation, respectively. For comparison, panels C, F and I show another control with added PBS-suspended zinc oxide nanoparticles (ZnO) [11]. Normal discocytic shapes of erythrocytes were observed after 1, 3 and 24 hours in all samples. Singular echinocytes were found in a positive control (ZnO-treated) samples ([11], Figure 2C, F and I). It can be seen in panels B, E and H that agglomerated CB nanomaterial adhered to the erythrocyte membrane. Large CB nanomaterial agglomerates have filled the discocyte dimple (Figure 2B, black arrow) but also agglomerates adhered to the membrane at any location on the erythrocyte surface (Figure 2E and H, black arrows). Large CB nanomaterial agglomerates can

adhere to two cells and therefore mediate the interaction between them (Figure 2B, white arrow). A close-up view shows that in the case of small contact area between CB agglomerate and membrane (Figure 3A, C and D, white arrows), the CB nanomaterial does not affect the local membrane curvature (the membrane shape is undisturbed at the site of contact of the CB nanomaterial agglomerate). Moreover, the global shape of the cell remains discocytic (Figure 3A). However, singular disintegrated cells were found in the sample (Figure 3B). The cell shown in Figure 3B adhered to huge CB nanomaterial agglomerate and the shape was distorted. Large pores in the membrane (Figure 3B, white arrow) caused the exchange of the inner and the outer cell solutions. A similar process of cell deformation upon the adhesion to a huge agglomerate is indicated in Figure 2H (white arrow), but in this case, the membrane integrity is still preserved.

Carbon black effects on erythrocyte shape in population of erythrocytes and platelets

Figure 4 shows the effect of CB nanomaterial on coalescence of washed erythrocytes in three subjects with no record of the disease (A, B and C) as observed by the phase contrast optical microscope. Washed erythrocytes were incubated with PBS for control and with CB nanomaterial (suspended in PBS) for 1, 3 and 24 hours, respectively. It can be seen that the control samples remained unchanged after 1, 3 and 24 hours while in the test samples, the erythrocytes have gathered around larger CB nanomaterial agglomerates already after 1 hour. The size of the complexes

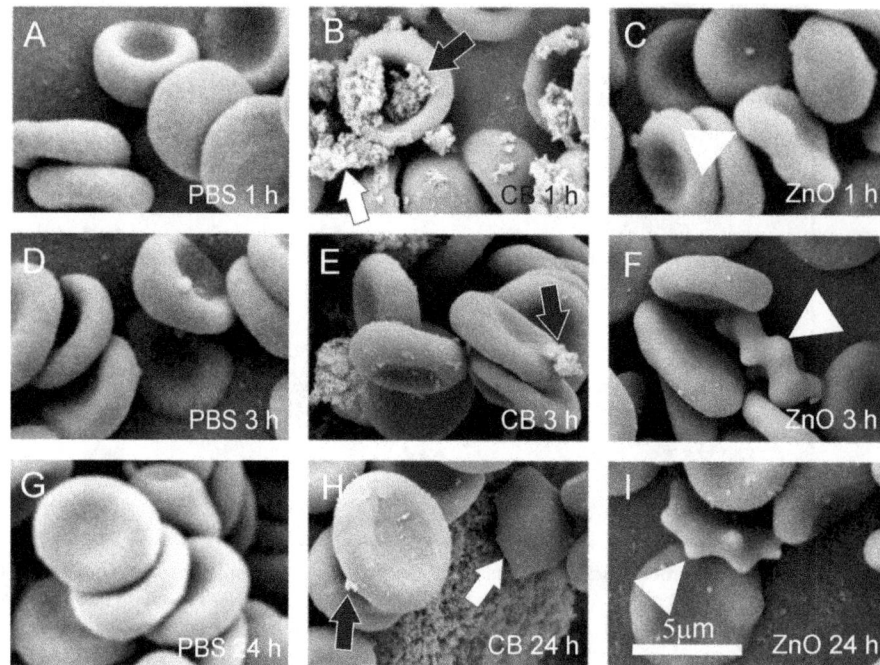

Figure 2 The effect of carbon black nanomaterial on washed human erythrocytes. The effect of carbon black (CB) nanomaterial on washed human erythrocytes as observed by scanning electron microscope. Panels **A**, **D** and **G** show the control samples with added citrated and phosphate buffered saline (PBS); panels **B**, **E** and **H** show the samples with added PBS-suspended CB nanomaterial after 1 hour, 3 hours and 24 hours of incubation, respectively and panels **C**, **F** and **I** show the samples with added PBS-suspended ZnO after 1 hour, 3 hours and 24 hours, respectively. Large CB agglomerates adhered to the erythrocyte surface (**B, E, H**, black arrows). Agglomerate which adhered to two erythrocytes mediated a bridging interaction between them (**B**, white arrow). In ZnO-treated samples, singular echinocytes were found (**C, F, I**, white triangles).

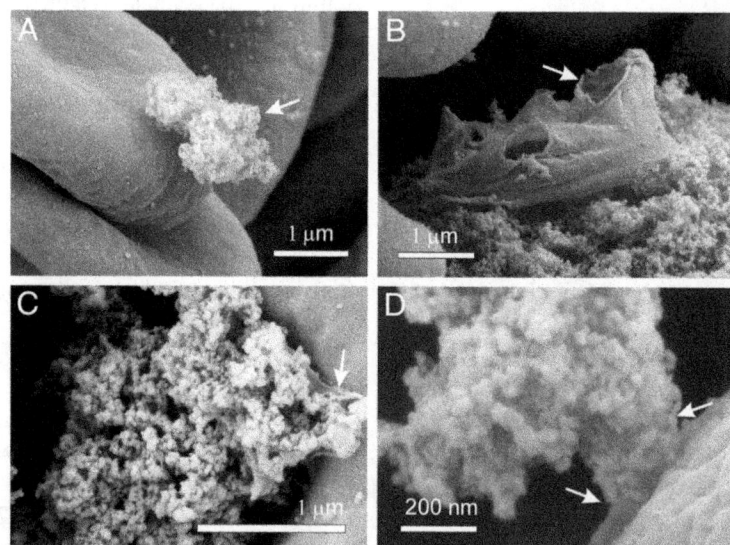

Figure 3 Adhesion of large agglomerates of carbon black nanomaterial to the erythrocyte membrane. In the case of a rather small contact area (panels **A, C** and **D**, white arrows) the local and the global membrane shape remained undisturbed. In the case of a large contact area **(B)** the membrane was considerably affected leading to formation of a ghost with large pores in the membrane (**B**, white arrow).

Effect of carbon black nanomaterial on biological membranes revealed by shape of human erythrocytes...

149

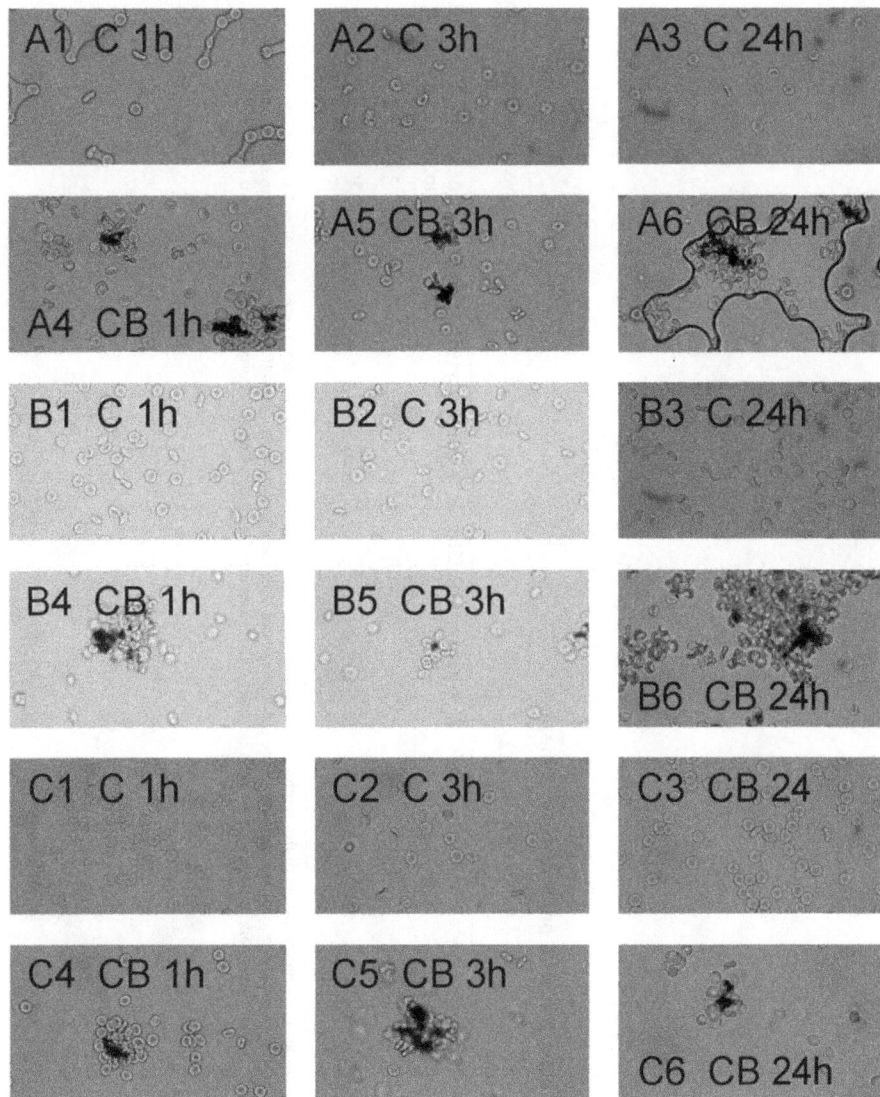

Figure 4 The effect of carbon black nanomaterial on coalescence between erythrocytes. The effect of carbon black (CB) nanomaterial on coalescence between erythrocytes in three subjects with no record of the disease (marked **A**, **B** and **C**) as observed by the phase contrast optical microscope. Washed erythrocytes were incubated with citrated and phosphate buffered saline (PBS) for control (denoted by **C**) and with PBS-suspended CB nanomaterial (denoted by **CB**) for 1, 3 and 24 hours, respectively, as indicated.

remained essentially the same also after 3 and 24 hours (Figure 4). Erythrocytes that were not in direct contact with the visible CB nanomaterial agglomerates did not adhere to each other.

We analyzed the effect of CB nanomaterial on erythrocyte shape at the population level (Table 1). The examined blood initially contained mostly discocytes (Table 1). With time the portion of discocytes in the test sample and in the control sample decreased at the expense of the portion of echinocytes (Table 1), but the effect was stronger in the control sample. The differences between the respective test and control samples were statistically significant (p < 0.05) (Table 1). We observed no decreasing trend of the number of erythrocytes in the samples with time.

Figure 5 shows the effect of CB nanomaterial on population of platelets as observed by SEM. Washed platelets were incubated with PBS (control; A, D, G) and with CB nanomaterial suspended in PBS (B, E, H) for 1, 3 and 24 hours, respectively. For comparison, we show also platelets incubated with a positive control (ZnO agglomerated nanomaterial [11]) for 1, 3 and 24 hours, (C, F, I, respectively). After 1 and 3 hours, the control and test samples showed disc-like shapes (A, B, D, E) characteristic for resting platelets. There were some platelets exhibiting protrusions and shape deformation characteristic for activated platelets after 24 hours in the control and in the test sample (G,H). For comparison, the effect of ZnO nanoparticles caused shape transformation in some platelets after

Table 1 A population study of the effect of carbon black nanomaterial on erythrocyte shape

Sample	Sample size (number of frames) Test/Control	Test				Control				Difference between test and control		
		D: % discocytes (SD)	E: % echinocytes (SD)	S: % stomatocytes (SD)		D: % discocytes (SD)	E: % echinocytes (SD)	S: % stomatocytes (SD)		ΔD (%) (p, P)	ΔE (%) (p, P)	ΔS (%) (p, P)
1 hour	56/54	99 (2)	1 (1)	0		90 (8)	9 (7)	1 (5)		9 ($<10^{-6}$, 1)	-8 ($<10^{-6}$, 1)	NA
3 hours	50/53	90 (18)	9 (18)	0		84 (11)	16 (11)	0		6 (0.05, 0.62)	-6 (0.04, 0.75)	NA
24 hours	49/51	79 (9)	21 (9)	0		62 (9)	38 (10)	0		17 ($<10^{-6}$, 1)	-17 ($<10^{-6}$, 1)	NA

A population study of the effect of carbon black nanomaterial on erythrocyte shape. The average values of the portion of discocytes (D), echinocytes (E) and stomatocytes (S) are given after 1 hour, 3 hours and 24 hours of incubation with the carbon black nanomaterial (test sample) and with citrated and phosphate buffered saline (control sample). SD: standard deviation. p: statistical significance of the difference between the average values of test and control, calculated by the t-test. P: statistical power. NA: non applicable as there were les than 1% of stomatocytes in all samples. Altogether 10712 cells were included in the analysis.

Figure 5 The effect of carbon black nanomaterial on population of platelets. The effect of carbon black (CB) nanomaterial on population of platelets was observed by the scanning electron microscope. Washed platelets were incubated with citrated and phosphate-buffered saline (PBS) for control **(A, D, G)**, with PBS-suspended CB nanomaterial **(B, E, H)** and for comparison, with ZnO [11] **(C, F, I)** for 1, 3 and 24 hours, respectively. After 1 and 3 hours the control and test samples contained disc-like platelets **(A, B, D, E)** characteristic for the resting state, while there were some platelets exhibiting deviation from the disc-like shape and tubular protrusions (characteristic for activated platelets) found after 24 hours **(E, F)**. For comparison, the effect of ZnO caused shape changes of some platelets after 3 hours **(F)** while after 24 hours all platelets attained swollen globular shape with remnants of tubular protrusions, indicating their activation **(G)**.

3 hours (F) and swollen globular shapes with remnants of tubular protrusions characteristic for activated platelets after 24 hours (I).

Carbon black effect on giant unilamellar phospholipid vesicles

Figure 6 shows the effect of CB nanomaterial on GUV abundance and shape. Incubation of GUVs with CB nanomaterial caused a considerable decrease in the number of GUVs already after 20 minutes (Figure 6A). The remaining GUVs in the test suspension exhibited higher eccentricity than the control GUVs (Figure 6B).

Carbon black effect on parameters of blood cell populations

Tables 2, 3, 4 show blood cell population parameters (erythrocyte parameters: erythrocyte concentration RBC, hematocrit HCT, red cell distribution width RDW and hemoglobin distribution width HDW (Table 2), platelet parameters: concentration of platelets PLT, plateletcrit PCT, mean platelet volume MPV, platelet distribution width PDW, mean platelet complement MPC and mean platelet mass MPM (Table 3) and concentrations of leukocyte populations: total leukocytes (WBC), neutrophils, lymphocytes, monocytes, eosinophils and

basophils, (Table 4) of samples incubated with two different concentrations of CB nanomaterial (0.5 mg/mL (A) and 1 mg/mL (B)), and control samples (incubated with equal volume of PBS). We found no differences between average values of parameters in test samples and in control samples with respect to number, mass and volume of erythrocytes and platelets and the shapes of their distributions (Tables 2 and 3). We found statistically significant difference in parameter MPC which measures refractive index of platelets (Table 3) and in the number of leukocytes; control samples had considerably less leukocytes than the control sample (20% (A) and 16% (B), respectively) (Table 4). Moreover, we found differences in distribution of events within the flow cytometer scatter diagram regions that are in blood samples attributed to different types of leukocytes (Table 4). The number of events within the eosinophils and basophils region was increased while the number of events within the monocyte region was decreased with respect to the control sample (Table 4). These differences were statistically significant and of sufficient statistical power (the power was considered sufficient if larger than 0.8). However, there were no differences between the respective samples incubated with different concentrations of nanomaterial (Table 4).

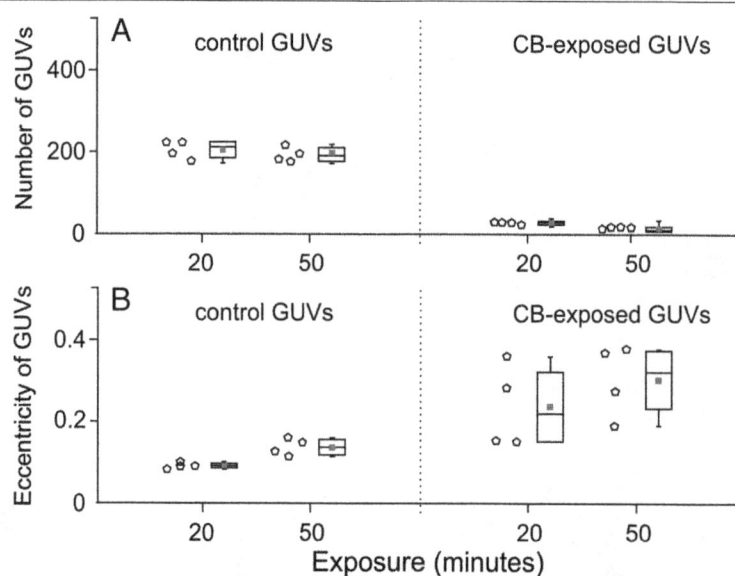

Figure 6 The effect of carbon black nanomaterial on giant unilamellar vesicles. Quantity of giant unilamellar vesicles (GUVs) incubated in chambers with 0.3 M glucose solution (control) and suspension of CB nanomaterial (0.05 mg/mL in 0.3 M glucose solution). Samples of GUVs were incubated with sugar solution for control and with carbon black (CB) nanomaterial suspended in sugar solution (0.05 mg/mL), for 20 minutes and 50 minutes, respectively. Each empty dot corresponds to a number of vesicles obtained in one sample, i.e. one replicate of each tested group. One sample of a population consists of all vesicles captured in two individual video tracks at different places in a chamber. Based on preliminary calculations, two such video tracks capture approximately 3% of the whole population of vesicles in the chamber. The red square in the center of the box plot represents the mean value of **(A)** the number of GUVs or **(B)** GUVs' eccentricity, the central line is the median value, and the lower and upper lines of the box represent the 25th and 75th percentiles. The whiskers are set at two standard deviations from the mean value. Addition of CB nanomaterial considerably diminished the number of GUVs in the sample while the remaining GUVs exhibited an increased eccentricity of the contour (elongation).

Discussion

We considered the effect of CB nanomaterial on biological membranes. We found that CB nanomaterial agglomerated in PBS and in platelet rich plasma (Figure 1). Agglomerates of CB nanomaterial were found to interact with human erythrocytes (Figures 2 and 3). Adhesion of agglomerates did not change the local membrane curvature (Figures 2 and 3). SEM micrographs indicate that the surface of large CB nanomaterial agglomerates in close contact with membrane can be relatively small and discretized (Figure 3D, white arrow). If attached over a large area to a CB nanomaterial agglomerate, erythrocytes may be damaged. Figure 3B shows remnants of a cell attached to huge CB nanomaterial complex. Large holes with bulby circular edge were formed in the membrane of the cell shown in Figure 3B (white arrow) causing the cell interior and exterior to exchange the solutions, and a considerable deformation of the shape. Similarly, a cell shown in Figure 2H was attached to huge CB nanomaterial complex over a large area. However, such events were rarely observed in the SEM micrographs.

Yet the interaction between CB nanomaterial agglomerates and the membrane was strong enough to mediate the

bridging interaction between erythrocyte membranes (Figure 2B, white arrow) and induce formation of agglomerate-cell complexes already after an hour of incubation with CB nanomaterial in PBS – diluted blood cell suspensions (Figure 4). Similar structures were formed also in blood plasma (Figure 1D2). Such formations may present mechanical occlusions in the bloodstream.

Incubation of the erythrocyte suspension with CB nanomaterial preserved the average number of intact cells after 24 hours. In the control and in the test sample we observed an increase of the portion of echinocytes with respect to discocytes with time but this process was less prominent in the test sample than in the control sample (Table 1).

In contrast, we observed a deleterious effect of agglomerated CB nanomaterial on GUVs. We found a considerable decrease in the number of GUVs already after 20 minutes of incubation with CB nanomaterial. Although the GUV artificial membrane presents a backbone of biological membranes and therefore shares important similarities with the membrane of an erythrocyte or a more complex cell, the difference between these systems may derive also from the properties of other parts of the experimental system (the composition

Table 2 Erythrocyte parameters in blood samples incubated with carbon black nanomaterial

Parameter	Sample	1	2	3	4	5	6	7	8	9	10	11	12	13	14	AVG	STD	p
RBC (10^{12} cells/L)	A	3.27	3.72	3.65	3.27	4.75	3.89	3.38	3.26	2.92	2.98	3.6	2.82	4.46	3.67	3.55	0.55	0.42 (A-C)
	B	3.51	3.56	3.5	3.41	4.46	3.84	3.39	3.44	3.35	2.5	3.95	3.23			3.51	0.46	0.27 (B-C)
	C	3.26	3.54	3.2	3.22	4.54	3.95	3.46	3.26	3.38	3.14	3.68	3.27	4.34	3.3	3.40	0.41	0.80 (A-B)
HCT (%)	A	0.28	0.32	0.30	0.28	0.40	0.34	0.31	0.28	0.26	0.26	0.32	0.24	0.37	0.31	0.31	0.04	0.96 (A-C)
	B	0.30	0.31	0.29	0.29	0.38	0.33	0.31	0.29	0.30	0.22	0.36	0.28			0.3	0.04	0.89 (B-C)
	C	0.28	0.31	0.27	0.27	0.39	0.34	0.32	0.27	0.30	0.27	0.33	0.28	0.36	0.28	0.3	0.04	0.76 (A-B)
RDW (%)	A	12.8	13.5	13	15.3	12.4	12.9	13.5	13.6	12.5	12.3	13	12.6	12.5	12.6	13.04	0.78	0.88 (A-C)
	B	12.7	13.5	13	15.4	12.4	12.8	13.4	13.6	12.8	12.4	12.9	12.6			13.13	0.82	0.90 (B-C)
	C	12.7	13.7	12.9	15.3	12.5	12.8	13.7	13.6	12.6	12.5	12.9	12.8	12.5	12.6	13.1	0.77	0.98 (A-B)
HDW (g/L)	A	24.5	22.5	27.1	26.3	24.6	24.0	22.6	25.7	27.2	24.1	23.4	22.7	25.9	25.9	24.8	1.62	0.95 (A-C)
	B	24.3	22.5	27.1	26.1	24.4	22.1	22.1	26.2	27.3	24.6	23.2	22.7			22.9	6.20	0.36 (B-C)
	C	24.6	22.7	26.8	26.2	24.7	23.9	22.3	26.2	27.5	24.3	23.5	22.6	25.9	25.9	24.8	1.66	0.37 (A-B)

Erythrocyte parameters - (Erythrocyte (RBC), Hematocrit (HCT), RBC distribution width (RDW) and hemoglobin distribution width (HDW)) in blood samples incubated with two different CB nanomaterial concentrations (A – 0.5 mg/mL, B - 1 mg/mL), and control suspension incubated with PBS (C). Average value (AVG), standard deviation (STD) and statistical significance of the difference between the test and the control samples (A-C and B-C, respectively) and between the two test samples (A-B), given by the respective probability (p).

of the solution, the interactions at the interfaces of contact of the sample and the laboratory material).

When GUVs are formed in the electroformation process, they are connected by a network of nanotubules [27]. This network is torn when GUVs are rinsed out of the electroformation chamber, but its remnants stay attached to the GUVs [28]. Under an optical microscope fresh GUVs appear spherical while the attached nanotubules cannot be observed due to their thinness. If a spontaneous process exists within the observation chamber which removes the lipid molecules from the outer layer, the average mean curvature of the membrane

Table 3 Platelet parameters in blood samples incubated with carbon black nanomaterial

Parameter	Sample	1	2	3	4	5	6	7	8	9	10	11	12	13	14	AVG	STD	p	P
PLT (10^9 cells/L)	A	106	71	91	62	56	69	169	-	114	146	114	109	38	37	89.4	39.0	0.16 (A-C)	
	B	61	56	99	80	49	104	176	56	123	184	92	119			99.9	44.9	0.21 (B-C)	
	C	117	82	129	86	56	117	182	104	121	209	141	129	49	59	113	45.9	0.91 (A-B)	
PCT (%)	A	0.1	0.07	0.09	0.07	0.05	0.07	0.16	0.06	0.11	0.13	0.1	0.1	0.04	0.03	0.08	0.04	0.21 (A-C)	
	B	0.06	0.06	0.1	0.08	0.05	0.1	0.17	0.05	0.12	0.16	0.08	0.11			0.10	0.04	0.31 (B-C)	
	C	0.11	0.08	0.13	0.08	0.05	0.11	0.16	0.09	0.11	0.17	0.12	0.12	0.05	0.05	0.10	0.04	0.87 (A-B)	
MPV (fL)	A	9.5	10.5	9.9	10.5	9.2	9.7	9.4	9.2	9.6	8.6	9	8.9	9.4	8.3	9.41	0.63	0.62 (A-C)	
	B	10	10.5	10.2	9.9	9.3	9.6	9.6	8.8	9.6	8.4	8.7	9.4			9.5	0.63	0.39 (B-C)	
	C	9.5	10.3	10.1	9.7	9	9.5	8.8	8.7	9	8.4	8.8	9.6	9.7	9	9.29	0.56	1.00 (A-B)	
PDW (%)	A	56	63.8	57.6	65.4	72.2	61.3	62.1	67.7	59.1	54	61.9	53.8	51.6	47.7	59.6	6.67	0.15 (A-C)	
	B	61.5	61.0	59.7	56.8	63.2	61.8	61.8	54.5	56.8	55.4	47.2	60			58.3	4.49	0.31 (B-C)	
	C	54.6	62.3	55.4	54	60.2	58.1	58.3	59.7	52.9	52.4	60	50.9	56.3	57.8	56.6	3.39	0.17 (A-B)	
MPC (g/L)	A	215	211	216	209	223	222	228	209	215	219	219	229	206	174	214	13.4	0.02 (A-C)	0.88
	B	189	206	219	215	218	236	218	212	220	220	209	229			216	11.7	0.00 (B-C)	0.84
	C	214	225	222	232	226	239	232	233	230	230	229	229	219	193	225	11.2	0.61 (A-B)	
MPM (pg)	A	1.89	1.99	1.95	0.96	1.81	1.97	1.97	1.74	1.88	2.74	1.79	1.89	1.78	1.38	1.84	0.38	0.45 (A-C)	
	B	1.72	1.96	2.01	1.94	1.84	2.06	1.91	1.73	1.92	1.71	1.72	1.89			1.87	0.12	0.10 (B-C)	
	C	1.88	2.09	2.04	2.07	1.86	2.1	1.89	1.87	1.92	1.77	1.86	2.03	1.92	1.6	1.92	0.14	0.90 (A-B)	

Platelet parameters (Platelet count (PLT), plateletcrit (PCT), platelet distribution width (PDW), mean platelet volume (MPV), mean platelet mass (MPM) and mean platelet component (MPC)) in blood samples incubated with two different carbon black nanomaterial concentrations (A – 0.5 mg/mL, B – 1 mg/mL), and control suspension incubated with citrated and phosphate-buffered saline (C). Average value (AVG), standard deviation (STD) and statistical significance of the difference between the test and the control samples (A-C and B-C, respectively) and between the two test samples (A-B), given by the respective probability (p). Statistically significant differences have also been evaluated for statistical power at α = 0.05 (P).

Table 4 Leukocyte parameters in blood samples incubated with carbon black nanomaterial

Parameter	Sample	1	2	3	4	5	6	7	8	9	10	11	12	13	14	AVG	STD	p	P
WBC (10^9 cells/L)	A	8.51	6.81	8.45	6.3	6.13	8.97	6.92	8.1	8.83	7.05	6.91	6.12	4.74	4.71	7.04	1.39	0.04 (A-B)	0.81
	B	9.04	5.97	8.12	5.78	6.33	8.28	7.12	7.29	8.94	8.55	8.3	7.46	4.78	4.57	7.60	1.13	0.00 (B-C)	1
	C	7.82	5.36	7.1	4.65	4.24	6.56	6.2	5.98	7.59	6.04	6.04	6.72	4.78		5.98	1.14	0.70 (A-B)	
Neutrophil region (10^9 cells/L)	A	3.04	4.82	5.81	3.18	4.59	3.99	4.62	3.05	2.57	3.97	3.53	3.24	3.21	2.37	3.71	0.97	0.25 (A-B)	0.86
	B	4.05	5.58	5.72	4.15	4.19	3.86	4.35	3.46	2.5	3.79	3.89	3.08	3.05	2.29	4.05	0.91	0.04 (B-C)	
	C	3.79	4.12	5.29	2.93	3.74	3.68	3.72	2.29	1.98	3.33	3.22	2.89			3.31	0.86	0.37 (A-B)	
Lymphocyte region (10^9 cells/L)	A	2.14	1.39	1.71	2.84	1.74	2.11	2.35	1.97	2.7	3.53	3.11	2.46	1.01	1.82	2.21	0.68	0.58 (A-B)	
	B	2.17	1.62	1.83	3.17	1.66	2.05	2.2	1.75	2.45	3.33	2.91	2.15			2.27	0.58	0.39 (B-C)	
	C	2.15	1.56	1.71	2.56	1.61	1.97	2.12	1.56	2.37	3.17	3.16	2.04	1.19	1.86	2.07	0.58	0.78 (A-B)	1
Monocyte region (10^9 cells/L)	A	0.14	0.19	0.23	0.27	0.22	0.23	0.24	0.24	0.08	0.2	0.18	0.15	0.25	0.15	0.20	0.05	0.00 (A-B)	1
	B	0.18	0.17	0.23	0.33	0.16	0.17	0.2	0.34	0.11	0.16	0.21	0.16			0.20	0.07	0.00 (B-C)	
	C	0.4	0.24	0.36	0.43	0.31	0.39	0.36	0.3	0.22	0.3	0.36	0.3	0.34	0.18	0.32	0.07	0.87 (A-B)	
Eosinophil region (10^9 cells/L)	A	0.69	0.38	0.95	0.76	1.41	0.44	1.61	0.72	0.8	0.63	1.6	0.9	0.2	0.3	0.81	0.45	0.00 (A-B)	1
	B	0.91	0.85	1.02	0.82	1.06	0.92	1.29	0.66	0.58	0.64	1.88	0.5			0.93	0.37	0.00 (B-C)	
	C	0.35	0.08	0.19	0.09	0.3	0.12	0.31	0.07	0.06	0.27	1.04	0.09	0.17	0.19	0.24	0.25	0.49 (A-B)	
Basophil region (10^9 cells/L)	A	0.1	0.14	0.14	0.28	0.13	0.14	0.14	0.15	0.15	0.1	0.1	0.06	0.06	0.07	0.13	0.06	0.00 (A-B)	1
	B	0.16	0.06	0.15	0.07	0.2	0.12	0.23	0.12	0.14	0.2	0.15	0.08			0.14	0.05	0.00 (B-C)	
	C	0.2	0.03	0.02	0.03	0.03	0.04	0.06	0.02	0.03	0.04	0.04	0.03	0.03	0.04	0.05	0.05	0.51 (A-B)	

Leukocyte parameters (Leukocyte count (WBC) and counts within Neutrophil region, Lymphocyte region, Monocyte region, Eosionphil region and Basophil region) in blood samples incubated with two different carbon black nanomaterial concentrations (A – 0.5 mg/mL, B – 1 mg/mL), and control suspension incubated with citrated and phosphate-buffered saline (C). Average value (AVG), standard deviation (STD) and statistical significance of the difference given by the probability (p) are given. Statistically significant differences have also been evaluated for statistical power at α = 0.05 (P).

diminishes. To keep the membrane free energy as low as possible the GUV adjusts its shape by integration of the nanotubule(s). Concomitantly, the globular part of the GUV becomes flaccid and its thermal shape fluctuations become visible [28] while the eccentricity of GUVs increases when the protrusions are integrated in the mother vesicle. It is important at which stage of this process the test substance (CB nanomaterial) was added to GUVs. Added CB nanomaterial interacts with solutes, and with surfaces of the observation chamber which affects the osmolarity of the suspension. Water migrates through the phospholipid membrane [29] to equalize its chemical potential. If inserted in hypotonic medium, this mechanism causes GUVs to swell while for large enough differences in chemical potential they burst [30]. Experiments with GUVs were performed immediately after the electroformation when the mother vesicle was almost spherical with hindered possibilities to adjust its shape. As the membrane poorly tolerates stretching, the vesicles which were almost spherical, bursted. This mechanism provides a possible explanation for the decrease of the number of GUVs in the first 20 minutes of observation (Figure 6). Measured 50 minutes after the addition of the CB nanomaterial, the number of GUVs remained more or less unchanged with respect to the measurements at 20 minutes. Also, the remaining GUVs in the test sample exhibited larger average eccentricity than the GUVs in the control sample. In agreement with the proposed mechanism we interpret that the vesicles which have prior to addition of CB nanomaterial integrated enough membrane from the remnants of the nanotubular network were flaccid enough and therefore able to adjust their shape and resist the osmotic shock. Similar osmotic effects took place also in erythrocyte suspension, however, erythrocytes have a relatively low relative volume (in discocytes which prevailed in the samples considered this ratio is below 0.7), so the cells were able to adjust their shape in attaining the osmotic equilibrium. As cells were far from the limiting spherical shapes they did not burst, instead, they adjusted their shape by minimizing the free energy of the membrane with underlying membrane skeleton [31] according to the new conditions. Bursting of GUVs in the experiment (Figure 6) therefore likely reflects the *in vitro* conditions.

Echinocytosis in *in vitro* conditions was observed in the control and in the test sample. The possible mechanisms include interaction of solutes with glass surface of the observation chamber [32], change of the difference between the outer and the inner membrane layer areas due to change in composition or shape of the membrane constituent molecules [33-36], change in the composition of the inner and the outer solution [37] and ATP depletion [38]. CB nanomaterial may interfere with these mechanisms. As the zeta potential is negative, the nanoparticles and their agglomerates attract positively charged molecules in the solution and thereby cause changes of the bulk concentration of solutes, with consequent changes in the Donnan equilibrium including pH and therefrom induced transformations of membrane protein conformation [39]. Also, CB nanomaterial may interact with the glass surfaces of the observation chamber and indirectly affect the glass – induced mechanism responsible for echinocytosis in the control sample. However, the effect of CB nanomaterial in our case suppressed the echincytogenic effect observed in the control sample (Table 1).

We found singular ghosts in the samples treated with CB nanomaterial (Figure 3B) but the membrane of the blood cells did not appear locally deformed due to the presence of CB nanomaterial (Figures 2 and 3). CB nanomaterial also did not cause dramatic changes in platelet shape (Figure 5). The treated and the control samples contained mostly resting platelets while the signs of the platelet activation after 24 hours of incubation with CB nanomaterial (roundness of shape, the presence of the tubular protrusions) were observed also in the control sample and were therefore not ascribed to the effect of CB nanomaterial. Figure 5 for comparison presents also platelets that were incubated with agglomerated ZnO where it can be seen that after 24 hours the platelets were considerably activated, but as previously found [11] this effect can be observed already 3 hours after the addition of ZnO nanoparticles to blood cells.

Further, no differences between CB-treated samples and control samples with respect to number, mass and volume of platelets and shapes of their distributions were found. This indicates lack of evidence for platelet activation which is in agreement with our other results. However, we found statistically significant differences in parameter MPC which measures refractive index of platelets. MPC was suggested as a potential measure of the activation of platelets, however the results are not yet decisive [40]. A change in refractive index of platelets due to incubation with CB nanomaterial (Table 2) could in our experiments be a consequence of adsorption of nanoparticles to the cell surface, (Figure 2).

It was found that the number of leukocytes was statistically significantly increased in samples incubated with CB nanomaterial. Peroxidase reaction staining showed an increase of the number of stained particles that are in blood samples ascribed to eosinophils. We suggest that higher number of leukocytes in CB-treated samples could be a consequence of detecting CB aggregates by the cytometer in the frame within the scatter diagram pertaining to leukocytes and of a decreased adsorption of leukocytes to epruvette walls in samples treated with CB nanomaterial. Higher number of peroxidase stained cells could be due to adsorption of CB nanomaterial to the cell surface, thereby influencing the position of events in the cytometer scatter diagram.

We have incubated blood with CB nanomaterial for 24 hours since this was the interval in which interaction of nanomaterial with membranes took place in our previous experiments with TiO_2 and ZnO [11]. We used the concentration of nanomaterial of the same order that showed an effect in previous work. In analysis of standard blood parameters we used two different concentrations of CB nanomaterial, however, we observed no difference between samples incubated with different concentrations of nanomaterial (Table 2).

In this work we have focused on the effect of CB nanomaterial on biological membranes as revealed by the change of the erythrocyte, platelet and phospholipid vesicle shape. Activation of platelets can be assessed also by other methods, therefore, further studies should be performed taking into account different aspects of blood coagulation.

Our results do not indicate a CB nanomaterial – induced risk for thromboembolic events due to an increased activation of platelets. However, the risk for occlusion of small vessels is indicated due to the formation of rather large agglomerates which may present mechanical occlusions in the vessels. Adhesion of blood cells to these complexes additionally aggravates the situation. Moreover, by mediating the attractive interaction between membranes, CB nanomaterial accumulated in the erythrocyte dimple can stabilize the shape of the cell that is unfavorable for flow through capillaries. Larger complexes of agglomerated CB nanoparticles present a risk for the occlusion of the vessels even with no cells attached.

The observed effects of interaction of nanomaterial with biological membranes on their biophysical properties present only one viewpoint on the effect of nanomaterial on cells. It was observed that PEGylated gold nanoparticles considerably affected function of erythrocytes (decrease of deformability, oxygen delivering ability, CD47 and ATP content, aggregation of band-3), although they did not cause haemolysis [41]. Moreover, these effects can *in vivo* last for weeks due to biocompatibility of nanoparticles [41]. It was suggested that decreased deformability of erythrocytes could be a consequence of adsorption of gold nanoparticles on the erythrocyte membrane [41]. As we have also observed adsorption of CB (Figure 2) as well as of TiO_2 and ZnO nanomaterial [11] on the erythrocyte membrane, similar long lasting effects of CB nanomaterial would also be probable *in vivo*. Activation of human platelets was induced by nanodiamonds [42], *in vivo* effect on immune system was induced by uptake of carbon ultrafine particles [43], drug carrier nanoparticles dendrimeres were found to induce procoagulant activity of leukocytes [44] and genotoxic effects of TiO_2 were reported in leukocytes [45]. On the other hand, carbon nanotubes did not induce acute immune response [46]. To better understand underlying

mechanisms, further studies considering different viewpoints and different types of nanomaterial are therefore indicated.

Blood cells are a convenient system for the study of the effect of different substances on cells and biological membranes. Monocytes were found an appropriate system for testing the pyrogeneicity of compounds alternative to rabbit immune response [47-49]. Erythrocytes, platelets and GUVs can also reveal information on the relevant effects. It is however necessary to continue improving the methods for quantification of these effects and expanding the possibilities of the use of blood cells.

Analysis of the sample in the observation chamber involves taking images of the sample at different locations. When comparing the images it can be clearly seen that the sample is not completely homogeneous with respect to the number of cells and cell shapes. Statistical analysis of the representative images of the sample should therefore be performed. We have previously developed a computer aided method to assess an ensemble of representative images and analyzed the effect of different substances on GUVs [50,51]. In this work a statistical approach is used also to compare the concentration and the shape of erythrocytes. We have distinguished between three different types of erythrocyte shapes (discocyte, echinocyte and stomatocyte). As besides the errors of this method, the source of the fluctuations in the average values are also systemic errors in preparing the sample, we presented the results in terms of the portions of cells of a certain type of the shape. A choice of relevant parameters and an iterative approach involving the construction of an ensemble of images representative for the system, calculating the average values and statistical significance of the differences and checking whether the power of the ensemble is large enough will lead to the optimization of the method with respect to the time used for analysis.

The analysis of the potentially prothrombogenic effect of exogenously added substance could be complemented by considering also the extracellular vesicles isolated from blood samples [13,52]. Currently, the most widely used markers for assessment of extracellular vesicles are the antibodies. However, development of ethically superior materials (aptameres) that are produced *in vitro* [53], and the corresponding technologically advanced methods [54] should be considered.

It was reported that various types of nanoparticles and in particular CB, can directly oxidize molecules by surface-mediating reactions and leakage of transition metals and redox-active substances, and thus damage the membranes [3]. CB nanomaterial was also observed to induce apoptosis in HUVECs [55] and in bronchial epithelial cells [56] which was interpreted by leakage of the enzymes due to (mitochondrial) membrane injury [56]. Biochemical approaches also indicate the importance of the integrity, composition and properties of the membrane in vital cell processes and support our indications that understanding basic membrane properties and relevant interactions are crucial in nanotoxicology studies.

Analysis of the shape and population characteristics of erythrocytes and GUVs provides a quantitative method suitable for the study of the effect of added substances on biological membranes and should be considered in the future paradigms of research and testing.

Conclusions

Carbon black nanomaterial suspended in citrated and phosphate buffered saline interacted with erythrocyte membrane but in 24 hours the direct interactions did not cause observable changes in local membrane shape or in global erythrocyte shape. Also we observed no changes in shape of platelets due to incubation of platelet rich plasma with the nanomaterial within 24 hour observation and in population parameters that consider concentration, volume and distribution shape. The carbon black nanomaterial however formed large agglomerates in sugar solution, in citrated and phosphate buffered saline and in blood plasma. Indirect effects on the artificial membranes were interpreted by changes in osmolarity induced by addition of carbon black nanomaterial to the giant phospholipid vesicle suspensions. Blood cells and giant phospholipid vesicles are appropriate systems for the study of the effects of nanomaterial on biological membranes.

Methods
Chemicals

The phospholipid 1-Palmitoyl-2-Oleoyl-sn-Glycero-3-Phosphocholine (POPC) and cholesterol were purchased from Avanti Polar Lipids, Inc. (Alabaster, Al, USA). Sucrose and glucose were purchased from Sigma–Aldrich (Steinheim, Germany). Stock solutions (1 mg/mL) of both POPC and cholesterol were prepared by dissolving the lipid in a mixture of $CHCl_3$ and MeOH (2:1, v/v). Sucrose, glucose, paraformaldehyde, $CHCl_3$ and MeOH were purchased from Sigma–Aldrich (Steinheim, Germany), NaCl, $NaH_2PO_4 \cdot 2H_2O$, $Na_2HPO_4 \cdot 2H_2O$ and Triton X 100 surfactant (OmniPur) from Merck KGaA (Darmstadt, Germany) and KCl and KH_2PO_4 from Kemika (Zagreb, Croatia). Citrated and phosphate buffered saline (PBS): 137 mM NaCl, 2.7 mM KCl, 7.8 mM $Na_2HPO_4 \cdot 2H_2O$, 1.5 mM KH_2PO_4, 10.9 mM $Na_3C_6H_5O_7$, pH 7.4) was prepared with ultrapure distilled H_2O and filtered before use through 0.22 µm pore filters. Glutaraldehyde and osmium tetroxide (OsO_4) were purchased from SPI Supplies (West Chester, PA USA).

Carbon black nanomaterial

CB nanopowder was from PlasmaChem GmbH (Berlin, Germany). The average size of primary particles was 13 nm. Stock suspensions of CB nanomaterial were prepared in PBS at 5 mg/ml.

Characterization and imaging of carbon black nanomaterial suspensions

Characterization of CB nanomaterial (primary characteristics of CB nanomaterial) and nanoparticle suspensions (secondary characteristics of CB nanomaterial) were assessed by SEM and TEM electron microscopy, by DLS and by zeta potential measurements. For TEM, water suspension of nanoparticles was applied on a copper-grid-supported transparent carbon foil at room temperature. TEM imaging of nanomaterial was performed using a JEOL 2100 transmission electron microscope (Tokyo, Japan) operated at 200 kV. SEM imaging of nanomaterial in PBS was performed using LEO Gemini 1530 (LEO, Oberkochen, Germany) scanning electron microscope.

Dispersed CB nanomaterial dissolved in PBS was inspected by DLS using 3D DLS-SLS spectrometer (LS Instruments, Fribourg, Switzerland). Zeta potential of CB nanomaterial in PBS was measured at the pH value of the suspension using ZetaPals (Brookhaven Instruments Corporation, Holtsville, NY, USA).

Blood sampling

For SEM imaging blood was collected from two authors (female and male, average age 40 years) and a volunteer (female, 30 years) with no record of the disease into 2.7 mL tubes containing 270 µL trisodium citrate at a concentration 0.109 mol/L. Blood was collected by vein puncture with free flow into tubes (BD Vacutainers, Becton Dickinson, CA) by using a 21-gauge needle (length 70 mm, inner radius 0.4 mm) (Microlance, Becton Dickinson, NJ, USA). Sampling was performed according to the Declaration of Helsinki and a written consent to take blood was given by the donors. The study was approved by the National Ethics Committee, No 117/02/10. No adverse effects on donors' health due to sampling were observed.

For direct observation under the optical microscope, a drop of blood was taken from an author (female, 54 years) with a pipette from a small incision on a fingertip. Blood was inserted into an Eppendorf tube previously filled with 1 mL of PBS and gently mixed by turning the sample upside down.

For the study of population of blood cells, 2.3 mL of blood was collected from 14 volunteers with no record of disease (13 female and 1 male, average age 22 years) into evacuated tubes (BD Vacutainers, Becton Dickinson, CA) containing 270 µL trisodium citrate at a concentration 0.109 mol/L. Blood was collected by vein puncture by using a 21-gauge needle (length 70 mm, inner radius 0.4 mm)

(Microlance, Becton Dickinson, NJ, USA). Each subject donated 3 tubes of blood.

Preparation of blood cells for electron microscopy

Blood was processed within 1 hour from sampling. Blood was centrifuged in a Centric 400R centrifuge (Domel d.o.o., Železniki, Slovenia) at 150 g and 37°C for 10 minutes to separate erythrocytes from platelet rich plasma. Erythrocytes were repeatedly washed with PBS by centrifugation at 1550 g and 37°C for 10 minutes. Washed erythrocytes or platelet rich plasma were aliquoted into equal parts (50 μL in case of erythrocytes and 200 μL in case of platelet rich plasma). CB nanomaterial suspended in PBS (or PBS alone for control) was added to aliquots in v/v ratio of 2:1 in case of erythrocytes and 3:1 in case of platelet rich plasma, 3 and 2 units corresponding to erythrocytes and platelet rich plasma, respectively. After incubation the samples were fixed in 0.1% glutaraldehyde, incubated for another hour at room temperature and centrifuged at 1550 g and 37°C for 10 minutes. Supernatant was exchanged for PBS samples were vortexed, centrifuged at 1550 g and 37°C for 10 minutes and fixed in 2% glutaraldehyde for an hour.

Phase contrast microscopy of erythrocytes

5 μL of erythrocyte suspension was placed into observation chamber composed of two cover glasses which were glued together with nail polish. Samples were observed under Olympus GWB BH-2 (Olympus Corporation, Tokyo, Japan) microscope with phase contrast optics at objective magnification 400×. Images were taken with Canon EOS 450D digital camera (Canon Inc., Tokyo, Japan).

Scanning electron microscopy of blood cells

Fixed samples were washed by exchanging supernatant with PBS and incubated for 20 minutes at room temperature. This procedure was repeated 4 times while the last incubation was performed over night at 8°C. Samples were then post-fixed for 60 minutes at 22°C in 1% OsO_4 dissolved in 0.9% NaCl, dehydrated in a graded series of acetone/water (50%–100%, v/v), critical-point dried, gold-sputtered, and examined using a LEO Gemini 1530 (LEO, Oberkochen, Germany) scanning electron microscope.

Carbon black nanomaterial effect on erythrocyte shape in populations of erythrocytes

PBS-diluted blood samples incubated with the test or control solution were observed under the Leitz Aristoplan (Leitz, Wetzlar, Germany) optical microscope equipped by the Watec (Model: 902DM3S), Watec Inc., New York, USA, camera and Pinnacle Studio HD, Version 15.0.0.7593 frame graber and software, Avid Technology Inc., USA. Final concentration of CB nanomaterial in samples was 0.5 mg/mL. The samples were observed after 1, 3 and 24 hours, in order to compare the effect with previously considered effect of TiO_2 and ZnO nanoparticles [11]. The sample was further diluted to obtain a required density of blood cells for observation. The required density rendered a monolayer of relatively closely packed cells in an observation frame under the microscope. An observation chamber 1.5×1 cm^2 was created on the glass by using silicon grease. 40 μL of the test suspension was placed in the observation chamber and closed by the cover glass. Care was taken not to leave voids in the grease boundary in order to prevent evaporation of liquid from the observation chamber. For quantitative analysis of the effect of CB nanomaterial on blood cell membranes we randomly imaged at least 50 frames over the sample. We avoided regions close to the silicon grease. Two independent experiments were performed.

Image analysis of blood cell populations

For each sample we created an ensemble of pictures from which we omitted pictures of invalid quality. We counted the number of discocytes, echinocytes and stomatocytes in each picture. It was assumed that each picture was representative for the sample and that the number of cells of each type fluctuated around the corresponding average value. The average values of the percent of discocytes, echinocytes and stomatocytes were calculated for each sample. Altogether 10712 cells were included in the final analysis.

Electroformation of giant unilamellar phospholipid vesicles

The electroformation of GUVs [49] was performed at room temperature. In brief, 40 μL of the lipid mixture of POPC (80%, v/v) and cholesterol (20%, v/v), both dissolved in 2:1 chloroform/methanol mixture, was spread over two platinum electrodes and the solvent was allowed to evaporate in low vacuum for two hours. The electrodes were then placed into the electroformation chamber (2 mL Eppendorf cup), filled with 2 mL of 0.3 M sucrose solution. An alternating electric field was applied as described in [57]. After the electroformation, 600 μL of 0.3 M sucrose solution containing GUVs was added to 1 ml of 0.3 M glucose solution.

Experiments with giant unilamellar phospholipid vesicles and carbon black nanomaterial

Immediately after the electroformation, the suspension of GUVs in sucrose/glucose solution was mixed by turning the Eppendorf cup upside down (five times). The suspension was diluted (300 μL of 0.3 M glucose solution was added to 100 μl of original vesicle solution) in order to obtain the desired concentration of GUVs, which was found convenient according to our preliminary experiments. Subsequently, the diluted suspension was aliquoted into 2 vials (80 μL in each vial). Two parallels of test suspension of CB nanomaterial (0.05 mg/mL in 0.3 M glucose solution) and two parallels of control suspension (0.3 M glucose solution) were added to GUV suspensions in volume ratio 9:1 (GUVs: test suspension) to reach the final concentration of nanomaterial 10 μg/mL. Vials were turned upside down five times to mix the test suspension with GUV suspension. Thus obtained samples (70 μL each) were separately transferred into four CoverWellTM Perfusion chambers PC4L-0.5 (Grace Bio-Labs Sigma-Aldrich, Steinheim, Germany).

The observation chambers with samples were mounted to the inverted phase contrast light microscope (Nikon Eclipse TE2000-S, Tokyo, Japan). After pre-defined durations of incubation (20 and 50 minutes), GUV populations in each chamber were recorded by a Sony CCD video camera module, model: XC-77 CE (Minato, Japan). These recordings consisted of two video sequences each approximately 2 minutes long. During the recording, the object glass was moved to capture images of thousands of GUVs. With the use of image processing algorithms, the video sequences were transformed into large mosaics [58] where each mosaic contained all vesicles acquired within the sample. GUVs in mosaics were segmented by using a computer-aided approach [58].

A sample of GUVs was incubated with sugar solution for control and another sample was incubated with CB nanomaterial dissolved in sugar solution (0.05 mg/mL). The vesicles were imaged after 20 minutes and after 50 minutes.

The mosaics were separated into images (micrographs) with the size of a single field of view with the microscope at 400× magnification. Using the GUVs' locations inside the mosaics, all vesicles were associated with the appropriate micrograph allowing us to extract information on vesicle density throughout the micrographs composing the mosaic. Numbers of GUVs in the micrograph were averaged over all micrographs within the test and control samples.

Vesicle shapes were characterized by their deviations from sphere. The nonspherical contours were approximated by fitting with an ellipse with smaller semiaxis b and larger semiaxis a. The eccentricity $\varepsilon = (1 - b^2/a^2)^{1/2}$ was assigned to each vesicle in the mosaic and the average value for the sample was calculated.

Analysis of blood cell populations with respect to concentration, volume and distribution

Each subject donated 3 samples of blood to create 3 populations of samples. Population A was incubated with 200 μl PBS-dissolved CB nanomaterial with final concentration 0.5 mg/mL, population B was incubated with equal volume of PBS-dissolved CB nanomaterial with final concentration 1 mg/mL, population C was incubated with equal volume of PBS. Samples were incubated for one hour during which they were continuously slowly turned upside down at room temperature. After one hour the samples were left in upright

position for another hour to allow sedimentation of nanomaterial to the bottom of the tubes. Upper milliliter of the samples was pipetted into plastic tubes. After additional 10 minutes, the samples were examined for presence of CB agglomerates. 40 μl of sample was taken with the pipette from the bottom of the tube and placed into an observation chamber created between two cover glasses by silicon grease. No CB agglomerates were observed under the optical microscope Leitz Aristoplan (Leitz, Wetzlar, Germany) equipped by the Watec (Model: 902DM3S), Watec Inc., New York, USA, camera and Pinnacle Studio HD, Version 15.0.0.7593 frame graber and software, Avid Technology Inc., USA. Blood cell populations were assesed by the ADVIA flow cytometer for erythrocyte parameters: erythrocyte concentration RBC, hematocrit HCT, red cell distribution width RDW and hemoglobin distribution width HDW, platelet parameters: concentration of platelets PLT, plateletcrit PCT, mean platelet volume MPV, platelet distribution width PDW, mean platelet complement MPC and mean platelet mass MPM, and leukocyte concentration (WBC).

Statistical analysis

Methods of descriptive statistics were used to compare the test samples and the respective control samples. We used two-sided pooled t-test with equal variance. Probability of the t-test below 0.05 was considered statistically significant. The software used for statistical analysis was from R Development Core Team: R: A language and environment for statistical computing [59]. The statistical significance of the t- test was given by the probability of the type I or type II error (i.e. by α or β). Power analysis was performed in order to validate the size of samples. Power larger than 0.8 at $\alpha = 0.05$ indicated the sample of a proper size. For calculation, we used Microsoft Excel software (Microsoft® Office Excel® 2007 SP3) and Power & Sample Size Calculator.

Abbreviations
CB: Carbon black; GUVs: Giant unilamellar phospholipid vesicles; PBS: Citrated and phosphate-buffered saline; TEM: Transmission electron microscope; SEM: Scanning electron microscope; DLS: Dynamic light scattering; RBC: Concentration of erythrocytes; HCT: Hematocrit; RDW: Red cell distribution width; HDW: Hemoglobin distribution width; PLT: Concentration of platelets; PCT: Plateletcrit; MPV: Mean platelet volume; PDW: Platelet distribution width; MPC: Mean platelet complement; MPM: Mean platelet mass; WBC: Concentration of leukocytes; POPC: 1-Palmitoyl-2-Oleoyl-sn-Glycero-3-Phosphocholine.

Competing interests
The authors declare that they have no competing interests.

Authors' contributions
MP and MŠ prepared blood samples, performed experiments with blood and contributed in writing the manuscript. VŠ and HH prepared samples with erythrocytes and platelets, performed imaging with scanning electron microscope at Abo Akademi University and critically reviewed the manuscript. JLK performed experiments with populations of erythrocytes and statistical analysis of these experiments and contributed in writing the manuscript. RŠ prepared samples for electron microscopy, performed flow cytometry analysis and contributed in writing the manuscript. DM performed

characterization of carbon black nanomaterial and critically reviewed the manuscript, BD and VK performed experiments with GUVs and statistical analysis of these experiments and critically reviewed the manuscript. DD participated in design and coordination of the study and contributed in writing the manuscript. VKI participated in design and coordination of the study, elaborated the method for analysis of population of erythrocytes and contributed in writing the manuscript. All authors have read and approved the final version of the manuscript.

Acknowledgements

Authors are indebted to Jernej Zupanc for guidance in the computer-aided analysis of giant phospholipid vesicle populations and to Sabina Boljte for technical support with preparation of giant phospholipid vesicles and with image analysis. The work was supported by ARRS grants P3-0388, J3-5499 and J1-6728.

Author details

[1]Laboratory of Clinical Biophysics, University of Ljubljana, Faculty of Health Sciences, Zdravstvena pot 5, Ljubljana SI-1000, Slovenia. [2]Group of Nanobiology and Nanotoxicology, University of Ljubljana, Biotechnical Faculty, Večna pot 111, Ljubljana SI-1000, Slovenia. [3]Lymphocyte Cytoskeleton Group, Institute of Biomedicine/Pathology, BioCity, University of Turku, Tykistökatu 6B, Turku SF-20520, Finland. [4]J. Stefan Institute, Jamova 39, Ljubljana SI-1000, Slovenia. [5]Department of Biosciences, BioCity, Åbo Akademi University, BioCity, Artillerigatan 6, Åbo/Turku SF-20520, Finland.

References

1. National Research Council of The National Academies. Toxicity testing in the 21st century: a vision and a strategy. Washington, DC: The National Academies Press; 2007.
2. Hartung T. Food for thought… on alternative methods for nanoparticle safety testing. Altex. 2010;27:87–95.
3. Møller P, Jacobsen NR, Folkmann JK, Danielsen PH, Mikkelsen L, Hemmingsen JG, et al. Role of oxidative damage in toxicity of particulates. Free Radic Res. 2009;44:1–46.
4. Knight A, Bailey J, Balcombe J. Animal carcinogenicity studies: 1. Poor human predictivity. Altern Lab Anim ATLA. 2006;34:19–27.
5. Knight A, Bailey J, Balcombe J. Animal carcinogenicity studies: 2. Obstacles to extrapolation of data to humans. Altern Lab Anim ATLA. 2006;34:29–38.
6. Knight A, Bailey J, Balcombe J. Animal carcinogenicity studies: 3. Alternatives to the bioassay. Altern Lab Anim ATLA. 2006;34:39–48.
7. Clift MJD, Gehr P, Rothen-Rutishauser B. Nanotoxicology: a perspective and discussion of whether or not in vitro testing is a valid alternative. Arch Toxicol. 2011;85:723–31.
8. Canham PB. The minimum energy of bending as a possible explanation of the biconcave shape of the human red blood cell. J Theor Biol. 1970;26:61–81.
9. Seifert U. Configurations of fluid membranes and vesicles. Adv Phys. 1997;46:13–137.
10. Gorbet MB, Sefton MV. Biomaterial-associated thrombosis: roles of coagulation factors, complement, platelets and leukocytes. Biomaterials. 2004;25:5681–703.
11. Šimundić M, Drašler B, Šuštar V, Zupanc J, Štukelj R, Makovec D, et al. Effect of engineered TiO2 and ZnO nanoparticles on erythrocytes, platelet-rich plasma and giant unilamelar phospholipid vesicles. BMC Vet Res. 2013;9:7.
12. Ghio A, Sobus J, Pleil J, Madden M. Controlled human exposures to diesel exhaust. Swiss Med Wkly. 2012;142:w13597.
13. Schara K, Janša V, Šuštar V, Dolinar D, Pavlič JI, Lokar M, et al. Mechanisms for the formation of membranous nanostructures in cell-to-cell communication. Cell Mol Biol Lett. 2009;14:636–56.
14. Sahu D, Kannan GM, Vijayaraghavan R. Carbon black particle exhibits size dependent toxicity in human monocytes. Int J Inflamm. 2014;2014:e827019.
15. Ema M, Naya M, Horimoto M, Kato H. Developmental toxicity of diesel exhaust: a review of studies in experimental animals. Reprod Toxicol. 2013;42:1–17.
16. Reijnders L. Human health hazards of persistent inorganic and carbon nanoparticles. J Mater Sci. 2012;47:5061–73.

17. Suwalsky M, Novoa V, Villena F, Sotomayor CP, Aguilar LF, Ronowska A, et al. Structural effects of Zn2+ on cell membranes and molecular models. J Inorg Biochem. 2009;103:797–804.
18. Suwalsky M, Hernández P. Aluminum enhances the toxic effects of amyloid β-peptide on cell membranes and a molecular model. Monatshefte Für Chem - Chem Mon. 2011;142:431–7.
19. Rojas-Aguirre Y, Hernández-Luis F, Mendoza-Martínez C, Sotomayor CP, Aguilar LF, Villena F, et al. Effects of an antimalarial quinazoline derivative on human erythrocytes and on cell membrane molecular models. Biochim Biophys Acta BBA - Biomembr. 2012;1818:738–46.
20. Suwalsky M, Belmar J, Villena F, Gallardo MJ, Jemiola-Rzeminska M, Strzalka K. Acetylsalicylic acid (aspirin) and salicylic acid interaction with the human erythrocyte membrane bilayer induce in vitro changes in the morphology of erythrocytes. Arch Biochem Biophys. 2013;539:9–19.
21. Manrique-Moreno M, Londoño-Londoño J, Jemioła-Rzemińska M, Strzałka K, Villena F, Avello M, et al. Structural effects of the Solanum steroids solasodine, diosgenin and solanine on human erythrocytes and molecular models of eukaryotic membranes. Biochim Biophys Acta BBA - Biomembr. 2014;1838:266–77.
22. Suwalsky M, Villena F, Norris B, Soto MA, Sotomayor CP, Messori L, et al. Structural effects of titanium citrate on the human erythrocyte membrane. J Inorg Biochem. 2005;99:764–70.
23. Rothen-Rutishauser BM, Schürch S, Haenni B, Kapp N, Gehr P. Interaction of fine particles and nanoparticles with red blood cells visualized with advanced microscopic techniques. Environ Sci Technol. 2006;40:4353–9.
24. Zupanc J, Drobne D, Drašler B, Valant J, Iglič A, Kralj-Iglič V, et al. Experimental evidence for the interaction of C-60 fullerene with lipid vesicle membranes. Carbon. 2012;50:1170–8.
25. Mocan T. Hemolysis as expression of nanoparticles-induced cytotoxicity in red blood cells. Biotechnol Mol Biol Nanomedicine BMBN. 2013;1:7–12.
26. Drašler B, Drobne D, Novak S, Valant J, Boljte S, Otrin L, et al. Effects of magnetic cobalt ferrite nanoparticles on biological and artificial lipid membranes. Int J Nanomedicine. 2014;9:1559–81.
27. Mathivet L, Cribier S, Devaux PF. Shape change and physical properties of giant phospholipid vesicles prepared in the presence of an AC electric field. Biophys J. 1996;70:1112–21.
28. Kralj-Iglič V, Gomišček G, Majhenc J, Arrigler V, Svetina S. Myelin-like protrusions of giant phospholipid vesicles prepared by electroformation. Colloids Surf Physicochem Eng Asp. 2001;181:315–8.
29. Vitkova V, Genova J, Bivas I. Permeability and the hidden area of lipid bilayers. Eur Biophys J. 2004;33:706–14.
30. Peterlin P, Arrigler V, Diamant H, Haleva E. Chapter ten - permeability of phospholipid membrane for small polar molecules determined from osmotic swelling of giant phospholipid vesicles. In: Iglič A, editor. Advances in planar lipid bilayers and liposomes. London: Academic; 2012. p. 301–35.
31. Iglič A. A possible mechanism determining the stability of spiculated red blood cells. J Biomech. 1997;30:35–40.
32. Wong P. A hypothesis of the disc–sphere transformation of the erythrocytes between glass surfaces and of related observations. J Theor Biol. 2005;233:127–35.
33. Sheetz MP, Singer SJ. Biological membranes as bilayer couples. A molecular mechanism of drug-erythrocyte interactions. Proc Natl Acad Sci. 1974;71:4457–61.
34. Gimsa J, Ried C. Do band 3 protein conformational changes mediate shape changes of human erythrocytes? Mol Membr Biol. 1995;12:247–54.
35. Iglič A, Kralj-Iglič V, Hägerstrand H. Amphiphile induced echinocyte-spheroechinocyte transformation of red blood cell shape. Eur Biophys J. 1998;27:335–9.
36. Iglič DA, Kralj-Iglič V, Hägerstrand H. Stability of spiculated red blood cells induced by intercalation of amphiphiles in cell membrane. Med Biol Eng Comput. 1998;36:251–5.
37. Wong P. A basis of echinocytosis and stomatocytosis in the disc–sphere transformations of the erythrocyte. J Theor Biol. 1999;196:343–61.
38. Gov NS, Safran SA. Red blood cell membrane fluctuations and shape controlled by ATP-induced cytoskeletal defects. Biophys J. 2005;88:1859–74.
39. Rudenko SV. Erythrocyte morphological states, phases, transitions and trajectories. Biochim Biophys Acta BBA - Biomembr. 2010;1798:1767–78.
40. Beard MJ, Jeewa Z, Bashir S, Cardigan R, Thomas S. Comparison of platelet activation in platelet concentrates measured by flow cytometry or ADVIA 2120. Vox Sang. 2011;101:122–30.
41. He Z, Liu J, Du L. The unexpected effect of PEGylated gold nanoparticles on the primary function of erythrocytes. Nanoscale. 2014;6:9017–24.

42. Kumari S, Singh MK, Singh SK, Grácio JJ, Dash D. Nanodiamonds activate blood platelets and induce thromboembolism. Nanomed. 2014;9:427–40.

43. Frampton MW, Stewart JC, Oberdörster G, Morrow PE, Chalupa D, Pietropaoli AP, et al. Inhalation of ultrafine particles alters blood leukocyte expression of adhesion molecules in humans. Environ Health Perspect. 2006;114:51–8.

44. Dobrovolskaia MA, Patri AK, Potter TM, Rodriguez JC, Hall JB, McNeil SE. Dendrimer-induced leukocyte procoagulant activity depends on particle size and surface charge. Nanomed. 2012;7:245–56.

45. Frenzilli G, Bernardeschi M, Guidi P, Scarcelli V, Lucchesi P, Marsili L, et al. Effects of in vitro exposure to titanium dioxide on DNA integrity of bottlenose dolphin (Tursiops truncatus) fibroblasts and leukocytes. Mar Environ Res. 2014;100:68–73.

46. Medepalli K, Alphenaar B, Raj A, Sethu P. Evaluation of the direct and indirect response of blood leukocytes to carbon nanotubes (CNTs). Nanomedicine Nanotechnol Biol Med. 2011;7:983–91.

47. Fischer M, Hartzsch K, Hartung T, Montag-Lessing T. First results in the prevaluation of the human whole blood assay for pyrogens in biological pharmaceuticals. ALTEX. 1998;15:10–3.

48. Schindler S, von Aulock S, Daneshian M, Hartung T. Development, validation and applications of the monocyte activation test for pyrogens based on human whole blood. ALTEX. 2009;26:265–77.

49. Hasiwa N, Daneshian M, Bruegger P, Fennrich S, Hochadel A, Hoffmann S, et al. Evidence for the detection of non-endotoxin pyrogens by the whole blood monocyte activation test. ALTEX. 2013;30:169–208.

50. Zupanc J, Valant J, Drobne D, Kralj-Iglič V, Iglič A. A new approach to analyse effects of nanoparticles on lipid vesicles. Int J Biomed Nanosci Nanotechnol. 2010;1:34.

51. Zupanc J, Dobnikar A, Drobne D, Valant J, Erdogmus D, Bas E. Biological reactivity of nanoparticles: mosaics from optical microscopy videos of giant lipid vesicles. J Biomed Opt. 2011;16:026003.

52. De Paoli SH, Diduch LL, Tegegn TZ, Orecna M, Strader MB, Karnaukhova E, et al. The effect of protein corona composition on the interaction of carbon nanotubes with human blood platelets. Biomaterials. 2014;35:6182–94.

53. Marolt U, Cencic A, Gorenjak M, Potrc S. Generating aptamers for cancer diagnosis and therapy. Clin Exp Pharmacol. 2012;2:111.

54. Miodek A, Castillo G, Hianik T, Korri-Youssoufi H. electrochemical aptasensor of human cellular prion based on multiwalled carbon nanotubes modified with dendrimers: a platform for connecting redox markers and aptamers. Anal Chem. 2013;85:7704–12.

55. Vesterdal LK, Mikkelsen L, Folkmann JK, Sheykhzade M, Cao Y, Roursgaard M, et al. Carbon black nanoparticles and vascular dysfunction in cultured endothelial cells and artery segments. Toxicol Lett. 2012;214:19–26.

56. Hussain S, Thomassen LC, Ferecatu I, Borot M-C, Andreau K, Martens JA, et al. Carbon black and titanium dioxide nanoparticles elicit distinct apoptotic pathways in bronchial epithelial cells. Part Fibre Toxicol. 2010;7:10.

57. Angelova MI, Soléau S, Méléard P, Faucon F, Bothorel P. Preparation of giant vesicles by external AC electric fields. Kinetics and applications. In: Helm C, Lösche M, Möhwald H, editors. Trends in colloid and interface science VI. Darmstadt: Steinkopff; 1992. p. 127–31.

58. Zupanc J, Drobne D, Ster B. Markov random field model for segmenting large populations of lipid vesicles from micrographs. J Liposome Res. 2011;21:315–23.

59. The R project for statistical computing. http://www.r-project.org/ (accessed 31 Jan2015).

14

Novel pegylated silver coated carbon nanotubes kill *Salmonella* but they are non-toxic to eukaryotic cells

Atul A Chaudhari[1], Shanese L Jasper[1], Ejovwoke Dosunmu[1], Michael E Miller[2], Robert D Arnold[3], Shree R Singh[1] and Shreekumar Pillai[1*]

Abstract

Background: Resistance of food borne pathogens such as *Salmonella* to existing antibiotics is of grave concern. Silver coated single walled carbon nanotubes (SWCNTs-Ag) have broad-spectrum antibacterial activity and may be a good treatment alternative. However, toxicity to human cells due to their physico-chemical properties is a serious public health concern. Although pegylation is commonly used to reduce metal nanoparticle toxicity, SWCNTs-Ag have not been pegylated as yet, and the effect of pegylation of SWCNTs-Ag on their anti-bacterial activity and cell cytotoxicity remains to be studied. Further, there are no molecular studies on the anti-bacterial mechanism of SWCNTs-Ag or their functionalized nanocomposites.

Materials and methods: In this study we created novel pegylated SWCNTS-Ag (pSWCNTs-Ag), and employed 3 eukaryotic cell lines to evaluate their cytotoxicity as compared to plain SWCNTS-Ag. Simultaneously, we evaluated their antibacterial activity on *Salmonella enterica* serovar Typhimurium (*Salmonella* Typhimurium) by the MIC and growth curve assays. In order to understand the possible mechanisms of action of both SWCNTs-Ag and pSWCNTs-Ag, we used electron microscopy (EM) and molecular studies (qRT-PCR).

Results: pSWCNTs-Ag inhibited *Salmonella* Typhimurium at 62.5 μg/mL, while remaining non-toxic to human cells. By comparison, plain SWCNTs-Ag were toxic to human cells at 62.5 μg/mL. EM analysis revealed that bacteria internalized either of these nanocomposites after the outer cell membranes were damaged, resulting in cell lysis or expulsion of cytoplasmic contents, leaving empty ghosts. The expression of genes regulating the membrane associated metabolic transporter system (*artP*, *dppA*, and *livJ*), amino acid biosynthesis (*trp* and *argC*) and outer membrane integrity (*ompF*) protiens, was significantly down regulated in *Salmonella* treated with both pSWCNTs-Ag and SWCNTs-Ag. Although EM analysis of bacteria treated with either SWCNTs-Ag or pSWCNTs-Ag revealed relatively similar morphological changes, the expression of genes regulating the normal physiological processes of bacteria (*ybeF*), quorum sensing (*sdiA*), outer membrane structure (*safC*), invasion (*ychP*) and virulence (*safC*, *ychP*, *sseA* and *sseG*) were exclusively down regulated several fold in pSWCNTs-Ag treated bacteria.

Conclusions: Altogether, the present data shows that our novel pSWCNTs-Ag are non-toxic to human cells at their bactericidal concentration, as compared to plain SWCNTS-Ag. Therefore, pSWCNTs-Ag may be safe alternative antimicrobials to treat foodborne pathogens.

Keyword: Pegylation, Silver coated carbon nanotubes, Toxicity, Gene regulation, Food borne

* Correspondence: spillai@alasu.edu
[1]Center for Nanobiotechnology Research, Alabama State University, Montgomery, AL, USA
Full list of author information is available at the end of the article

Background

Antibiotic resistance of foodborne pathogens is a matter of serious concern as it has a detrimental effect not only on human health, but also results in food deterioration thus requiring immediate attention [1]. The use of nanoparticles appears to be one of the most promising strategies for overcoming microbial resistance as development of resistance to these nanoparticles is not likely [2]. Antimicrobial nanoparticles (NP) have been reported to be effective antibacterial agents due to their high surface area to volume ratios, unique physicochemical properties and multiple mechanisms of antimicrobial action that prevent development of microbial resistance to these nanoparticles [3,4]. Of relevance, silver nanoparticles (AgNPs) have garnered attention due to their ability to kill both gram-positive and gram-negative bacteria [5-8]. However, due to their high surface energy, AgNPs tend to aggregate into large particles which may affect their bactericidal property due to instability in the growth medium [9]. Alternatively, the use of carbon nanotubes (CNTs) as a host nanomaterial [10] for AgNPs creates a stable nanocomposite [8,11-13]. CNTs have been shown to be a suitable vehicle for the efficacious and target specific delivery of molecules [10,14]. Silver coated CNTs (AgCNTs) have demonstrated stronger antibacterial activity against gram-positive as well as gram-negative bacteria compared to commercially available AgNPs [8,13].

However, AgNPs or CNTs are toxic to human cells, with several known mechanisms of toxicity to eukaryotic cells [2,15-21]. Toxicity of AgNPs or CNTs can be reduced using various functionalization strategies, which also improves their solubility and dispersability [20-26]. More recently, neutral amphiphiles of single walled carbon nanotubes have been shown to be less toxic and stable in various media [10,14]. Specifically, pegylation of CNTs using polyethylene glycol (PEG) has been shown to increase biocompatibility and reduce the toxicity of CNTs administered intravenously in mice as compared to non-pegylated CNTs [24]. As yet, most research has focused on how the toxicity of either plain AgNPs or CNTs could be reduced by pegylation. However, AgCNTs have not been pegylated as yet, and the effect of pegylation on their toxicity to human cell has not been investigated. Further, it also remains to be determined whether pegylation of AgCNTs affect their antibacterial activity, as PEG molecules may cover the silver coating on CNTs, thus reducing their antibacterial activity. Another gap in this area is that very little is known about the antibacterial mechanism of AgCNTs and pegylated AgCNTs. Some of the proposed mechanisms include silver ion dissolution, direct contact with cell membranes and generation of reactive oxygen species [9]. A recent study demonstrated that the antibacterial activity of AgCNTs is mediated through generation of reactive oxygen species via its direct contact with the bacterial cells [9]. However, the molecular mechanisms of action remain to be explored.

Accordingly, in the present study we created novel pegylated silver coated single walled carbon nanotubes (pSWCNTs-Ag) using PEG and employed 3 eukaryotic cell lines to evaluate their cytotoxicity as compared to plain SWCNTS-Ag. Simultaneously, we evaluated their antibacterial activity on *Salmonella* Typhimurium, a gram-negative foodborne pathogen of serious public health concern, using the MIC and growth curve assays. Further, to understand the possible mechanisms of action of both SWCNTs-Ag and pSWCNTs-Ag, we performed electron microscopy (EM) and molecular studies using quantitative reverse transcriptase polymerase chain reaction (qRT-PCR).

Results

Characterization of pSWCNTs-Ag and SWCNTs-Ag

As indicated by the manufacturer, dispersion of SWCNTs-Ag in NanoSperse AQ® resulted in a relatively homogenous, yet insoluble suspension, whereas pegylation of SWCNTs-Ag produced a homogenous and highly stable water soluble suspension as reported elsewhere (Figure 1) [27]. The zeta potential value of pSWCNTs-Ag (Table 1) was positive (+8.99) compared to the negatively charged SWCNTs-Ag (−41.9), indicating that phospholipid-poly (ethylene glycol)-amine (PL-PEG-amine) molecules were strongly anchored over SWCNTs-Ag, which imparted the positive charge and made them water-soluble. Fourier transform infrared spectroscopy (FT-IR) analysis showed the presence of characteristic peaks of PL-PEG (alkane C-H, carbonyl

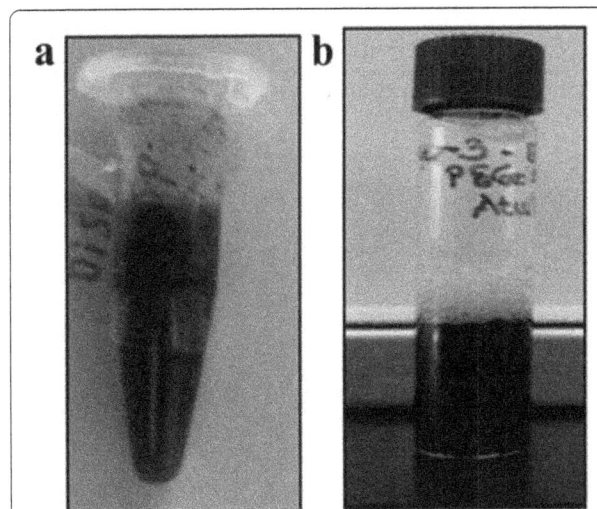

Figure 1 Suspensions of silver coated single walled carbon nanotubes (SWCNTs-Ag). (a) Homogenously dispersed SWCNTs-Ag in NanoSperse AQ®. **(b)** Water soluble pSWCNTs-Ag.

Table 1 Zeta potential measurements of the nanocomposites

Nanocomposite	Zeta potential (mV)	STDEV
SWCNTs-Ag	−41.9	1.30
pSWCNTs-Ag	8.99	0.52

c = o etc.) on pSWCNTs-Ag, whereas SWCNTs-Ag did not possess any similar peaks (Figure 2). Further, SEM imaging of pSWCNTs-Ag clearly indicated a cloudy hazy coating of PEG around the SWCNTs-Ag (Figure 3b, silver can be seen as spherical deposits, indicated by arrowheads) as compared to SWCNTs-Ag with no coating around them (Figure 3a). TEM images clearly verified the pegylation as evidenced by increase in size of pSWCNTs-Ag (54 nm) (Figure 3d & f) compared to SWCNTs-Ag (6 nm) (Figure 3c & e). The amount of PL-PEG that was deposited on 10 mg/mL of SWCNTs-Ag was observed to be 2 μM equivalent to 10 mg/mL as measured by the inorganic phosphate assay.

Antibacterial activity of SWCNTs-Ag

Next, we thoroughly examined the antibacterial activity of SWCNTs-Ag and pSWCNTs-Ag. The MIC values for both, SWCNTs-Ag and pWCNTs-Ag, were between 31.25 μg/mL and 62 μg/mL for *Salmonella* Typhimurium, *Escherichia coli*, *Staphylococcus aureus* and *Streptococcus pyogenes* (Figure 4; Additional file 1: Figures S1-S3).

Additionally, the Kirby Bauer disc diffusion assay demonstrated strong antibacterial activity against all four pathogens characterized by zones of inhibition on the MH agar plates (Figure 5; Additional file 1: Table S1). Further, the growth curve assay of *Salmonella* Typhimurium exposed to various concentrations of SWCNTs-Ag showed that bacterial growth was dramatically inhibited in a time and concentration dependent manner (Figure 6). Similarly, pSWCNTs-Ag exhibited strong antibacterial activity at 125, 62.5 and 31.25 μg/mL as evidenced by reduced bacterial numbers and impeded growth as the time progressed (Figure 6). Based on the bacterial growth curve analysis, we further analyzed live/dead staining of bacteria upon exposure to 12.5 μg/mL of SWCNTs-Ag and pSWCNTs-Ag for 16 h. The live/dead proportion was approximately 7–8 fold lower when bacteria were exposed to SWCNTs-Ag and pSWCNTs-Ag compared to non-treated controls (Figure 7).

Pegylated SWCNTs-Ag are non-toxic to human cells at their bactericidal concentrations

The cytotoxicity of SWCNTs-Ag and pSWCNTs-Ag was compared by the MTT assay using 3 cell lines including the A549 (human lung carcinoma), Hep2 (hepatocellular carcinoma) and J774 (murine macrophages) cells. As represented in Figure 8, SWCNTs-Ag showed dose dependent toxicity in all the cell lines, whereas pSWCNTs-Ag were observed not to be toxic at similar concentrations. Further,

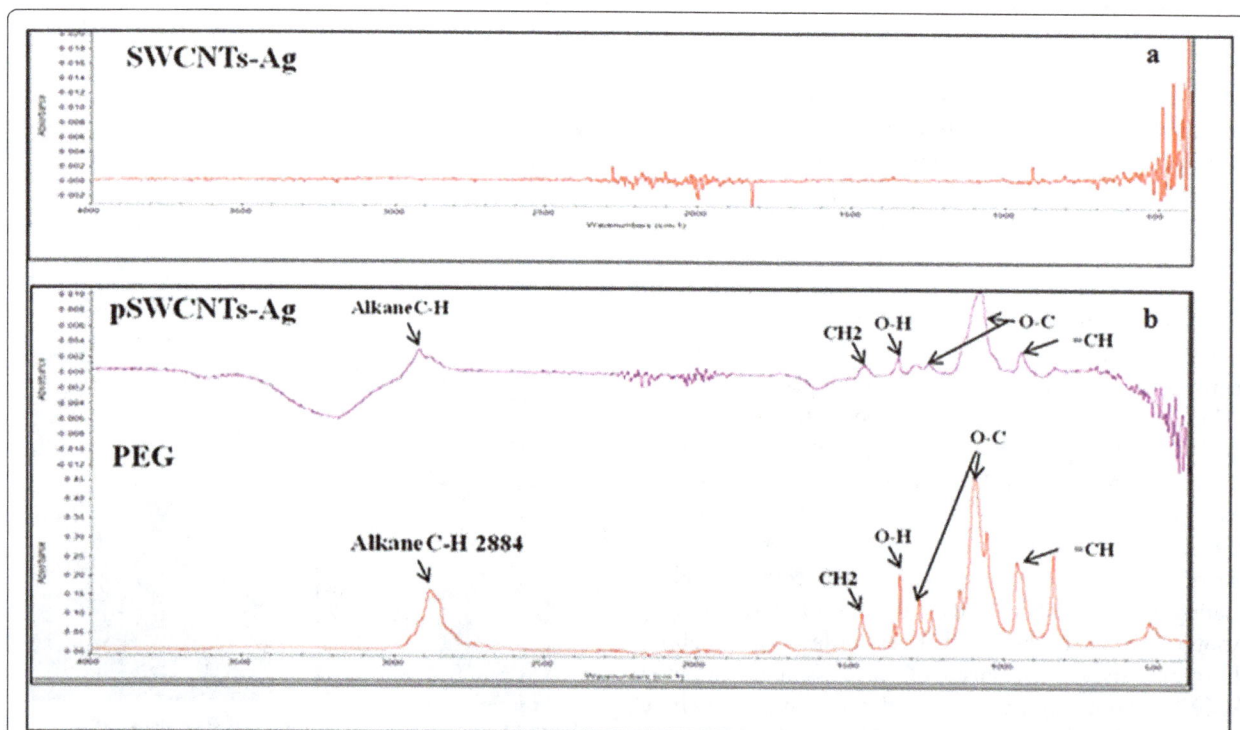

Figure 2 FT-IR pattern of (a) SWCNTs-Ag. (b) pSWCNTs-Ag and PL-PEG. The characteristic peaks on PL-PEG and pSWCNTs-Ag such as alkane C-H, carbonyl c = o, hydroxyl O-H and methylene CH$_2$ are indicated by arrows.

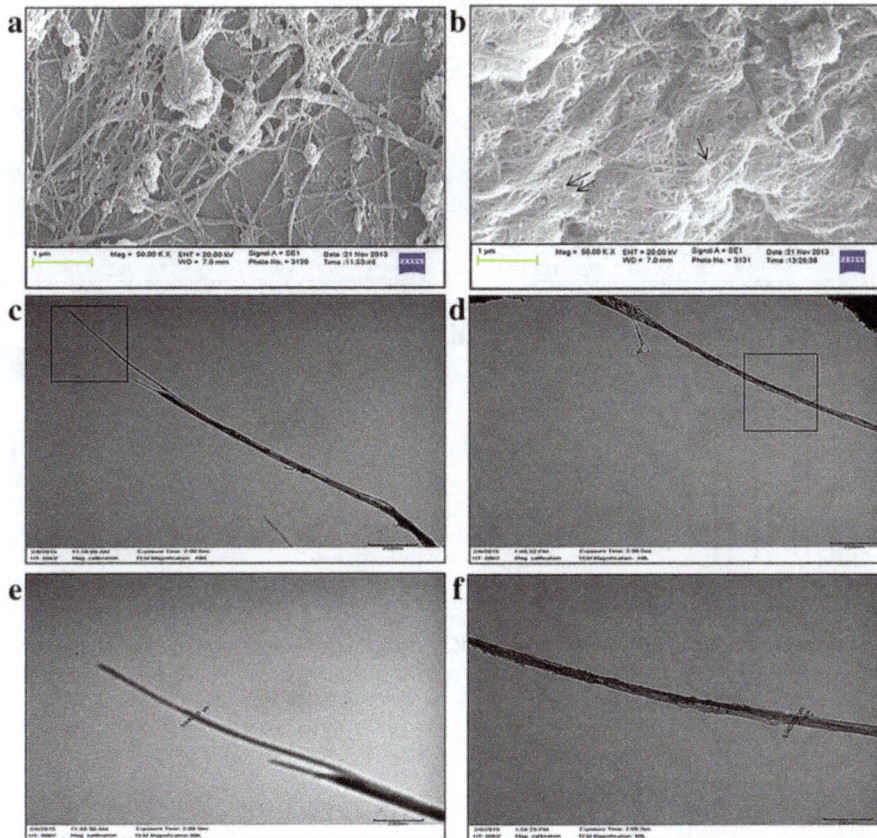

Figure 3 Characterization of silver coated single walled carbon nanotube formulations by SEM (a & b) and TEM (c & d). (a) SEM image of silver coated single walled carbon nanotubes dispersed in NanoSperse AQ® dispersant (SWCNTs-Ag). **(b)** SEM image of pegylated SWCNTs (pSWCNTs-Ag). The hazy coating around the SWCNTs-Ag is apparent and silver deposition is indicated by an arrow. **(c)** TEM image of SWCNTs-Ag. **(d)** TEM of pSWCNTs-Ag. **(e)** Magnified inset of image C showing the diameter of SWCNT-Ag as 6 nm. **(f)** Magnified inset of image D showing pSWCNTs-Ag of 54 nm.

A549 cells treated with plain SWCNTs-Ag (32.25 & 15.6 μg/mL) showed damaged morphology such as loss of nucleus, cell organelles and presence of apoptotic vesicles upon TEM analysis (Figure 9b & c). On the other hand, cells exposed to pSWCNTs-Ag at similar concentrations as that of SWCNTs-Ag showed normal morphology including intact nuclei and cell organelles (Figure 9d & e) similar to non-treated control cells (Figure 9a). Additionally, the DNA fragmentation assay showed fragmented DNA, a characteristic of apoptotic cell death, when treated with SWCNTs-Ag whereas pSWCNT-Ag treated cells did not show DNA fragmentation similar to the non-treated control cells (Figure 9f).

EM analysis of *Salmonella* Typhimurium exposed to SWCNTs-Ag and pSWCNTs-Ag

The SEM and TEM images of either healthy non-treated bacterial cells or nanocomposites-treated cells are shown in Figure 10. SEM analysis revealed that treatment with the SWCNTs-Ag or pSWCNTs-Ag resulted in the complete lysis of the bacterial cells (indicated by a black solid arrow), or the cells appeared as hollow disrupted entities (indicated by a white solid block arrow) compared to non-treated healthy bacterial cells (Figure 10a). Some of the bacterial cells also showed the presence of holes in either a central division region (Figure 10b), or a polar division region (Figure 10c indicated by white line arrows). These findings were further supported by TEM imaging (Figure 10d-j), which showed damaged bacterial cells upon treatment with the nanocomposites (Figure 10e & j) compared to healthy untreated bacteria (Figure 10d). Upon interaction with the nanocomposites (indicated by double headed black arrows), bacterial cell membranes were ruptured causing the expulsion of cytoplasmic contents outside the cells, leaving behind either an empty ghost cell (Figure 10h, indicated by a dotted line black arrow) or debris of the cell (indicated by a white arrowhead). By comparison, non-treated bacteria had intact cell membranes and cytoplasmic contents (Figure 10d).

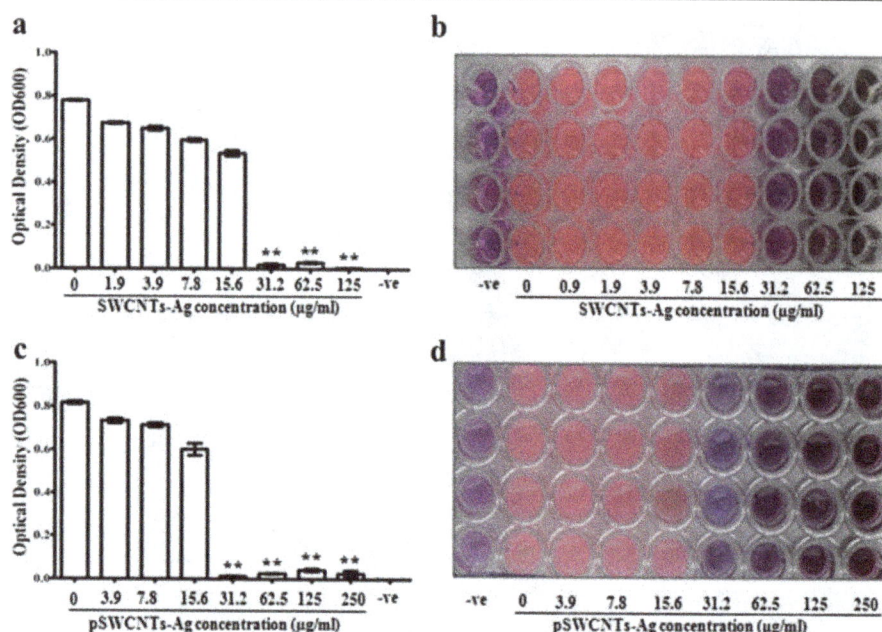

Figure 4 Evaluation of the minimum inhibitory concentrations (MICs) using the redox resazurin dye-based microtiter broth dilution assay. 1×10^5 cfu/mL bacteria were exposed to doubling concentrations of nanocomposites. **(a)** SWCNTs-Ag without resazurin. **(b)** SWCNTs-Ag with resazurin. **(c)** pSWCNTs-Ag without resazurin. **(d)** pSWCNTs-Ag with resazurin. All the plates were incubated at 37°C and the optical density at 600 nm (OD600) was determined after 24 h. All values were considered to be significant at $p \leq 0.05$ or 0.01 versus the controls (0 µg/mL of SWCNTs-Ag present in bacterial culture). **$p \leq 0.01$ indicating highly significant differences. Error bars represent standard deviations determined from at least six replicates.

Molecular studies of *Salmonella* Typhimurium exposed to nanocomposites

The expression of genes associated with bacterial metabolism, structural integrity and virulence was investigated to further explore the antibacterial action of nanocomposites. Bacteria treated with plain SWCNTs-Ag showed down-regulation of the *ompF* gene and up-regulation of the *ychP* gene whereas there was no significant change in expression of the genes such as *cigR* and *safC* (Figure 11). On the other hand, pSWCNTs-Ag treated cells showed significant down regulation of *ompF*, *safC* and *ychP* genes whereas the *cigR* gene was up regulated (Figure 11). Further, the genes such as *dppA*, *livJ*, *artP*, *trpA* and *argC* were significantly down-regulated several-fold upon exposure to SWCNTs-Ag and pSWCNTs-Ag (Figure 11c-f) whereas the expression of the *fliH* gene was significantly up-regulated upon treatment with both the nanocomposites (Figure 11g & h). The expression of *ybeF* and *sdiA* was down-regulated significantly in pSWCNTs-Ag treated bacteria, whereas the expression of these two genes remained unchanged upon SWCNTs-Ag treatment (Figure 11g & h). Also, the bacteria treated with plain SWCNTs-Ag showed significant up regulation of genes such as *sseA sseB* and *ssaH* whereas the expression of gene *sseG* remained unchanged (Figure 11i). Conversely, pSWCNTs-Ag treated bacteria showed down regulation of *sseA* and

sseG genes with no change in expression of *sseB* and *ssaH* genes (Figure 11h).

Discussion

Resistance of food borne pathogens to antibiotics is a serious problem and the development of resistance-free antibacterial agents is necessary to treat bacterial infections effectively. Although the use of SWCNTs-Ag is gaining popularity due to their excellent antibacterial properties, toxicity to human cells remains a major concern. Results of our study show that pegylation of SWCNTs-Ag reduces their toxicity to eukaryotic cells, compared to plain non-pegylated SWCNTs-Ag, and our results are in agreement with the previous findings that show pegylation reduces the toxicity of plain SWCNTs [22,28]. Additionally, our findings also corroborate with the recent findings wherein biotinylated amphiphiles of SWCNTs have been shown to reduce the toxicity to various mammalian cell lines [10,14]. However, to our knowledge, this is the first attempt to pegylate SWCNTs-Ag and to report that the pSWCNTs-Ag were non-toxic to human cells. Further, the effect of pegylation on the antibacterial activity of SWCNTs-Ag also remains to be reported. As the SWCNTs used in the present study were silver coated, it is possible that the anti-bacterial activity of SWCNTs-Ag could have been compromised by pegylation, as PL-PEG-amine molecules

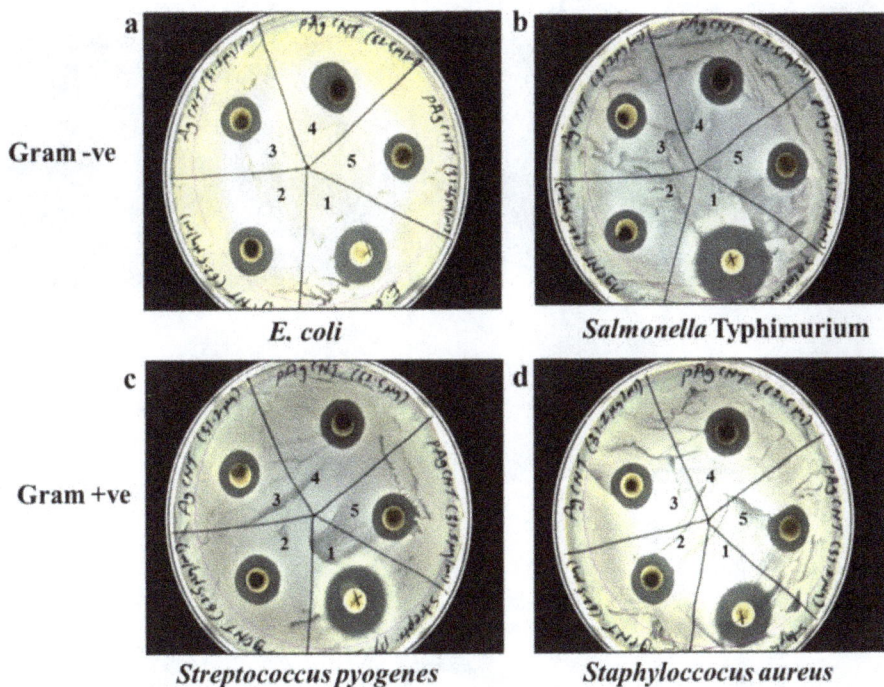

Figure 5 Zone of inhibition test using Kirby–Bauer disc diffusion assay against gram- negative and gram-positive organisms. The zone of inhibition around discs containing MIC concentrations of SWCNTs-Ag and pSWCNTs-Ag and the broad spectrum antibiotic amoxicillin–clavulanic acid (30 µg) can be clearly seen for both, the gram- negative pathogens such as **(a)** *Escherichia coli* and **(b)** *Salmonella* Typhimurium; and gram-positive pathogens such as **(c)** *Streptococcus pyogenes* and **(d)** *Staphylococcus aureus*. The numbers indicated are; 1: antibiotic control; 2: SWCNTs-Ag (62.5 µg/mL); 3: SWCNTs-Ag (31.25 µg/mL); 4: pSWCNTs-Ag (62.5 µg/mL); 5: pSWCNTs-Ag (31.25 µg/mL).

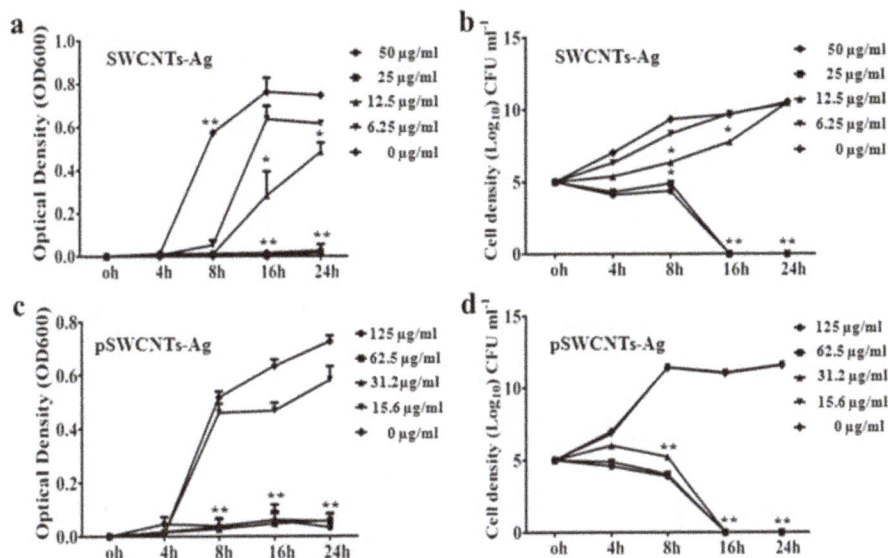

Figure 6 Growth curve and quantitative analysis of *Salmonella* Typhimurium exposed to various concentrations of nanocomposites. **(a)** Bacterial growth curve exposed to SWCNTs-Ag using optical density measurements. **(b)** cfu/mL of surviving bacteria upon exposure to SWCNTs-Ag. **(c)** Growth curve upon exposure to pSWCNTs-Ag using optical density measurements. **(d)** Quantification of bacteria upon exposure to pSWCNTs-Ag. Bacteria were grown in LB broth containing various concentrations of nanocomposites and all the cultures were incubated at 37°C with shaking at 250 rpm and the optical density measurements at 600 nm (OD600) and cfu/mL counts were done at 0, 4, 8, 16 and 24 h. The results are means of three experiments, with $p \leq 0.05$ indicating significant * differences, or $p \leq 0.01$ indicating highly significant ** differences. Error bars represent standard deviations determined from at least three replicates.

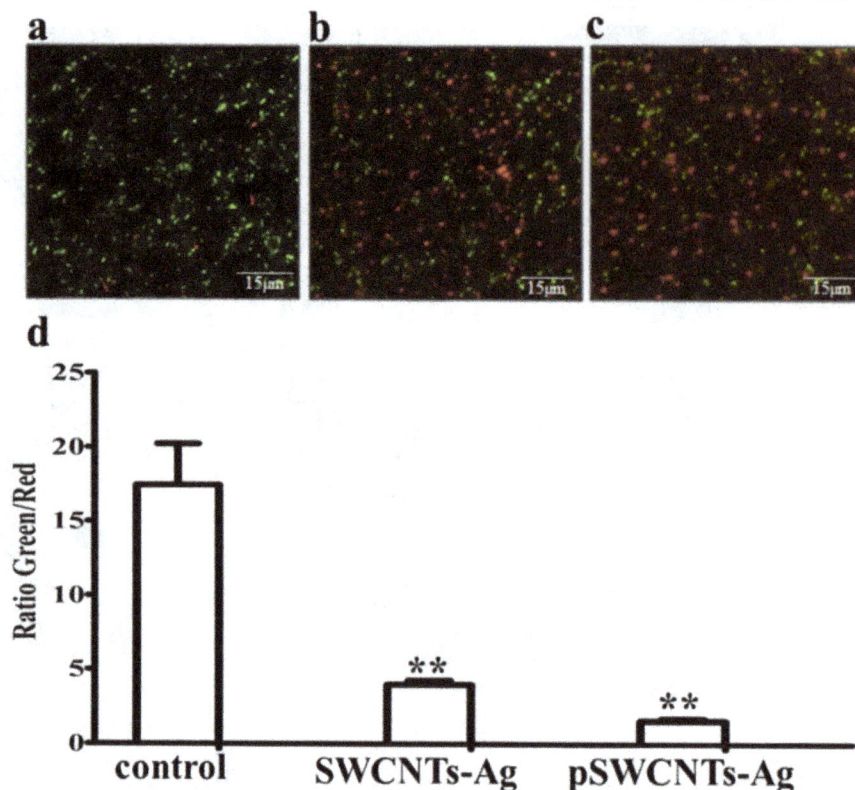

Figure 7 Live/ dead staining of bacteria exposed to different nanocomposites. 1×10^5 cfu/mL bacteria were exposed to 12.5 µg/mL of nanocomposites for 16 h and bacteria were stained using Baclight bacterial viability kits. **(a)** Non-treated bacteria. **(b)** Bacteria exposed to SWCNTs-Ag **(c)** pSWCNTs-Ag treated bacteria. **(d)** Ratio of green/red colored bacteria upon exposure to nanocomposites. All values were considered to be significant * at $p \leq 0.05$ * or ** highly significant at $p \leq 0.01$. Error bars represent standard deviations of the results determined with at least six replicates. The images are at 10× magnification. Scale bars: ~15 µm

may have covered the silver coating on the SWCNTs. However, our results clearly demonstrated that pegylated SWCNTs-Ag destroy *Salmonella* as effectively as plain SWCNTs-Ag. The bacterial MIC values for pSWCNTs-Ag did not differ from that of plain SWCNTs-Ag (62.5 & 31.2 µg/mL). Sequential monitoring of quantitative bacterial growth over time, and live/dead staining of bacteria upon exposure to various concentrations of nanocomposites provided further evidence that the antibacterial activity of SWCNTs-Ag was not reduced by pegylation. At 31.2 µg/mL concentration, both the nanocomposites exhibited bactericidal activity, however, the uncoated SWCNTs-Ag were cytotoxic. By contrast, pSWCNTs-Ag were non-toxic to all the cell lines at bactericidal concentrations of 62.5 µg/mL, which is at least twice the MIC value. The higher toxicity of plain non-pegylated SWCNTs-Ag may be attributed to their lower surface area as a result of their tendency to agglomerate and their surface chemistry [17]. Due to their insolubility and agglomerative properties, SWCNTs-Ag may not be able to enter cells thus inducing contact mediated cell toxicity similar to CNTs [22]. On the contrary, pSWCNTs-Ag were soluble in water and therefore showed reduced toxicity, or no

toxicity, possibly due to their appropriate distribution in the solution, lessening the contact mediated toxicity due to their ability to enter in to human cells similar to pegylated CNTs [22].

Next, we also investigated the morphological changes seen in *Salmonella* upon exposure to either plain or pegylated SWCNTs-Ag using electron microscopy. SEM and TEM clearly showed that nanocomposites damaged the bacterial cells by either creating a hole on the bacterial cell surface, or causing lysis of the bacterial cells. This mechanism of action results from the Ag coating of SWCNTs, as AgNPs are well known for contact-mediated membrane damage [5,6,29]. Further, SWCNTs by themselves have been shown to cause membrane damage in bacteria thereby regulating their antibacterial activity [30,31]. However, all these studies have shown the membrane damaging effects of AgNPs or CNTs by SEM analysis and lack detailed information about further validation by TEM. More recently, the antibacterial activity of Ag-doped multi-walled CNTs has been associated with the generation of reactive oxygen species through direct contact of the nanocomposites with bacteria causing membrane damage [9]. Our study indicated

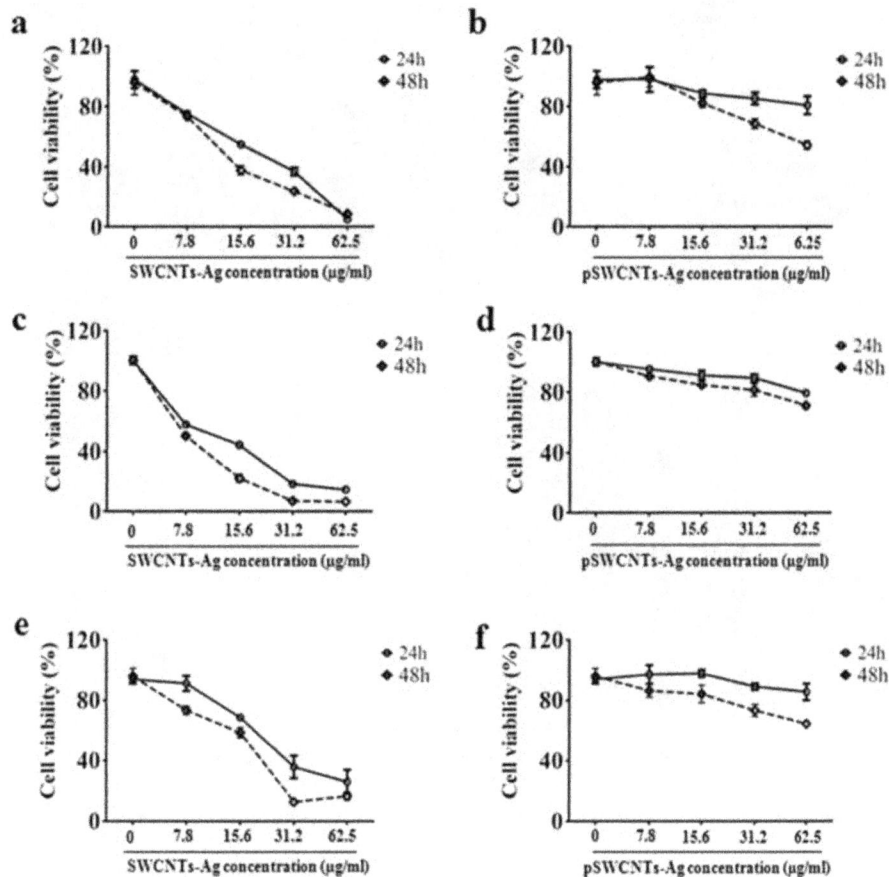

Figure 8 *In vitro* **cytotoxicity assay for SWCNTs-Ag and pSWCNTs-Ag. (a)** Viability of A549 cells exposed to SWCNTs-Ag. **(b)** Cell viability of A549 cells exposed to various concentrations of pSWCNTs-Ag. **(c)** J774 cell viability exposed to SWCNTs-Ag. **(d)** Cell viability of J774 treated with pSWCNTs-Ag. **(e)** Hep-2 cells viability treated with SWCNTsAg. **(f)** Viability of Hep-2 cells exposed to pSWCNTs-Ag. All the cells were treated with nanocomposites for 24 and 48 h and statistical differences were calculated compared to controls. All values were expressed as fold change expressed compared to non-treated bacteria. *when $p \leq 0.05$ indicate significant differences; **when $p \leq 0.01$ indicate highly significant differences. Error bars represent standard deviations of the results determined with at least 3 biological replicates.

additional possible mechanisms for the antibacterial activity of SWCNTs-Ag, perhaps linked with the well-known ghost formation phenomenon, as characterized by tunnel formation on bacterial cell membranes, resulting in the expulsion of the cytoplasmic contents [32]. However this may require further validation by more advanced analyses using atomic force microscopy [33] or in- lens field emission scanning electron microscopy [34] to capture the detailed morphological changes induced by SWCNTs-Ag in a sequential manner.

As EM studies revealed that SWCNTs-Ag caused membrane damage and formation of ghost cells, we further investigated the expression of a few, yet major, genes associated with invasion (*ychP*), outer membrane proteins (*ompF* and *safC*), inner membrane protein (*cigR*), amino acid biosynthesis (*trpA*, tryptophan and *argC*, arginine), ABC transport (*dppA*, *artP* and *livJ*), DNA transcriptionactivators (*fliH*, *sdiA* and *ybeF*) and *Salmonella* pathogenicity island-2 (SPI-2) associated

genes (*sseA*, *sseB*, *sseG* and *ssaH*) using qRT-PCR [35,36]. These genes were selected either on the basis of their common existence and physiological roles in a wide range of bacterial species, or their significant role in *Salmonella* virulence. The expression of genes associated with membrane integrity (*ompF*, *cigR*, *ychP*) was significantly down regulated by treatment with plain silver or pegylated silver nanocomposites. However, another outer membrane associated gene *safC* (associated with SPI-6) and the *ychP* invasin gene were exclusively down regulated only in pSWCNTs-Ag treated cells. These results support our hypothesis that both the nanocomposites cause bacterial cell membrane damage, as evidenced by change in gene expression that regulates proteins associated with the bacterial membranes. Our results also indicate that pSWCNTs-Ag interfere with the expression of additional genes compared to plain SWCNTs-Ag, which can be attributed to the solubility of pSWCNTs-Ag in water. Further, the genes regulating the bacterial

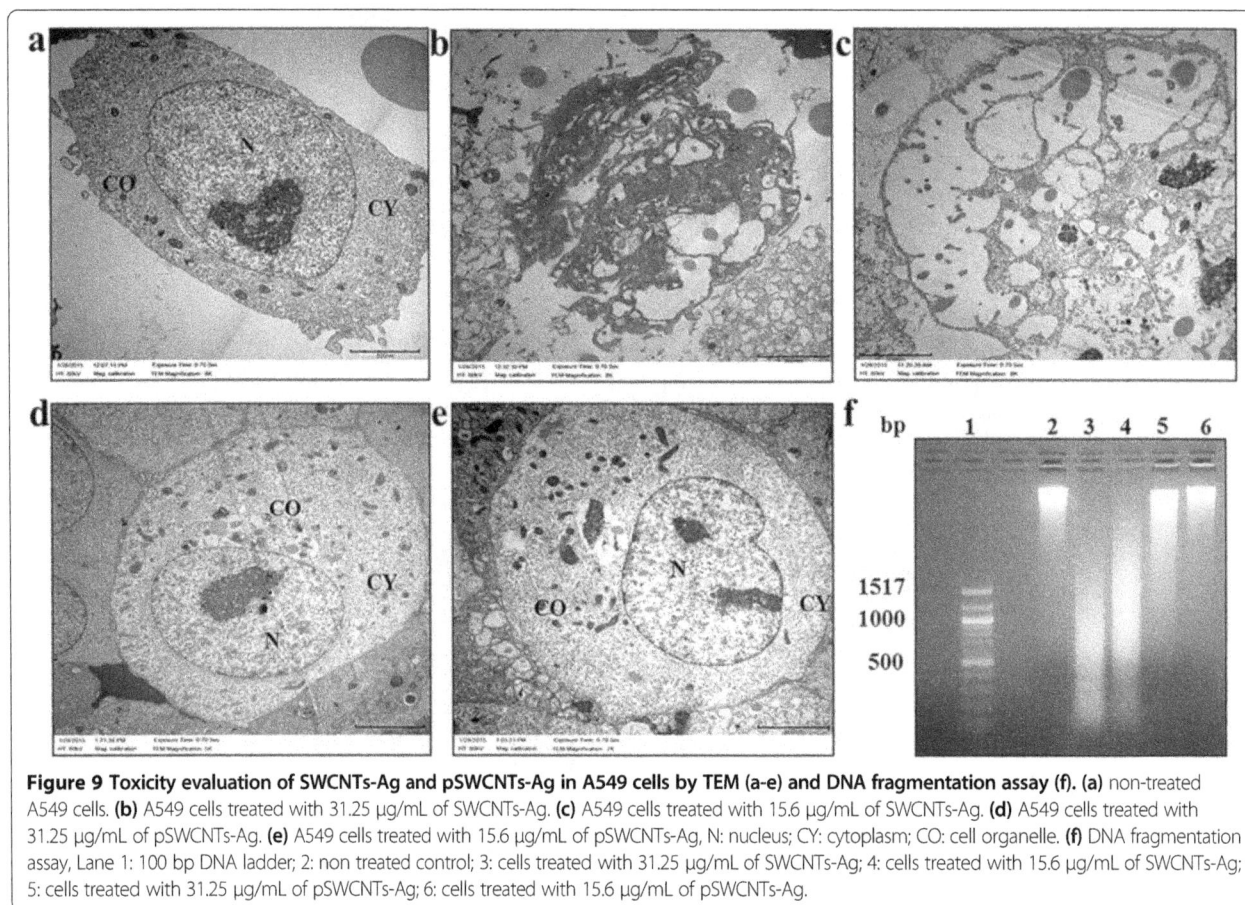

Figure 9 Toxicity evaluation of SWCNTs-Ag and pSWCNTs-Ag in A549 cells by TEM (a-e) and DNA fragmentation assay (f). (a) non-treated A549 cells. **(b)** A549 cells treated with 31.25 µg/mL of SWCNTs-Ag. **(c)** A549 cells treated with 15.6 µg/mL of SWCNTs-Ag. **(d)** A549 cells treated with 31.25 µg/mL of pSWCNTs-Ag. **(e)** A549 cells treated with 15.6 µg/mL of pSWCNTs-Ag, N: nucleus; CY: cytoplasm; CO: cell organelle. **(f)** DNA fragmentation assay, Lane 1: 100 bp DNA ladder; 2: non treated control; 3: cells treated with 31.25 µg/mL of SWCNTs-Ag; 4: cells treated with 15.6 µg/mL of SWCNTs-Ag; 5: cells treated with 31.25 µg/mL of pSWCNTs-Ag; 6: cells treated with 15.6 µg/mL of pSWCNTs-Ag.

membrane associated ABC transporter proteins and amino acid synthesis were down regulated significantly by either nanocomposites treatment. ABC transporters are extremely vital in cell viability, virulence and pathogenicity as they function as protein systems that counteract any undesirable changes occurring in the cell [37]. The periplasmic DppA [38] (initial receptor for dipeptides transport), artP [39] (an important component of the arginine binding and transport system) and livJ [40] (regulator of three branched-chain amino acids system) proteins are some of the major ABC transporters that allow transport of a wide variety of materials across cytoplasmic membranes, and play an important role in bacterial metabolism under conditions of stress. Down regulation of the genes regulating these proteins indicates the inhibition of these bacterial recovery mechanisms. Similarly, the down- regulation of tryptophan (an essential amino acid) and arginine biosynthesis genes further suggests that Ag nanocomposites may also interfere with the amino acid synthesis machinery of bacteria.

On the other hand, the *fliH* gene, which encodes the flagellar FliH protein was significantly up-regulated upon exposure to both nanocomposites. FliH is an effector protein involved in the flagellar export apparatus which

regulates the transcription of *flaB* and thus the motility of bacteria [41]. Up-regulation of *fliH* in treated bacteria thus, may indicate damage to the flagellar assembly, and a stress recovery response from the damage caused by the Ag nanocomposites.

In addition to these molecular changes, only the bacteria treated with pegylated SWCNTs-Ag showed differential expression of the genes associated with biofilm formation, quorum sensing and virulence. The expression of *ybeF* (a transcriptional regulator of the lysR family) and *sdiA* (a receptor for N-acyl-L-homoserine lactones), was down regulated significantly in pSWCNTs-Ag treated bacteria, whereas the expression of these two genes remained unchanged upon SWCNTs-Ag treatment (Figure 11g & h). The lysR family of protein regulators play a diverse role in cellular functions such as transport, response to oxidative stress, nitrogen fixation, biofilm formation and bacterial virulence, whereas sdiA proteins play a key role in quorum sensing mechanisms of pathogenic bacteria [42,43]. Our results thus indicate that pSWCNTs-Ag may have an additional capability of interfering with normal physiological processes and quorum sensing mechanisms of bacteria compared to non-pegylated plain SWCNTs-Ag. Similarly, the genes associated with the SPI-2 mediated type three-

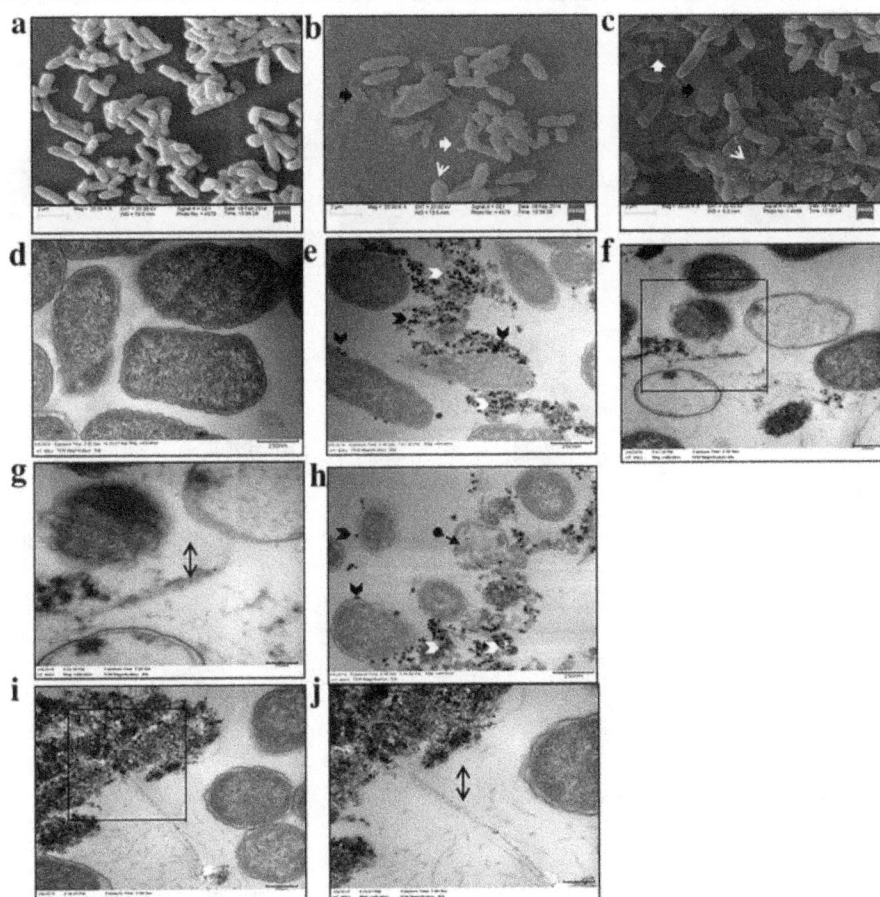

Figure 10 Evaluation of morphological changes in bacteria upon their interaction with SWCNTs-Ag and pSWCNTs-Ag by SEM (a-c) and TEM (d-j). (a) Non treated *Salmonella* Typhimurium. **(b)** *Salmonella* Typhimurium treated with SWCNTs-Ag. **(c)** Treatment with pSWCNTs-Ag. **(d)** Non-treated bacteria. **(e-g)** Bacteria exposed to SWCNTs-Ag; **(g)** is magnified inset of **(f)**. **(h-j)** Bacteria exposed to pSWCNTs-Ag; **(j)** is magnified inset of **(i)**. Black solid arrows indicate lysis of the cells. White solid block arrows indicate dissolved entities. White line arrows indicate pore formation. Black arrow-heads represent nanocomposites, whereas white arrow-heads indicate cell debris. Dotted line black arrows show empty ghost cells. Two headed arrow indicates the presence of nanocomposites.

secretion system (T3SS) showed different expression patterns upon exposure to either nanocomposite. The expression of *sseA* was significantly down regulated in bacteria treated with pSWCNTs-Ag, which resulted in unchanged expression of the *sseB* and *ssaH* genes, and down regulation of the *sseG* gene. Conversely, the expression of *sseA* was up-regulated, followed by several fold increase in *sseB* and *ssaH* expression when the bacteria were treated with SWCNTs-Ag. SPI-2 T3SS is required for intracellular survival of *Salmonella*, and development of systemic disease [44]. The *sseA* gene of SPI-2 T3SS plays an important role in pathogenesis and acts as a chaperone to regulate the stabilization and export of the *sseB* filament protein and other effector proteins such as *sseG* and *ssaH* [45].Thus, our findings indicate that pSWCNTs-Ag may affect several pathways to exert their antibacterial effect, as compared to plain SWCNTs-Ag. Further studies are required to fully elucidate the antibacterial effect of both nanocomposites on the bacterial genome using more advanced

methods such as DNA microarray or whole genome sequencing.

Conclusions

In conclusion, our results demonstrate that pSWCNTs-Ag are non-toxic to human cells at their bactericidal concentration, and down-regulate the genes associated with quorum sensing and virulence mechanisms in *Salmonella* (Figure 12). Although our EM results show that exposure to SWCNTs and pSWCNTs-Ag produce relatively similar morphological changes in bacteria, qRT-PCR findings clearly underline the differences in gene regulation mechanisms, indicating that pSWCNTs-Ag are more efficient than plain SWCNTs-Ag. Intracellular survival of *Salmonella* and their quorum sensing mechanisms are very important for their pathogenesis [43,44,46]. Treatment of *Salmonella* with pSWCNTs-Ag interrupts these two important mechanisms and thus has an additional advantage over plain SWCNTs-Ag as

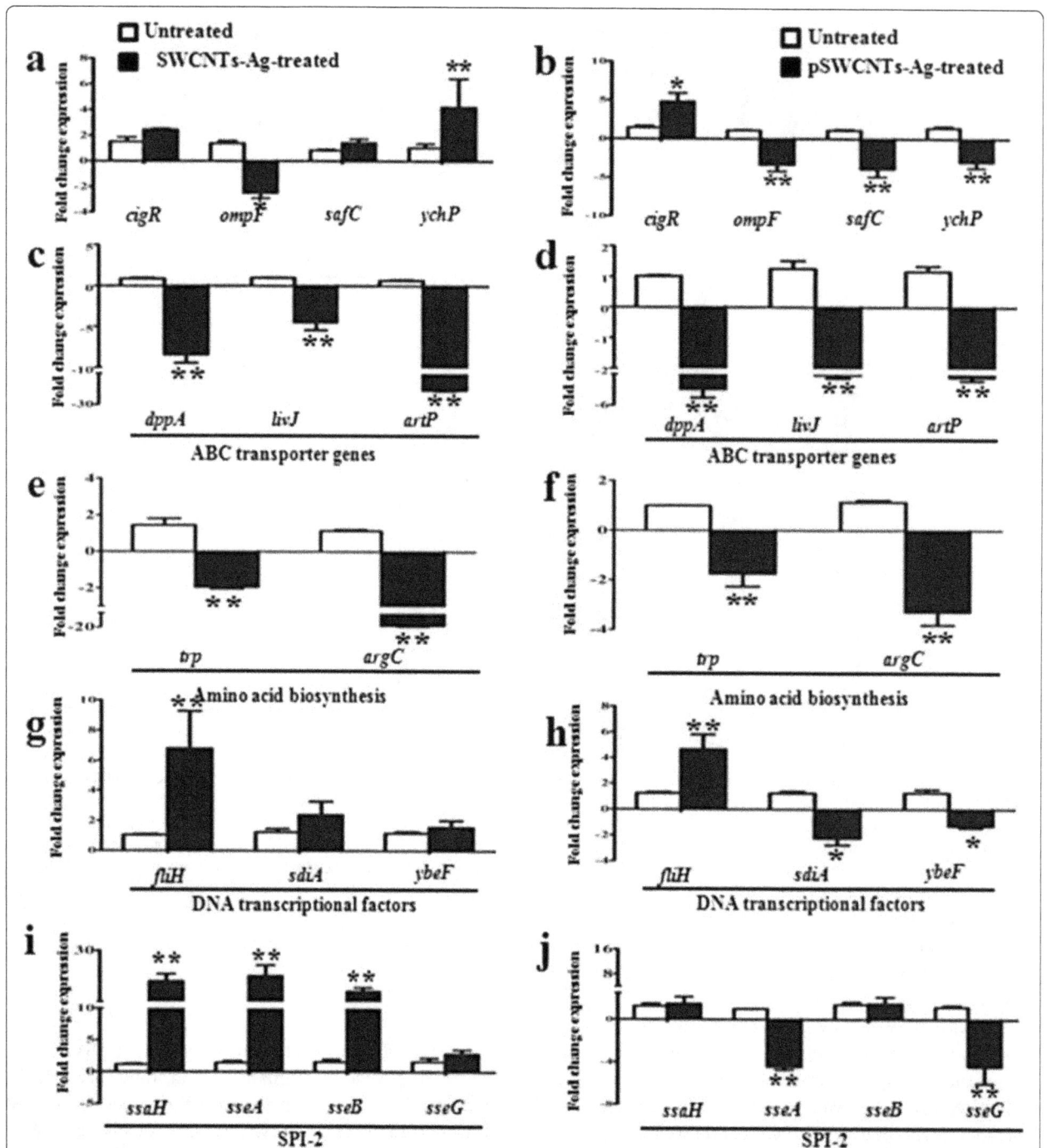

Figure 11 Gene expression studies in *Salmonella* Typhimurium exposed to either SWCNTs-Ag or pSWCNTs-Ag nanocomposites. (a) and
(b) represent gene expression of *cigR*, *ompF*, *safC* and *ychP* upon exposure to SWCNTs-Ag and pSWCNTs-Ag, respectively. **(c)** and **(d)** represent
ABC transporter systems-associated gene expression in bacteria treated with SWCNTs-Ag and pSWCNTs-Ag, respectively. **(e)** and **(f)** represent the
expression of amino acid biosynthesis genes in bacteria exposed to SWCNTs-Ag and pSWCNTs-Ag, respectively. **(g)** and **(h)** represent expression
of genes related to DNA transcription in bacteria treated with SWCNTs-Ag and pSWCNTs-Ag, respectively. **(i)** and **(j)** indicate the expression of
genes associated with the SPI-2 type three secretion system in bacteria exposed to SWCNTs-Ag and pSWCNTs-Ag, respectively. All values were
expressed as fold change expression compared to non-treated bacteria. *when $p \leq 0.05$ indicate significant differences; **when $p \leq 0.01$ indicate
highly significant differences. Error bars represent standard deviations of the results determined with at least 3 biological replicates.

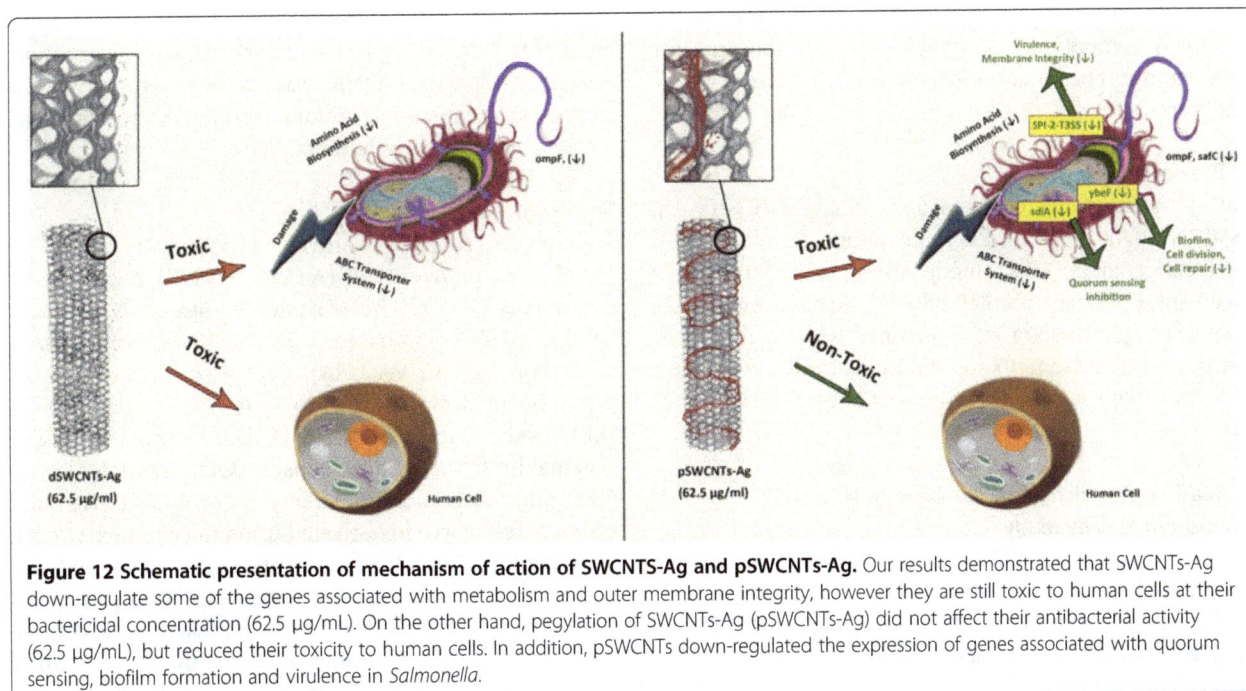

Figure 12 Schematic presentation of mechanism of action of SWCNTS-Ag and pSWCNTs-Ag. Our results demonstrated that SWCNTs-Ag down-regulate some of the genes associated with metabolism and outer membrane integrity, however they are still toxic to human cells at their bactericidal concentration (62.5 µg/mL). On the other hand, pegylation of SWCNTs-Ag (pSWCNTs-Ag) did not affect their antibacterial activity (62.5 µg/mL), but reduced their toxicity to human cells. In addition, pSWCNTs down-regulated the expression of genes associated with quorum sensing, biofilm formation and virulence in *Salmonella*.

far as the antibacterial mechanism is concerned. These results show promise towards developing resistance free and safe antimicrobials for the effective treatment of foodborne infections.

Methods

Preparation of SWCNTs-Ag solutions

SWCNTs-Ag, with purity greater than 95%, were purchased from NanoLab, Inc. Waltham, MA, USA. The SWCNTs-Ag were produced by catalytic chemical vapor deposition, and had an outer diameter of 1–5 nm and a length of 1–2 µm. SWCNTs-Ag were dispersed in 1 mL of specifically formulated dispersant- NanoSperse AQ® from NanoLab Inc. Briefly, 1 mg of SWCNTs-Ag were suspended in 1 mL of NanoSperse AQ® dispersant solution (according to manufacturer's instructions) and the suspensions were immediately sonicated for 1-2 h, and shaken for 30 min to obtain the SWCNTs-Ag dispersion (1 mg/mL).

Functionalization of SWCNTs-Ag using phospholipid-poly (ethylene glycol) (PL-PEG)

The SWCNTs-Ag were non-covalently functionalized using phospholipid-poly(ethylene glycol) PL-PEG5000-Amine (NOF Corporation, White Plains, NY, USA) as described previously [27]. Briefly, hydrophobic SWCNTs-Ag were mixed with water solutions of the amphiphilic polymer PL-PEG5000-Amine in a ratio of 1:5, sonicated for 60 min at room temperature (~22°C) and centrifuged for 6 h (24000 × g), at room temperature. After centrifugation, the supernatant solution containing functionalized

SWCNTs- Ag was run through an Amicon centrifugal filter (molecular weight cutoff of 100 kDa Millipore, Billerica, MA, USA) and centrifuged for 10 min (4,000 × g) at room temperature and the sediment was discarded. The functionalized SWCNTs were washed 5–6 times with water to remove excess PL-PEG molecules. Functionalized SWCNTs-Ag were solubilized in sterile nuclease-free water.

Determination of SWCNTs-Ag concentration in pSWCNTs-Ag solution

To measure the concentration of SWCNTs-Ag in pSWCNTs-Ag solution, the dry weight of the solution was calculated after vacuum drying of the suspension. Simultaneously, the PL-PEG concentration in a solution was determined using an inorganic phosphate assay following acid hydrolysis [47]. Once the concentration of PL-PEG was known, the concentration of SWCNTs-Ag was calculated by subtracting the concentration of PL-PEG from the total dry weight of pSWCNTs-Ag.

Characterization of SWCNTs-Ag and pSWCNTs-Ag

Both the nanocomposite formulations were characterized using zeta potential and fourier transform infrared spectroscopy (FT-IR) analysis [48].

Determination of zeta potential

The zeta potential of SWCNTs-Ag and pSWCNTs-Ag was measured using a Zetasizer (Nano-ZS; Malvern Instruments Ltd, Malvern, UK). The samples were diluted in distilled water to 1/10 (v/v), sonicated, and placed in a disposable cuvette for zeta potential measurements. All the measurements were carried out in triplicates for each sample. The values are reported as means of triplicate samples.

FT-IR spectroscopy

FTIR spectra were recorded for SWCNTs-Ag, pSWCNTs-Ag and PL-PEG5000-amine in attenuated total reflectance (ATR) mode using an infrared (IR) spectrophotometer (Nicolet 380 FT-IR; Thermo Fisher Scientific). The spectra were obtained with 64 scans per sample, ranging from 400 to 4000 cm^{-1} and a resolution of 4 cm^{-1}. The sample chamber was purged with dry N2 gas.

Evaluation of toxicity to eukaryotic cells

In vitro cell toxicity assay

The cell toxicity to SWCNTs-Ag and pSWCNTs-Ag was determined using Cell Titer 96® Non-Radioactive cell proliferation kits (Promega, Madison, WI). A549 (human lung carcinoma), Hep2 (hepatocellular carcinoma) cells and J774 (murine macrophages) cell lines were used for the cytotoxicity assay. The assay is a colorimetric assay based on the reduction of the tetrazolium dye MTT [3-(4,5-dimethylthiazol-2yl)-2,5-diphenyltetrazolium bromide). As per the manufacturer's protocol, 1×10^4 cells/well in 100 μL of minimum essential medium-10 (MEM-10, Gibco, Life technologies, Grand Island, NY) were seeded into a 96-well plate. After overnight incubation at 37°C in 5% CO_2 humidified atmosphere, the media from the 96-well plates were replaced with the MEM-10 media containing 62.5, 31.25, 15.12 or 7.8 μg/mL of either SWCNTs-Ag or pSWCNTs-Ag. The treated cells were further incubated at 37°C under 5% CO_2 for 24 and 48 h. At the end of the each incubation period, 15 μL of MTT dye were added into each well and the plate was allowed to incubate again for the next 4 h, in darkness. The reaction was then stopped using 100 μL of the stop solution. The absorbance of the plate was measured at 570 nm using a TECAN Sunrise™ enzyme-linked immunosorbent assay (ELISA) plate reader (Tecan US, Inc., Morrisville, NC, USA). Non-treated cells, in growth media, were used as a control.

DNA fragmentation assay

The DNA fragmentation assay was performed as described elsewhere [49]. Briefly, A549 cells were treated with SWCNTs-Ag and pSWCNTs-Ag at concentrations of 31.25 and 15.6 μg/mL for 48 h, trypsinized and re-suspendeded in 0.5 mL of lysis buffer [10 mM Tris–HCl buffer (pH 8.0) containing 2% (v/v) Triton-X100, 0.5 mM EDTA)], treated with RNAse (10 mg/mL) at 37°C for 2 h and; digested with 200 μg/mL proteinase K for 4 h at 50°C. Upon digestion, DNA was extracted with phenol/chloroform/Iso-amyl alchohol (25:24:1, v/v), and precipitated with 70% ethanol at –20°C overnight. Electrophoresis of the purified DNA was carried out using 1.5% agarose gels containing ethidium bromide and visualized using ultraviolet transillumination.

Bactericidal experiments

Salmonella enterica serovar Typhimurium (ATCC® 13311™), *Escherichia coli* (ATCC® 25922™), *Staphylococcus aureus* (ATCC® 9144™) and *Streptococcus pyogenes* (ATCC® 8135™) purchased from American Type Culture Collection (ATCC, VA USA), were used for the bacterial experiments. Bacteria were grown at 37°C in Luria-Bertani (LB) broth (Difco, Sparks, MD, USA) with continuous shaking until the optical density (OD) was 0.6–0.8 (at 600 nm). Bactericidal activity of SWCNTs-Ag and pSWCNTs-Ag was investigated using the parameters such as minimum inhibitory concentrations (MIC) and the Kirby-Bauer disk diffusion assay against all four pathogens. Further detailed evaluation of the antibacterial effects on *Salmonella* Typhimurium was done by growth curve analysis, and live/dead staining. The morphological changes that occurred in *Salmonella* by treatment with SWCNTs-Ag and pSWCNTs-Ag were examined by scanning electron microscopy (SEM) and transmission electron microscopy (TEM). Gene expression studies were carried out using the quantitative real time reverse transcriptase polymerase chain reaction (qRT-PCR).

Determination of MIC

The MIC values of SWCNTs-Ag and pSWCNTs-Ag were evaluated in quadruplet wells of sterile 96-well microtiter plates using the broth microdilution assay with or without using the redox reagent resazurin, following a previously described broth microdilution procedure [50,51]. Briefly, early log phase suspensions of the bacteria (1×10^5 cfu/mL) were exposed to doubling concentrations of SWCNTs-Ag and pSWCNTs-Ag starting at 1.9 μg/mL in the presence or absence of resazurin. Two-fold serial dilutions of pSWCNTs-Ag were performed in sterile nuclease free water, whereas the NanoSperse AQ dispersant solution was used as a diluent for the plain SWCNTs-Ag. All plates were sealed lightly (with ventilation) and then incubated at 37°C for 24 h. Each plate consisted of 8 dilutions of the SWCNTs-Ag dispersions/solutions, 1 negative control (no SWCNTs-Ag or no bacterial culture), and 1 positive control (only bacterial culture without SWCNTs-Ag). After incubation, the MIC was determined by the turbidity of the culture media in the wells. The concentration of the first well without turbidity was considered as the minimum inhibitory concentration. The inhibition of bacterial growth was determined by measuring absorbance at 600 nm with a TECAN Sunrise™ enzyme-linked immunosorbent assay (ELISA) plate reader (Tecan US, Inc Morrisville, NC, USA). While testing plates where resazurin dye was added, MIC values were identified at 24 h as the lowest concentration of antimicrobial agent with color shift from blue to red. All experiments were repeated at least three times.

Kirby–Bauer disc diffusion assay

The antibacterial activity of SWCNTs-Ag and pSWCNTs-Ag at their MICs such as 62.5 & 31.25 µg/mL was further tested using the Kirby–Bauer disc diffusion assay against all four pathogens as described earlier [52]. Bacterial suspensions of each bacterial strain (10^8 cfu/mL) were swabbed on the surface of Mueller–Hinton agar plates and filter paper discs (Fisher Scientific, MO) containing 62.5 & 31.25 µg of SWCNTs-Ag and pSWCNTs-Ag, were placed on the plate. A broad spectrum combination of amoxicillin and clavulanic acid (30 µg) was used as a positive control (BD, BBL™, USA). Plates were incubated at 37°C overnight and the diameters of the clear zones around the discs, called 'zones of inhibition', were recorded.

Quantitative growth analysis of *Salmonella*

Salmonella growth was quantified sequentially at time points of 0, 4, 8, 16 and 24 h post-exposure to SWCNTs-Ag or pSWCNTs-Ag. Twenty milliliters of cultures containing 1×10^5 cfu/mL of bacteria were exposed to 50, 25, 12.5 and 6.25 µg/mL of SWCNTs-Ag, or 125, 62.5, 31.2 and 15.7 µg/mL of pSWCNTs-Ag (these concentrations were selected based on the MICs of pSWCNTs-Ag and plain SWCNTs-Ag). The cultures were incubated at 37°C with shaking at 250 rpm and the optical density at 600 nm (OD600) was determined at 0, 4, 8, 16 and 24 h. The graph was plotted as O.D vs. time point on the Y and X axis, respectively. Similarly, at each time point, 1 mL aliquots of bacterial culture were collected, subjected to serial 10-fold dilution in sterile LB broth, and each dilution was then spread on PCA to determine the cfu/mL. Each sample was analyzed in triplicate and the analysis was repeated at least three times. The average cfu/mL value of each sample was then reported as described above.

Live/dead staining of bacteria

The viability of bacteria was examined by live/dead staining using the Baclight bacterial viability kit (L13152, Molecular probes, USA) according to manufacturer's instructions. Briefly, 1×10^5 cfu/mL bacteria were treated with 12.5 µg/mL of SWCNTs-Ag and pSWCNTs-Ag for 16 h in a shaking incubator (250 rpm) at 37°C. Post-treatment, the bacterial cells were harvested, washed with 0.85% sodium chloride (NaCl, Fischer Thermo scientific, New Jersey, USA), incubated further in 0.85% NaCl for 1 h and finally re-suspended in 200 µl of 0.85% Nacl. Bacterial samples were then incubated in the dark for 30–45 min with an equal amount of 2× stock solution of the LIVE/DEAD BacLight staining reagent containing a final concentration of 6 µM of the SYTO 9 dye and 30 µM of propidium iodide. The stained bacterial suspensions were trapped (5 µl) between a slide and an 18 mm square coverslip and the images were captured by Nikon Eclipse TE200 microscope (Nikon, Melville, NY, USA) using FITC-HYQ (Ex 450–500) and TRITC HYQ (Ex 530–550) filters. Viable cells had green fluorescence, while non-viable cells had red fluorescence. All the experiments were carried out in triplicate sets and the treated samples were compared with non-treated controls. The live and dead cells (of at least three images of treated and non-treated bacteria) were counted manually using NIS-Elements AR 3.1 microscope imaging software and the live/dead ratio for each sample was calculated.

Electron microscopy

Scanning electron microscopy (SEM, Zeiss EVO 50, Carl Zeiss Meditec, Oberkochen, Germany) and transmission electron microscopy (TEM, Zeiss EM 10C 10CR, Carl Zeiss Meditec, Oberkochen, Germany) were used to examine the morphology and size of SWCNTs-Ag and pSWCNTs-Ag. Additionally, SEM and TEM were used to observe the morphological features of *Salmonella* Typhimurium treated with SWCNTs-Ag and pSWCNTs-Ag. The non-treated bacterial cells were used as a control. Similarly, TEM analysis was performed to investigate the cytotoxicity of the Ag nanocomposites in A549 cells compared to non-treated cells. Samples of SWCNTs-Ag and pSWCNTs-Ag for SEM were prepared as described previously with minor modifications [8]. Briefly, the samples were sonicated, diluted 10 times and placed on a SEM stub with a conductive silver paint and sputter-coated with gold/palladium using a sputter coat device (EMS 550×, Carl Zeiss Meditec, Oberkochen, Germany). For bacterial sample preparation, cells (1×10^5 cfu/mL) were treated with 12.5 µg/mL of SWCNTs-Ag and pSWCNTs-Ag separately for 16 h in a shaking incubator (250 rpm) at 37°C. For eukaryotic sample preparation, A549 cells were treated with 31.25 & 15.6 µg/mL of SWCNTs-Ag and pSWCNTs-Ag separately for 48 h and incubated at 37°C in 5% CO_2 humidified atmosphere. Post treatment, the bacterial cells were harvested by centrifugation (13000 × **g** for 10 min), fixed in 2.5% glutaraldehyde and 1% formaldehyde overnight, followed by fixation in 1% aqueous osmium tetroxide for 1 h and then subjected to serial dehydration in increasing percentages of ethanol (30, 50, 70, 80, 90, 95 and 100%). After the final dehydration, step, 5 μl of each sample was placed on the SEM stub, air dried and sputter coated with a gold-palladium alloy for subsequent observation by SEM. For TEM, the 10 fold diluted samples of SWCNTs-Ag and pSWCNTs-Ag were dispersed onto the carbon-coated copper grids (200 mesh). The bacterial and A549 cells samples were passed through propylene oxide, polymerized in Embed 812 resin, followed by ultrathin gold colored sectioning on to copper grids and observation by TEM.

Molecular studies using qRT-PCR

The mRNA levels of several metabolically essential genes and virulence factor-associated genes of *Salmonella* Typhimurium were investigated using qRT-PCR. The primer sets and the function of each gene are described in Table 2. The bacteria (1×10^5 cfu/mL) were treated with 12.5 μg/mL of SWCNTs-Ag and pSWCNTs-Ag for 16 h in a shaking incubator (250 rpm) at 37°C, followed by total RNA extraction using RNeasy Mini kit (Qiagen, Germany). The RNA was quantified by using a NanoVue Plus spectrophotometer at 260 nm/280 nm (GE Healthcare Life Sciences, Pittsburg, PA), and cDNA synthesis was carried out in a 40 μl reaction volume using the Applied Biosystems High Capacity cDNA Reverse Transcriptase Kit (Life Technologies, Grand Island, NY). The expression of metabolically essential genes such as genes associated with ABC transporter system (*dppA*, *artP* and *livJ*), amino acid biosynthesis (*trp* and *argC*) and DNA transcription (*sdiA*, *fliH* and *ybeF*); virulence factor associated genes such as *Salmonella* pathogenicity island-2 (SPI) (*ssaH*, *sseA*, *sseB* and *sseG*), SPI-3 (*cigR*), SPI-6 (*safC*), outer membrane protein (*ompF*) and invasion (*ychP*) genes were quantified by qRT-PCR using the SYBR® Select Mastermix (Life Technologies, Grand Island, NY) according to the manufacturer's instructions. DNA was amplified and quantified in the Applied Biosystems® ViiA™ 7 real-time PCR system (Life Technologies). PCR conditions consisted of an initial denaturation at 95°C for 2mins, followed by 40 cycles of 95°C for 5 s, 56°C for 25 s and 72°C for 30s. Amplification efficiency of each primer set with respect to the endogenous control gene (*16srRNA*) was in between 95-98% (data not shown). The data obtained from three-independent experiments were used to analyze the relative gene expression compared to non-treated samples by the 2 – ΔΔCt method [53].

Statistical analyses

All data are expressed as the mean ± standard deviation (SD) unless otherwise specified. Analyses were performed using the GraphPad Prism Version 4 software (GraphPad Software, Inc., La Jolla, CA). Statistical differences for growth curves were evaluated by using two way ANOVA. Fold change expression of genes were analyzed by the Student's t test. Differences were considered to be statistically significant when the P-values were ≤0.05 or 0.01.

Table 2 Primers used in this study

Gene	Forward (5'-3')	Reverse (5'-3')	Function
sseA	GGGCTAAGGTGAGTCAACA	TGAAGAATACTCTCTGTCTCTCTG	Virulence (SPI-2 associated)
sseB	CATCTTATGGGGAAGTCAAAACC	GATAAGTCATCCTGGCTCCC	Virulence (SPI-2 associated)
ssah	TGATTTCCCAGGTACATGCGATG	GCCAACAATAATGCCAGACATACC	Virulence (SPI-2 associated)
sseG	GCCGATCCCAGTGTGTTATT	GCATTTGGGCTAACAGGTTTC	Virulence (SPI-2 associated)
ychP	GCGACCTTAGAACAGCGAAT	CCAAAGTACTGCTCCAGACTAAC	Invasion
cigR	CCCACATCAGAAGGGCAATATC	GCCATTACCATTTCCCGGATT	Inner membrane protein
safC	ATGGTAGCGCCATTCCTTTC	CCGCCAAACCAGTGAGATAAA	Outer membrane protein
ompF	CGTTGGCGCGAAATATGATG	GTTTGCCATTTCCACGGTATC	Outer membrane protein F
ybeF	TCCCGATCCGCTGTTTATTC	GCTTATCATAGCTGCCCGTAA	DNA transcription activator
FliH	TCGTCTGTCTCAGGTGGAA	CTTGCCACGGCAATTCATTAG	DNA transcription activator
sdiA	GCCCAGCGTTTCGGATTA	GTAAACGAGGAGCAGCGTAAA	DNA transcription, quorum sensing
trpA	CAGCCATTGTCAAGATTATCG	CTGAGACAAAGGACCTGAG	Tryptophan biosynthesis
argC	CAGTTTCTGTGAAGTGAGTTT	CAGATGGACGGCGATTTC	Arginine biosynthesis
dppA	ATCAAAGCCGTTTATCAG	TTAATATCGTCGTTGTAGC	ABC transporter system
artP	GTCTTTCAGCAATATAATCTTTGG	TCTTTTGTCAGACCCAGTA	ABC transporter system
livJ	CAACGGCGGCAAAGTATA	CGTACTGCTGCTTATCGT	ABC transporter system
16 s rRNA	CAGAAGAAGCACCGGCTAAC	AATGCAGTTCCCAGGTTGAG	Endogenous control

Additional file

Additional file 1: Antibacterial activity of SWCNTs-Ag and pSWCNTs-Ag against gram-positive and gram- negative bacteria.

Abbreviations
AgNPs: Silver nanoparticles; AgCNTs: Silver coated CNTs; ANOVA: Analysis of variance; ATR: Attenuated total reflectance; CNTs: Carbon nanotubes; ELISA: Enzyme-linked immunosorbent assay; EM: Electron microscopy; FT-IR: Fourier transform infrared spectroscopy; IR: Infrared; MIC: Minimum inhibitory concentrations; MTT: 3-(4,5-dimethylthiazol-2yl)-2,5-diphenyltetrazolium bromide; NP: Nanoparticles; OD: Optical density; PEG: Polyethylene glycol; PL-PEG: Phospholipid-poly(ethylene glycol); SWCNTs-Ag: Silver coated single walled carbon nanotubes; pSWCNTs-Ag: Pegylated SWCNTS-Ag; qRT-PCR: Quantitative reverse transcriptase polymerase chain reaction; SEM: Scanning electron microscopy; TEM: Transmission electron microscopy.

Competing interests
The authors declare that they have no competing interests.

Authors' contributions
AC and SP conceived the concept. AC conducted most of the experiments, and SJ, ED, MM and RA, conducted part of the experiments. AC, SP, SS and RA, analyzed the data and co-wrote the paper. The manuscript was written through contributions of all authors. All authors have given approval to the final version of the manuscript.

Acknowledgements
This research was supported by grants from the National Science Foundation-CREST (HRD-1241701), NSF-HBCU-UP (HRD-1135863) and National Institutes of Health-MBRS-RISE (1R25GM106995-01).

Author details
[1]Center for Nanobiotechnology Research, Alabama State University, Montgomery, AL, USA. [2]Research Instrumentation Facility, Auburn University, Auburn, AL, USA. [3]Department of Drug Discovery and Development, Auburn University, Auburn, AL, USA.

References
1. Durso LM, Cook KL. Impacts of antibiotic use in agriculture: what are the benefits and risks? Curr Opin Microbiol. 2014;19C:37–44.
2. Pelgrift RY, Friedman AJ. Nanotechnology as a therapeutic tool to combat microbial resistance. Adv Drug Deliv Rev. 2013;65:1803–15.
3. Chatterjee S, Bandyopadhyay A, Sarkar K. Effect of iron oxide and gold nanoparticles on bacterial growth leading towards biological application. J Nanobiotechnol. 2011;9:34.
4. Huh AJ, Kwon YJ. "Nanoantibiotics": a new paradigm for treating infectious diseases using nanomaterials in the antibiotics resistant era. J Control Release. 2011;156:128–45.
5. Dallas P, Sharma VK, Zboril R. Silver polymeric nanocomposites as advanced antimicrobial agents: classification, synthetic paths, applications, and perspectives. Adv Colloid Interface Sci. 2011;166:119–35.
6. Guzman M, Dille J, Godet S. Synthesis and antibacterial activity of silver nanoparticles against gram-positive and gram-negative bacteria. Nanomedicine. 2012;8:37–45.
7. Kurek A, Grudniak AM, Kraczkiewicz-Dowjat A, Wolska KI. New antibacterial therapeutics and strategies. Pol J Microbiol. 2011;60:3–12.
8. Rangari VK, Mohammad GM, Jeelani S, Hundley A, Vig K, Singh SR, et al. Synthesis of Ag/CNT hybrid nanoparticles and fabrication of their nylon-6 polymer nanocomposite fibers for antimicrobial applications. Nanotechnology. 2010;21:095102.
9. Su R, Jin Y, Liu Y, Tong M, Kim H. Bactericidal activity of Ag-doped multi-walled carbon nanotubes and the effects of extracellular polymeric substances and natural organic matter. Colloids Surf B Biointerfaces. 2013;104:133–9.
10. Brahmachari S, Duttaa S, Das PK. Biotinylated amphiphile-single walled carbon nanotube conjugate for target-specific delivery to cancer cells. J Mater Chem B. 2014;2:1160–73.
11. Jung JH, Hwang GB, Lee JE, Bae GN. Preparation of airborne Ag/CNT hybrid nanoparticles using an aerosol process and their application to antimicrobial air filtration. Langmuir. 2011;27:10256–64.
12. Li Z, Fan L, Zhang T, Li K. Facile synthesis of Ag nanoparticles supported on MWCNTs with favorable stability and their bactericidal properties. J Hazard Mater. 2011;187:466–72.
13. Brahmachari S, Mandal SK, Das PK. Fabrication of SWCNT-Ag nanoparticle hybrid included self-assemblies for antibacterial applications. PLoS One. 2014;9:e106775.
14. Brahmachari S, Das D, Shome A, Das PK. Single-walled nanotube/amphiphile hybrids for efficacious protein delivery: rational modification of dispersing agents. Angew Chem Int Ed Engl. 2011;50:11243–7.
15. Ahamed M, Alsalhi MS, Siddiqui MK. Silver nanoparticle applications and human health. Clin Chim Acta. 2010;411:1841–8.
16. Faedmaleki F, Shirazi F, Salarian AA, Ahmadi Ashtiani H, Rastegar H. Toxicity effect of silver nanoparticles on mice liver primary cell culture and hepg2 cell line. Iran J Pharm Res. 2014;13:235–42.
17. Kovvuru P, Mancilla PE, Shirode AB, Murray TM, Begley TJ, Reliene R. Oral ingestion of silver nanoparticles induces genomic instability and DNA damage in multiple tissues. Nanotoxicology. 2014. Epub ahead of print.
18. Park EJ, Yi J, Kim Y, Choi K, Park K. Silver nanoparticles induce cytotoxicity by a Trojan-horse type mechanism. Toxicol In Vitro. 2010;24:872–8.
19. Park S, Lee YK, Jung M, Kim KH, Chung N, Ahn EK, et al. Cellular toxicity of various inhalable metal nanoparticles on human alveolar epithelial cells. Inhal Toxicol. 2007;19:59–65.
20. Jain S, Singh SR, Pillai S. Toxicity issues related to biomedical applications of carbon nanotubes. J Nanomed Nanotechol. 2012;3:5.
21. Vardharajula S, Ali SZ, Tiwari PM, Eroglu E, Vig K, Dennis VA, et al. Functionalized carbon nanotubes: biomedical applications. Int J Nanomedicine. 2012;7:5361–74.
22. Dumortier H, Lacotte S, Pastorin G, Marega R, Wu W, Bonifazi D, et al. Functionalized carbon nanotubes are non-cytotoxic and preserve the functionality of primary immune cells. Nano Lett. 2006;6:1522–8.
23. Haberl N, Hirn S, Wenk A, Diendorf J, Epple M, Johnston BD, et al. Cytotoxic and proinflammatory effects of PVP-coated silver nanoparticles after intratracheal instillation in rats. Beilstein J Nanotechnol. 2013;4:933–40.
24. Schipper ML, Nakayama-Ratchford N, Davis CR, Kam NW, Chu P, Liu Z, et al. A pilot toxicology study of single-walled carbon nanotubes in a small sample of mice. Nat Nanotechnol. 2008;3:216–21.
25. Tejamaya M, Romer I, Merrifield RC, Lead JR. Stability of citrate, PVP, and PEG coated silver nanoparticles in ecotoxicology media. Environ Sci Technol. 2012;46:7011–7.
26. Thorley AJ, Tetley TD. New perspectives in nanomedicine. Pharmacol Ther. 2013;140:176–85.
27. Liu Z, Tabakman SM, Chen Z, Dai H. Preparation of carbon nanotube bioconjugates for biomedical applications. Nat Protoc. 2009;4:1372–82.
28. Chou CC, Hsiao HY, Hong QS, Chen CH, Peng YW, Chen HW, et al. Single-walled carbon nanotubes can induce pulmonary injury in mouse model. Nano Lett. 2008;8:437–45.
29. Chen M, Yang Z, Wu H, Pan X, Xie X, Wu C. Antimicrobial activity and the mechanism of silver nanoparticle thermosensitive gel. Int J Nanomedicine. 2011;6:2873–7.
30. Chen H, Wang B, Gao D, Guan M, Zheng L, Ouyang H, et al. Broad-spectrum antibacterial activity of carbon nanotubes to human gut bacteria. Small. 2013;9:2735–46.
31. Kang S, Herzberg M, Rodrigues DF, Elimelech M. Antibacterial effects of carbon nanotubes: size does matter! Langmuir. 2008;24:6409–13.
32. Szostak MP, Hensel A, Eko FO, Klein R, Auer T, Mader H, et al. Bacterial ghosts: non-living candidate vaccines. J Biotechnol. 1996;44:161–70.
33. Stukalov O, Korenevsky A, Beveridge TJ, Dutcher JR. Use of atomic force microscopy and transmission electron microscopy for correlative studies of bacterial capsules. Appl Environ Microbiol. 2008;74:5457–65.
34. Pawley J. The development of field-emission scanning electron microscopy for imaging biological surfaces. Scanning. 1997;19:324–36.
35. Blair JM, Richmond GE, Bailey AM, Ivens A, Piddock LJ. Choice of bacterial growth medium alters the transcriptome and phenotype of *Salmonella enterica* Serovar Typhimurium. PLoS One. 2013;8:e63912.

36. Gong H, Vu GP, Bai Y, Chan E, Wu R, Yang E, et al. A *Salmonella* small non-coding RNA facilitates bacterial invasion and intracellular replication by modulating the expression of virulence factors. PLoS Pathog. 2011;7:e1002120.

37. Davidson AL, Dassa E, Orelle C, Chen J. Structure, function, and evolution of bacterial ATP-binding cassette systems. Microbiol Mol Biol Rev. 2008;72:317–64.

38. Giaouris E, Samoilis G, Chorianopoulos N, Ercolini D, Nychas GJ. Differential protein expression patterns between planktonic and biofilm cells of *Salmonella enterica* serovar Enteritidis PT4 on stainless steel surface. Int J Food Microbiol. 2013;162:105–13.

39. Wissenbach U, Six S, Bongaerts J, Ternes D, Steinwachs S, Unden G. A third periplasmic transport system for L-arginine in *Escherichia coli*: molecular characterization of the artPIQMJ genes, arginine binding and transport. Mol Microbiol. 1995;17:675–86.

40. Matsubara K, Ohnishi K, Sadanari H, Yamada R, Fukuda S. A portion of the nucleotide sequence corresponding to the N-terminal coding region of *livJ* is essential for its transcriptional regulation. Biochim Biophys Acta. 2000;1494:54–62.

41. Guyard C, Raffel SJ, Schrumpf ME, Dahlstrom E, Sturdevant D, Ricklefs SM, et al. Periplasmic flagellar export apparatus protein, FliH, is involved in post-transcriptional regulation of *FlaB*, motility and virulence of the relapsing fever spirochete *Borrelia hermsii*. PLoS One. 2013;8:e72550.

42. Espinosa E, Casadesus J. Regulation of *Salmonella enterica* pathogenicity island 1 (SPI-1) by the LysR-type regulator LeuO. Mol Microbiol. 2014;91:1057–69.

43. Smith JL, Fratamico PM, Yan X. Eavesdropping by bacteria: the role of SdiA in *Escherichia coli* and *Salmonella enterica* serovar Typhimurium quorum sensing. Foodborne Pathog Dis. 2011;8:169–78.

44. Marcus SL, Brumell JH, Pfeifer CG, Finlay BB. *Salmonella* pathogenicity islands: big virulence in small packages. Microbes Infect. 2000;2:145–56.

45. Zurawski DV, Stein MA. The SPI2-encoded SseA chaperone has discrete domains required for SseB stabilization and export, and binds within the C-terminus of SseB and SseD. Microbiology. 2004;150:2055–68.

46. Gnanendra S, Mohamed S, Natarajan J. Identification of potent inhibitors for *Salmonella* typhimurium quorum sensing via virtual screening and pharmacophore modeling. Comb Chem High Throughput Screen. 2013;16:826–39.

47. Bartlett GR. Phosphorus assay in column chromatography. J Biol Chem. 1959;234:466–8.

48. Eroglu E, Tiwari PM, Waffo AB, Miller ME, Vig K, Dennis VA, et al. A nonviral pHEMA + chitosan nanosphere-mediated high-efficiency gene delivery system. Int J Nanomedicine. 2013;8:1403–15.

49. Seo JS, Moon MH, Jeong JK, Seol JW, Lee YJ, Park BH, et al. SIRT1, a histone deacetylase, regulates prion protein-induced neuronal cell death. Neurobiol Aging. 2012;33:1110–20.

50. Palomino JC, Martin A, Camacho M, Guerra H, Swings J, Portaels F. Resazurin microtiter assay plate: simple and inexpensive method for detection of drug resistance in *Mycobacterium tuberculosis*. Antimicrob Agents Chemother. 2002;46:2720–2.

51. Soehnlen MK, Kunze ME, Karunathilake KE, Henwood BM, Kariyawasam S, Wolfgang DR, et al. *In vitro* antimicrobial inhibition of *Mycoplasma bovis* isolates submitted to the Pennsylvania Animal Diagnostic Laboratory using flow cytometry and a broth microdilution method. J Vet Diagn Invest. 2011;23:547–51.

52. Bauer AW, Kirby WM, Sherris JC, Turck M. Antibiotic susceptibility testing by a standardized single disk method. Am J Clin Pathol. 1966;45:493–6.

53. Schmittgen TD, Livak KJ. Analyzing real-time PCR data by the comparative C (T) method. Nat Protoc. 2008;3:1101–8.

15

Bioactive magnetic near Infra-Red fluorescent core-shell iron oxide/human serum albumin nanoparticles for controlled release of growth factors for augmentation of human mesenchymal stem cell growth and differentiation

Itay Levy[1†], Ifat Sher[2†], Enav Corem-Salkmon[1], Ofra Ziv-Polat[1], Amilia Meir[3], Avraham J Treves[3], Arnon Nagler[4], Ofra Kalter-Leibovici[5], Shlomo Margel[1] and Ygal Rotenstreich[2*]

Abstract

Background: Iron oxide (IO) nanoparticles (NPs) of sizes less than 50 nm are considered to be non-toxic, biodegradable and superparamagnetic. We have previously described the generation of IO NPs coated with Human Serum Albumin (HSA). HSA coating onto the IO NPs enables conjugation of the IO/HSA NPs to various biomolecules including proteins. Here we describe the preparation and characterization of narrow size distribution core-shell NIR fluorescent IO/HSA magnetic NPs conjugated covalently to Fibroblast Growth Factor 2 (FGF2) for biomedical applications. We examined the biological activity of the conjugated FGF2 on human bone marrow mesenchymal stem cells (hBM-MSCs). These multipotent cells can differentiate into bone, cartilage, hepatic, endothelial and neuronal cells and are being studied in clinical trials for treatment of various diseases. FGF2 enhances the proliferation of hBM-MSCs and promotes their differentiation toward neuronal, adipogenic and osteogenic lineages in vitro.

Results: The NPs were characterized by transmission electron microscopy, dynamic light scattering, ultraviolet–visible spectroscopy and fluorescence spectroscopy. Covalent conjugation of the FGF2 to the IO/HSA NPs significantly stabilized this growth factor against various enzymes and inhibitors existing in serum and in tissue cultures. IO/HSA NPs conjugated to FGF2 were internalized into hBM-MSCs via endocytosis as confirmed by flow cytometry analysis and Prussian Blue staining. Conjugated FGF2 enhanced the proliferation and clonal expansion capacity of hBM-MSCs, as well as their adipogenic and osteogenic differentiation to a higher extent compared with the free growth factor. Free and conjugated FGF2 promoted the expression of neuronal marker Microtubule-Associated Protein 2 (MAP2) to a similar extent, but conjugated FGF2 was more effective than free FGF2 in promoting the expression of astrocyte marker Glial Fibrillary Acidic Protein (GFAP) in these cells.

Conclusions: These results indicate that stabilization of FGF2 by conjugating the IO/HSA NPs can enhance the biological efficacy of FGF2 and its ability to promote hBM-MSC cell proliferation and trilineage differentiation. This new system may benefit future therapeutic use of hBM-MSCs.

Keywords: IO/HSA NPs, FGF2, BM-MSCs

* Correspondence: Ygal.Rotenstreich@sheba.health.gov.il
†Equal contributors
2Goldschleger Eye Institute, Sackler Faculty of Medicine, Tel Aviv University, Sheba Medical Center, Tel-Hashomer 52621, Israel
Full list of author information is available at the end of the article

Background

Magnetic nanoparticles (NPs) which are known for their very large surface area and magnetic properties (size up to 0.1 μm) have a wide range of potential applications such as drug delivery, MRI, diagnostics, hyperthermia, specific cell labeling and separation, cell tracking and bio-catalysis [1-6]. Iron oxide (IO) NPs of sizes less than approximately 50 nm are superparamagnetic (possess magnetic properties when they are exposed to external magnetic field, and lose their magnetic properties when the magnetic field is removed), allowing therefore the separation of these NPs by using high gradient magnetic columns. IO NPs are also known for their non-toxicity and biodegradability, therefore ideal for biomedical applications [7,8]. Previous studies showed that it is also possible to mark the IO NPs with a fluorescent dye, e.g., near IR (NIR) dye, which further improves the probe capabilities [9].

NIR light of 700 to 1000 nm, achieves the highest tissue penetration due to minimal absorbency of the surface tissue in this spectral region. In vivo fluorescence imaging has experienced substantial growth with the "opening" of the NIR "window" because of the development of novel NIR fluorescence probes and optical imaging instruments [10-12]. In previous studies we described the generation of IO NPs coated with Human Serum Albumin (HSA) [13]. HSA exhibits an average blood half-life of 19 days and is emerging as a versatile protein carrier for drug targeting and improving the pharmacokinetic profile of peptide or protein-based drugs [14]. These properties combined with lack of toxicity, easy availability, biodegradability and preferential uptake in tumor and inflamed tissues make the core-shell IO/HSA NPs an ideal candidate for drug targeting and delivery. In addition, another important property of the HSA coating onto the IO NPs is that its various functional groups, e.g., carboxylates, amines, hydroxyls and thiols, can easily be used through different activation methods for conjugation of the IO/HSA NPs to various biomolecules such as proteins, amino acids, antibodies, oligonucleotides, etc. [15,16]. Conjugation of proteins to IO/HSA NPs is predicted to reduce their susceptibility to chemical, enzymatic and thermal degradation, thus enhancing the protein biological efficacy [17-20]. Furthermore, it may provide a mean for sustained release of the conjugated proteins.

Bone marrow mesenchymal stem cells (BM-MSCs) are multipotent cells that can differentiate into mesenchymal and non-mesenchymal lineages. They can give rise to osteogenic, chondrogenic, adipogenic, myoegenic, hepatogenic, endothelial and neurogenic cells both *in vitro* and *in vivo* [21-26]. BM-MSCs secrete trophic factors that can promote the survival of damaged cells, as well as immuno-modulatory cytokines that can suppress T-cell proliferation

and function [27-31]. Because of their good proliferation, differentiation and paracrine potential, as well as their relative ease of isolation and low immunogenicity, BM-MSCs have become a main source for tissue engineering of bone, cartilage, muscle, marrow stroma, tendon, fat, and other connective tissues [32-34]. Furthermore, we and others have shown that hBM-MSC transplantation has the potential to ameliorate the symptoms of various neurodegenerative diseases, including retinal degeneration, Alzheimer's disease, Parkinson, familial amyotrophic lateral sclerosis and multiple sclerosis [29,35-37] as well as other disease such as acute liver failure [38] and pulmonary emphysema [39]. These and other successful animal studies have led to numerous clinical trials using hBM-MSC as a source for cellular therapy for treatment of heart, liver, bone and cartilage repair, foot ulcers, spinal cord injuries, peripheral nerve injuries and acute graft-versus-host disease [40-46]. Since mesenchymal stem cells comprise only 0.001-0.01% of the bone mononuclear cells, extensive *in vitro* expansion is required to obtain sufficient number of cells for clinical use [47]. Although the cells have high proliferation potential, prolonged culture expansion may reduce the cell differentiation potential. In addition, proliferation and differentiation potential varies between donors [48]. Hence enhancing cell proliferation and differentiation potential could improve their yields for clinical applications.

In addition, following transplantation of hBM-MSc there is a need to repeatedly monitor the cells in vivo in a non-invasive manner. This cannot be achieved using histological and immunohistochemical techniques that require tissue removal. We have previously shown that pre-labeling of mesenchymal stem cells with IO NPs enables noninvasive *in-vivo* tracking following cell transplantation using Magnetic Resonance Imaging (MRI, [49]).

Several studies have demonstrated that supplementation of basic FGF (also known as FGF2) to BM-MSC culture medium increases cell proliferation rate and cell differentiation [50,51]. However, as the cells are cultivated at 37 degrees, rapid enzymatic degradation and protein denaturation leads to short time life of FGF2 of about 3–10 minutes and reduces its biological activity and functions [52,53]. In a previous study we showed that conjugation of FGF2 to IO/HSA NPs stabilized the factor and significantly improved its ability to promote rat nasal olfactory mucosa cell migration, growth and differentiation [54]. The present article describes a method of preparing FGF2-conjugated IO/HSA NIR fluorescent core-shell NPs that significantly stabilized the FGF2 through its covalent conjugation to the nanoparticle's surface [55,56]. We also show that FGF2 conjugated to IO/HSA NPs is internalized by hBM-MSCs and promotes the growth and trilineage (neuronal, bone, fat) differentiation capacity of the cells at a higher extent compared with the free FGF.

Results and discussion

The NIR fluorescent IO/HSA NPs were prepared by nucleation followed by stepwise growth of IO thin films onto the gelatin/IO nuclei as described in the "Methods" section.

Nanoparticles' characterization: dry and hydrodynamic size and size distribution

IO core NPs and IO/HSA core-shell NPs were both diluted with H_2O to a concentration of 1 mg/ml and dried over a TEM grid. TEM measurements (Figure 1A,B) indicate that the size and size distribution of the core and core/shell NPs are 17 ± 1 nm and 21 ± 3 nm, respectively. Samples of IO and IO/HSA were dispersed in H_2O and their hydrodynamic diameters were determined (using DLS) to be 103 ± 14 nm and 43 ± 5 nm, respectively (Figure 1C). These hydrodynamic measurements demonstrate that the albumin coating decreased the hydrodynamic size of the IO NPs, as clearly shown in Figure 1C. These size differences between TEM and DLS are attributed to the fact that TEM measures the dry diameter, while DLS determines the hydrodynamic

diameter, which takes the hydrated layers on the particle surface into account. In addition, the difference in the hydrodynamic size of the IO core NPs and the HSA/IO core-shell NPs may indicate that the heat denatured albumin coating is more hydrophobic than the core IO NPs.

Fluorescence spectroscopy

The excitation and emission spectrum of the Cy7-IO/HSA NPs and the free Cy7 in PBS are shown in Figure 2. The maximum fluorescence excitation of the Cy7-IO/HSA NPs and free Cy7 occurs at approximately 769 and 749 nm, respectively. The maximum fluorescence emission intensity of the Cy7-IO/HSA NPs and free Cy7 occurs at approximately 780 and 766 nm, respectively. The red-shift in the absorbance spectrum of the NIR fluorescent IO/HSA NPs compared with the free Cy7 dye is probably due to its binding to the gelatin within the IO core NPs that affects the dipole moment of the dye [20].

Photobleaching stabilization

To study the fluorescence stability of the Cy7 encapsulated NPs, a photobleaching experiment was performed

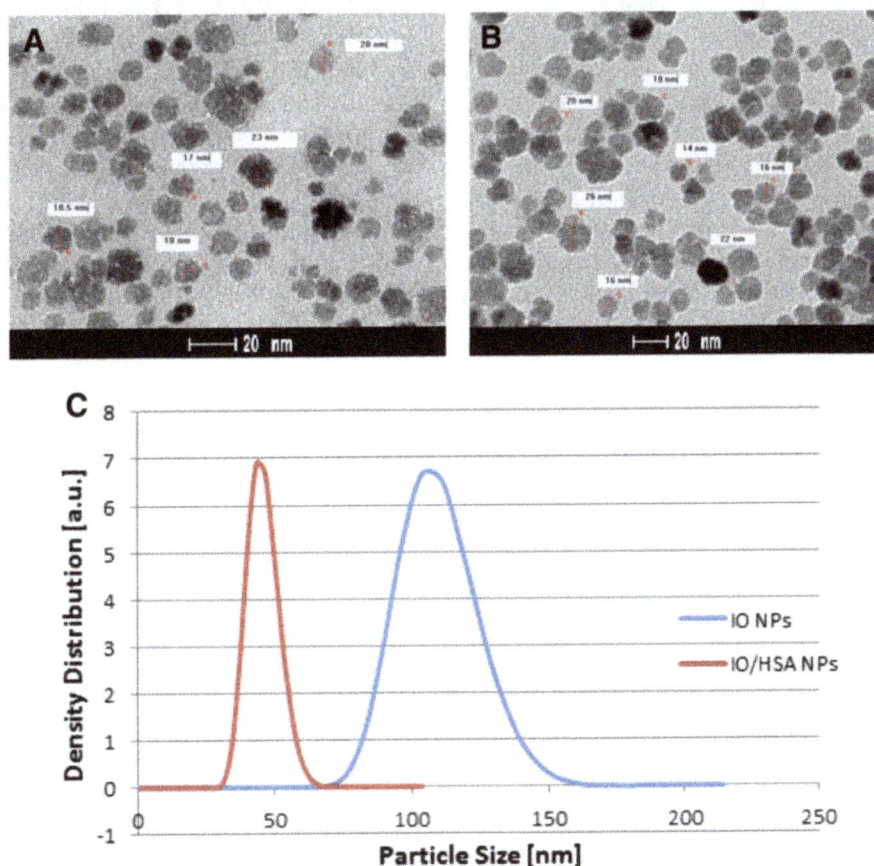

Figure 1 Dry and Hydrodynamic Size and Size Distribution of IO Nanoparticles. TEM images of the dry core IO (**A**) and the core-shell IO/HSA (**B**) NPs; (**C**) Hydrodynamic size and size distribution of the core IO & the core-shell IO/HSA NPs dispersed in an aqueous phase.

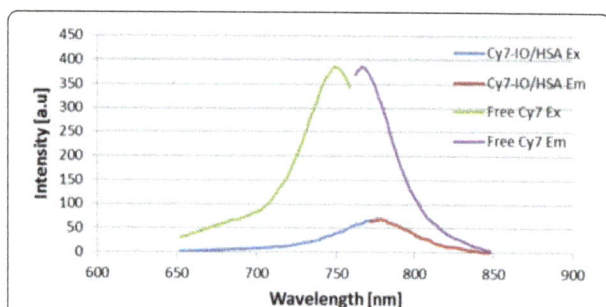

Figure 2 Excitation and emission of Free Cy7 and Cy7-IO/HSA NPs. The excitation and emission spectrum of Cy7-IO/HSA NPs and the free Cy7 in PBS at a final concentration of 250 ng/ml were determined using spectrofluorometer.

for free Cy7 dye and Cy7-IO/HSA, as described in the literature [14]. Both samples were illuminated at 800 nm, and their fluorescence intensities were measured, 5 cycles of 20 min each with 10 min recovery between cycles. The fluorescence intensity of the NIR fluorescent IO/HSA NPs decreased by 5% after the first cycle (t = 20 min) and by 15% after all 5 cycles (t = 140 min), while the fluorescence of free Cy7 decreased by 38% after the first cycle (t = 20 min) and by 87% after all cycles (t = 140 min), as shown in Figure 3.

The irreversible light-induced destruction of the fluorophore also known as photobleaching is affected by factors such as temperature, exposure time, oxygen, oxidizing or reducing agents and illumination levels [56]. The encapsulation of Cy7 within the NPs significantly reduced the photobleaching as demonstrated in Figure 3. Encapsulation of the dye probably protects the dye against reactive oxygen species, thereby reducing photobleaching [55,57]. Previous work in our lab showed similar results with the dye RITC conjugated to NPs [58].

Long term stability of free versus conjugated neurotrophic factors

FGF2 was chosen to serve as a model for neurotrophic factors. The stability of the free and conjugated FGF2 against various enzymes and inhibitors existing in serum and in tissue culture was examined. The stability was tested in various concentrations of serum. Figure 4A and B indicates that the concentrations of the free and the conjugated-FGF2 decreased with time and with increasing concentration of the serum. However, the concentration of the free FGF2 decreased significantly more rapidly than that of the conjugated factor (p < 0.004). For example, the residual concentration of the conjugated factor following incubation for one day in medium containing 40, 80 and 100% serum was 101 ± 4.5, 75 ± 6.2 and $51 \pm 4.2\%$ from the initial concentrations, respectively, while the residual concentrations of the free factor were only 62 ± 2.5, 23 ± 3.5 and $9.0 \pm 0.8\%$, respectively (Figure 4A). Examination of the residual concentrations of FGF2 remaining after incubation for one week in a medium containing 20, 40 and 60% of serum, demonstrated a dose–response relationship wherein increasing serum concentration resulted in reduced concentration of FGF2. Thus, the concentration of free factor was reduced to 29 ± 2.4, 5 ± 0.5 and 0% from the initial concentration, respectively. By contrast the concentration of the conjugated-FGF2 was significantly higher (66.9 ± 3.0, 49.1 ± 2.4 and $20.2 \pm 0.9\%$, respectively, Figure 4B). These results indicate that the conjugated-FGF2 is significantly more stable in serum than the free factor.

Conjugated-FGF2 promotes hBM-MSC cell expansion

To examine the effect of conjugated FGF2 on hBM-MSC expansion, the cells were subcultured for 3 passages in growth media supplemented with 0.1 ng/ml free or

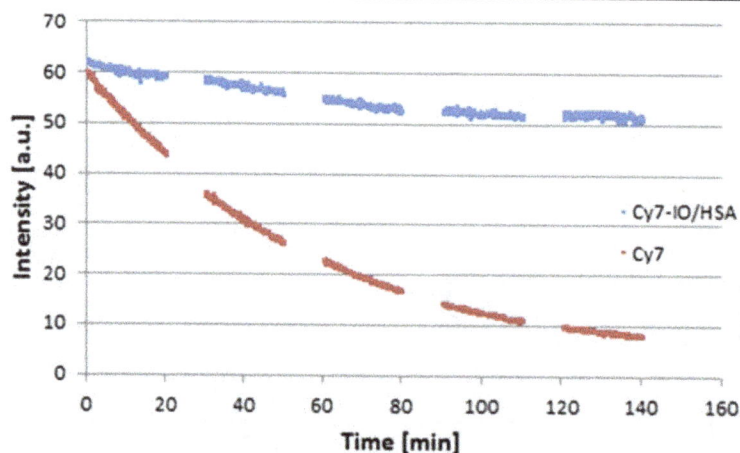

Figure 3 Photobleaching Stabilization of Cy7 by encapsulation. Fluorescence intensity as function of time of free Cy7 and Cy7-IO/HSA NPs following illuminated at 800 nm was measured using spectrofluorometer.

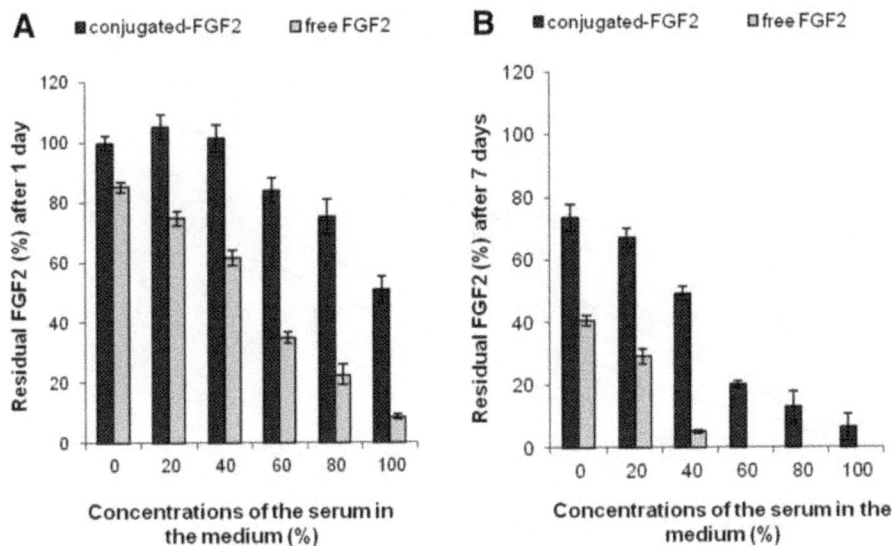

Figure 4 Stability of free versus conjugated-FGF2. Ten ng/ml of free or conjugated FGF2 were incubated in various concentrations of fetal calf serum (0–100 %) in the medium at 37°C according to the experimental section. The residual concentrations of the growth factors following 1 and 7 days of incubation are shown in **A** and **B**, respectively.

conjugated FGF2. As shown in previous studies [50,59], cells grown in the presence of free FGF2 demonstrated higher proliferation rates compared to control cultures (Figure 5A). Moreover, the population doubling (PD) time for hBM-MSCs expanded in the presence of conjugated FGF2 was consistently shorter than that of cells expanded in the presence of free FGF2 or control conditions at all passages (Figure 5A). Following 3 passages in the presence of conjugated FGF2, the cells reached on average 7 ± 0.14 population doublings. By contrast cells cultivated with free FGF2, non-conjugated NPs or control conditions, reached only 5.2 ± 0.3, 3.5 ± 0.3 or 4 ± 0.5 PDs, respectively (Figure 5A). The yield of cells obtained from expanding the cells in the presence of conjugated FGF2 was over 3 fold higher than that of free FGF2 and nearly 8 fold higher than the yield obtained from cells grown under control conditions (Additional file 1: Figure S1). Immunofluorescence analysis using antibody directed against the proliferating cell nuclear antigen (PCNA), demonstrated that all cells under all growth conditions were positive for this proliferation marker (data not shown). Trypan blue staining was performed in every subculturing. No dead cells were identified in any of treatments in any of passages or donors. As was previously demonstrated by others for free FGF2 [60,61], hBM-MSCs cultured in the presence of FGF2 conjugated to IO NPs were smaller compared with cells cultured under control conditions or in the presence of non-conjugated NPs (Figure 5B-F). Cell surface antigen phenotyping clearly demonstrated that hBM-MSCs cultivated for 3 passages in the presence of NP-FGF2

maintained the expression of mesenchymal cell markers CD73, CD90, CD105 and were negative for hematopoietic markers CD14, CD34 and CD45 (Additional file 2: Figure S2). Furthermore the cells maintained low levels of expression of HLA-DR, suggesting that expansion in the presences of 0.1 ng/ml conjugated FGF2 will not increase their immunogenicity in vivo. Taken together our findings suggest that expansion of hBM-MSCs in the presence of 0.1 ng/ml conjugated FGF2 substantially increases cell yields with no adverse effects on expression of mesenchymal surface markers.

Uptake of Nanoparticles by hBM-MSCs

To examine the uptake of conjugated FGF2, hBM-MSCs were incubated with increasing concentration of conjugated-FGF2 and analyzed by flow cytometry to detect the Cy7 fluorescence signal. Figure 6 demonstrates efficient uptake of the conjugated FGF2 by the cells following 48 h incubation with Cy7-IO/HSA-FGF2 NPs. Quantification of Cy7 positive cells revealed that 35.6% of the cells incubated with 10 ng/ml conjugated FGF2 (9 µg/ml Cy7-IO/HSA NPs) were Cy7 positive (Figure 6A). In cells cultured in the presence of 50 ng/ml conjugated FGF2 (45 µg/ml Cy7-IO/HSA NPs), 97% of the cells were Cy7-positive. Similar results were obtained in cells cultured with 100 ng/ml conjugated FGF2 (90 µg/ml Cy7-IO/HSA NPs), with 96% of the cells positive for Cy7 (Figure 6A). Only 26% or 28.4% of the cells were positive for Cy7 following incubation with 9 or 45 µg/ml free Cy7-IO/HSA NPs, respectively (Figure 6B and data not shown). However, when cells were cultured with 90 µg/ml Cy7-IO/

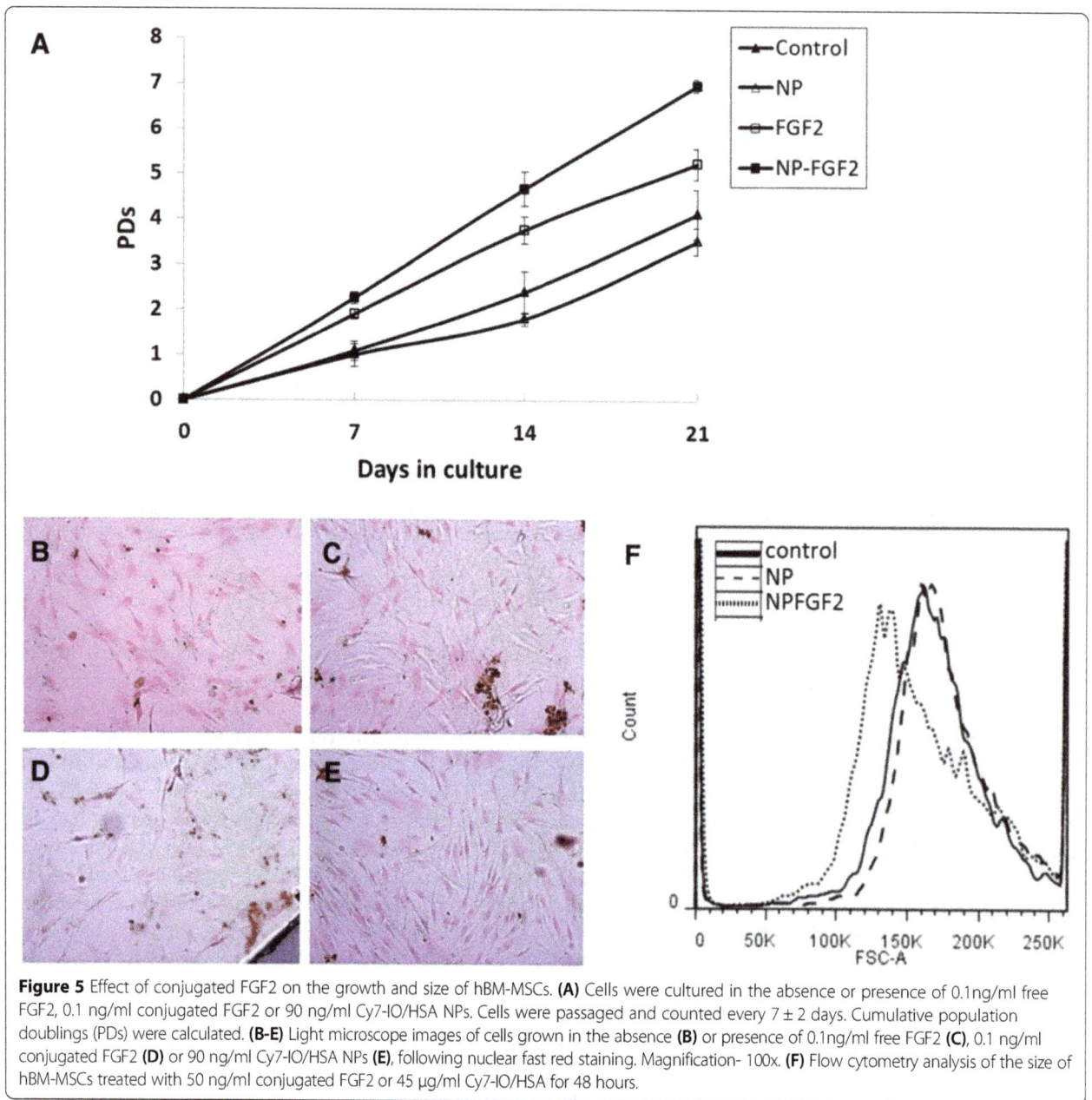

Figure 5 Effect of conjugated FGF2 on the growth and size of hBM-MSCs. **(A)** Cells were cultured in the absence or presence of 0.1ng/ml free FGF2, 0.1 ng/ml conjugated FGF2 or 90 ng/ml Cy7-IO/HSA NPs. Cells were passaged and counted every 7 ± 2 days. Cumulative population doublings (PDs) were calculated. **(B-E)** Light microscope images of cells grown in the absence **(B)** or presence of 0.1ng/ml free FGF2 **(C)**, 0.1 ng/ml conjugated FGF2 **(D)** or 90 ng/ml Cy7-IO/HSA NPs **(E)**, following nuclear fast red staining. Magnification- 100x. **(F)** Flow cytometry analysis of the size of hBM-MSCs treated with 50 ng/ml conjugated FGF2 or 45 μg/ml Cy7-IO/HSA for 48 hours.

HSA NPs, 90% of the cells were Cy7 positive, suggesting that at low concentrations the Cy7-IO/HSA internalization was mediated at least in part by endocytosis of FGF2, but at high concentrations the cells can internalize the NPs in a non-FGF2 dependent pathway.

The cellular uptake of these nanoparticles into the hBM-MSCs was also confirmed by Prussian Blue iron staining as shown in Figure 7. This figure demonstrates blue granules inside the cells incubated with 10–100 ng/ml FGF2 conjugated to Cy7-IO/HSA NPs for 48 h, indicating the accumulation of conjugated NPs in the cells. Increasing the concentration of conjugated FGF2 enhanced the amount of

positively stained cells. Thus, at 10 ng/ml conjugated FGF2, nearly 23% of the cells were positively stained with Prussian Blue (Figure 7C). By contrast, 100% of cells incubated with 50 or 100 ng/ml conjugated FGF2 were positively stained with this dye (Figure 7F,I).

Cells incubated with non-conjugated IO/HSA NPs were also positive for Prussian Blue staining, but to a lower extent compared with the FGF2-conjugated IO/HSA NPs (Figure 7B,E,H). This data strongly suggest that at concentration of 50 ng/ml and lower, the majority of FGF2-conjuagted NPs were most probably internalized by the cells via receptor-mediated endocytosis.

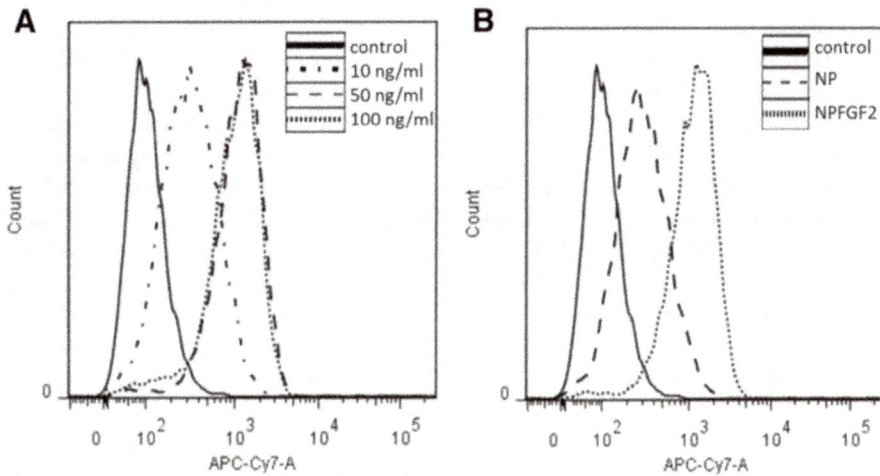

Figure 6 Determination of uptake of Cy7-IO/HSA NPs by hBM-MSCs using flow cytometry. Flow cytometry analysis of the uptake of Cy7-IO/HSA by hBM-MSCs non-treated (control) or treated with **(A)** FGF2 conjugated Cy7-IO/HSA in three FGF2 concentrations (10, 50 and 100 ng/ml) for 48h or with **(B)** 50 ng/ml FGF2-Cy7-IO/HSA or 45 µg/ml non-conjugated NPs.

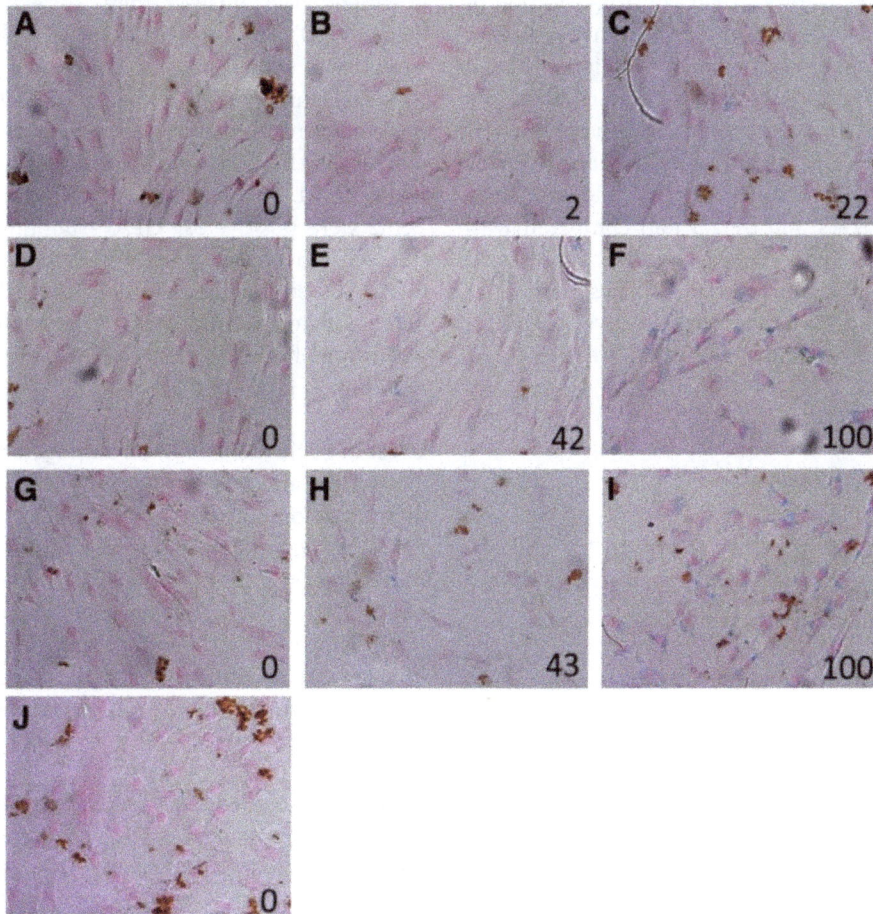

Figure 7 Determination of uptake of Cy7-IO/HSA NPs by hBM-MSCs using Prussian Blue iron staining. Cells were grown in the absence (control, **J**) or presence of 10 ng/ml **(A,C)**, 50 ng/ml **(D,F)** or 100 ng/ml **(G,I)** free FGF2 **(A, D, G)** or conjugated-FGF2 NPs **(C, F, I)** , or non-conjugated IO/HSA NPs at concentration of 9 µg/ml **(B)**, 45 µg/ml **(E)** or 90 µg/ml **(H)**. Following fixation in 4% PFA, cells were stained with Prussian Blue iron stain (blue) and counter stained with Nuclear Fast Red (red). The percentage of Prussian Blue positive cells was calculated from 3 microscopic fields and is indicated at the bottom of each picture. Magnification (×200).

Endocytosis of FGF2-conjugated NPs could be mediated by FGF receptor 1 (FGFR1) that is expressed in hBM-MSCs and mediates FGF2 internalization and signaling [62,63]. Our findings are supported by various studies that demonstrated that conjugating a variety of ligands to NP surfaces facilitates receptor-mediated endocytosis of the NPs [64]. TUNEL staining revealed that there were no apoptotic cells in cultures supplemented with 100 ng/ml conjugated or free FGF2, 90 µg/ml non-conjugated NPs or control cells, supporting the biocompatibility of the IO/HSA NPs (data not shown).

Effect of conjugated FGF2 on the clonal expansion capacity of hBM-MSCs

One of the characteristic features of hBM-MSCs is generation of colonies when plated at low densities and the efficacy of colony formation is indicative of the cell proliferation potential [65]. We compared the effect of supplementing the growth media with free or conjugated FGF2 on the clonal expansion capacity of the cells. Non-conjugated NPs were added as control. Addition of 45 µg/ml free IO/HSA NPs to the growth media had no significant effect on cloning efficiency (Figure 8 and Additional file 3: Figure S3), further demonstrating the biocompatibility of these nanoparticles. Supplementation of conjugated FGF2 significantly enhanced colony formation by the hBM-MSC by nearly 2 fold compared with free IO/HSA NPs and by 1.5 fold compared with free FGF2. The difference between the 3 supplements was statistically significant ($p < 0.001$). These data suggest that conjugated FGF2 is more effective in enhancing cloning efficiency of hBM-MSC than free FGF2.

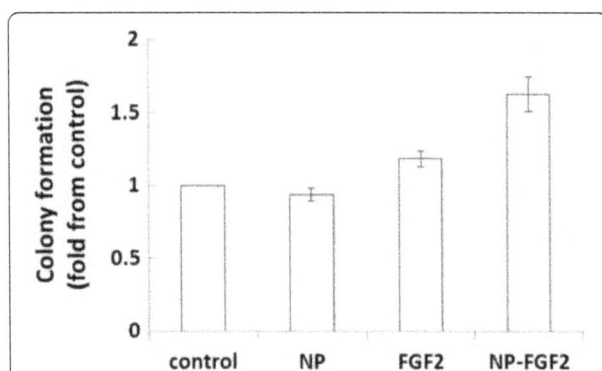

Figure 8 Enhanced clonal expansion capacity of hBM-MSCs in the presence of conjugated FGF2. Human BM-MSCs were seeded in 6 well plates and incubated in growth media alone (control) or growth media supplemented with 50 ng/ml free or conjugated FGF2, or free NPs. Media was change every 3 days. Seven days post seeding, colonies were counted following extensive washing and Giemsa staining. Data is presented as fold increase in colony number compared with control (mean ± SE of 3 experiments in duplicates or triplicates using cells from 3 different donors).

Effect of conjugated FGF2 on neurogenic differentiation

Human BM-MSCs are multipotent cells that can differentiate into a variety of cell types, including neuronal, bone and fat cells. We first examined the effect of conjugated FGF2 on neurogenic capacity of hBM-MSCs by monitoring cell morphology as well as the expression of two neuronal differentiation markers - Microtubule-Associated Protein 2 (MAP2), a marker of neuronal differentiation that is expressed exclusively during the neuronal differentiation of neural precursor cells, and the astrocyte marker Glial Fibrillary Acidic Protein (GFAP). As shown in Figure 9, cells incubated for 14 days in neuronal differentiation media with no supplementation of FGF2 (control) or in the presence of free IO/HSA NPs failed to present the morphological changes characteristic of neuronal cells or to express the MAP2 and GFAP markers (Figure 9 A,B,E,F,I,J). By contrast, cells supplemented with either free FGF2 or conjugated growth factor at 50ng/ml presented morphologic changes typical to neuronal cells, with bipolar morphology and elongated processes (arrows in bright field images point to elongated processes, Figure 9C,D). Furthermore, 100% of cells incubated for 2 weeks with either free or conjugated FGF2 expressed MAP2 (Figure 9G,H). Conjugated FGF2 was over 3 fold more efficient in inducing the expression of the astrocyte marker GFAP in these cells compared with the free FGF2 (Figure 9K,L). Thus, 11.4% of cells incubated with free FGF2 were positive for GFAP staining whereas 35% of the cells incubated with conjugated FGF2 expressed GFAP. Taken together our data suggest that conjugated FGF2 enhances hBM-MSC neurogenic potential and was more efficient than free FGF2 in promoting differentiation to glial cells.

Effect of conjugated FGF2 on adipogenic differentiation

To test the effect of conjugated FGF2 on the adipogenic differentiation potential of hBM-MSCs, cells were grown for 14 days in adipogenic media with or without supplementation of 3 ng/ml free FGF2 or conjugated FGF2, or free IO/HSA NPs. To evaluate adipogenesis, cultures were stained with Oil red that stains lipid droplets in adipocytes. As shown in Figure 10, conjugated FGF2 was over 2 fold more efficient in promoting adipogenic differentiation compared with free FGF2 ($p < 0.0001$).

Effect of conjugated FGF2 on Osteogenic differentiation

The effect of conjugated FGF2 on osteogenic differentiation was examined by treating the cells for 14 days in osteogenic induction media supplemented with 3 ng/ml or 10 ng/ml free FGF2 or conjugated FGF2 or free IO/HSA NPs, as control. Following fixation, cells were stained with Alizarin Red S solution to visually detect the presence of mineralization. Conjugated FGF2 at 3 ng/ml was 2 fold more efficient in promoting osteogenic differentiation

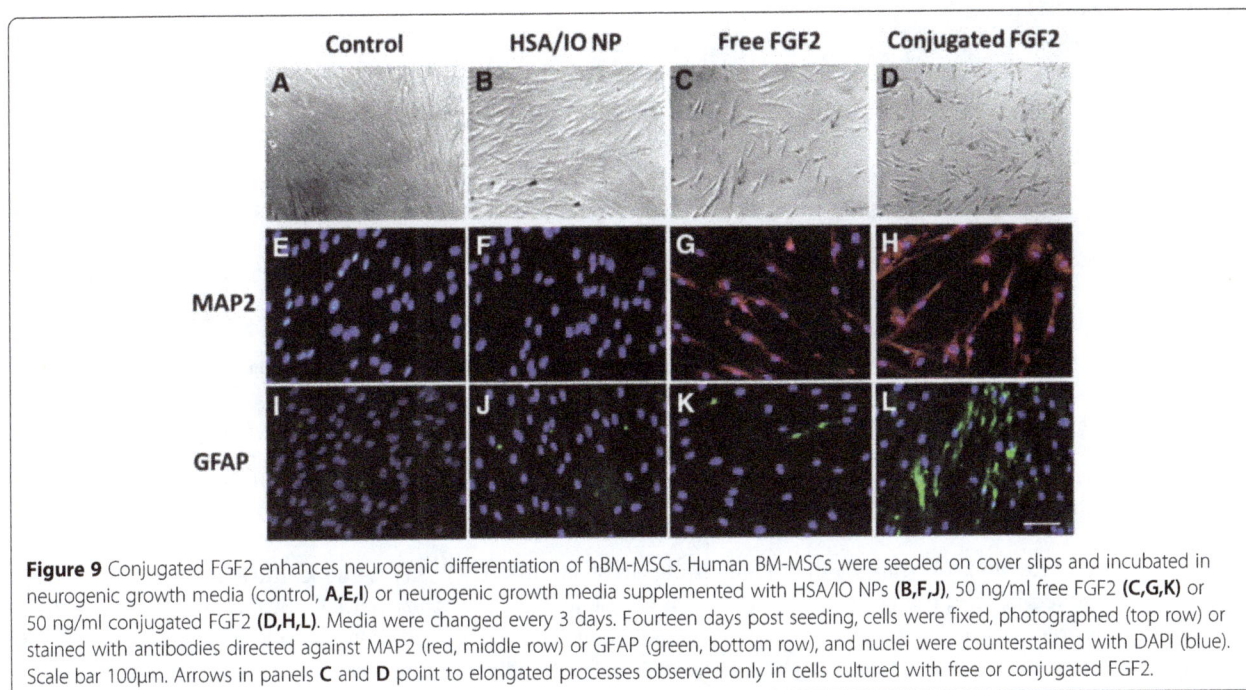

Figure 9 Conjugated FGF2 enhances neurogenic differentiation of hBM-MSCs. Human BM-MSCs were seeded on cover slips and incubated in neurogenic growth media (control, **A,E,I**) or neurogenic growth media supplemented with HSA/IO NPs (**B,F,J**), 50 ng/ml free FGF2 (**C,G,K**) or 50 ng/ml conjugated FGF2 (**D,H,L**). Media were changed every 3 days. Fourteen days post seeding, cells were fixed, photographed (top row) or stained with antibodies directed against MAP2 (red, middle row) or GFAP (green, bottom row), and nuclei were counterstained with DAPI (blue). Scale bar 100μm. Arrows in panels **C** and **D** point to elongated processes observed only in cells cultured with free or conjugated FGF2.

compared with free FGF2 at this concentration, and slightly more efficient than 10 ng/ml free FGF2 (Figure 11). Statistical analysis suggested a highly significant difference between the free and conjugated FGF2 treatments (p < 0.001) and between the different treatment concentrations (p = 0.025). Together, the effects of concentration and treatment explained a high proportion of the observed osteogenic differentiation (R squared = 0.864). There was no interaction between the concentration and treatment type parameters.

Conclusions

Taken together we have shown here that IO/HSA NPs are biocompatible and that FGF2-conjugated IO/HSA NPs significantly enhanced hBM-MSC growth and trilineage differentiation compared with the same concentration of free FGF2. Our findings suggest that these FGF-coupled NPs may possibly be used for expanding hBM-MSCs and enhancing their differentiation potential for future therapeutic use. As the cells endocytose the FGF2-IO/HSA NPs, it is very likely that these NPs will

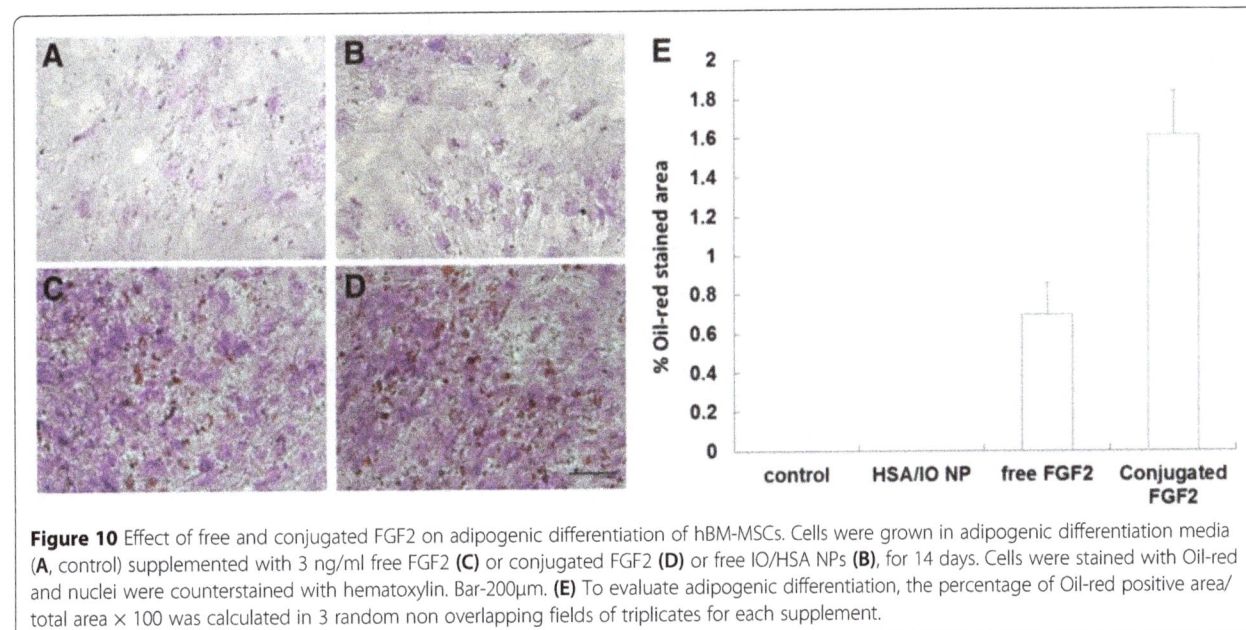

Figure 10 Effect of free and conjugated FGF2 on adipogenic differentiation of hBM-MSCs. Cells were grown in adipogenic differentiation media (**A**, control) supplemented with 3 ng/ml free FGF2 (**C**) or conjugated FGF2 (**D**) or free IO/HSA NPs (**B**), for 14 days. Cells were stained with Oil-red and nuclei were counterstained with hematoxylin. Bar-200μm. (**E**) To evaluate adipogenic differentiation, the percentage of Oil-red positive area/ total area × 100 was calculated in 3 random non overlapping fields of triplicates for each supplement.

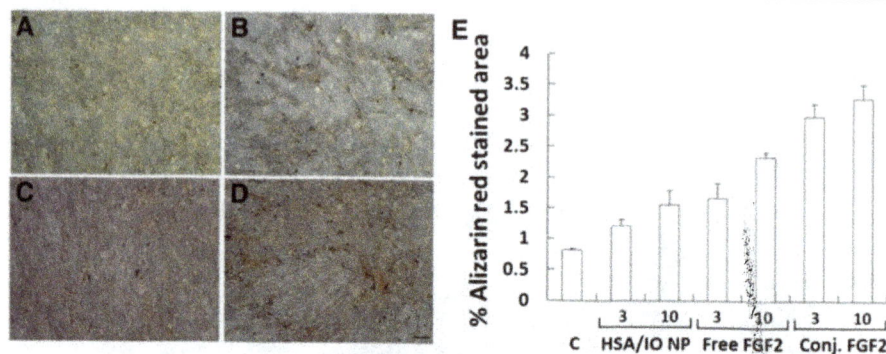

Figure 11 Effect of free and conjugated FGF2 on osteogenic differentiation of hBM-MSCs. Cells were grown in osteogenic differentiation media (**A**, control) supplemented with 10 ng/ml free FGF2 (**C**) or conjugated FGF2 (**D**) or free IO/HSA NPs (**B**), for 14 days. Cells were stained with Alizarin-red. Bar-200µm. (**E**) To evaluate the osteogenic differentiation capacity of each supplement, the Alizarin-red O positive area/total area x100 was calculated in 3 areas from triplicate samples for control (c) or each supplement at 3 or 10 ng/ml as indicated in the graph.

facilitate in-vivo detection of transplanted cells using MRI and as an added benefit, the labeled cell may be imaged by NIR fluorescence using optical coherence tomography (OCT) and other in vivo imaging systems equipped with NIR fluorescence. In this work we tested the biological activity of conjugated FGF2 as there is a vast amount of literature supporting the use of this growth factor for enhancing growth and differentiation of hBM-MSCs. The effect of supplementing the growth media of hBM-MSCs with FGF2-conjugated IO/HSA NPs on the cell therapeutic effect in vivo in animal models of neuroretinal degeneration will be investigated in future studies.

In future work we will also test the effect of other conjugated factors such as Bone Morphogenetic Protein 1 (BMP1) for bone defect applications and Heparin binding Epidermal Growth Factor-like Growth Factor (HB-EGF) for neuroretinal degeneration applications.

Methods

Materials

The following analytical-grade chemicals were purchased from commercial sources and used without further purification: bicarbonate buffer (BB; 0.1M, pH 8.4), ferric chloride hexahydrate, hydrochloric acid (1 M), sodium hydroxide (1 M), sodium nitrate, Triton X-100, gelatin from porcine skin, human serum albumin (HSA), NHS-Cy7, rhodamine isothiocyanate (RITC), divinyl sulfone (DVS), triethylamine (TEA), D-glucose from Sigma (Israel); FGF2 ELISA kit and recombinant human FGF2 from PeproTech (Israel); Midi-MACS magnetic columns from Almog Diagnostic (Israel); phosphate-buffered saline (PBS free of Ca+2 and Mg+2; 0.1 M, pH 7.4) from Biological-Industries (Israel); tissue culture plates (96 wells) and plastic tips from Greiner bio-one (Germany); Water was purified by passing deionized water through an Elgastat Spectrum reverse osmosis system (Elga, High

Wycombe, UK). All tissue culture reagents were from Biological Industries (Israel). B27 and DAPI were from Invitrogen. Dexamethasone, insulin, β-glycophostphate, ascorbate phosphate, neuron-specific microtubule-associated protein 2 mouse monoclonal antibody and dyes were from Sigma. Glial Fibrillary Acidic Protein rabbit monoclonal antibody was from Cell Signaling. TUNEL TMR Red was from Roche. Secondary antibodies were from Jackson ImmunoResearch. PCNA (pc1-0) mouse monoclonal IgG2a was from Santa Cruz, USA.

Preparation of the non-fluorescent and fluorescent core IO nanoparticles

Core IO NPs of narrow size distribution were prepared by nucleation in the initial part, followed by stepwise controlled growth of IO thin films onto gelatin/IO nuclei. Briefly, IO NPs of 18 ± 1 nm diameter were prepared by adding $FeCl_2$ solution (10 mmol/5 ml H_2O, 1 N HCl 0.5ml) to 80 ml aqueous solution containing 240 mg gelatin (during the whole procedure, the aqueous suspension is agitated at 60°C), followed by $NaNO_3$ solution (7 mmol/5 ml H_2O). Next, 1N NaOH aqueous solution was added up to pH 9.5. This procedure was repeated three times with 10 min intervals. The formed magnetic NPs were then washed from excess reagents with water using high gradient magnetic field (HGMF) technique. As soon as the washing step was completed, the column was removed from the magnetic field and the NPs were eluted by adding an aqueous bicarbonate buffer (BB, 0.1M, pH = 8.3) [66]. NIR core IO NPs were prepared similarly, by substituting the gelatin for gelatin covalently conjugated with NHS Cy7 to obtain NIR-IO NPs [20].

HSA coating onto the fluorescent IO core nanoparticles

HSA coating was performed by shaking the aqueous suspension of the fluorescent IO NPs with 10% HSA

(MW ~66,000) at 75°C for 12h. The HSA coated NPs were then washed from excess reagents by magnetic columns with PBS (pH = 7.4).

Activation of the fluorescent IO/HSA nanoparticles

Activation of the fluorescent IO/HSA NPs was performed by functionalization of these NPs with excess DVS. One double bond created a covalent bond with the amino groups of the HSA coating onto the fluorescent IO NPs. The residual activated double bond was then used for covalent binding of ligands containing primary amino groups. Briefly, 20 µl of DVS were added to 1 ml of the fluorescent IO/HSA NPs (5 mg/ml) dispersed in the BB continuous phase. The dispersion was then shaken for 12h at 60°C and the remaining free DVS was then washed from the obtained DVS-conjugated NPs using magnetic columns with BB.

Conjugation of FGF2 to the activated fluorescent IO/HSA nanoparticles

Bioactive ligands such as amino acids, proteins, antibodies and more can be easily conjugated to the DVS activated NPs. Briefly, 200 µl of dissolved FGF2 (0.1 mg/ml) were mixed with 200 µl of the DVS activated fluorescent IO/HSA NPs (5 mg/ml) dispersed in BB (0.1M, pH = 8.3). Next, the dispersion was shaken at room temperature for 60 min in order to allow the nucleophilic attack of primary amino groups (from the bioactive ligand) on the DVS-IO/HSA NPs. Blocking of residual activated DVS groups was then performed with glycine, by adding glycine (1% w/v) and mixing the dispersion for additional 30 min at room temp. Excess of unbound ligands were then removed by magnetic columns and the FGF2 conjugated IO/HSA NPs were then eluted with PBS (pH = 7.4).

Transmission Electron Microscopy (TEM)

The TEM image provides direct information on the dry particle shape and size, in which approximately 200 NPs were measured to determine its average size. The core IO NPs and the core-shell IO/HSA NPs were diluted with H_2O to a concentration of 1 mg/ml, dripped on a TEM grid and then dried.

Dynamic Light Scattering (DLS)

Dynamic light scattering measures Brownian motion and relates the intensity fluctuations in the scattered light to the size and size distribution of the particles in its hydrated shape. The fluorescent core IO NPs and core-shell IO/HSA NPs were diluted with H_2O inside a cuvette and the average diameter was then measured, while each measurement was repeated 5 times.

Spectrofluorometer

Spectrofluorometer uses the fluorescent properties of a molecule to provide information about their concentration and fluorescence intensity, both excitation and emission, in different wavelength. Cy7 and Cy7-IO NPs were diluted with PBS to a concentration of 250 ng/ml of the dye, followed by emission, excitation and stability measurements.

Enzyme-Linked ImmunoSorbent Assay (ELISA)

Enzyme-linked immunosorbent assay (ELISA) is commonly used to determine if a particular protein is present in a sample and its concentration [67]. In the present work the concentration of the free and conjugated FGF2 was determined by FGF2 ELISA kit (Pepro-Tech, Israel) based on a calibration curve of known concentrations of free FGF2, according to the literature and following manufacturer's instructions [67]. Samples of the Cy7-IO/HSA-DVS-FGF NPs were diluted with ELISA assay diluent to 3 different NPs' concentrations, each concentration was tested in triplicates and the mean value was calculated. The concentration of the bound FGF2 was determined from a calibration curve of free FGF2 and found to be 1.1 µg/mg core-shell NPs.

Comparative stability studies of free versus conjugated FGF2

For stability measurements, free or conjugated-FGF2 were incubated (10 ng/ml) in various concentrations of fetal calf serum in the medium (0–100% of non-heated serum diluted in medium) at 37°C for 1 and 7 days. The concentration of the residual free and conjugated-FGF2 was then determined by FGF2 ELISA kit.

Production of BM-MSCs

Fresh human Bone Marrows Mesenchymal Stromal Cells (hBM-MSCs) were collected from 6 healthy donors in the operating room at The Sheba Medical Center, Tel-Hashomer, under sterile conditions. The research was approved by the institutional review board at the Sheba Medical Center. Bone marrow mononuclear cells were separated by Ficoll gradient (1.077g/dl) according to the manufacturer instructions and were seeded in tissue culture flasks with culture media containing low-glucose Dulbecco's Modified Eagle's Medium (DMEM) supplemented with 15% FCS, 100U/ml penicillin, 100 ug/ml streptomycin and 2mM L-Glutamine. Tissue culture media was changed after 48 h and then twice a week until 70-80% confluence was reached. Trypan blue staining was performed in every subculturing.

Expansion of hBM-MSCs

Cell expansion in the presence of different supplements was tested in cells derived from 3 BM donors. Cells were grown for 3 passages between passages 2–5. Cells, at concentration of 5×10^3 cell/well, were seeded in 6 well plates in the presence of 0.1 ng/ml free FGF2 or conjugated FGF2 or 90 ng/ml Cy7-IO/HSA NPs in duplicates. Cells were subcultured every 7 ± 2 days and reseeded at 5×10^3 cells per well. Since all plates for each donor were subcultured at the same time for each passage, control and non-conjugated NPs cultures were less confluent than their FGF-treated counterparts during subculturing. To evaluate cell morphology and size, cells were seeded on cover slips and stained with Nuclear Fast Red or analyzed by flow cytometry.

Flow cytometry analysis

Cell surface antigen phenotyping was performed by flow cytometry (FACSCalibur, Becton-Dickinson) using antibodies directed against CD14, CD34, CD45, CD73, CD90, CD105 and HLA-DR to confirm mesenchymal cell phenotype [29,35-37].

The uptake of Cy7 within cells was evaluated by FACSAria III (BD) cell sorting. In order to maximize cell viability and minimize mechanical perturbations, we set the flow rate to 1.1 (minimum). For Cy7 analysis 633nm excitation laser was used with a filter. Data were processed by FlowJo v7.6.4.

Cellular uptake of NPs

Cells were seeded on coverslips precoated with MSC- attachment solution following manufacturer instructions (Biological Industries, Israel). After 24h, NPs were added to cell growth media for 48hr followed by fixation with 4% paraformaldehyde (PFA). Cells were stained with Prussian Blue iron stain and Nuclear Fast Red and visualized by light microscopy (Olympus BX51).

Colony formation assay

For Colony Forming Unit-Fibroblasts (CFU-F) assay, cells were seeded in 6 well plates (250 cell/well) in growth medium. Medium was changed every 3 days. Colonies were formed, analyzed and counted within 7 days after seeding. Cells were washed to remove non adherent colonies. Colonies were fixed in methanol, stained with Giemsa stain, and manually counted. All counting were done in a masked fashion.

Adipogenic differentiation assay

For induction of adipogenic differentiation, cells were cultured for two weeks in growth medium supplemented with 0.6M dexamethasone and 10 mg/l Insulin. Oil-Red-O staining was performed to identify the adipogenic cells followed by hematoxylin counter staining.

Osteogenic differentiation assay

For induction of osteogenic differentiation, cells were cultured for two weeks in growth medium supplemented with 0.1M dexamethasone, 10mM β-glycophostphate and 50 ng/ml ascorbate phosphate. Alizarin red staining was performed to identify the osteogenic cells.

Neurogenic differentiation assay

To induce neurogenic differentiation, cells were grown in DMEM supplemented with 0.5% B27, 1% fetal bovine serum, 5% horse serum, 0.5mM retinoic acid, 20 ng/ml epidermal growth factor and 50 ng/ml nerve growth factor. After two weeks, cells were fixed in 4% PFA, and immunostained with neuron-specific Microtubule-Associated Protein 2 (MAP2) mouse monoclonal antibody or Glial Fibrillary Acidic Protein (GFAP) rabbit monoclonal antibody.

Statistical analysis

A general linear model was used using the donor, replicate and treatment type as independent parameters and differentiation score or colony number as the dependent variables. In the analysis of adipose differentiation the score was normalized after logarithmic transformation. Equality of variance was tested and maintained in all analyzes. We used the Bonferroni correction for post hoc analyses.

Additional files

Additional file 1: Figure S1. Growth curves for hBM-MSCs grown with conjugated FGF2. Cells were cultured in the presence of absence of 0.1 ng/ml free FGF2, 0.1 ng/ml conjugated FGF2 or 90 ng/ml Cy7-IO/HSA NPs. Cells were passaged and counted every 7 ± 2 days and number of live cells was calculated following Trypan blue staining.

Additional file 2: Figure S2. Cell-Surface markers of hBM-MSCs following expansion in the presence of conjugated FGF2. Human BM-MSCs were expanded in the absence or presence of 0.1 ng/ml conjugated FGF2 for 3 passages. Flow cytometry analysis was performed using antibodies directed against CD14, CD34, CD45, CD73, CD90, CD105 and HLA-DR. Non labeled cells (non shaded), Isotype-matched IgG controls (dash-lined) and labeled cells (shaded) curves are shown. A minimum of 10,000 events was recorded.

Additional file 3: Figure S3. Clonal expansion capacity of hBM-MSCs in the presence of conjugated FGF2. Human BM-MSCs from 3 donors were seeded in duplicates or triplicates in 6 well plates and incubated in growth media alone (control) or growth media supplemented with free IO/HSA NPs, free or conjugated FGF2 at concentration of 50 ng/ml. Media was change every 3 days. Seven days post seeding, colonies were counted following extensive washing and Giemsa staining. Data is presented as colony number per 100 cells (mean ± SE). For donor B colony expansion with non conjugated NPs was not determined.

Competing interests

The authors declare that they have no competing interests.

Authors' contributions

IL- Preparation and characterization of the non-conjugated and FGF2-conjugated IO/HSA nanoparticles, drafting of manuscript, critical revision. IS – Study conception and design, carried out and analyzed all the BM-MSC experiments, drafting of manuscript, critical revision. ECS – Preparation and characterization of the non-conjugated and FGF2-conjugated IO/HSA nanoparticles. OZP – Analysis of the activity of the free and conjugated FGF2. AM- Preparation of BM-MSC from donors. AJV- Study design, critical revision. AN- BM aspiration, critical revision. OKL- Statistical analysis. SM- Supervising the preparation and characterization of the non-conjugated and FGF2-conjugated IO/HSA nanoparticles, and revision of the article. YR- Study conception and design, critical revision. All authors read and approved the final manuscript.

Grant information

This study was supported by a grant from the Claire and Amedee Maratier Institute for the Study of Blindness and Visual Disorders, Sackler Faculty of Medicine, Tel-Aviv University, and a grant from the Israeli Ministry of Trade and Industry KAMIN–Yeda Program (to YR). IS was partially supported by the Israeli Ministry of Absorption and Immigration. The supporting organizations had no role in the design or conduct of this research.

Author details

[1]Department of Chemistry, Bar-Ilan Institute of Nanotechnology and Advanced Materials, Ramat-Gan 52900, Israel. [2]Goldschleger Eye Institute, Sackler Faculty of Medicine, Tel Aviv University, Sheba Medical Center, Tel-Hashomer 52621, Israel. [3]Center for Stem Cells and Regenerative Medicine, Cancer Research Center, Sheba Medical Center, Tel-Hashomer 52621, Israel. [4]Hematology Division, Sheba Medical Center, Tel-Hashomer 52621, Israel. [5]Unit of Cardiovascular Epidemiology, Gertner Institute for Epidemiology and Health Policy Research, Ramat Gan, Israel, Sackler Faculty of Medicine, Tel-Aviv University, Tel-Aviv, Israel.

References

1. Hergt R, Hiergeist R, Hilger I, Kaiser W, Lapatnikov Y, Margel S, et al. Maghemite nanoparticles with very high AC-losses for application in RF-magnetic hyperthermia. J Magn Magn Mater. 2004;270:345–57.

2. Liong M, Lu J, Kovochich M, Xia T, Ruehm SG, Nel AE. Multifunctional inorganic nanoparticles for imaging, targeting, and drug delivery. ACS Nano. 2008;2:889–96.

3. Rogers WJ, Meyer CH, Kramer CM. Technology insight: in vivo cell tracking by use of MRI. Nat Clin Pract Cardiovasc Med. 2006;3:554–62.

4. Pankhurst QA, Connolly J, Jones S, Dobson J. Applications of magnetic nanoparticles in biomedicine. J Phys D Appl Phys. 2003;36:R167–81.

5. Qiao R, Yang C, Gao M. Superparamagnetic iron oxide nanoparticles: from preparations to in vivo MRI applications. J Mater Chem. 2009;19:6274–93.

6. Park YI, Piao Y, Lee N, Yoo B, Kim BH, Choi SH, et al. Transformation of hydrophobic iron oxide nanoparticles to hydrophilic and biocompatible maghemite nanocrystals for use as highly efficient MRI contrast agent. J Mater Chem. 2011;21:11472–7.

7. Jin X, Chen K, Huang J, Lee S, Wang J, Gao J, et al. PET/NIRF/MRI triple functional iron oxide nanoparticles. Biomaterials. 2010;31:3016–22.

8. Arruebo M, Fernández-Pacheco R, Ibarra MR, Santamaría J. Magnetic nanoparticles for drug delivery. Nano Today. 2007;2:22–32.

9. Lee H, Yu MK, Park S, Moon S, Min JJ, Jeong YY, et al. Thermally cross-linked superparamagnetic iron oxide nanoparticles: synthesis and application as a dual imaging probe for cancer in vivo. J Am Chem Soc. 2007;129:12739–45.

10. Altınoğlu El, Adair JH. Near infrared imaging with nanoparticles. WIREs Nanomed Nanobiotechnol. 2010;2:461–77.

11. Kinsella JM, Jimenez RE, Karmali PP, Rush AM, Kotamraju VR, Gianneschi NC, et al. X-Ray computed tomography imaging of breast cancer by using targeted peptide-labeled bismuth sulfide nanoparticles. Angew Chem Int Ed Engl. 2011;50:12308–11.

12. He X, Wang K, Cheng Z. In vivo near-infrared fluorescence imaging of cancer with nanoparticle-based probes. Wiley Interdiscip Rev Nanomed Nanobiotechnol. 2010;2:349–66.

13. Corem-Salkmon E, Ram Z, Daniels D, Perlstein B, Last D, Salomon S, et al. Convection-enhanced delivery of methotrexate-loaded maghemite nanoparticles. Int J Nanomedicine. 2011;6:1595–602.

14. Kratz F. Albumin as a drug carrier: design of prodrugs, drug conjugates and nanoparticles. J Controlled Release. 2008;132:171–83.

15. Galperin A, Margel S. Synthesis and characterization of radiopaque magnetic core-shell nanoparticles for X-ray imaging applications. J Biomed Mater Res B Appl Biomater. 2007;83:490–8.

16. Boguslavsky Y, Margel S. Synthesis and characterization of poly (divinylbenzene)-coated magnetic iron oxide nanoparticles as precursor for the formation of Air-stable carbon-coated iron crystalline nanoparticles. J Colloid Interface Sci. 2008;317:101–14.

17. MacDonald K, Murrell WG, Bartlett P, Bushell GR, Mackay-Sim A. FGF2 promotes neuronal differentiation in explant cultures of adult and embryonic mouse olfactory epithelium. J Neurosci Res. 1996;44:27–39.

18. Huang Y-C, Huang Y-Y. Tissue engineering for nerve repair. Biomed Eng Appl Basis Commun. 2006;18:100–10.

19. Wu Y, Cai W, Chen X. Near-infrared fluorescence imaging of tumor integrin Avβ3 expression with Cy7-labeled RGD multimers. Mol Imaging Biol. 2006;8:226–36.

20. Zhang S, Uludağ H. Nanoparticulate systems for growth factor delivery. Pharm Res. 2009;26:1561–80.

21. Pittenger MF, Mackay AM, Beck SC, Jaiswal RK, Douglas R, Mosca JD, et al. Multilineage potential of adult human mesenchymal stem cells. Science. 1999;284:143–7.

22. Kopen GC, Prockop DJ, Phinney DG. Marrow stromal cells migrate throughout forebrain and cerebellum, and they differentiate into astrocytes after injection into neonatal mouse brains. Proc Natl Acad Sci U S A. 1999;96:10711–6.

23. Oswald J, Boxberger S, Jørgensen B, Feldmann S, Ehninger G, Bornhäuser M, et al. Mesenchymal stem cells can be differentiated into endothelial cells in vitro. Stem Cells. 2004;22:377–84.

24. Jiang Y, Jahagirdar BN, Reinhardt RL, Schwartz RE, Keene CD, Ortiz-Gonzalez XR, et al. Pluripotency of mesenchymal stem cells derived from adult marrow. Nature. 2002;418:41–9.

25. Mezey E, Key S, Vogelsang G, Szalayova I, Lange GD, Crain B. Transplanted bone marrow generates New neurons in human brains. Proc Natl Acad Sci U S A. 2003;100:1364–9.

26. Arien-Zakay H, Lazarovici P, Nagler A. Tissue regeneration potential in human umbilical cord. Best Pract Res Clin Haematol. 2010;23:291–303.

27. Krampera M, Glennie S, Dyson J, Scott D, Laylor R, Simpson E, et al. Bone marrow mesenchymal stem cells inhibit the response of naive and memory antigen-specific T cells to their cognate peptide. Blood. 2003;101:3722–9.

28. Di Nicola M, Carlo-Stella C, Magni M, Milanesi M, Longoni PD, Matteucci P, et al. Human bone marrow stromal cells suppress T-lymphocyte proliferation induced by cellular or nonspecific mitogenic stimuli. Blood. 2002;99:3838–43.

29. Sadan O, Melamed E, Offen D. Bone-marrow-derived mesenchymal stem cell therapy for neurodegenerative diseases. Expert Opin Biol Ther. 2009;9:1487–97.

30. Resnick IB, Barkats C, Shapira MY, Stepensky P, Bloom AI, Shimoni A, et al. Treatment of severe steroid resistant acute GVHD with mesenchymal stromal cells (MSC). Am J Blood Res. 2013;3:225–38.

31. Michael M, Shimoni A, Nagler A. Recent compounds for immunosuppression and experimental therapies for acute graft-versus-host disease. Isr Med Assoc J. 2013;15:44–50.

32. Petite H, Viateau V, Bensaid W, Meunier A, de Pollak C, Bourguignon M, et al. Tissue-engineered bone regeneration. Nat Biotechnol. 2000;18:959–63.

33. Phinney DG, Prockop DJ. Concise review: mesenchymal stem/multipotent stromal cells: the state of transdifferentiation and modes of tissue repair–current views. Stem Cells. 2007;25:2896–902.

34. Caplan AI. Adult mesenchymal stem cells for tissue engineering versus regenerative medicine. J Cell Physiol. 2007;213:341–7.

35. Tzameret A, Sher I, Belkin M, Treves AJ, Meir A, Nagler A, et al. Transplantation of human bone marrow mesenchymal stem cells as a thin subretinal layer ameliorates retinal degeneration in a Rat model of retinal dystrophy. Exp Eye Res. 2014;118:135–44.

36. Xiong N, Yang H, Liu L, Xiong J, Zhang Z, Zhang X, et al. bFGF promotes the differentiation and effectiveness of human bone marrow mesenchymal stem cells in a rotenone model for Parkinson's disease. Environ Toxicol Pharmacol. 2013;36:411–22.

37. Lee JK, Jin HK, Endo S, Schuchman EH, Carter JE, Bae J. Intracerebral transplantation of bone marrow-derived mesenchymal stem cells reduces amyloid-beta deposition and rescues memory deficits in Alzheimer's disease mice by modulation of immune responses. Stem Cells. 2010;28:329–43.

38. Yu J, Yin S, Zhang W, Gao F, Liu Y, Chen Z, et al. Hypoxia preconditioned bone marrow mesenchymal stem cells promoted liver regeneration in a Rat massive hepatectomy model. Stem Cell Res Ther. 2013;4:83.

39. Zhao Y, Xu A, Xu Q, Zhao W, Li D, Fang X, et al. Bone marrow mesenchymal stem cell transplantation for treatment of emphysemic rats. Int J Clin Exp Med. 2014;7:968–72.

40. Chen SL, Fang WW, Qian J, Ye F, Liu YH, Shan SJ, et al. Improvement of cardiac function after transplantation of autologous bone marrow mesenchymal stem cells in patients with acute myocardial infarction. Chin Med J (Engl). 2004;117:1443–8.

41. Dabiri G, Heiner D, Falanga V. The emerging use of bone marrow-derived mesenchymal stem cells in the treatment of human chronic wounds. Expert Opin Emerg Drugs. 2013;18:405–19.

42. Gopal K, Amirhamed HA, Kamarul T. Advances of human bone marrow-derived mesenchymal stem cells in the treatment of cartilage defects: a systematic review. Exp Biol Med (Maywood). 2014;239:663–9.

43. Pal R, Venkataramana NK, Bansal A, Balaraju S, Jan M, Chandra R, et al. Ex vivo-expanded autologous bone marrow-derived mesenchymal stromal cells in human spinal cord injury/paraplegia: a pilot clinical study. Cytotherapy. 2009;11:897–911.

44. Muroi K, Miyamura K, Ohashi K, Murata M, Eto T, Kobayashi N, et al. Unrelated allogeneic bone marrow-derived mesenchymal stem cells for steroid-refractory acute graft-versus-host disease: a phase I/II study. Int J Hematol. 2013;98:206–13.

45. Bianco Martinez AM, de Oliveira GC, dos Santos RB, Teixeira Oliveira J, Martins Almeida F. Neurotrauma and mesenchymal stem cells treatment: from experimental studies to clinical trials. World J Stem Cells. 2014;6:179–94.

46. ClinicalTrials.gov, https://clinicaltrials.gov/. Accessed 2014.

47. Ringden O, Uzunel M, Rasmusson I, Remberger M, Sundberg B, Lonnies H, et al. Mesenchymal stem cells for treatment of therapy-resistant graft-versus-host disease. Transplantation. 2006;81:1390–7.

48. DiGirolamo CM, Stokes D, Colter D, Phinney DG, Class R, Prockop DJ. Propagation and senescence of human marrow stromal cells in culture: a simple colony-forming assay identifies samples with the greatest potential to propagate and differentiate. Br J Haematol. 1999;107:275–81.

49. Amsalem Y, Mardor Y, Feinberg MS, Landa N, Miller L, Daniels D, et al. Iron-oxide labeling and outcome of transplanted mesenchymal stem cells in the infarcted myocardium. Circulation. 2007;116:I38–45.

50. Auletta JJ, Zale EA, Welter JF, Solchaga LA. Fibroblast growth factor-2 enhances expansion of human bone marrow-derived mesenchymal stromal cells without diminishing their immunosuppressive potential. Stem Cells Int. 2011;2011:235176.

51. Bianchi G, Banfi A, Mastrogiacomo M, Notaro R, Luzzatto L, Cancedda R, et al. Ex vivo enrichment of mesenchymal cell progenitors by fibroblast growth factor 2. Exp Cell Res. 2003;287:98–105.

52. Edelman ER, Nugent MA, Karnovsky MJ. Perivascular and intravenous administration of basic fibroblast growth factor: vascular and solid organ deposition. Proc Natl Acad Sci U S A. 1993;90:1513–7.

53. Whalen GF, Shing Y, Folkman J. The fate of intravenously administered bFGF and the effect of heparin. Growth Factors. 1989;1:157–64.

54. Skaat H, Ziv-Polat O, Shahar A, Last D, Mardor Y, Margel S. Magnetic scaffolds enriched with bioactive nanoparticles for tissue engineering. Adv Healthc Mater. 2012;1:168–71.

55. Ziv-Polat O, Topaz M, Brosh T, Margel S. Enhancement of incisional wound healing by thrombin conjugated iron oxide nanoparticles. Biomaterials. 2010;31:741–7.

56. Green-Sadan T, Kuttner Y, Lublin-Tennenbaum T, Kinor N, Boguslavsky Y, Margel S, et al. Glial cell line-derived neurotrophic factor-conjugated nanoparticles suppress acquisition of cocaine self-administration in rats. Exp Neurol. 2005;194:97–105.

57. Ziv-Polat O, Skaat H, Shahar A, Margel S. Novel magnetic fibrin hydrogel scaffolds containing thrombin and growth factors conjugated iron oxide nanoparticles for tissue engineering. Int J Nanomedicine. 2012;7:1259–74.

58. Perlstein B, Lublin-Tennenbaum T, Marom I, Margel S. Synthesis and characterization of functionalized magnetic maghemite nanoparticles with fluorescent probe capabilities for biological applications. J Biomed Mater Res BAppl Biomater. 2010;92:353–60.

59. Ahn HJ, Lee WJ, Kwack K, Kwon YD. FGF2 stimulates the proliferation of human mesenchymal stem cells through the transient activation of JNK signaling. FEBS Lett. 2009;583:2922–6.

60. Solchaga LA, Penick K, Porter JD, Goldberg VM, Caplan AI, Welter JF. FGF-2 enhances the mitotic and chondrogenic potentials of human adult bone marrow-derived mesenchymal stem cells. J Cell Physiol. 2005;203:398–409.

61. Sotiropoulou PA, Perez SA, Salagianni M, Baxevanis CN, Papamichail M. Characterization of the optimal culture conditions for clinical scale production of human mesenchymal stem cells. Stem Cells. 2006;24:462–71.

62. Dombrowski C, Helledie T, Ling L, Grunert M, Canning CA, Jones CM, et al. FGFR1 signaling stimulates proliferation of human mesenchymal stem cells by inhibiting the cyclin-dependent kinase inhibitors P21 and P27. Stem Cells. 2013;31:2724–36.

63. Burgess WH, Maciag T. The heparin-binding (fibroblast) growth factor family of proteins. Annu Rev Biochem. 1989;58:575–602.

64. Swami A, Shi J, Gadde S, Votruba AR, Kolishetti N, Farokhzad OC. Nanoparticles for targeted and temporally controlled drug delivery. In: Nanoparticles for targeted and temporally controlled drug delivery. Multifunctional Nanoparticles for Drug Delivery Applications: Springer, 2012: 9–29

65. Phinney DG. Building a consensus regarding the nature and origin of mesenchymal stem cells. J Cell Biochem. 2002;85:7–12.

66. Margel S, Tennenbaum T, Gura S. Synthesis and characterization of nano- and micron-sized iron oxide and iron particles for biomedical applications. Lab Tech Biochem Mol Biol. 2007;32:119–62.

67. Lequin RM. Enzyme immunoassay (EIA)/enzyme-linked immunosorbent assay (ELISA). Clin Chem. 2005;51:2415–8.

16

Detection and quantification of bacterial biofilms combining high-frequency acoustic microscopy and targeted lipid microparticles

Pavlos Anastasiadis[1,2,3], Kristina D A Mojica[4,5], John S Allen[3*] and Michelle L Matter[1*]

Abstract

Background: Immuno-compromised patients such as those undergoing cancer chemotherapy are susceptible to bacterial infections leading to biofilm matrix formation. This surrounding biofilm matrix acts as a diffusion barrier that binds up antibiotics and antibodies, promoting resistance to treatment. Developing non-invasive imaging methods that detect biofilm matrix in the clinic are needed. The use of ultrasound in conjunction with targeted ultrasound contrast agents (UCAs) may provide detection of early stage biofilm matrix formation and facilitate optimal treatment.

Results: Ligand-targeted UCAs were investigated as a novel method for pre-clinical non-invasive molecular imaging of early and late stage biofilms. These agents were used to target, image and detect *Staphylococcus aureus* biofilm matrix *in vitro*. Binding efficacy was assessed on biofilm matrices with respect to their increasing biomass ranging from $3.126 \times 10^3 \pm 427$ UCAs per mm^2 of biofilm surface area within 12 h to $21.985 \times 10^3 \pm 855$ per mm^2 of biofilm matrix surface area at 96 h. High-frequency acoustic microscopy was used to ultrasonically detect targeted UCAs bound to a biofilm matrix and to assess biofilm matrix mechanoelastic physical properties. Acoustic impedance data demonstrated that biofilm matrices exhibit impedance values (1.9 MRayl) close to human tissue (1.35 - 1.85 MRayl for soft tissues). Moreover, the acoustic signature of mature biofilm matrices were evaluated in terms of integrated backscatter ($0.0278 - 0.0848$ $mm^{-1} \times sr^{-1}$) and acoustic attenuation (3.9 Np/mm for bound UCAs; 6.58 Np/mm for biofilm alone).

Conclusions: Early diagnosis of biofilm matrix formation is a challenge in treating cancer patients with infection-associated biofilms. We report for the first time a combined optical and acoustic evaluation of infectious biofilm matrices. We demonstrate that acoustic impedance of biofilms is similar to the impedance of human tissues, making *in vivo* imaging and detection of biofilm matrices difficult. The combination of ultrasound and targeted UCAs can be used to enhance biofilm imaging and early detection. Our findings suggest that the combination of targeted UCAs and ultrasound is a novel molecular imaging technique for the detection of biofilms. We show that high-frequency acoustic microscopy provides sufficient spatial resolution for quantification of biofilm mechanoelastic properties.

Keywords: Targeted therapy, Lipid particles, Biofilm matrix, Targeted ultrasound contrast agents, Cancer, Acoustic microscopy, Molecular imaging, Microbubbles

* Correspondence: alleniii@hawaii.edu; matter@hawaii.edu
[3]Mechanical Engineering, University of Hawaii at Manoa, Honolulu, HI 96822, USA
[1]University of Hawaii Cancer Center, Honolulu, HI 96813, USA
Full list of author information is available at the end of the article

Background

Bacterial biofilms are three-dimensional extracellular matrices composed of carbohydrates, proteins and exopolysaccharides [1-6] that develop on solid–liquid or solid-air interfaces in the body [3,7]. Biofilms consist of bacterial cells and matrix proteins. The majority of biofilms contain 10% or less of bacterial cells and over 90% matrix [8]. Biofilm matrices are highly conserved dynamic structures. Initiation of a biofilm matrix occurs by a transient interaction of bacteria with a surface followed by an adhesive stage that allows for microcolony formation and a subsequent growth and maturation stage. The complexity of biofilms allows bacteria cells to survive a multitude of environments and promotes cell dispersion to colonize new areas. These matrices may form on medical devices or fragments of dead tissue [3,9-12]. Clinically, biofilms may occur during chemotherapy and infectious diseases such as endocarditis [6,13-16].

Biofilm associated infections are resistant to treatment and recur even after repeated antibiotic therapy. One primary issue is that established biofilm matrices act as diffusion barriers and actively bind up antibiotics and antibodies thereby providing increased resistance. Overall killing bacteria that are surrounded by a microbial biofilm require up to 1000 times higher concentrations of antibiotics than those without a surrounding biofilm [17-20]. Thus, detecting, treating and inhibiting biofilm formation inside the body is a key medical challenge.

Moreover, in the clinical setting antibiotic therapy efficacy is decreased in the presence of an established biofilm making early detection critical. For example, infective endocarditis may occur due to chronic infection, has a poor prognosis and is associated with high mortality rates [14-16,21]. Indeed, there are significant diagnostic challenges for endocarditis that are attributed to the inaccessibility of intra-cardiac biofilms and the non-specific nature of the clinical symptoms [22]. Although echocardiography permits non-invasive detection of biofilms [23] it has significant limitations in the detection of early biofilm matrix formation. In addition, clinical diagnosis primarily occurs after biofilm matrices are fully established, thereby significantly decreasing available treatment options. Therefore, early detection is a crucial component of diagnosis; however no current diagnostic methodology is available that clearly delineates early and late stage matrices.

Ultrasound is an effective method for imaging biofilms in vitro [24-28]. One method used to enhance biofilm detection is the addition of UCAs (encapsulated gas bubbles), which provide a unique acoustic scattering signature thereby significantly enhancing imaging capabilities [29]. Furthermore, linking a ligand to a contrast agent's outer membrane aids in UCAs binding to tissue and is crucial in delineating disease specific regions from surrounding healthy tissue [30-35].

In this study, ligand-targeted UCAs were used as a novel method for pre-clinical non-invasive molecular imaging of early and late stage biofilms. These agents were used to target and detect Staphylococcus aureus (S. aureus) biofilm formation. Binding efficacy was assessed on established biofilms as a function of surface area. A combination of acoustic and optical microscopy was used to quantify the mechanical and structural properties of a three dimensional biofilm matrix. We show that high-frequency scanning acoustic microscopy (SAM) provides sufficient high spatial resolution for imaging and quantification of biofilm thickness and mechanoelastic properties.

Results

Biofilm formation occurs when bacterial cells enter the body and attach to the underlying endothelium or tissues. Over time, biofilms form a protective three dimensional matrix that results in lower antibody efficacy in vivo (Figure 1). Biofilm surface areas were assessed by epifluorescence microscopy images of stained S. aureus biofilms at various time points (Figure 2A). Biofilm matrix surface area doubled during the first 12 hours after inoculation (growing from 26.85 mm^2 ± 6.72 mm^2 to 51.7 mm^2 ± 2.12 mm^2 at 12 h and 24 h respectively; p < 0.05). Similar growth patterns were observed through 96 hours (68.95 mm^2 ± 4.6 mm^2, 122.2 mm^2 ± 8.56 mm^2 and 179.2 mm^2 ± 2.97 mm^2 for 48 h, 72 h and 96 h respectively; p < 0.05). These data suggest that biofilm matrices are produced over time in our in vitro culture system.

To determine whether targeted UCAs bind to a biofilm matrix in vitro, we next examined whether targeted ultrasound contrast agents (UCAs) bound to the biofilm matrix over time. We observed an increase in the binding rate of targeted UCAs to the biofilm matrix (Figure 2B). We tested whether labeled targeted UCAs were detectable upon a labeled biofilm matrix. Tetramethylrhodamine isothiocyanate (TRITC)-streptavidin conjugated UCAs (red staining) were detectable from fluorescein isothiocyanate (FITC) anti-WGA labeled matrix (green staining). At the 12 h time point 1.109 × 10^3 ± 142 UCAs were bound to the biofilm. The number of bound bubbles significantly increased to 3.126 × 10^3 ± 427 over the following 12 h. Between 24 h and 72 h labeled UCAs binding increased (5.042 × 10^3 ± 285 UCAs at 48 h, 7.563 × 10^3 ± 142 at 72 h; p < 0.05). Between 72 h and 96 h a significant increase in targeted UCAs was observed (7.563 × 10^3 ± 142 to 21.985 ± 855 at 96 h; p < 0.05) suggesting that binding increases in correlation with biofilm matrix surface area. Fluorescence images stained for S. aureus biofilm matrix at various time points (Figure 2C) confirms that targeted UCAs bound more as the biofilm matrix increased over 96 hours.

Developing a non-invasive diagnostic method to detect biofilm matrices early (or at initial stages) would be a valuable clinical tool if the targeted agents could be detected

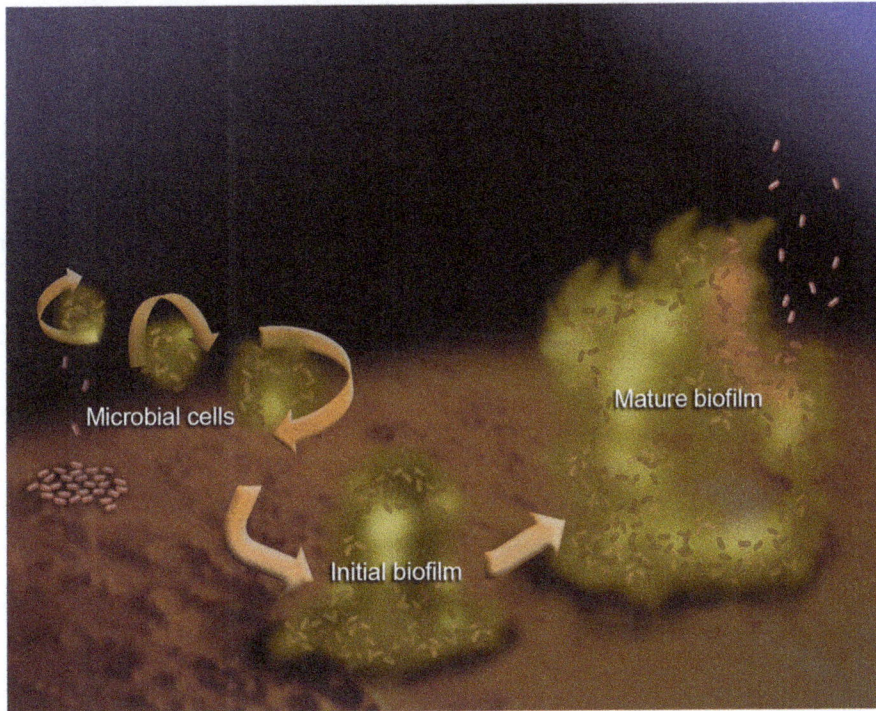

Figure 1 Biofilm matrix formation. Individual bacterial cells gain entrance into the bloodstream and attach at favorable sites. As they continue growing, they form a protective biofilm matrix against hostile agents, the immune system or fluid turbulences caused by hemodynamic forces. As the biofilm matrix matures, individual cells are dispersed into the bloodstream where they travel to distant sites in the body forming colonies. Figure adapted from [6].

acoustically. Because we determined that targeted UCAs bound proportionately to biofilm matrix mass we next assessed whether ultrasound could be used to detect targeted UCAs *in vitro*. The center frequency for the ultrasonic evaluation was 100 MHz [26] allowing for a rigorous quantification of biofilm matrix mechanoelastic properties in our *in vitro* biofilm culture system (Table 1).

For the physical evaluation of biofilm matrix properties (density, acoustic attenuation, ultrasound velocity, acoustic impedance and bulk modulus) a time-resolved high-frequency scanning acoustic microscope was used (Fraunhofer IBMT, St. Ingbert, Germany; Table 2). For imaging, an acoustic lens is triggered by a piezoelectric transducer that emits and receives highly focused sound waves and resolves them along the time axis (Figure 3A). The echoes reflected off the sample surface, the substrate and the interface between the sample and the substrate were taken into consideration for mechanoelastic quantification. The acoustic lens is mounted on top of the stage of a Zeiss Axiovert M200, inverted light microscope (Figure 3B). This custom arrangement [36,37], where the optical microscope objective and the acoustic lens are confocally aligned allows for corresponding optical (or fluorescence) imaging and therefore facilitating novel simultaneous acoustical and optical evaluation of specimen.

S. aureus mature biofilms at day five were ultrasonically and fluorescently evaluated (Figure 4). UCAs were conjugated with TRITC-labeled streptavidin that allowed for the detection of the corresponding fluorescent signal. Because the acoustic lens is confocally aligned we were able to overlay the corresponding acoustic and fluorescent signals. In each case three or more different regions were scanned covering a total surface of 1 mm^2 for each independent acquisition. For the same time point, fluorescence and optical images were acquired from the identical regions (Figure 4). The grey colored area corresponds to the acquired acoustic dataset of the biofilm matrix while the inset depicts a fluorescent image of a partial area within that ultrasonically acquired region. The epifluorescent images were comparable in terms of specimen location and provided different complementary information on the biofilm structure and mechanical properties (Table 1). Microbubbles (Targeson, San Diego, CA, USA) were 2–3 μm in diameter and were detected based on fluorescence and acoustic signals (Figure 4).

We next examined whether regions of biofilm matrices can be delineated based on targeted and bound or non-targeted, non-bound UCAs. When no biofilm mass was present the targeted UCAs remain unbound as there is no ligand for them to bind to (Figure 5A). As biofilm matrix formation progresses, targeted UCAs bind to the ligand

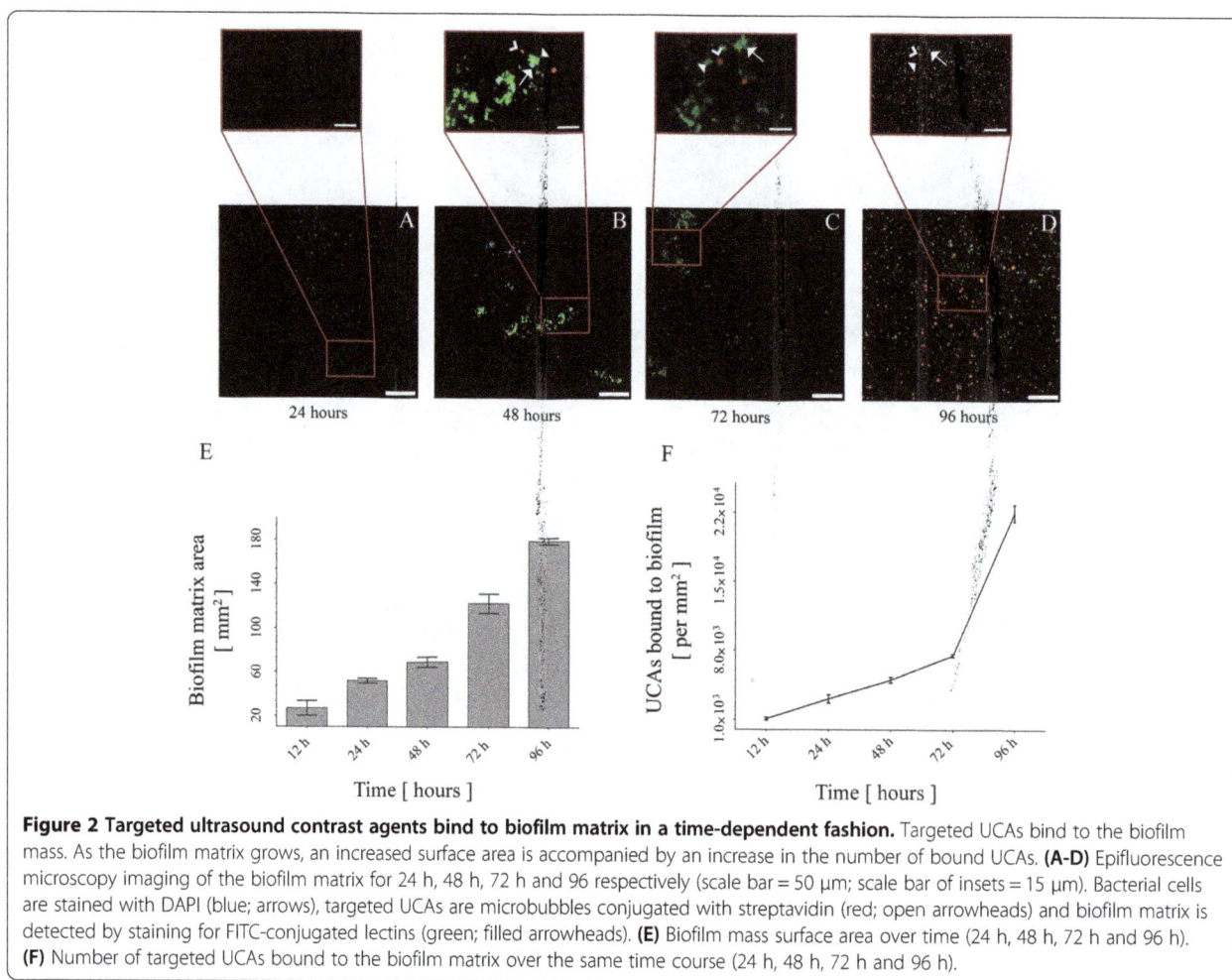

Figure 2 Targeted ultrasound contrast agents bind to biofilm matrix in a time-dependent fashion. Targeted UCAs bind to the biofilm mass. As the biofilm matrix grows, an increased surface area is accompanied by an increase in the number of bound UCAs. **(A-D)** Epifluorescence microscopy imaging of the biofilm matrix for 24 h, 48 h, 72 h and 96 respectively (scale bar = 50 μm; scale bar of insets = 15 μm). Bacterial cells are stained with DAPI (blue; arrows), targeted UCAs are microbubbles conjugated with streptavidin (red; open arrowheads) and biofilm matrix is detected by staining for FITC-conjugated lectins (green; filled arrowheads). **(E)** Biofilm mass surface area over time (24 h, 48 h, 72 h and 96 h). **(F)** Number of targeted UCAs bound to the biofilm matrix over the same time course (24 h, 48 h, 72 h and 96 h).

present in the biofilm matrix (Figure 5B). Targeted UCAs bound to the biofilm matrix scattered sound and produced a detectable acoustic signature [38,39], which correlates with a biofilm matrix. The images shown in Figure 4 depict both the corresponding optical and acoustic images of the UCAs. The acoustic image in Figure 6A depicts UCAs reflectivity in backscatter intensity and their spatial location is depicted as red signals in the fluorescent image.

Furthermore, Figure 6A demonstrates that bound UCAs provide a stronger backscatter intensity as compared to the regions of biofilm matrix alone.

Based on linear acoustics [40] the integrated backscatter coefficient and acoustic attenuation were calculated for regions of bound and unbound UCAs. The mean integrated backscatter coefficient (IBSC; Figure 6A) was determined for frequencies ranging from 97 MHz to 104 MHz from

Table 1 Mechanical and elastic parameters of an *S. aureus* biofilm at 96 hours as determined by time-resolved high-frequency scanning acoustic microscopy

Structural and physical properties	Biofilm matrix
Thickness [μm]	127.23 ± 2.87
Ultrasound velocity [m/s]	1523.14 ± 12.01
Attenuation [Np/mm]	4.2 ± 0.18
Acoustic impedance [MRayl]	1.9 ± 0.01
Density [kg/m^3]	1246.58 ± 11.48
Bulk modulus [GPa]	$2.8 ± 2.9 \times 10^{-5}$

Table 2 Physical properties of the high-frequency acoustic lens used in this study

Acoustic lens properties	
Excitation center frequency [MHz]	100
Max PRF [kHz]	100
Gain [dB]	40
Sampling rate [MSamples/s]	400
Focal resolution [μm]	10
Aperture [μm]	950
Aperture angle [°]	55
Working distance [μm]	900

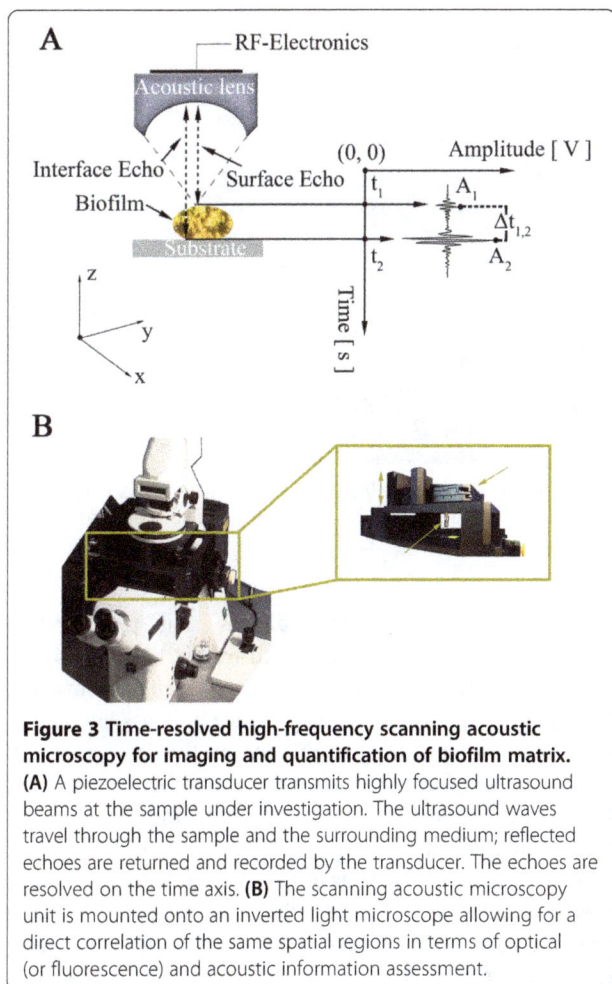

Figure 3 Time-resolved high-frequency scanning acoustic microscopy for imaging and quantification of biofilm matrix. **(A)** A piezoelectric transducer transmits highly focused ultrasound beams at the sample under investigation. The ultrasound waves travel through the sample and the surrounding medium; reflected echoes are returned and recorded by the transducer. The echoes are resolved on the time axis. **(B)** The scanning acoustic microscopy unit is mounted onto an inverted light microscope allowing for a direct correlation of the same spatial regions in terms of optical (or fluorescence) and acoustic information assessment.

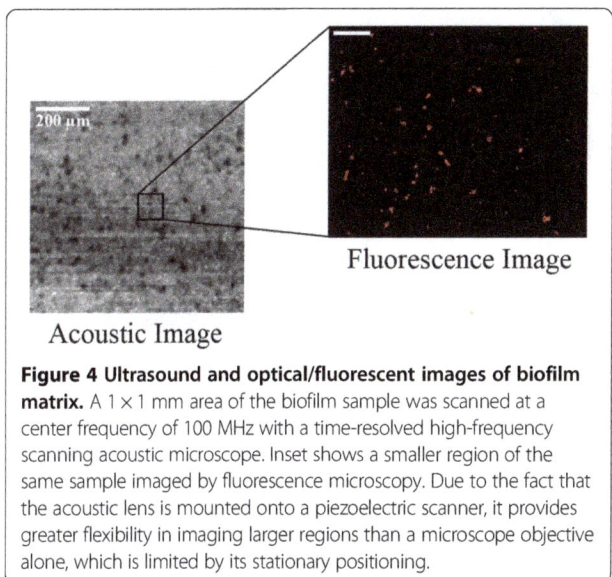

Figure 4 Ultrasound and optical/fluorescent images of biofilm matrix. A 1×1 mm area of the biofilm sample was scanned at a center frequency of 100 MHz with a time-resolved high-frequency scanning acoustic microscope. Inset shows a smaller region of the same sample imaged by fluorescence microscopy. Due to the fact that the acoustic lens is mounted onto a piezoelectric scanner, it provides greater flexibility in imaging larger regions than a microscope objective alone, which is limited by its stationary positioning.

the measured biofilm regions in which targeted UCAs were bound versus biofilm matrix alone. This frequency range corresponded to 36 points at a sampling frequency of 400 MHz while the scanning of the region of interests (ROIs) was performed with step sizes in the order of 10 μm in the x- and y-direction respectively. The acquired mean values of the IBSCs for the ROIs that remained bound to targeted UCAs range from 0.0278 $mm^{-1} \times sr^{-1}$ to 0.0848 $mm^{-1} \times sr^{-1}$ while the values for the standard deviation (StDev) vary from 0.0016 $mm^{-1} \times sr^{-1}$ to 0.0043 $mm^{-1} \times sr^{-1}$. The evaluation of the ROIs corresponding to the matrix without UCAs yielded for the IBSCs mean values in the range from 0.0167 $mm^{-1} \times sr^{-1}$ to 0.0694 $mm^{-1} \times sr^{-1}$ with StDev values ranging from 0.0012 $mm^{-1} \times sr^{-1}$ to 0.0024 $mm^{-1} \times sr^{-1}$ respectively.

The same ROIs used for the quantification of the IBSC were further evaluated with regard to sound attenuation (Figure 6B). ROIs were analyzed in which UCAs were either bound to the matrix or not bound. Each of the fifteen ROIs per condition consisted of nine pixels corresponding to 135 raw radio-frequency (RF) time-signals for the UCA ROIs and similarly fifteen ROIs for the matrix ROIs corresponding to another 135 raw RF time-signals. The frequency-dependent attenuation was calculated over the frequency range from 97 MHz to 104 MHz. Equivalent to our IBSC findings this frequency range consisted of 36 points at a sampling frequency of 400 MHz. The attenuation graphs for the ROIs with bound targeted UCAs and the ROIs with plain matrix over the selected frequency range are shown in Figure 6B. Taken together, detection of targeted bound UCAs is significant compared with unbound UCAs. Our data highlight the potential of targeted UCAs as a means of molecular imaging to detect the early stages of biofilm matrix formation.

Conclusions

In this study, we report for the first time a combined optical and acoustic imaging method of infectious biofilm matrices. Ligand-targeted UCAs were used as a novel method for pre-clinical non-invasive molecular imaging of early to late stage biofilms. These agents were used to target *S. aureus* biofilm formation and assess the binding efficacy on early to late stage biofilm matrices with respect to their surface area. A combination of acoustic and optical microscopy was used to quantify *S. aureus* biofilm mechanoelastic properties. We show that time-resolved high-frequency SAM is a viable method for ultrasonic imaging in addition to quantifying mechanical and elastic properties in soft materials (eg. tissues, cells, biofilm matrices) in a non-invasive setting. Moreover, the use of targeted UCAs with high-frequency SAM allow for UCAs detection at higher frequencies other than their resonance frequency. The mechanoelastic properties of the *S. aureus* biofilm matrix are summarized in Table 1.

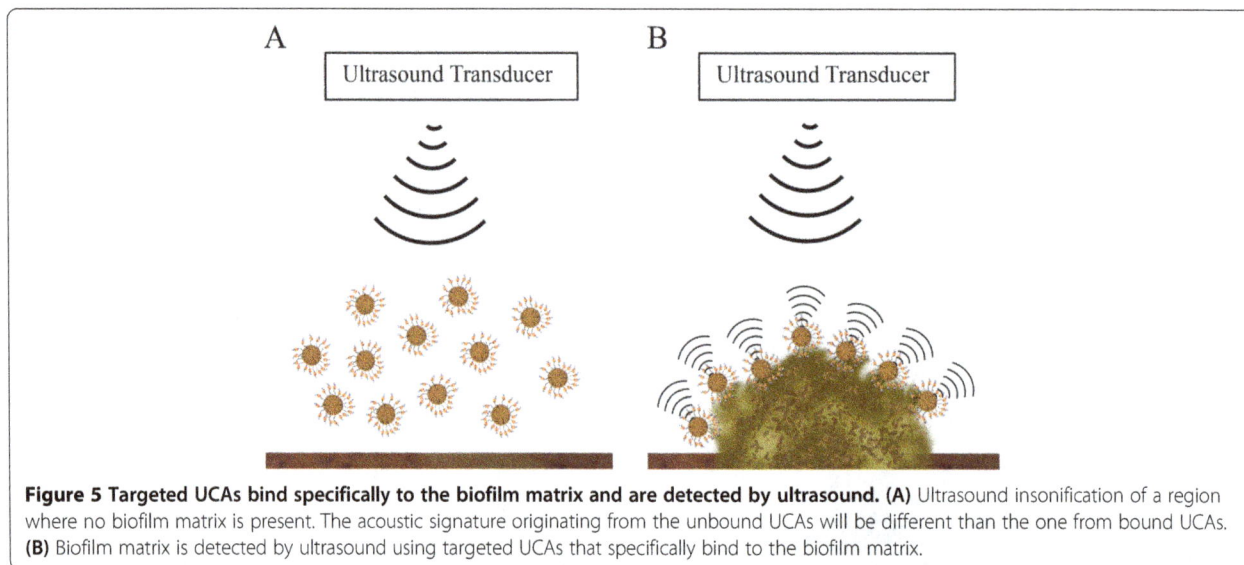

Figure 5 Targeted UCAs bind specifically to the biofilm matrix and are detected by ultrasound. (A) Ultrasound insonification of a region where no biofilm matrix is present. The acoustic signature originating from the unbound UCAs will be different than the one from bound UCAs. **(B)** Biofilm matrix is detected by ultrasound using targeted UCAs that specifically bind to the biofilm matrix.

Biofilms occurring from infections pose a challenge to current medicine because of the difficulty for early detection and diagnosis. Biofilms protect bacteria and promote resistance to antibiotics and chemotherapeutic agents. Moreover, detecting early and late biofilm formation may be problematic due to their dynamic profile. Individual bacterial cells may detach from the biofilm to colonize other niches or an entire biofilm colony may move as a whole across a region [6]. Thus, biofilm-mediated rippling effects that occur during detachment and transmigration pose biomedical challenges. One example is ventilator-associated pneumonia in immunologically compromised patients that may occur due to biofilm rippling (e.g. cancer patients) [41]. Our data indicates that the binding efficiency of UCAs correlates with matrix biomass. Thus, we propose that the rolling and rippling effects observed during biofilm maturation may reduce biomass and therefore decrease imaging capabilities at the late stages of biofilm matrix formation. It may be that there is a critical timeframe where UCAs bind well and imaging is enhanced and that as the biofilm grows, detaches and ripples, binding is decreased. Developing better detection methodologies and hence diagnostic clinical imaging methods are needed to assess biofilm formation early. By detection of biofilm infections at their earlier stages, this method will potentially offer more treatment options.

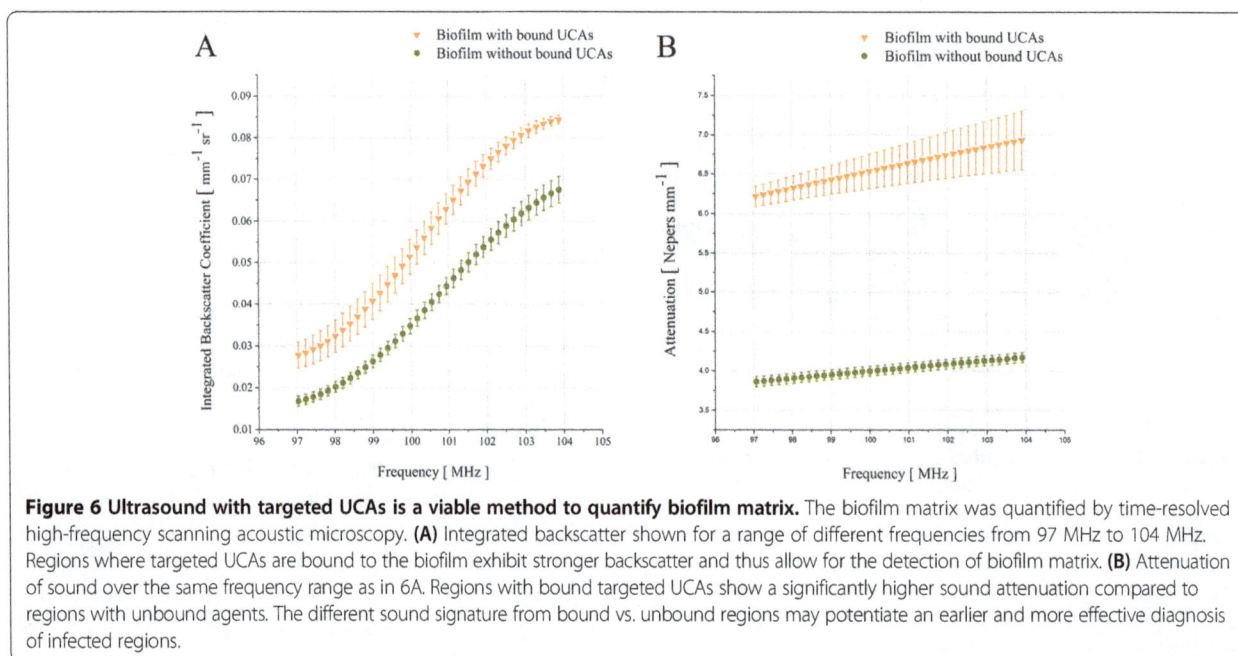

Figure 6 Ultrasound with targeted UCAs is a viable method to quantify biofilm matrix. The biofilm matrix was quantified by time-resolved high-frequency scanning acoustic microscopy. **(A)** Integrated backscatter shown for a range of different frequencies from 97 MHz to 104 MHz. Regions where targeted UCAs are bound to the biofilm exhibit stronger backscatter and thus allow for the detection of biofilm matrix. **(B)** Attenuation of sound over the same frequency range as in 6A. Regions with bound targeted UCAs show a significantly higher sound attenuation compared to regions with unbound agents. The different sound signature from bound vs. unbound regions may potentiate an earlier and more effective diagnosis of infected regions.

Due to the complex structure of biofilm matrices, we focused on the lectins concanavalin A and WGA [5,42] because biofilms may switch between these two polysaccharides during growth (Table 3). Targeting carbohydrate epitopes that are present in early biofilm matrices may provide novel biofilm markers that will enhance a more optimal molecular imaging, particularly at early stage formation.

Ultrasound imaging devices are readily available in clinical settings and the application of ultrasound techniques for biofilm-relevant infections is familiar to hospital personnel for diseases such as infective endocarditis [12,23,43] and cancer [44]. Targeted UCAs have the potential to recognize and bind to early stage biofilm matrix and thus, facilitate an early diagnosis. Our study demonstrated that while more targeted UCAs bound to larger biofilm matrix mass, a significant number of targeted UCAs also bound to less developed biofilm matrices. Targeted UCAs with ultrasound imaging may provide a means for early detection of biofilm formation within a non-invasive setting. This would include detecting endocarditis and biofilm matrix formation at the site of medical implants such as prosthetic devices, and catheters. Moreover, cancer patients rely on catheterization for chemotherapy treatment and biofilms are prevalent at the catheter interface. In immunologically compromised patients, rippling effects of the late stage biofilms have been reported to promote biofilm transmigration to the lungs causing additional complications in treatment [45,46]. Thus, the use of targeted UCAs may provide a rapid method to facilitate early diagnosis in a number of diseases.

High-frequency ultrasound may be used to assess biofilm development *in vitro*. Currently clinical biomicroscopy uses frequencies in the range of 15–50 MHz including intravascular ultrasound spectroscopy (IVUS), cardiovascular and ocular applications [47-54]. In particular, clinical imaging applications, such as detecting metastases in the eye, are in the range of 15–50 MHz, although research has been performed to measure at the higher frequency of 75 MHz [50]. The use of higher frequencies in the range of 100 MHz has been used for the imaging of choroidal metastasis [55]. Clinical ultrasound applications have focused on higher frequencies in the range of 100–200 MHz [56]. Low resonance frequencies are used clinically for drug delivery applications in conjunction with UCAs and targeted drug delivery as these low frequencies induce microbubble rupture [57]. Thus, in terms of imaging the higher frequencies provide enhanced imaging capabilities whereas lower resonant frequencies allow for more efficient targeted drug delivery. We report for the first time a method to quantify backscatter intensity and mechanoelastic properties of biofilms [58-64]. With regard to the integrated backscatter and the acoustic attenuation, considering differences in the frequency domain, similar values have been previously reported for cancer cells and tissues [58,61,63-66]. A more in-depth understanding of the three-dimensional biofilm matrix structural and mechanoelastic parameters will enhance biofilm imaging and subsequent treatment. Targeted UCAs potentially provide a novel means of imaging for the diagnosis of biofilm infections *in vivo*.

Materials and methods
Bacterial strains and cultivation of biofilms
We used a penicillin-resistant mutant of *S. aureus*. *S. aureus* and coagulase-negative staphylococci account for the majority of device-related infections [67].

S. aureus cultures were stored frozen at $-80°C$ in 10% glycerol and 90% tryptic soy broth (TSB, T8907, Sigma-Aldrich, St. Louis, USA) solution dissolved in sterile ultra-pure water (Alfa Aesar, Ward Hill, MA, USA).

A vial of frozen bacterial culture was thawed at room temperature (RT) and added to 250 mL of TSB. The inoculum was propagated and incubated overnight on an incubator shaker at 37°C and 160 rotations per minute (RPMs). The bacterial cultures were harvested after standardization to an optical density at 600 nm (OD_{600}) of 0.05 relative to the TSB culture medium (Beckman Coulter, Inc., Fullerton, CA, USA).

Biofilm assays were conducted by adding three milliliters of the standardized bacterial culture solution to the pretreated 35 mm glass (World Precision Instruments, Inc., Sarasota, FL, USA) and polystyrene petri dishes (Greiner Bio-One, Monroe, NC, USA). Glass and polystyrene petri dishes were treated in a previous step with Collagen IV (BD Biosciences, San Jose, CA, USA) for twenty minutes and rinsed in three washing steps with sterile distilled water. Prior to the addition of the inoculum, a 22 × 22 mm sterile micro cover glass (VWR International, LLC, West Chester, PA, USA) was placed into each of the polystyrene petri dishes. The glass and polystyrene petri dishes were then kept inside an incubator shaker at 37°C and 120 RPMs for up to 96 hours without replacement and addition of fresh culture medium in the interim.

Lectins, antibodies and immunofluorescence
Fluorescently-labeled lectins, concanavalin A (conA; binds to α-Man, α-Glc) [5,42] and wheat germ agglutinin

Table 3 Carbohydrate-binding specificity of lectins employed for staining of *S. aureus* biofilms

Lectin (source)	Abbreviation	Conjugate	Main specificity	Reference
Concanavalin A (Canavalia ensiformis)	ConA	FITC, TRITC	α-Man, α-Glc	Goldstein and Hayes [42]
Wheat germ agglutinin (Triticum vulgaris)	WGA	FITC, TRITC	(β-GlcNAc)₂, NeuNAc	

(WGA; binds to (β-GlcNAc)$_2$ and NeuNAc; Sigma-Aldrich Corp., St. Louis, MO, USA) [5,42] conjugated with FITC were used for the visualization of carbohydrate-containing extracellular polymeric substances in biofilms of *S. aureus* (Table 3) [5,42]. Stock solution of ConA at a concentration of 1 mg/mL in 0.1 M sodium bicarbonate (pH 8.3) and WGA at a concentration of 1 mg/mL in phosphate buffered saline (PBS; pH 7.4) were prepared, aliquoted and stored at –20°C. Prior to use, thawed portions of ConA and WGA aliquots were diluted with 0.1 M sodium bicarbonate (pH 8.3) and PBS (pH 7.4) respectively to a lectin final concentration of 10 μg/mL.

The blue-fluorescent nucleic acid stain 4′,6-diamidino-2-phenylindole, dihydrochloride (DAPI, Sigma-Aldrich Corp., St. Louis, MO, USA) was used to visualize bacterial cell distribution in the biofilms. A DAPI stock solution at a concentration of 5 mg/mL and 14.3 mM in ultrapure water was prepared, aliquoted and stored at –20°C. An aliquot was diluted to 300 nM in PBS immediately before use.

The monoclonal immunoglobulin antibody to protein A of *S. aureus* was used as a conjugation agent for the UCA particles. Anti-Protein A (APA) was developed in rabbit using protein A purified from *S. aureus* (Sigma-Aldrich Corp., St. Louis, MO, USA). Protein A localizes on the surface of staphylococcal bacterial strains and its distribution is inhomogeneous [68,69]. The lyophilized content of the vial was reconstituted in 2 mL PBS (pH 7.4) yielding a solution with a protein concentration of 23.7 mg/mL. The lectin from *P. aeruginosa* (PA-IL, Sigma-Aldrich Corp., St. Louis, MO, USA) was used, similarly to APA, to conjugate the surface of the UCA particles. The lyophilized content was diluted in 1 mL PBS (pH 7.4) yielding a protein concentration of 1 mg/mL. Following the reconstitution of APA and PA-IL, the proteins were biotinylated and conjugated onto the surface of the UCAs according to the method that will be described in more detail.

Sulfo-NHS-LC-Biotin (Thermo Scientific, Rockford, IL, USA) was applied to label APA and PA-IL with biotin. The vial of Sulfo-NHS-LC-Biotin was stored at –20°C and equilibrated to RT before opening to avoid condensation. For the biotin labeling reaction 2.2 mg of Sulfo-NHS-LC-Biotin were dissolved in 400 μL ultrapure water immediately before use yielding a 10 mM solution. A 20-fold molar excess of biotin reagent to label APA and PA-IL, resulting in 4–6 biotin groups per antibody molecule, was found to be suitable. For APA with a concentration of 23.7 mg/mL, a volume of 320 μL Sulfo-NHS-LC-Biotin was used for the biotinylation reaction while 13.5 μL Sulfo-NHS-LC-Biotin were used to label PA-IL with a concentration of 1 mg/mL. Following the incubation on ice for two hours at RT a Zeba® desalt spin column (Thermo Scientific,

Rockford, IL, USA) was applied to remove the excess non-reacted and hydrolyzed Sulfo-NHS-LC-Biotin reagent from the APA and PA-IL solutions. The column was placed into a sterile 15 mL falcon tube and centrifuged at 1000 × G for two minutes. After centrifugation the storage buffer collected at the bottom of the falcon tube was discarded, the column placed back into the same falcon tube and equilibrated by adding 2.5 mL of PBS (Thermo Scientific, Rockford, IL, USA) to the top of the resin bed and centrifuging at 1000 × G for two minutes. Next, the flow-through was discarded and the same step was repeated for a total of three times. Subsequently the column was placed into a new sterile 15 mL falcon tube and the antibody solution was applied onto the center of the resin bed. Finally, the column was centrifuged at 1000 × G for two minutes. The collected purified flow-through antibody solutions were aliquoted and stored appropriately.

Targeted ultrasound contrast agents

Biotin-conjugated lipid-encapsulated perfluorocarbon UCAs with a mean diameter of 3.02 μm ± 0.05 μm (Targeson, San Diego, CA, USA) were removed from a sealed vial using a four-way stopcock-syringe combination with a 22 G needle while simultaneously venting the vial with an additional needle. Using a 22 G needle, targestar conjugation buffer (TCB, Targeson, San Diego, CA, USA) was withdrawn into the syringe containing the UCA particles to a total volume of 3.5 mL and centrifuged at 400 × G for three minutes to remove excess free unincorporated lipids from the UCA particle solution. After centrifugation the infranatant was drained drop-wise and the UCAs were re-suspended in 1.0 mL TCB. Afterward, UCAs were incubated with 150 μL FITC-streptavidin (Invitrogen, Carlsbad, CA, USA) at a concentration of 1 mg/mL for twenty minutes at RT with occasional gentle shaking of the vial. The unreacted FITC-streptavidin was removed in a centrifugal washing at 400 × G for three minutes similarly to the previous step and re-suspended in 1.0 mL TCB. Finally, the UCA particles were incubated on ice with either APA or PA-IL for 30 minutes while the unconjugated targeting antibody and ligand molecules were removed by a final centrifugal washing as was described in the previous steps.

Lectin-staining for biofilms

For the *S. aureus* biofilm matrix, a double staining approach with ConA and WGA was chosen [5,42]. After incubation for twenty minutes in the dark at RT, excess staining solution was removed by rinsing three times with sterile distilled water.

Epifluorescence microscopy

Epifluorescence microscopy was carried out with a Zeiss Axioskop 2 microscope equipped with an AxioCam MRc Rev 3. Negative and positive controls were conducted using epifluorescence microscopy. For each case four biofilm samples were used to image five random positions on every sample. Each position was imaged by applying the respective filter for FITC, TRITC and DAPI. The negative control did not use any dyes to control for any possible autofluorescence effects.

Ultrasonic investigation of biofilms

Ultrasonic imaging and RF data acquisition was performed with a high-frequency scanning acoustic microscope (Fraunhofer IBMT, St. Ingbert, Germany). A detailed description of the acoustic lens used is shown in Table 2.

The recorded RF data were stored for further processing. The post-processing was conducted using custom-written scripts in MATLAB (The Math-Works, Natick, MA, USA). The scripts were applied for the visual reconstruction in 2D and 3D of the raw RF data for the selected ROIs. RF raw signals were gated by applying a rectangular window function. The window length at 100 MHz excitation center frequency was set such as it corresponded to 10 wavelengths. The wavelength estimation is based on the center frequency of the lens (100 MHz).

The gating of RF raw time-signals allows estimations of scattering properties to be related to distinct ROIs in the volume under interrogation [70]. However, the gating process also allows unwanted frequency content to be added into the backscattered power spectrum which subsequently, leads to inaccurate estimates of scatterer properties. In order to minimize such unwanted effects, a Hamming window was applied [70]. The tapered windows reduced the high-frequency content added into the gated RF time-signals by smoothing the edges.

When the acoustic lens is moved over a ROI where the substrate is covered by an EPS layer under investigation, two echoes are received. One echo originates from the top surface of the layer, $S_0(t)$, and the second, $S(t)$, from the interface between the layer and the substrate. These signals can be written as follows [71-74]:

$$S_0(t) = A_0 \, s(t{-}t_0) \otimes g(t, z_0)$$
$$S(t) = A_1 s(t{-}t_1) \otimes g(t, z_1) + A_2 \, s(t{-}t_2) \otimes g(t, z_2)$$

$S(t)$ is the reflected signal from the top and the sample-substrate interface of the EPS matrix. Provided that the defocus is positive meaning that the acoustic lens is elevated above the maximum focus position, it is adequate to constrain the function g to be independent

of t and to be a real function of z only. The optimum value of z was found experimentally, by scanning along the z axis and finding the minimum positive value at which the shape of the waveform remained approximately constant as a function of z. The z value was experimentally determined to be 900 μm. Within the approximation of the independence of the waveform shape on z, the signals can be written respectively as [74]:

$$S_0(t) = A_0 \, s(t{-}t_0) \times g(z_0)$$
$$S(t) = A_1 s(t{-} t_1) \times g(t, z_1) + A_2 \, s(t{-} t_2) \times g(z_2)$$

From the height in amplitude and position on the time axis of each maximum, the following parameters were measured:

$$\Delta T_{0/1} = t_0{-}t_1$$
$$\Delta T_{0/1} = t_0{-}t_1$$

where t_1 and t_2 are the arrival times of the sample and the interface echo respectively (Figure 3A) and t_0 is the time arrival of the reference signal (not shown) when no sample is placed in between the acoustic lens and the substrate. The velocity of the coupling medium, which in this case was degassed biofilm medium at 25°C, was approximated to the velocity of distilled water and set to be 1497 m/s [75,76] while the attenuation of the same medium was set equal to the attenuation of distilled water at 25°C, 2 dB/mm [77]. The density of the medium was calculated with a microbalance and a micropipette at 25°C. From the density of the coupling medium, denoted as ρ_{cm}, and the respective ultrasonic velocity, denoted as v_{cm}, the acoustic impedance, Z_{cm}, of the coupling medium was deduced:

$$Z_{cm} = \rho_{cm} \times V_{cm}$$

From the difference in time between the reference signal, t_0, and the reflection from the sample surface, t_1, and by applying the velocity, v_0, of the coupling medium, the thickness of the layer is:

$$d = \frac{1}{2}(t_0{-} t_1)V_0$$

From the ratio of magnitude of the reflection A_1 from the surface of the layer to the magnitude of the reference signal A_0, and by applying the impedance Z_0 of the coupling medium as has been calculated in and the acoustic impedance of the substrate Z_s, the acoustic impedance of the biofilm sample is:

$$Z_{bf} = Z_0 \frac{A_0 + A_1}{A_0{-}A_1}$$

Finally, from the amplitude A_2 of the echo from the interface between the layer and the substrate, the amplitude of the substrate echo A_0, the attenuation in the cell, in units of Nepers per unit length, can be calculated as follows:

$$\alpha = \alpha_0 + \frac{1}{2d}\log_e\left(\frac{A_0}{A_2}\frac{Z_s\text{-}Z_{bf}}{Z_s+Z_{bf}}\frac{4Z_c\times Z_0}{(Z_c+Z_0)^2}\frac{Z_s+Z_0}{Z_s\text{-}Z_0}\right)$$

Abbreviations
APA: Anti-Protein A; DAPI: 4',6-diamidino-2-phenylindole, dihydrochloride; IBSC: Integrated backscatter coefficient; FITC: Fluorescein isothiocyanate; OD: Optical density; PBS: Phosphate buffered saline; RF: Radio-frequency; ROI: Region of interest; RPM: Revolutions per minute; RT: Room temperature; StDev: Standard deviation; TCB: Targeson conjugation buffer; TRITC: Tetramethylrhodamine isothiocyanate; TSB: Tryptic soy broth.

Competing interests
The authors declare that they have no competing interests.

Authors' contributions
PA, KDAM, JSA, MLM: Designed the study. PA, KDAM: Conducted the experiments, performed data analysis, performed statistical analysis. PA, MLM: Prepared and edited the manuscript. JSA, KDAM: Edited the manuscript. All authors added intellectual content, read and approved the final version.

Acknowledgements
We thank Dr. Tung T. Hoang (Microbiology Dept., University of Hawaii at Manoa) for providing us with the bacterial strain. We would like to thank Dr. Terry Matsunaga (Dept. of Radiology, University of Arizona, Tucson, AZ, USA), Dr. Joshua Rychak (Targeson, San Diego, CA, USA) and Dr. Alexander L. Klibanov (Dept. of Biomedical Engineering, University of Virginia, Charlottesville, VA, USA) for sharing their technical expertise with us and for valuable discussions. This work was supported by: National Institutes of Health (NCRR P20-RR016453 and RO1GM104984 to M.L.M).

Author details
[1]University of Hawaii Cancer Center, Honolulu, HI 96813, USA. [2]Molecular Biosciences and Bioengineering, University of Hawaii at Manoa, Honolulu, HI 96822, USA. [3]Mechanical Engineering, University of Hawaii at Manoa, Honolulu, HI 96822, USA. [4]Department of Oceanography, School of Ocean and Earth Sciences and Technology, University of Hawaii at Manoa, Honolulu, HI, USA. [5]Current address: Department of Biological Oceanography, Royal Netherlands Institute for Sea Research (NIOZ), P.O. Box 59, 1790 AB Den Burg, Texel, The Netherlands.

References
1. Liu D, Lau YL, Chau YK, Pacepavicius G: Simple technique for estimation of biofilm accumulation. Bull Environ Contam Toxicol 1994, 53(6):913–918.
2. Allison DG, Ruiz B, SanJose C, Jaspe A, Gilbert P: Extracellular products as mediators of the formation and detachment of Pseudomonas fluorescens biofilms. Fems Microbiol Letters 1998, 167(2):179–184.
3. Costerton JW, Stewart PS, Greenberg EP: Bacterial biofilms: a common cause of persistent infections. Science 1999, 284(5418):1318–1322.
4. Wingender J, Neu TR, Flemming HC: What are Bacterial Extracellular Substances? In Microbial Extracellular Polymeric Substances: Characterization, Structure and Function. 1st edition. Edited by Wingender J, Neu TR, Flemming HC. Berlin: Springer; 1999:1–19.
5. Strathmann M, Wingender J, Flemming HC: Application of fluorescently labelled lectins for the visualization and biochemical characterization of polysaccharides in biofilms of pseudomonas aeruginosa. J Microbiol Methods 2002, 50(3):237–248.
6. Hall-Stoodley L, Costerton JW, Stoodley P: Bacterial biofilms: from the natural environment to infectious diseases. Nat Rev Microbiol 2004, 2(2):95–108.
7. Davey ME, GA O'T: Microbial biofilms: from ecology to molecular genetics. Microbiol Mol Biol Rev 2000, 64(4):847–867.
8. Flemming HC, Wingender J: The biofilm matrix. Nat Rev Microbiol 2010, 8(9):623–633.
9. Lambe DW, Ferguson KP, Mayberry-Carson KJ, Tober-Meyer B, Costerton JW: Foreign-body-associated experimental osteomyelitis induced with bacteroides fragilis and staphylococcus epidermidis in rabbits. Clin Orthop Relat Res 1991, 266:285–294.
10. Harris LG, Richards RG: Staphylococci and implant surfaces: a review. Injury-Int J Care Injured 2006, 37:3–14.
11. Baldassarri L, Montanaro L, Creti R, Arciola CR: Underestimated collateral effects of antibiotic therapy in prosthesis-associated bacterial infections. Int J Artif Organs 2007, 30:786–791.
12. Kaplan JB: Methods for the treatment and opinion prevention of bacterial biofilms. Expert Opinion Therap Patents 2005, 15(8):955–965.
13. Sullam PM, Drake TA, Sande MA: Pathogenesis of endocarditis. Am J Med 1985, 78(6B):110–115.
14. Caldwell DA, Lovasik D: Endocarditis in the immunocompromised. Am J Nurs 2002, (Suppl):32–36.
15. Donlan RM, Costerton JW: Biofilms: survival mechanisms of clinically relevant microorganisms. Clin Microbiol Rev 2002, 15(2):167–193.
16. Petti CA, Fowler VG Jr: Staphylococcus aureus bacteremia and endocarditis. Cardiol Clin 2003, 21(2):219–233. vii.
17. Nickel JC, Wright JB, Ruseska I, Marrie TJ, Whitfield C, Costerton JW: Antibiotic-resistance of pseudomonas-aeruginosa colonizing a urinary catheter invitro. Eur J Clin Microbiol Infect Dis 1985, 4(2):213–218.
18. Allison DG, Gilbert P: Modification by surface association of antimicrobial susceptibility of bacterial-populations. J Ind Microbiol 1995, 15(4):311–317.
19. Stewart PS, Costerton JW: Antibiotic resistance of bacteria in biofilms. Lancet 2001, 358(9276):135–138.
20. Parsek MR, Singh PK: Bacterial biofilms: an emerging link to disease pathogenesis. Annu Rev Microbiol 2003, 57:677–701.
21. Furuya EY, Lowy FD: Antimicrobial strategies for the prevention and treatment of cardiovascular infections. Curr Opin Pharmacol 2003, 3(5):464–469.
22. Durack DT, Lukes AS, Bright DK, Alberts MJ, Bashore TM, Corey GR, Douglas JM, Gray L, Harrell FE, Harrison JK, Heinle SA, Morris A, Kisslo JA, Nicely LM, Oldham N, Penning LM, Sexton DJ, Towns M, Waugh RA: New criteria for diagnosis of infective endocarditis - utilization of specific echocardiographic findings. Am J Med 1994, 96(3):200–209.
23. Baddour LM, Bettmann MA, Bolger AF, Epstein AE, Ferrieri P, Gerber MA, Gewitz MH, Jacobs AK, Levison ME, Newburger JW, Pallasch TJ, Wilson WR, Baltimore RS, Falace DA, Shulman ST, Tani LY, Taubert KA: Infective endocarditis: diagnosis, antimicrobial therapy, and management of complications. Circulation 2005, 112(15):2373.
24. Shemesh H, Goertz DE, van der Sluis LW, de Jong N, Wu MK, Wesselink PR: High frequency ultrasound imaging of a single-species biofilm. J Dent 2007, 35(8):673–678.
25. Vaidya K, Osgood R, Ren D, Pichichero ME, Helguera M: Ultrasound imaging and characterization of biofilms based on wavelet de-noised radiofrequency data. Ultrasound Med Biol 2014, 40(3):583–595.
26. Good MS, Wend CF, Bond LJ, McLean JS, Panetta PD, Ahmed S, Crawford SL, Daly DS: An estimate of biofilm properties using an acoustic microscope. IEEE Trans Ultrason Ferroelectr Freq Control 2006, 53(9):1637–1646.
27. Holmes AK, Laybourn-Parry J, Parry JD, Unwin ME, Challis RE: Ultrasonic imaging of biofilms utilizing echoes from the biofilm/air interface. IEEE Trans Ultrason Ferroelectr Freq Control 2006, 53(1):185–192.
28. Kujundzic E, Fonseca AC, Evans EA, Peterson M, Greenberg AR, Hernandez M: Ultrasonic monitoring of early-stage biofilm growth on polymeric surfaces. J Microbiol Methods 2007, 68(3):458–467.
29. Sbeity F, Menigot S, Charara J, Girault JM: Contrast improvement in sub- and ultraharmonic ultrasound contrast imaging by combining several hammerstein models. Int J Biomed Imag 2013, 2013:270523.
30. Anderson CR, Hu X, Zhang H, Tlaxca J, Decleves A-E, Houghtaling R, Sharma K, Lawrence M, Ferrara KW, Rychak JJ: Ultrasound molecular imaging of tumor angiogenesis with an integrin targeted microbubble contrast agent. Investig Radiol 2011, 46(4):215–224.
31. Klibanov AL: Microbubble contrast agents - targeted ultrasound imaging and ultrasound-assisted drug-delivery applications. Investig Radiol 2006, 41(3):354–362.
32. Klibanov AL: Ultrasound molecular imaging with targeted microbubble contrast agents. J Nucl Cardiol 2007, 14(6):876–884.

33. Klibanov AL: **Preparation of targeted microbubbles: ultrasound contrast agents for molecular imaging.** *Med Biol Eng Comput* 2009, **47**(8):875–882.

34. Klibanov AL, Hughes MS, Villanueva FS, Jankowski RJ, Wagner WR, Wojdyla JK, Wible JH, Brandenburger GH: **Targeting and ultrasound imaging of microbubble-based contrast agents.** *Magnet Reson Mat Biol Physics Med* 1999, **8**(3):177–184.

35. Unnikrishnan S, Klibanov AL: **Microbubbles as ultrasound contrast agents for molecular imaging: preparation and application.** *Am J Roentgenol* 2012, **199**(2):292–299.

36. Weiss EC, Anastasiadis P, Pilarczyk G, Lemor RM, Zinin PV: **Mechanical properties of single cells by high-frequency time-resolved acoustic microscopy.** *Ultrason Ferroelectr Frequen Contr IEEE Trans* 2007, **54**(11):2257–2271.

37. Weiss EC, Lemor RM, Pilarczyk G, Anastasiadis P, Zinin PV: **Imaging of focal contacts of chicken heart muscle cells by high-frequency acoustic microscopy.** *Ultrasound Med Biol* 2007, **33**(8):1320–1326.

38. Zinin PV, Allen JS III: **Deformation of biological cells in the acoustic field of an oscillating bubble.** *Phys Rev E* 2009, **79**(2 Pt 1):021910.

39. Zinin PV, Allen JS 3rd, Levin VM: **Mechanical resonances of bacteria cells.** *Phys Rev E Stat Nonlin Soft Matter Phys* 2005, **72**(6 Pt 1):061907.

40. Moran CM, Watson RJ, Fox KAA, McDicken WN: *In vitro* **acoustic characterisation of four intravenous ultrasonic contrast agents at 30 MHz.** *Ultrasound Med Biol* 2002, **28**(6):785–791.

41. Inglis TJ: **Evidence for dynamic phenomena in residual tracheal tube biofilm.** *Br J Anaesth* 1993, **70**(1):22–24.

42. Goldstein IJ, Hayes CE: **The lectins: carbohydrate-binding proteins of plants and animals.** *Adv Carbohydr Chem Biochem* 1978, **35**:127–340.

43. Archibald LK, Gaynes RP: **Hospital-acquired infections in the United States. The importance of interhospital comparisons.** *Infect Dis Clin North Am* 1997, **11**(2):245–255.

44. Azzopardi EA, Ferguson EL, Thomas DW: **The enhanced permeability retention effect: a new paradigm for drug targeting in infection.** *J Antimicrob Chemother* 2013, **68**(2):257–274.

45. Inglis TJJ, Titmeng L, Mahlee N, Eekoon T, Kokpheng H: **Structural features of tracheal tube biofilm formed during prolonged mechanical ventilation.** *Chest* 1995, **108**(4):1049–1052.

46. Al Akhrass F, Al Wohoush I, Chaftari AM, Reitzel R, Jiang Y, Ghannoum M, Tarrand J, Hachem R, Raad I: **Rhodococcus bacteremia in cancer patients is mostly catheter related and associated with biofilm formation.** *Plos ONE* 2012, **7**(3):e32945.

47. Hejsek L, Pasta J: **Ultrasonographic biomicroscopy of the eye.** *Ultrazvukova? Biomikroskopie oka* 2005, **61**(4):273–280.

48. Liang HD, Blomley MJK: **The role of ultrasound in molecular imaging.** *Br J Radiol* 2003, **76**(suppl_2):S140–S150.

49. Silverman RH: **High-resolution ultrasound imaging of the eye - a review.** *Clinical Exper Ophthalmol* 2009, **37**(1):54–67.

50. Silverman RH, Cannata J, Shung KK, Gal O, Patel M, Lloyd HO, Feleppa EJ, Coleman DJ: **75 MHz ultrasound biomicroscopy of anterior segment of eye.** *Ultrason Imaging* 2006, **28**(3):179–188.

51. Snetkova YV, Levin VM, Petriniuk YS, Denisov AF, Bogachenkov AN, Denisova LA, Khramtzova YA: **Application of the method of acoustic microscopy for the study of eye tissues.** *Morfologiya* 2005, **127**(2):72–75.

52. Goertz DE, Frijlink ME, De Jong N, Steen A: **Nonlinear intravascular ultrasound contrast imaging.** *Ultrasound Med Biol* 2006, **32**(4):491–502.

53. Goertz DE, Frijlink ME, Tempel D, Bhagwandas V, Gisolf A, Krams R, de Jong N, van der Steen AFW: **Subharmonic contrast intravascular ultrasound for vasa vasorum imaging.** *Ultrasound Med Biol* 2007, **33**(12):1859–1872.

54. Phillips LC, Klibanov AL, Wamhoff BR, Hossack JA: **Intravascular ultrasound detection and delivery of molecularly targeted microbubbles for gene delivery.** *IEEE Trans Ultrason Ferroelectr Freq Control* 2012, **59**(7):1596–1601.

55. Witkin AJ, Fischer DH, Shields CL, Reichstein D, Shields JA: **Enhanced depth imaging spectral-domain optical coherence tomography of a subtle choroidal metastasis.** *Eye* 2012, **26**(12):1598–1599.

56. Knspik DA, Starkoski B, Pavlin CJ, Foster FS: **A 100–200 MHz ultrasound biomicroscope.** *IEEE Trans Ultrason Ferroelectr Freq Control* 2000, **47**(6):1540–1549.

57. Liu Y, Miyoshi H, Nakamura M: **Encapsulated ultrasound microbubbles: therapeutic application in drug/gene delivery.** *J Control Release* 2006, **114**(1):89–99.

58. Saijo Y, Sasaki H, Okawai H, Nitta S-I, Tanaka M: **Acoustic properties of atherosclerosis of human aorta obtained with high-frequency ultrasound.** *Ultrasound Med Biol* 1998, **24**(7):1061–1064.

59. Saijo Y, Jorgensen CS, Falk E: **Ultrasonic tissue characterization of collagen in lipid-rich plaques in apoE-deficient mice.** *Atherosclerosis* 2001, **158**(2):289–295.

60. Saijo Y, Santos E, Sasaki H, Yambe T, Tanaka M, Hozumi N, Kobayashi K, Okada N: **Ultrasonic tissue characterization of atherosclerosis by a speed-of-sound microscanning system.** *IEEE Trans Ultrason Ferroelectr Frequen Contr* 2007, **54**(8):1571–1577.

61. Saijo Y, Miyakawa T, Sasaki H, Tanaka M, Nitta SI: **Acoustic properties of aortic aneurysm obtained with scanning acoustic microscopy.** *Ultrasonics* 2004, **42**(1–9):695–698.

62. Brand S, Solanki B, Foster DB, Czarnota GJ, Kolios MC: **Monitoring of cell death in epithelial cells using high frequency ultrasound spectroscopy.** *Ultrasound Med Biol* 2009, **35**(3):482–493.

63. Brand S, Weiss EC, Lemor RM, Kolios MC: **High frequency ultrasound tissue characterization and acoustic microscopy of intracellular changes.** *Ultrasound Med Biol* 2008, **34**(9):1396–1407.

64. Strohm EM, Czarnota GJ, Kolios MC: **Quantitative measurements of apoptotic cell properties using acoustic microscopy.** *IEEE Transactions Ultrason Ferroelectr Frequen Contr* 2010, **57**(10):2293–2304.

65. Baddour RE, Sherar MD, Hunt JW, Czarnota GJ, Kolios MC: **High-frequency ultrasound scattering from microspheres and single cells.** *J Acoust Soc Am* 2005, **117**(2):934–943.

66. Saijo Y, Tanaka M, Okawai H, Sasaki H, Nitta SI, Dunn F: **Ultrasonic tissue characterization of infarcted myocardium by scanning acoustic microscopy.** *Ultrasound Med Biol* 1997, **23**(1):77–85.

67. Baddour LM, Bettmann MA, Bolger AF, Epstein AE, Ferrieri P, Gerber MA, Gewitz MH, Jacobs AK, Levison ME, Newburger JW, Pallasch TJ, Wilson WR, Baltimore RS, Falace DA, Shulman ST, Tani LY, Taubert KA: **Nonvalvular cardiovascular device-related infections.** *Circulation* 2003, **108**(16):2015–2031.

68. DeDent AC, McAdow M, Schneewind O: **Distribution of protein A on the surface of staphylococcus aureus.** *J Bacteriol* 2007, **189**(12):4473–4484.

69. Schneewind O, Fowler A, Faull KF: **Structure of the cell wall anchor of surface proteins in staphylococcus aureus.** *Science* 1995, **268**(5207):103–106.

70. Oelze ML, O'Brien WD: **Improved scatterer property estimates from ultrasound backscatter for small gate lengths using a gate-edge correction factor.** *J Acoust Soc Am* 2004, **116**(5):3212–3223.

71. Briggs A: *Advances in Acoustic Microscopy.* New York: Plenum Press; 1995.

72. Briggs A, Kolosov O: *Acoustic Microscopy.* New York: Clarendon; 2010.

73. Briggs GAD, Rowe JM, Sinton AM, Spencer DS: **Quantitative Methods in Acoustic Microscopy.** In *Ultrasonics Symposium Proceedings: 1988.* Chicago, IL, USA: Publ by IEEE; 1988:743–749.

74. Briggs GAD, Wang J, Gundle R: **Quantitative acoustic microscopy of individual living human cells.** *J Microsc* 1993, **172**(Pt1):3–12.

75. Martin G, Carroll ET: **Tables of the speed of sound in water.** *J Acoustic Soc Am* 1959, **31**(1):75–76.

76. Greenspan M, Tschiegg CE: **Tables of the speed of sound in water.** *J Acoustic Soc Am* 1959, **31**(1):75–76.

77. Akashi N, Kushibiki J, Dunn F: **Acoustic properties of egg yolk and albumen range 20–400 MHz.** *J Acoust Soc Am* 1997, **102**(6):3774–3778.

Permissions

All chapters in this book were first published in JN, by BioMed Central; hereby published with permission under the Creative Commons Attribution License or equivalent. Every chapter published in this book has been scrutinized by our experts. Their significance has been extensively debated. The topics covered herein carry significant findings which will fuel the growth of the discipline. They may even be implemented as practical applications or may be referred to as a beginning point for another development.

The contributors of this book come from diverse backgrounds, making this book a truly international effort. This book will bring forth new frontiers with its revolutionizing research information and detailed analysis of the nascent developments around the world.

We would like to thank all the contributing authors for lending their expertise to make the book truly unique. They have played a crucial role in the development of this book. Without their invaluable contributions this book wouldn't have been possible. They have made vital efforts to compile up to date information on the varied aspects of this subject to make this book a valuable addition to the collection of many professionals and students.

This book was conceptualized with the vision of imparting up-to-date information and advanced data in this field. To ensure the same, a matchless editorial board was set up. Every individual on the board went through rigorous rounds of assessment to prove their worth. After which they invested a large part of their time researching and compiling the most relevant data for our readers.

The editorial board has been involved in producing this book since its inception. They have spent rigorous hours researching and exploring the diverse topics which have resulted in the successful publishing of this book. They have passed on their knowledge of decades through this book. To expedite this challenging task, the publisher supported the team at every step. A small team of assistant editors was also appointed to further simplify the editing procedure and attain best results for the readers.

Apart from the editorial board, the designing team has also invested a significant amount of their time in understanding the subject and creating the most relevant covers. They scrutinized every image to scout for the most suitable representation of the subject and create an appropriate cover for the book.

The publishing team has been an ardent support to the editorial, designing and production team. Their endless efforts to recruit the best for this project, has resulted in the accomplishment of this book. They are a veteran in the field of academics and their pool of knowledge is as vast as their experience in printing. Their expertise and guidance has proved useful at every step. Their uncompromising quality standards have made this book an exceptional effort. Their encouragement from time to time has been an inspiration for everyone.

The publisher and the editorial board hope that this book will prove to be a valuable piece of knowledge for researchers, students, practitioners and scholars across the globe.

List of Contributors

Zhiwei Ma
State Key Laboratory of Military Stomatology, Department of Periodontology and Oral Medicine, The School of Stomatology, Fourth Military Medical University, Xi-an, China

Chuanxu Yang
Interdisciplinary Nanoscience Center (iNANO), Department of Molecular Biology and Genetics, Aarhus University, Gustav Wiedsvej 14, 8000 Aarhus C, Denmark

Wen Song
State Key Laboratory of Military Stomatology, Department of Prothodontics, The School of Stomatology, Fourth Military Medical University, Xi-an, China

Qintao Wang
State Key Laboratory of Military Stomatology, Department of Periodontology and Oral Medicine, The School of Stomatology, Fourth Military Medical University, Xi-an, China

Jørgen Kjems
Interdisciplinary Nanoscience Center (iNANO), Department of Molecular Biology and Genetics, Aarhus University, Gustav Wiedsvej 14, 8000 Aarhus C, Denmark

Shan Gao
State Key Laboratory of Military Stomatology, Department of Periodontology and Oral Medicine, The School of Stomatology, Fourth Military Medical University, Xi-an, China

Syeda Sohaila Naz
Nanoscience and Catalysis Division, National Centre for Physics, Quaid-i-Azam University Campus, Islamabad 44000, Pakistan

Muhammad Raza Shah
H.E.J. Research Institute of Chemistry, International Centre for Chemical and Biological Sciences, University of Karachi, Karachi 75270, Pakistan

Nazar Ul Islam
Institute of Chemical Sciences, University of Peshawar, Peshawar 25120, Pakistan
Sarhad University of Science and Information Technology, Peshawar 2500, Pakistan

Ajmal Khan
H.E.J. Research Institute of Chemistry, International Centre for Chemical and Biological Sciences, University of Karachi, Karachi 75270, Pakistan

Samina Nazir
Nanoscience and Catalysis Division, National Centre for Physics, Quaid-i-Azam University Campus, Islamabad 44000, Pakistan

Sara Qaisar
Nanoscience and Catalysis Division, National Centre for Physics, Quaid-i-Azam University Campus, Islamabad 44000, Pakistan

Syed Sartaj Alam
Department of Plant Pathology, Khyber Pakhtunkhwa Agricultural University Peshawar, Peshawar, Pakistan

Qing Wang
School of Life Sciences and Biotechnology, Shanghai Jiao Tong University, Shanghai 200240, China
Institute of Nano Biomedicine and Engineering, Key Laboratory for Thin Film and Microfabrication Technology of the Ministry of Education, Department of Instrument Science & Engineering, School of Electronic Information and Electrical Engineering, Shanghai Jiao Tong University, 800 Dongchuan RD, Shanghai 200240, China

Chunlei Zhang
Institute of Nano Biomedicine and Engineering, Key Laboratory for Thin Film and Microfabrication Technology of the Ministry of Education, Department of Instrument Science & Engineering, School of Electronic Information and Electrical Engineering, Shanghai Jiao Tong University, 800 Dongchuan RD, Shanghai 200240, China

Guangxia Shen
Institute of Nano Biomedicine and Engineering, Key Laboratory for Thin Film and Microfabrication Technology of the Ministry of Education, Department of Instrument Science & Engineering, School of Electronic Information and Electrical Engineering, Shanghai Jiao Tong University, 800 Dongchuan RD, Shanghai 200240, China

Huiyang Liu
Institute of Nano Biomedicine and Engineering, Key Laboratory for Thin Film and Microfabrication Technology of the Ministry of Education, Department of Instrument Science & Engineering, School of Electronic Information and Electrical Engineering, Shanghai Jiao Tong University, 800 Dongchuan RD, Shanghai 200240, China

F Hualin
Institute of Nano Biomedicine and Engineering, Key Laboratory for Thin Film and Microfabrication Technology of the Ministry of Education, Department of Instrument Science & Engineering, School of Electronic Information and Electrical Engineering, Shanghai Jiao Tong University, 800 Dongchuan RD, Shanghai 200240, China

Daxiang Cui
School of Life Sciences and Biotechnology, Shanghai Jiao Tong University, Shanghai 200240, China
Institute of Nano Biomedicine and Engineering, Key Laboratory for Thin Film and Microfabrication Technology of the Ministry of Education, Department of Instrument Science & Engineering, School of Electronic Information and Electrical Engineering, Shanghai Jiao Tong University, 800 Dongchuan RD, Shanghai 200240, China

Ludmilla Regina de Souza
Institute of Biological Sciences, Molecular Biology Programme, University of Brasília, Brasília, DF 70910-900, Brazil

Luis Alexandre Muehlmann
Institute of Biological Sciences, Molecular Biology Programme, University of Brasília, Brasília, DF 70910-900, Brazil

Mayara Simonelly Costa dos Santos
Institute of Biological Sciences, Molecular Biology Programme, University of Brasília, Brasília, DF 70910-900, Brazil

Rayane Ganassin
Institute of Biological Sciences, University of Brasília, Brasília, DF 70910-900, Brazil

Rosana Simón-Vázquez
Biomedical Research Center (CINBIO), Institute of Biomedical Research of Vigo, University of Vigo, Vigo, Pontevedra 36310, Spain

Graziella Anselmo Joanitti
Institute of Biological Sciences, Molecular Biology Programme, University of Brasília, Brasília, DF 70910-900, Brazil

Ewa Mosiniewicz-Szablewska
Institute of Physics, Polish Academy of Sciences, Warsaw 02-668, Poland

Piotr Suchocki
Department of Bioanalysis and Drugs Analysis, Warsaw Medical University, Warsaw 02-097, Poland
Department of Pharmaceutical Chemistry, National Medicines Institute, Warsaw 00-725, Poland

Paulo César Morais
Institute of Physics, University of Brasília, Brasília, DF 70910-900, Brazil
School of Automation, Huazhong University of Science and Technology, Wuhan, Hubei 430074, China

África González-Fernández
Biomedical Research Center (CINBIO), Institute of Biomedical Research of Vigo, University of Vigo, Vigo, Pontevedra 36310, Spain

Ricardo Bentes Azevedo
Institute of Biological Sciences, Molecular Biology Programme, University of Brasília, Brasília, DF 70910-900, Brazil

Sônia Nair Báo
Institute of Biological Sciences, Molecular Biology Programme, University of Brasília, Brasília, DF 70910-900, Brazil

Ling Zhang
State Key Laboratory of Bioelectronics, Southeast University, Nanjing 210096, China
School of Biomedical Engineering, Hubei University of Science and Technology, Xianning 437000, China

Xin Wang
State Key Laboratory of Bioelectronics, Southeast University, Nanjing 210096, China

Jinglu Zou
State Key Laboratory of Bioelectronics, Southeast University, Nanjing 210096, China

Yingxun Liu
State Key Laboratory of Bioelectronics, Southeast University, Nanjing 210096, China

Jinke Wang
State Key Laboratory of Bioelectronics, Southeast University, Nanjing 210096, China

Laura Andolfi
IOM-CNR, Area Science Park, Basovizza, Trieste, Italy

Elisa Trevisan
Department of Life Sciences University of Trieste, Trieste, Italy

Barbara Troian
A.P.E. Research Srl, AREA Science Park, Basovizza, Trieste, Italy

Stefano Prato
A.P.E. Research Srl, AREA Science Park, Basovizza, Trieste, Italy

Rita Boscolo
Institute for Maternal and Child Health, IRCCS Burlo Garofolo, Trieste, Italy

Elena Giolo
Institute for Maternal and Child Health, IRCCS Burlo Garofolo, Trieste, Italy

Stefania Luppi
Institute for Maternal and Child Health, IRCCS Burlo Garofolo, Trieste, Italy

Monica Martinelli
Institute for Maternal and Child Health, IRCCS Burlo Garofolo, Trieste, Italy

Giuseppe Ricci
Institute for Maternal and Child Health, IRCCS Burlo Garofolo, Trieste, Italy
Department of Medicine, Surgery and Health Sciences, University of Trieste, Italy

Marina Zweyer
Department of Medicine, Surgery and Health Sciences, University of Trieste, Italy

Lynn M Murray
The MacDiarmid Institute for Advanced Materials and Nanotechnology, Department of Electrical and Computer Engineering, University of Canterbury, Christchurch 8140, New Zealand

Volker Nock
The MacDiarmid Institute for Advanced Materials and Nanotechnology, Department of Electrical and Computer Engineering, University of Canterbury, Christchurch 8140, New Zealand

John J Evans
The MacDiarmid Institute for Advanced Materials and Nanotechnology, and Centre for Neuroendocrinology, Department of Obstetrics and Gynaecology, University of Otago, Christchurch 8011, New Zealand

Maan M Alkaisi
The MacDiarmid Institute for Advanced Materials and Nanotechnology, Department of Electrical and Computer Engineering, University of Canterbury, Christchurch 8140, New Zealand

David C Kennedy
Department of Biomolecular Systems, Max Planck Institute of Colloids and Interfaces (MPIKG), 14476 Potsdam, Germany
National Research Council Canada (CNRC), 100 Sussex Drive, Ottawa, Ontario, Canada

Guillermo Orts-Gil
Department of Biomolecular Systems, Max Planck Institute of Colloids and Interfaces (MPIKG), 14476 Potsdam, Germany

Chian-Hui Lai
Department of Biomolecular Systems, Max Planck Institute of Colloids and Interfaces (MPIKG), 14476 Potsdam, Germany

Larissa Mller
Division 1.1 Inorganic Trace Analysis, Federal Institute for Materials Research and Testing (BAM), Richard-Willsttter-Strae 11, 12489 Berlin, Germany

Andrea Haase
Departments Chemical and Product Safety, German Federal Institute for Risk Assessment (BfR), 10589 Berlin, Germany

Andreas Luch
Departments Chemical and Product Safety, German Federal Institute for Risk Assessment (BfR), 10589 Berlin, Germany

Peter H Seeberger
Department of Biomolecular Systems, Max Planck Institute of Colloids and Interfaces (MPIKG), 14476 Potsdam, Germany
Institute for Chemistry and Biochemistry, Free University Berlin, Arnimallee 22, 14195 Berlin, Germany

Milagros Ramos-Gómez
Centre for Biomedical Technology, Polytechnic University of Madrid, 28223 Madrid, Spain
Biomedical Research Networking Center in Bioengineering Biomaterials and Nanomedicine (CIBER-BBN), Madrid, Spain

Emma G Seiz
Department of Molecular Biology and Center of Molecular Biology "Severo Ochoa", Autonomous University of Madrid-C.S.I.C, 28049 Madrid, Spain

Alberto Martínez-Serrano
Department of Molecular Biology and Center of Molecular Biology "Severo Ochoa", Autonomous University of Madrid-C.S.I.C, 28049 Madrid, Spain

Paul Schlinkert
Department of Molecular Biology, Paris Lodron-University of Salzburg, Hellbrunnerstr. 34A-5020 Salzburg, Austria

Eudald Casals
Institute Catalá de Nanotecnologia, Barcelona, Spain

Matthew Boyles
Department of Molecular Biology, Paris Lodron-University of Salzburg, Hellbrunnerstr. 34A-5020 Salzburg, Austria

Ulrike Tischler
Department of Molecular Biology, Paris Lodron-University of Salzburg, Hellbrunnerstr. 34A-5020 Salzburg, Austria

Eva Hornig
Department of Molecular Biology, Paris Lodron-University of Salzburg, Hellbrunnerstr. 34A-5020 Salzburg, Austria

Ngoc Tran
Institute Catalá de Nanotecnologia, Barcelona, Spain

Jiayuan Zhao
Institute for Work and Health, Lausanne, Switzerland

Martin Himly
Department of Molecular Biology, Paris Lodron-University of Salzburg, Hellbrunnerstr. 34A-5020 Salzburg, Austria

Michael Riediker
Institute for Work and Health, Lausanne, Switzerland Institue for Occupational Medicine (IOM) Singapore, Downtown Core, Singapore

Gertie Janneke Oostingh
Department of Molecular Biology, Paris Lodron-University of Salzburg, Hellbrunnerstr. 34A-5020 Salzburg, Austria Biomedical Sciences, Salzburg University of Applied Sciences, Puch, Salzburg, Austria

Victor Puntes
Institute Catalá de Nanotecnologia, Barcelona, Spain

Albert Duschl
Department of Molecular Biology, Paris Lodron-University of Salzburg, Hellbrunnerstr. 34A-5020 Salzburg, Austria

Simona Bancos
Center for Drug Evaluation and Research, Food and Drug Administration, Building 51 Room 4159, 10903 New Hampshire Ave., Silver Spring, MD 20993, USA

Katherine M Tyner
Center for Drug Evaluation and Research, Food and Drug Administration, Building 51 Room 4159, 10903 New Hampshire Ave., Silver Spring, MD 20993, USA

Yekaterina I Brandt
Department of Cell Biology and Physiology, University of New Mexico Health Sciences Center, Albuquerque, New Mexico 87131-0001, USA

Therese Mitchell
Department of Cell Biology and Physiology, University of New Mexico Health Sciences Center, Albuquerque, New Mexico 87131-0001, USA

Gennady A Smolyakov
Center for High Technology Materials, University of New Mexico, 1313 Goddard SE, Albuquerque, New Mexico 87106-4343, USA

Marek Osiński
Center for High Technology Materials, University of New Mexico, 1313 Goddard SE, Albuquerque, New Mexico 87106-4343, USA

Rebecca S Hartley
Department of Cell Biology and Physiology, University of New Mexico Health Sciences Center, Albuquerque, New Mexico 87131-0001, USA

Manca Pajnič
Laboratory of Clinical Biophysics, University of Ljubljana, Faculty of Health Sciences, Zdravstvena pot 5, Ljubljana SI-1000, Slovenia

Barbara Drašler
Group of Nanobiology and Nanotoxicology, University of Ljubljana, Biotechnical Faculty, Večna pot 111, Ljubljana SI-1000, Slovenia

Vid Šuštar
Lymphocyte Cytoskeleton Group, Institute of Biomedicine/Pathology, BioCity, University of Turku, Tykistökatu 6B, Turku SF-20520, Finland

Judita Lea Krek
Laboratory of Clinical Biophysics, University of Ljubljana, Faculty of Health Sciences, Zdravstvena pot 5, Ljubljana SI-1000, Slovenia

Roman Štukelj
Laboratory of Clinical Biophysics, University of Ljubljana, Faculty of Health Sciences, Zdravstvena pot 5, Ljubljana SI-1000, Slovenia

Metka Šimundić
Laboratory of Clinical Biophysics, University of Ljubljana, Faculty of Health Sciences, Zdravstvena pot 5, Ljubljana SI-1000, Slovenia

Veno Kononenko
Group of Nanobiology and Nanotoxicology, University of Ljubljana, Biotechnical Faculty, Večna pot 111, Ljubljana SI-1000, Slovenia

Darko Makovec
J. Stefan Institute, Jamova 39, Ljubljana SI-1000, Slovenia

Henry Hägerstrand
Department of Biosciences, BioCity, Åbo Akademi University, BioCity, Artillerigatan 6, Åbo/Turku SF-20520, Finland

Damjana Drobne
Group of Nanobiology and Nanotoxicology, University of Ljubljana, Biotechnical Faculty, Večna pot 111, Ljubljana SI-1000, Slovenia

Veronika Kralj-Iglič
Laboratory of Clinical Biophysics, University of Ljubljana, Faculty of Health Sciences, Zdravstvena pot 5, Ljubljana SI-1000, Slovenia

Atul A Chaudhari
Center for Nanobiotechnology Research, Alabama State University, Montgomery, AL, USA

Shanese L Jasper
Center for Nanobiotechnology Research, Alabama State University, Montgomery, AL, USA

Ejovwoke Dosunmu
Center for Nanobiotechnology Research, Alabama State University, Montgomery, AL, USA

Michael E Miller
Research Instrumentation Facility, Auburn University, Auburn, AL, USA

Robert D Arnold
Department of Drug Discovery and Development, Auburn University, Auburn, AL, USA

Shree R Singh
Center for Nanobiotechnology Research, Alabama State University, Montgomery, AL, USA

Shreekumar Pillai
Center for Nanobiotechnology Research, Alabama State University, Montgomery, AL, USA

Itay Levy
Department of Chemistry, Bar-Ilan Institute of Nanotechnology and Advanced Materials, Ramat-Gan 52900, Israel

Ifat Sher
Goldschleger Eye Institute, Sackler Faculty of Medicine, Tel Aviv University, Sheba Medical Center, Tel-Hashomer 52621, Israel

Enav Corem-Salkmon
Department of Chemistry, Bar-Ilan Institute of Nanotechnology and Advanced Materials, Ramat-Gan 52900, Israel

Ofra Ziv-Polat
Department of Chemistry, Bar-Ilan Institute of Nanotechnology and Advanced Materials, Ramat-Gan 52900, Israel

Amilia Meir
Center for Stem Cells and Regenerative Medicine, Cancer Research Center, Sheba Medical Center, Tel-Hashomer 52621, Israel

Avraham J Treves
Center for Stem Cells and Regenerative Medicine, Cancer Research Center, Sheba Medical Center, Tel-Hashomer 52621, Israel

Arnon Nagler
Hematology Division, Sheba Medical Center, Tel-Hashomer 52621, Israel

Ofra Kalter-Leibovici
Unit of Cardiovascular Epidemiology, Gertner Institute for Epidemiology and Health Policy Research, Ramat Gan, Israel, Sackler Faculty of Medicine, Tel-Aviv University, Tel-Aviv, Israel

Shlomo Margel
Department of Chemistry, Bar-Ilan Institute of Nanotechnology and Advanced Materials, Ramat-Gan 52900, Israel

Ygal Rotenstreich
Goldschleger Eye Institute, Sackler Faculty of Medicine, Tel Aviv University, Sheba Medical Center, Tel-Hashomer 52621, Israel

Pavlos Anastasiadis
University of Hawaii Cancer Center, Honolulu, HI 96813, USA
Molecular Biosciences and Bioengineering, University of Hawaii at Manoa, Honolulu, HI 96822, USA
Mechanical Engineering, University of Hawaii at Manoa, Honolulu, HI 96822, USA

Kristina D A Mojica
Department of Oceanography, School of Ocean and Earth Sciences and Technology, University of Hawaii at Manoa, Honolulu, HI, USA
Department of Biological Oceanography, Royal Netherlands Institute for Sea Research (NIOZ), P.O. Box 59, 1790 AB Den Burg, Texel, The Netherlands.

John S Allen
Mechanical Engineering, University of Hawaii at Manoa, Honolulu, HI 96822, USA

Michelle L Matter
University of Hawaii Cancer Center, Honolulu, HI 96813, USA